The Urban Book Series

The Urban Book Series is a resource for urban studies and geography research worldwide. It provides a unique and innovative resource for the latest developments in the field, nurturing a comprehensive and encompassing publication venue for urban studies, urban geography, planning and regional development.
The series publishes peer-reviewed volumes related to urbanization, sustainability, urban environments, sustainable urbanism, governance, globalization, urban and sustainable development, spatial and area studies, urban management, transport systems, urban infrastructure, urban dynamics, green cities and urban landscapes. It also invites research which documents urbanization processes and urban dynamics on a national, regional and local level, welcoming case studies, as well as comparative and applied research.
The series will appeal to urbanists, geographers, planners, engineers, architects, policy makers, and to all of those interested in a wide-ranging overview of contemporary urban studies and innovations in the field. It accepts monographs, edited volumes and textbooks.

Indexed by Scopus.

More information about this series at https://link.springer.com/bookseries/14773

Emanuele Giorgi • Tiziano Cattaneo
Alfredo Mauricio Flores Herrera
Virginia del Socorro Aceves Tarango
Editors

Design for Vulnerable Communities

Springer

Editors
Emanuele Giorgi
School of Architecture, Art and Design
Tecnológico de Monterrey
Chihuahua, Chihuahua, Mexico

Alfredo Mauricio Flores Herrera
School of Architecture, Art and Design
Tecnológico de Monterrey
Chihuahua, Chihuahua, Mexico

Tiziano Cattaneo
Department of Civil Engineering
and Architecture
University of Pavia, Italy

College of Design and Innovation
Tongji University
Shanghai, China

Virginia del Socorro Aceves Tarango
School of Hummanities
Tecnológico de Monterrey
Chihuahua, Chihuahua, Mexico

ISSN 2365-757X ISSN 2365-7588 (electronic)
The Urban Book Series
ISBN 978-3-030-96865-6 ISBN 978-3-030-96866-3 (eBook)
https://doi.org/10.1007/978-3-030-96866-3

This Springer imprint is published by the registered company Springer Nature Switzerland AG
The registered company address is: Gewerbestrasse 11, 6330 Cham, Switzerland

For all the residents of Paso del Norte in Chihuahua

Acknowledgments

All the residents of the Paso del Norte community made possible this research, which was funded by the *Fondo de Investigación 2020–2021* of *Observatorio de Ciudades del Tecnológico of Monterrey* (Mexico), through the research project '*Design for Vulnerables*', with a duration from October 2020 to December 2021.

As authors, we are grateful to all contributors, who participated in this stimulating discussion process, and to all the anonymous reviewers, who provided constructive comments for all the chapters of this book.

We are especially thankful to those Directors and Deans of the Tecnológico de Monterrey who encouraged and supported this project at different scales: Dr. Roberto Iñiguez Flores (National Dean, School of Architecture, Art, and Design), Dr. Rodolfo Julio Castello Zetina (Director of campus Chihuahua), Alfredo Henry Hidalgo Rasmussen (Dean of Research, School of Architecture, Art, and Design), Rául Juárez Perezlete (Western Regional Dean, School of Architecture, Art, and Design), and Pablo Hernández Quiñones (Dean of Division, School of Architecture, Social Science, and Humanities). Moreover, we are particularly grateful to Dr. Miguel Angel Montoya Bayardo (Director of Graduate and Research, School of Architecture, Art, and Design) and Prof. Pablo Alberto Rentería Rodríguez (Director of Architecture, campus Chihuahua), who provided valuable comments on methodology and the earlier structure of this book.

A special thank goes to Karol Serrato (research assistant) and to all those students of Architecture at Tecnológico de Monterrey, who participated in the research activities: Lizbeth Salas, Deborah Garay, Eduardo Olivas, Erika Flores, Alejandra Concha, Tania Cuilty, Ixchel Hinojosa, Ximena Arreola. Several other designers, students, and professors must be thanked because they made possible the project "Design for Vulnerables." In particular, we would like to thank all those 25 international experts—from 17 universities, 7 countries, and 5 schools of the Tecnológico de Monterrey—who enriched the discussion, thanks to their participation in the round tables.

We are especially grateful also to Bruno Fiuza and Hemalatha Velarasu from the Springer publishing team, who constantly helped us in the publication process.

Finally, great and sincere gratitude goes to all the residents of Paso del Norte in the city of Chihuahua (Mexico).

Contents

Contributors

Hocine Alioauane-Shaw ENSAP Bordeaux, Talence, France

Ardeth Barnhart Arlington, MA, USA

Carlo Berizzi Department of Civil Engineering and Architecture, University of Pavia, Pavia, Italy

Christina Boyes Centro de Investigacion y Docencia Economicas, A. C., DEI, Ciudad de México, Mexico

Chris Butters GAIA Architects, Oslo, Norway

Margherita Capotorto Department of Civil Engineering and Architecture, University of Pavia, Pavia, Italy

Tiziano Cattaneo Department of Civil Engineering and Architecture, University of Pavia, Italy

College of Design and Innovation, Tongji University, Shanghai, China

Paolo Ceccarelli International Laboratory of Architecture & Urban Design (ILAUD), Italy

Ali Cheshmehzangi Department of Architecture and Built Environment, University of Nottingham Ningbo China, Ningbo, China

Urban Innovation Lab, University of Nottingham Ningbo China, Ningbo, China

Network for Education and Research for Peace and Sustainability (NERPS), Hiroshima University, Hiroshima, Japan

Amir Cheshmehzangi School of Engineering, The University of Warwick, Warwick, UK

Moisés Gerardo Contreras Ruiz Esparza Instituto de Ingeniería, Universidad Nacional Autónoma de Mexico, Mexico City, Mexico

Gabriela Amado Chetto Coutinho Faculty of Architecture, Universidade Federal da Bahia, Salvador, Brazil

Ayotunde Dawodu Department of Architecture and Built Environment, University of Nottingham Ningbo China, Ningbo, China

Urban Innovation Lab, University of Nottingham Ningbo China, Ningbo, China

Corinna Dean University of Westminster, London, UK

Nivaldo Vieira de Andrade Junior Faculty of Architecture, Federal University of Bahia, Salvador, Brazil

Edgar Paul Martínez Ludert Muñoz de Cote Tecnologico de Monterrey, Escuela de Arquitectura, Arte y Diseño, Hermosillo, Mexico

Roberto De Lotto Department of Civil Engineering and Architecture, Urban Project Laboratory—UPLab, University of Pavia, Pavia, Italy

Virginia del Socorro Aceves Tarango Tecnologico de Monterrey, Escuela de Humanidades, Chihuahua, Mexico

Viviana Barquero Diaz-Barriga Tecnologico de Monterrey, Escuela de Arquitectura, Arte y Diseño, Querétaro, Mexico

Pius Leuba dit Galland Grolimund+Partner AG, Bern, Switzerland

Emanuele Giorgi Tecnologico de Monterrey, Escuela de Arquitectura, Arte y Diseño, Chihuahua, Mexico

Carlos Gotlieb ENSAP Bordeaux, Talence, France

Pilar Maria Guerrieri Politecnico di Milano, DASTU, Milano, Italy

Alfredo Mauricio Flores Herrera Tecnologico de Monterrey, Escuela de Arquitectura, Arte y Diseño, Chihuahua, Mexico

Carmelo Ignaccolo Department of Urban Studies and Planning, Massachusetts Institute of Technology, Cambridge, MA, USA

Ruoyu Jin Built Environment and Architecture/Construction, Property and Surveying, London South Bank University, London, UK

Ersela Kripa Texas Tech University, El Paso, TX, USA

Aleksandra Krstikj Tecnologico de Monterrey, School of Architecture, Art and Design, Ciudad López Mateos, State of Mexico, Mexico

Simone Lucatello Instituto Mora, CONACYT, Mexico City, Mexico

Greta Lukoseviciute Tecnologico de Monterrey, School of Architecture, Art and Design, Puebla, Mexico

Annie Malcolm University of California, Berkeley, CA, USA

Eugenio Mangi Department of Architecture and Built Environment, University of Nottingham Ningbo China, Ningbo, China

Gustavo Merino Escuela de Gobierno y Transformación Pública, Tecnológico de Monterrey, Mexico City, Mexico

Christiane Molina EGADE Business School, Tecnologico de Monterrey, Mexico City, Mexico

Marco Morandotti Department of Civil Engineering and Architecture (DICAr), University of Pavia, Pavia, Italy

Committee for International Cooperation and Development (CICOPS), University of Pavia, Pavia, Italy

Stephen Mueller Texas Tech University, El Paso, TX, USA

Minqing Ni College of Design and Innovation, Tongji University, Shanghai, China

Hiroyuki Shinohara The Chinese University of Hong Kong, Hong Kong, China

Gaia Nerea Terlicher Department of Civil Engineering and Architecture, University of Pavia, Pavia, Italy

Luca Trabattoni Department of Architecture and Planning, University of Technology of Opole, Opole, Poland

Giulio Verdini University of Westminster, London, UK

Yuqi Zhai College of Architecture and Urban Planning, Tongji University, Shanghai, China

Xiaocun Zhu College of Design and Innovation, Tongji University, Shanghai, China

Chapter 1
Introduction on Design for Vulnerable Communities

Tiziano Cattaneo, Emanuele Giorgi, Alfredo Mauricio Flores Herrera, and Virginia del Socorro Aceves Tarango

1.1 Background and Definitions

Vulnerability is essentially a planetary problem recurrent in all places worldwide. Actually, it is a permanent human condition in the history of human being and all of us could experience or are experiencing right now a certain grade of being vulnerable. Nevertheless it can be difficult to define what we mean by the term "vulnerable." Dictionaries are quite uniform in providing definitions of the terms "vulnerable" and "vulnerability" as follows, and we think interesting to include as first the etymology of the word "vulnerable":

T. Cattaneo (✉)
Department of Civil Engineering and Architecture, University of Pavia, Italy

College of Design and Innovation, Tongji University, Shanghai, China
e-mail: tiziano.cattaneo@unipv.it

E. Giorgi
Tecnologico de Monterrey, Escuela de Arquitectura, Arte y Diseño, Chihuahua, Mexico

A. M. F. Herrera
Tecnologico de Monterrey, Escuela de Arquitectura, Arte y Diseño, Campus Chihuahua, Chihuahua, Mexico

V. del Socorro Aceves Tarango
Tecnologico de Monterrey, School of Humanities, Tecnológico de Monterrey, Monterrey, Mexico

E. Giorgi et al. (eds.), *Design for Vulnerable Communities*, The Urban Book Series,
https://doi.org/10.1007/978-3-030-96866-3_1

1.1.1 Etymology of Vulnerable

Vulnerable (adj.). c. 1600, from Late Latin vulnerabilis "wounding," from Latin vulnerare "to wound, hurt, injure, maim," from vulnus (genitive vulneris) "wound," perhaps related to vellere "pluck, to tear" (see svelte), or from PIE *wele-nes-, from *wele- (2) "to strike, wound" (see Valhalla). (Etymology dictionary, 2021a).

1.1.2 Definition of Vulnerable

/ˈvʌln(ə)rəb(ə)l/. adjective: vulnerable. Exposed to the possibility of being attacked or harmed, either physically or emotionally. (of a person) in need of special care, support, or protection because of age, disability, or risk of abuse or neglect. (Google's English dictionary, 2021b).

1.1.3 Definition of Vulnerability

/vʌln(ə)rəˈbɪlɪti/. noun: vulnerability; plural noun: vulnerabilities. The quality or state of being exposed to the possibility of being attacked or harmed, either physically or emotionally. (Google's English dictionary, 2021).

Vulnerability can be thus a very wide concept. There are people with evident and undeniable more obvious vulnerabilities—such as elderly, children, and people who are facing disadvantages in many aspects of their lives. However, in our society there are also more concealed vulnerabilities that harm either physically or emotionally people of all walks of life, as well as there are conditions of vulnerability inside more common challenges such as mobility, technology, healthcare, and many more. In other words, vulnerability invades all the sphere of life in terms of economic, social, and environmental threats.

So, what are the sources of vulnerability?

Climate change hazards, environmental crisis, and social and economical inequity are notions entered by the time in our daily existence, and both media and scientists remember us that we are far to control and to totally prevent those risks impacting our society.

We should note that the term "risk" is omnipresent along with the term vulnerability.[1] Moreover, the term "vulnerable" also includes the word "risk" in its

[1] In this regards the Chap. 19 of The Fifth Assessment Report of the Intergovernmental Panel on Climate Change (IPCC) see: Oppenheimer, M., M. Campos, R. Warren, J. Birkmann, G. Luber, B. O'Neill, and K. Takahashi, 2014: Emergent risks and key vulnerabilities. In: Climate Change 2014: Impacts, Adaptation, and Vulnerability. Part A: Global and Sectoral Aspects. Contribution of Working Group II to the Fifth Assessment Report of the Intergovernmental Panel on Climate Change [Field, C.B., V.R. Barros, D.J. Dokken, K.J. Mach, M.D. Mastrandrea, T.E. Bilir,

definition. The notion of risk society, developed by two sociologists Ulrich Beck and Anthony Giddens, confirms that "our society had become a risk society."

According to the sociologist Ulrich Beck (1992), a risk society is "a systematic way of dealing with hazards and insecurities induced and introduced by modernization itself," while the British sociologist Anthony Giddens (1999) defines it as "a society where we increasingly live on a high technological frontier which absolutely no one completely understands and which generates a diversity of possible futures."

In this regards, Giddens (1999) stated:

> The origins of risk society can be traced to two fundamental transformations which are affecting our lives today. Each is connected to the increasing influence of science and technology, although not wholly determined by them. The first transformation can be called *the end of nature*; and the second *the end of tradition.*

About this aspect Giddens (1999) describes the End of Nature as world:

> [...] in which the natural environment disappears. It means that there are now few if any aspects of the physical world untouched by human intervention. The end of nature is relatively recent. It isn't something, of course, which can be precisely dated, but we can nevertheless roughly plot when the end of nature happened. It happened when a transition came about from the sort of anxieties people used to have about nature to a new set of worries. For hundreds of years, people worried about what nature could do to us—earthquakes, floods, plagues, bad harvests and so on. At a certain point, somewhere over the past 50 years or so, we stopped worrying so much about what nature could do to us, and we started worrying more about what we have done to nature. The transition makes one major point of entry in risk society. It is a society which lives 'after nature'.

Consequentially for Giddens (1999) the End of Tradition is.

> [...] essentially to be in a world where life is no longer lived as fate. For many people—and this is still a source of class division in modern societies—diverse aspects of life were established by tradition as fate. It was the fate of a woman to be involved in a domestic milieu for much of her life, to have children and look after the house. It was the fate of men to go out to work, to work until they retired and then—quite often soon after retirement—essentially

M. Chatterjee, K.L. Ebi, Y.O. Estrada, R.C. Genova, B. Girma, E.S. Kissel, A.N. Levy, S. MacCracken, P.R. Mastrandrea, and L.L. White (eds.). Cambridge University Press, Cambridge, United Kingdom and New York, NY, USA, pp. 1039–1099], has the objective to assess new literature published since the Fourth Assessment Report (AR4) on emergent risks and key vulnerabilities to climate change from the perspective of the distribution of risk over geographic location, economic sector, time period, and socioeconomic characteristics of individuals and societies. The Report, which is of course centered on the issue on climate change provide several interesting definition of 'vulnerability' and 'risk' and their interdependencies. We include here their definition and for more deepen information see the full Report. "*Vulnerability*: The propensity or predisposition to be adversely affected. Vulnerability encompasses a variety of concepts and elements including sensitivity or susceptibility to harm and lack of capacity to cope and adapt.

A broad set of factors such as wealth, social status, and gender determine vulnerability and exposure to climate-related risk. *Risk*: The potential for consequences where something of value is at stake and where the outcome is uncertain, recognizing the diversity of values. Risk is often represented as probability of occurrence of hazardous events or trends multiplied by the impacts if these events or trends occur. Risk = (Probability of Events or Trends) × Consequences. Risk results from the interaction of vulnerability, exposure, and hazard. In this report, the term risk is used primarily to refer to the risks of climate-change impacts".

to fade away. We no longer live our lives as fate, in a process which Ulrich Beck calls individualization.

Within these two transformations that determine the Risk Society according to Giddens (1999) exist also a distinction between two kinds of risk: the *external risk* which is expressed in down-to-earth terms and lasted for the last 200 years on industrial society and the other might be called the *manufactured risk*:

> Manufactured risk is risk created by the very progression of human development, especially by the progression of science and technology. Manufactured risk refers to new risk environments for which history provides us with very little previous experience. We often don't really know what the risks are, let alone how to calculate them accurately in terms of probability tables. Manufactured risk is expanding in most dimensions of human life.
>
> [...]
>
> Manufactured uncertainty intrudes directly into personal and social life—it isn't confined to more collective settings of risk. In a world where one can no longer simply rely on tradition to establish what to do in a given range of contexts, people have to take a more active and risk-infused orientation to their relationships and involvements"

Thus, in a world dominated by technology and modernization with the aspiration to improve wellness and well-being, paradoxically we can observe that a rise of vulnerability has been produced at all levels and strata of society. To be more specific, Giddens has explained that the advance of science and technology that allowed us the "domination" over nature has brought to the actual environmental crisis. Our worries should not be about *what nature could do to us, but what we can do for nature*. In other words, the root of our crisis (environmental, social, and economical) is that human being is separated from Nature: our lives are not embedded in any ecosystem. It means that while natural hazards are still a major threat in many parts of the world, the Manufactured risk has been also spread everywhere in the so-called developed world with the result that we are all substantially more vulnerable.

1.2 New Vision of Design

These aspects bring us to consider that our discipline of Design (architectural, urban, planning, landscape, etc.) could be now the cure of this crisis only if we fully reconsider our design approach. In 1972, Victor Papanek (Papanek, 2005) in its book *Design for the Real World. Human Ecology and Social Change* wrote:

> We all live in a designed world. Design is basic to all human activity… In this age of mass production, when everything must be planned and designed, design has become the most powerful tool with which man shapes his tools and environment (and, by extension, society and himself).

To provide a first basic shelter for protecting human being from nature has been the main goal of architecture since its origin. This radical simplicity of architectural design should be radically changed in order to face the appearance of the Manufactured risk as earlier defined.

As argued by Chinchilla Izaskun (2011).

> International standards of design procedures are still having Modernity as a main conceptual and methodological framework. [...] the definition of vulnerability will be therefore inevitably linked to the evolution and influence of Modernity. We first postulate Modernity has been an important source of vulnerability [...] But this production of vulnerability was not a part of Modernity's set of goals. Therefore, while being a major cause, Modernity was not including vulnerability as a language, as a cultural principle or as a sensibility. Maybe only the technical rise of lighter materials and the use of previously considered weak techniques could suppose an attempt to face vulnerability as a main characteristic of the inhabitable world after Modernity. But this shy flirt with weakness is not meaning Modernity could easily evolve to care vulnerable realities. The first reason we will uphold for this is that in fact Modern traditions are still fighting for a first set of objectives that has never been fully achieved. The second reason is that Modernity is not only lacking anthropological and ecological clues on its route but including logics and thought principles that are directly incompatible with those perspectives.

Accordingly, we have to recognize together with Van der Ryn and Cowan (2007) that.

> In many ways, the environmental crisis is a design crisis. It is a consequence of how things are made, buildings are constructed, and landscapes are used. Design manifests culture, and culture rests firmly on the foundation of what we believe to be true about the world.

Now, having delineated this context, we can back to the main object of our research: Design for vulnerable communities. As a general condition affecting many parts of our urbanized world, we understand that is possible to talk about vulnerability in many different ways. Trying to avoid mixed-rambling conclusions and taking a stance on architectural design we want to focus on how the protection of vulnerable realities would require a change in our design approach. Actually, the built environment is also a primary part of the current challenges, as it is, for example, a major consumer of energy and natural resources, and a major producer of waste. Moreover, the built environment is the place where the human life can thrive or can decline (Cattaneo et al., 2020).

So, how to contribute to sustainable communities and quality of life development? What kind of design strategies and principles we can propose for redesigning the built environment? How does architecture [architectural design] intensify social inclusion and health by taking care of vulnerable realities? What kind of design narratives and methods we should adopt for re-Naturing the built environment?

As mentioned earlier, we are in a time of crisis that someone consider a civilization crisis, others talk about the crisis of the Anthropocene and so forth (Giorgi, 2020). According to the Colombian-American anthropologist Arturo Escobar the roots of crisis is in the fact that the Capitalist society has created a dualistic competitive world in which human is separated from nature and is living in constant dichotomy between elements such as the separation between body and spirit. Escobar in his book *Designs for the Pluriverse: Radical Interdependence, Autonomy, and the Making of Worlds* (Escobar, 2018) introduced the concept of interdependences between human and non-human and between human and nature. His suggestion is to look at the world and the environment without separation and he put Design at the center of his argument. The design in fact was and is still at the center of the actual "devastation" but design could be now at the center of the environmental restoration and reparation. What is needed is an ontological idea of Design (Willis, 2006). Following the Papanek's words ontological design is the design of human experience.

> It does so by operating under one essential assumption: that by designing objects, spaces, tools and experiences, we are in fact designing the human being itself. And the ability to design human beings is going to be central to survive the technological shifts of the coming decades with even a semblance of agency (Fraga, 2020).

In short, we can synthetize that Design [including of course architecture and urban planning] is ways of being and ways of doing. Hence, design is very important domain for making life and for making story. Design should stop to be anthropocentric but, in the words of Escobar, should build *radical interdependencies* and *radical relationalities* starting from thinking the relationships of human and non-human with the Cosmos. Same as Bruno Munari (Munari, 1971) defined Design in his *Codice ovvio* (Obvious code) in 1971:

> Design as creation of objects, events and environments inspired by the observation of everyday life and natural phenomena.

In order to conclude this section we back to the Arturo Escobar's new vision of design theory and practice by reporting here the six strategies (principles) for redesigning the cities, or as he claims *re-earthing the city*.

The first principle is called Re-communalization that basically advocates to re-communalize the social life by a process of sharing of resources (basic needs of life, territories, neighborhoods, etc.) and emphasize the human need of sense of belonging to a community. Re-localization is another important principle that means to bring back to local vary kind of productions: from materials production to energy production according to the local resources and in respect of the place. This principle also includes one of the most important aspects, which is to re-localize the food production and food chain starting from cultural identity, food as heritage and understanding of nutrition feeding within a community. The third principle is linked to the concept of "autonomy" and it is connected to the previous one. It is necessary to underline that Escobar does not refer to individual autonomy but to a "collective autonomy." Local autonomy refers to the participation of all members of the community in activities and decisional processes of and for the community. This principle is strictly related to the politics activities associated with the very local governance and it is aimed at improving community's life. Fourth one is divided in three objectives: de-patriarchalization, de-racialization, and de-colonization of the society. This principle is connected to the previous one and it refers to important concepts such as equity between the community itself and between community and the world. All the Escobar's principles are related to design and vulnerability; however the fifth principle is perhaps the most important for architectural design and planning: re-integration of all the elements of the community (tangible and intangible, material and spiritual) with the Earth rules and natural mechanism. This is the base of the concept of *re-earthing the city* in which architecture should create interdependencies and rationalities with ecosystems. As last but not less important is the sixth principle that advocates to radical changes of the actual trends by starting the "re-construction" of vulnerable communities from the bottom, that is to say from local level by creating convergences of alternatives networks and organizations in order to benefit as much as possible a mutual rising of people.

1.3 Transitional Narrative of Vulnerable

While the focus is on Design for vulnerable communities, our research highlight important multiple goals for human health, quality of life, well-being, and so forth, with the aim to give evidence to a multiplicity of knowledge and interdisciplinary approaches. Thus, in conclusion of this section it necessary to reframe the concept of vulnerable. This need arises from the consciousness that more and more often designers will be called to propose solutions for vulnerable realities, and, at the same time, these realities will become more and more complex to analyze and to intervene. In fact, what designer can do is to create stories around vulnerabilities through the system thinking of the discipline of Design using its own tools of representation such as texts and a vast range of images (i.e., collages, photomontages, drawings, storyboards, comic strips, animations). The main aim of this design narrative is to create stories that not necessarily provide solutions but also ignite new knowledge visualization from a transition theory perspective and practice on vulnerabilities. In other words, we would propose a *transitional narrative of vulnerable*. In order to draft this transitional narrative of vulnerable we borrow a set of definition and insight from the social and environmental studies by Samantha Winter (2020). Winter's research focuses on the social and environmental determinants driving inequities in women's health and access to health-related services including changes in these factors related to climate change (Winter et al., 2019). In other words, Winter's investigation provides a clear meaning of vulnerability through the lens of healthy community. While the focus is on healthy community in informal settlements, her research and definitions highlight manifold characteristics on vulnerabilities of other communities. In the presentation titled *Healthy Communities: Exploring the Multilevel and Multidimensional Nature of Health and Well-Being in Informal Settlement Communities* Samantha Winter stated:

> Globally, an estimated one billion people live in informal settlements—defined as densely populated residential areas lacking durable housing; sufficient living and public spaces; access to basic infrastructure and services; and secure tenancy. Settlement residents often face discrimination and disadvantage, lack of recognition by governance and formal service frameworks, threats of eviction and demolition, precarious livelihoods, and high exposure to disease and violence. Research suggests health in these communities is unique from other communities. It is tied to the shared physical and social environment within settlements. It is important to explore the multilevel and multidimensional nature of health in these communities using a socio-ecological approach that accounts for communalities between settlements, honors the contextual nature of health within each settlement, and takes a broader perspective of health--focusing not only on negative health outcomes, but on the well-being, agency, and resilience of communities and the residents who live there.

Beside the informal settlements definition and residents vulnerabilities description we can here find the first fundamental proposition: vulnerability is contextual.

Social and environmental features of one place are unique and should be analyzed and investigated starting from its specific characteristics. Analyzing the specificity of vulnerable neighborhoods, which are deeply contextual, helps to explain why the effectiveness of interventions varies between vulnerable communities like informal settlements and other, "formal" communities. This drives us to understand that it is not possible to apply models (standardized) of interventions and every design project or planned intervention for vulnerable communities should start from site-specific uniqueness. That is to say, interventions that are effective in non-informal settlement communities are often not transferrable to informal settlements and vice versa (Winter, 2020).

A second important proposition (which is a key issue for designers, architectural designers, urban designers, and planners) is to consider vulnerability not only as a function of poverty,[2] but also strictly linked to the inappropriateness of the built environment, such as lack of accessibility to services and buildings, lack of public spaces, toxic materials for constructions, lack of safety of the street and urban environment in general, and many more, that inevitably conduct to both social risks and high environmental risks such as those that arise from lacking access to formal sanitation, sewerage systems, or solid waste disposal in densely populated communities (Winter, 2020).

Despite uniqueness and site-specific contextual problematic of vulnerable communities it is possible to identify some persisting communalities challenging both informal and non-informal settlements. Common living conditions, such as densely populated neighborhoods, reliance on precarious, informal job sector, diffuse poverty, inadequate public and green spaces, inadequate water, sanitation, sewerage, solid waste disposal, electricity, unstable structures and living spaces, located in ecologically sensitive areas (floodplains, steep slopes, areas with limited water availability, poor soil) with limited vegetation that are vulnerable to natural disasters, e.g., flooding, landslides, drought, heavy rains, and fire, and higher level of interpersonal violence and insecurity are all common threats in vulnerable communities. For these reasons the third proposition is to explore vulnerabilities using a

[2] Luis R. Fernández Carril, Vicepresidencia de Inclusión, Impacto social y Sostenibilidad, Tecnologico de Monterrey (Mexico), in his presentation titled "*Biopolitics and Necropolitics of the Anthropocene*" at the 'Design for vulnerables' workshop held online on Thursday, seventh of January 2021 has stated: "Vulnerability is not exclusively a condition of exposure to environmental risk. It is embedded in socio-political relations at all scales. Furthermore, in many cases, vulnerability is created by political decisions, whether by imposing unethical living conditions or exposing populations to death, willingly or unwillingly. These decisions that create or exacerbate vulnerability could be described through the conceptual framework of biopolitics and necropolitics. Biopolitical and necropolitical dynamics are embedded in the very design of social strategies on climate change coping mechanisms, creating or exacerbating vulnerability. In this manner, we claim the need to look beyond environmental risks, at the power dynamics in place to find additional (or the main) sources of vulnerability. Furthermore, we claim the importance of necessarily including social justice and inequity in design in order to reduce the vulnerability of populations to environmental risks.

multilevel and multidimensional socio-ecological model that accounts for communalities between settlements (Winter, 2020).

The last proposition here presented perhaps is the most interesting and it could be called "*learning from vulnerable*." As Samantha Winter said: while people living in vulnerable communities often face higher environmental risks, social threats, and health challenges, they may also experience important factors for well-being by living in these communities. For examples: strong social ties, mutual support, and networks; shared experiences and understanding with community members tribal or ethnic connection; religious or spiritual connection; connection to place; access to opportunities; independence for freedom to live in one's truth (for example, a marginalized identity); and access to services (e.g., health). So, what we should learn, let us better say, what vulnerable people living in these informal settlements can teach to us, is that in many ways despite they are living in risky conditions they are still able to live in interdependencies and in connection with nature, human, and non-human components. In other words, approaching vulnerability we cannot focus only on negative aspects and outcomes, but on the well-being, agency, and resilience of communities and the residents who live there (Winter, 2020).

Aware to be far from a fully exhaustive new transitional narrative the outcomes of our investigation is the draft of a set of proposals for design for vulnerable communities that do not have a "manifesto format" but are instead build up through others' contributions aimed to achieve mutual benefit. By keeping as the background the Agenda 2030 and the 17 Sustainable Development Goals established in 2015 by the United Nations General Assembly as 17 interlinked global goals designed to be a "blueprint to achieve a better and more sustainable future for all," this research project presented in this book explores mainly, but not only, two aspects:

- How design (architectural, urban, planning, etc.) can improve well-being of vulnerable communities.
- The role of the built and natural environment in vulnerable communities.

Vulnerable communities are places and spaces of transitions: sustainable solutions based on a new transitional narrative of vulnerable can be created, tested, and scaled.

1.4 Scope, Structure, and Contents of the Book

The book "Design for Vulnerable Communities" is an edited book, which collects thoughts of international experts from different fields of knowledge about the challenges that the designers will have to face in the next few years when designing for vulnerable communities. These contributions do not present only a professional or personal research because as they would be too narrow, but they should be read as different perspectives by the authors' originality, sensitivity, and experience, and which are open and inclusive of possible further insights. Obviously, the challenges in designing a proper solution for a community will strictly depend by the community's context. Anyway, there are some main issues that represent some challenges for vulnerable communities all around the world: social segregation, environmental pollution, economic and energy poverty, lack of education, and healthy emergences just to list some of them. Even if several national and international programs, mainly led by the SDGs, are doing a lot in term to solve these challenges and to push toward a sustainable future, some unexpected and overwhelming changes impact our society and territories. Covid-19 pandemic, technological inequalities, A.I., climate changes, political crisis, and economic uncertainties are setting very important moments of thought for all the societies. While some sectors of our contemporary society can manage more easily these challenges, the most vulnerated ones seems to be even more vulnerable to these changes. "Vulnerables" is a plural noun referring to all those societies, community, and persons who will see their lives fronting these unexpected and overwhelming changes without having the sufficient material and immaterial resources to face them. Since the strong responsibility of design practice, designers are called to act and to think about proper solutions. Of course, for the nature of these new challenges, design solutions can't be based on those strategies and methodologies that nowadays are commonly considered best practices: some new design solutions must be studied, developed, and implemented.

It's from the consciousness of the relevance and of the complexity of this "call to act" that the idea of this book arises from: more and more often designers will be called to propose solutions for vulnerable realities, and, at the same time, these realities will become more and more complex to analyze and to intervene. It's for this reason that this book collects research from scholars of different disciplines which can help to give interpretation, with a point of view different form the designers' one, the phenomena that are characterizing the vulnerable communities in the next few years. This will help the designers to have a general view of the variables to be considered during a development of a project.

To test the interaction among different disciplines in a design activity, all the scholars have been invited to discuss their research and points of view during the design workshop "Design for Vulnerables," which has been held digitally between January and February 2021. The workshop, organized by Tecnológico de Monterrey and Tongji University, aimed to propose solutions for the vulnerable community "Paso del Norte" in the city of Chihuahua (Mexico). During the workshop, design groups (composed by students, architects, engineers and residents of the community) interacted with the chapters' authors in round tables and project revisions.

This book is divided into five main parts developing the topic of vulnerable communities. In addition, mutual benefits and the co-implementation of well-being fostered by the Design as important domain for making better life through architecture, planning and landscape deign for vulnerable communities are presented.

- Part I: Sustainability and Vulnerabilities in Time of the Anthropocene.
- Part II: Approaches, Principles and Paradigms to Contemporary Research and Practice for Vulnerable Communities Goals and Strategies.
- Part III: Designing for Vulnerabilities. Applications and Actions.

- Part IV: Social Engagement in Vulnerable Communities Between Digital and Humanist Visions.
- Part V: Vulnerabilities in Context: Analysis and Projects in the US–Mexico Border Region.

The various chapters provide up-to-date scientific background information identifying knowledge gaps, and addressing vary aspects of vulnerabilities and vulnerable communities and lay out pressing challenges along with applications, actions, experiences, and questions. Chapters provide also specific examples of interventions in vulnerable communities with case studies, from Europe, China, North America, and Latin America including specific contextual challenges at the US–Mexico Border Region. These topics were intensively discussed at the "Design for Vulnerables" workshop held from January 4 to February 17, 2021. The workshop was organized by the Tecnologico de Monterrey Campus Chihuahua within international cooperation with the College of Design and Innovation of Tongji University and the Department of Civil Engineering and Architecture of the University of Pavia, Italy.

Relevant and up-to-date topics along with the main contributions of this book divided in five parts are described in detailed in the next section. This book contributes to an increased understanding on vulnerable communities and how design can be the means for the creation of new narratives of vulnerable people.

1.4.1 Part I: Sustainability and Vulnerabilities in Time of the Anthropocene

The first part presents an overview on sustainability and vulnerabilities from the perspective of politics, economics, and technology in the era of great climate changes. The part discusses how different understandings of vulnerability result in multiple interpretations describing the challenges that our contemporary society has to face in the time of Anthropocene.

To frame the concept of Anthropocene with the philosophical position that considers technology as the environment of humanity *Emanuele Giorgi* explores the important role of technology development in contemporary ages for vulnerable communities with the aims to understand the challenges for design, architecture, and politics in understanding how technology can help or harm vulnerable communities in the next decades. This chapter is based on a deep analysis of the existing literature accompanied with numerous case studies to highlight, from the perspective of urban-architectural design, the potential risks (technology illiteracy, data production, and exclusion) and benefits (reduction of distances, self-production, transportation, monitoring, and engagement) of technological development for vulnerable communities. Finally, the author highlights issues for designers, architects, and policy makers generated by technological development and its possible impacts on vulnerable communities. Regarding an economics and policy perspective for vulnerable communities, *Gustavo Merino* discuss the evolution of poverty and its multidimensional nature, as well as some global trends affecting the livelihoods of people in poverty, exploring how design and innovation can contribute to the eradication of poverty and improve the livelihoods and opportunities for those facing this condition. It considers not just the design of goods, services, and infrastructure but also of policies and programs. The analysis takes a multidisciplinary perspective, drawing particularly from economics and public policy, and considers the role of governments, civil society, and businesses. Within the actual scenario of natural disasters, climate change, war conflicts, unequal socioeconomic growth, technological changes, and massive urban migrations *Paolo Ceccarelli and Pilar Maria Guerrieri* argued that this condition continuously produces vulnerable environments in local communities all over the world. This thesis is supported with numerous cases in different contexts, such as the new settlements after the Fukushima explosion, the war refugee cities of Za'atar in Jordan and Dadaab in Kenya, shrinking obsolete industrial cities in Poland, migrant farmers camps in Spain and Italy, and urban-rural migration in India, in order to identify significant contemporary vulnerability issues and highlight the potential of design, imagining correct and durable solutions.

1.4.2 Part II: Approaches, Principles, and Paradigms to Contemporary Research and Practice for Vulnerable Communities

Chapters in the second part of the book discuss the facets approaches and principles on contemporary research and practice for vulnerable communities from the perspective of different but complementary disciplines: anthropology and ethnographic studies, environmental design, urban planning design, and business, in particular, highlighting the economic, energy, social, and technological issues that impact vulnerable communities.

In a first paper Annie Malcolm introduces the method of ethnography as a way of doing anthropology to produce knowledge that engages and produces something called the social, new ways of thinking about research, human interaction and place. By referring to main literature on the field the author highlights several lines of thought to promote a conversation about ethnographic engagement and meaning-making applying the method through the examination of creative work on the outskirts of the city of Shenzhen in China in a village called Wutong Shan. In other words, the interesting assumption is: context create meaning. In their chapter Xiaocun Zhu and Pius Leuba dit Galland departing from evolutionary psychology theory for which humans are suffering from a blind spot, caused by evolutionary inertia, which is masking distant/indirect/invisible interconnections advocate that systems thinking, which pays more attention to interdependence than independence, can contribute to establishing this new sensitivity to compensate our cognitive blind spot. Within this context they argue that design is the foundation of all human activity and every design defines interactions between Human and Natural systems. Vulnerable communities represent the extreme of the human system, thus rendering design positive to the environment in the urban context will make a significant and far-reaching contribution to saving our species and civilization. With the aim to propose principle for design in vulnerable communities Christiane Molina provides interesting approaches starting from the concept of consciousness in the field of business. In this chapter, the author describes this concept and three of the main elements of this stream of business literature and practice with the aim to advance consciousness in the professional exercise of urban and architectural design. Complementing the previous chapter which aims to propose principles for urban design, Ali Cheshmehzangi and coauthors take stance from an intriguing assumption: all the cities and urban environments are vulnerable due to "climate change impacts." In order to enable planners and policy makers to seek for innovative models of development, the authors provide a set of analytical view points on 12 key principles for urban design. To conclude this part, the chapter by Roberto De Lotto starts from the assumption that the term "vulnerability" is quite clear when it is described in simple words; it is an intuitive concept that everyone perceives in a qualitative way. It is more difficult to describe vulnerability with numbers, in a quantitative way. This dichotomy between qualitative and quantitative aspects, among the storytelling and the numbers, is visible in 2030 SDGs. To quantify in quantitative way, the vulnerability could be an interesting line of research.

1.4.3 Part III: Designing for Vulnerabilities. Applications and Actions

The third part of this book presents visions on how to understand, map, and address issues related with vulnerability presenting also multidisciplinary intervention strategies in vulnerable communities, stressing the main characteristics that are independent from a single context. The first two chapters have in common to see the vulnerability of architectural heritage as a base for the fragility of the local communities. The first paper by *Nivaldo Vieira de Andrade Junior and coauthor* provides a comprehensive overview on the scientific literature social housing experiences in Latin American historic centers. Considering housing a key point for historic centers rehabilitation, both for the heritage preservation and for overcoming the critical social framework, this paper will analyze social housing projects carried out in historic centers in Argentina, Brazil, Ecuador, Mexico, Peru, and Uruguay in the last decades. These initiatives demonstrate how the good architecture has the potential to serve social justice by establishing quality of life, decent housing, and empowerment of the most vulnerable social groups. *Eugenio Mangi and Hiroyuki Shinohara* propose the research by design framework to define a conceptual model that addresses vulnerabilities through the definition of public spaces and collective facilities in historical area using the case study the *hutong* areas in Beijing. Rapid urban development pressure, as it happened for decades in Shanghai, often exacerbates existing vulnerabilities and generates unresolved spatial tensions between pre-existing neighborhood features and new forms of development. *Carmelo Ignaccolo and Yuqi Zhai*, through a comparative analysis of new datasets, engage with the recent attempt of the city of Shanghai to repurpose the outdated commercial typologies built in the early '90s during the government-led urban renewal. More specifically, this study portrays the morphological transitions of Hongkou district through the lens of commercial activities and sheds light on the relationship between people's vulnerability and street commerce. Starting from an opposite method from the previous chapter *Giulio Verdini and Corinna Dean* propose an approach to understand and map city social and environmental vulnerabilities, alongside bottom-up and emerging community experiences in response to the pandemic, with the aim to provide a critical understanding on whether the goal of developing climate resilient and socially inclusive post pandemic neighborhoods can be achievable. Coherently with the aim and scope of Part III of this book the last three chapters provide description of projects that combine theoretical approach with applied design research in different context and different types of vulnerabilities. *Marco Morandotti* presents three projects focused on healthcare sustainable design in the global south: a medical dispensary in Chakama, Malindi, Kenya; a mid-size project focused on the Ayamè General Hospital, in Ivory Coast; and design of the new campus of the University Foundation Juan de Castellanos of Tunja, Colombia, that includes a new mother and child hospital. *Carlo Berizzi and coauthors* present projects on the vulnerable coastal areas in Italy in order to combine the need to preserve and strengthen the existing landscape with the opportunity to reactivate the coastal territories in which tourism represents the leading economy.

1.4.4 Part IV: Social Engagement in Vulnerable Communities Between Digital and Humanist Visions

The fourth part of this book deals with an important issue at the time of digital and ecological transitions and how design can boost social engagement in vulnerable communities using digital innovation but also in contrary by implementing humanist visions as responsible process of close and dynamic interaction between man and his environment. The first paper of this part written by *Emanuele Giorgi, Mauricio Flores, Tiziano Cattaneo, and Edgar Ledarth* investigate the contemporary deep interlacing between real and virtual worlds. This chapter presents a number of interesting case studies to understand which technological applications allow AR and VR to take on a key role in empowering vulnerable communities. This chapter makes evident that these technological tools are particularly crucial when applied to vulnerable contexts that are at risk of falling into a chasm of technological illiteracy that could become difficult to overcome. Digital transitions became a mantra of many government and intergovernmental institutions and along a political narrative of digital participation for inclusive growth. Moreover, digitalization is omnipresent in today's social and urban life. *Minqing Ni* chapter attempts to discuss how to make inclusive growth happen in the city by fostering digital transition process. The paper focuses on a case study of Singapore's digital collaborative governance innovation model, which brings citizen perspectives especially vulnerable groups into the design of public policies and services co-creation. Through the case study *Ni* paper offers a revised definitions of inclusive digital social innovation toward a new paradigm of hybrid resilient community. On the other edge of this discussion is the paper by *Carlos Gotlieb and coauthor* which describes the Bordeaux School of Architecture and Landscape (ENSAP Bordeaux) post-master program entitled "Rebuilding the World," which advocates a new model of training urban planners following a humanist vision opposed to the one resulting from productivism. This vision, invested in an ethical dimension, is conceived as an instance of collective creation and sharing of meaning prior to any intervention in space to foster responsible process of close and dynamic interaction between man and his environment.

1.4.5 Part V: Vulnerabilities in Context: Analysis and Projects in the US–Mexico Border Region

The last part focuses on a specific context. As we said the vulnerability is contextual, in particular this part presents a series of analysis, good practice examples, projects in the territories of US–Mexico Border Region. Within this part, a last chapter introduces the outputs of the "Design for Vulnerables" Real-life Web Laboratory, which is a design-based research project in Chihuahua (Mexico). *Simone Lucatello* introduces challenges (environmental changes, ecological disruption, and human impacts), focusing on the northern part Mexico, and particularly the US–Mexico border. Starting from available studies showing how vulnerable communities in the US–Mexico border region will soon be affected by important changes in drinking water supply, air quality, and heat islands-waves among others, the author explores possible solutions for climate change adaptation actions and policies in the Metropolitan area of El Paso del Norte. *Aleksandra Krstikj and coauthors* analyze factors that determine accessibility, availability, and acceptability of food; this chapter investigates weaknesses of the food system in Mexico's metropolitan periphery. Krstikj's study highlights the need for an integrated context-specific approach for adaptive urban planning that can support community health and sustainable development. Among other environmental risks, Ersela Kripa and Stephen Mueller underline in their chapter that populations living close to the US–Mexico international border are particularly vulnerable to health risks, including heightened incidence of respiratory disease from exposure to high levels of airborne dust. To cope this problem new methods are needed to gather and broadcast environmental health data across the international divide. The authors propose

"Nephelometry," which is a distributed network of dust sensors in Ciudad Juárez and El Paso, transcribing a shared aerial environment across the national boundary. The project documents the scattered territory of airborne particulate through a series of experiments in collective "drawing." The project provides a model for trans-boundary environmental assessment and action capable of addressing issues of environmental justice in the borderland. Approaching vulnerability through the assessment of energy poverty is substantial to this region, as it is alone one of the most complex and unfair threats society faces today. Climate change and energy poverty together represent a critical and irreversible global social, economic, and environmental crisis, affecting equity in all its representations. With this premise the paper of *Ardeth Barnhart and Viviana Barquero* discusses on whether the spatial design of a community may impact vulnerability, and whether this affects its ability to adapt to climate change and overcome energy poverty. Current assessment methods and planning tools for energy access are reviewed, concluding with thoughtful insights of design opportunities for adaptation for these communities. The *editors* complement the book with a conclusion chapter. The last chapter critically reviews the outputs of the "Design for Vulnerables" Real Life Web Laboratory, which is a research project based in Chihuahua (Mexico). The Real Life Web Lab has been an opportunity to discuss about challenges for vulnerable communities bringing together community members, designers, students, and international experts in a series of events and workshops. A set of five strategies to empower vulnerable communities were the results of these dialogues: (1) empowering connections and public spaces to increase communitarian resilience; (2) enhance the local sense of belonging to innovate the social relations and to strength the urban local phenomena; (3) design effective strategies of management and social cohesion, also optimizing shared public spaces; (4) enhance the natural heritage as cohesive common good for the community and opportunity to connect with the environment; and (5) tourism activities can offer business opportunities, enhance the sense of belonging, and, finally, provide instruments to reduce community's vulnerability.

We hope this book besides proposing several propositions (and narratives) on how design can impact specific vulnerable communities, it makes evident that a shift of paradigms is also needed that starts from "Design for Vulnerable" to "Design with Vulnerable." Many of the themes have applications beyond vulnerable communities highlighting important multiple goals for human health, quality of life, well-being, and so forth, with the aim to give evidence to a multiplicity of knowledge and interdisciplinary approaches.

Acknowledgments We thank the external reviewers for providing critical and helpful comments on earlier versions of this chapter and on the book proposal. This work was funded by Fondo de Investigación 2020–2021 del Observatorio de Ciudades del Tecnologico de Monterrey (Mexico), through the research project "Design for vulnerables," project duration from October 2020 to December 2021.

References

Beck, U. (1992). *Risk Society: Towards a New Modernity*. Translated by Ritter, Mark. London: Sage Publications.

Cattaneo, T., Giorgi, E., Flores, M., & Barquero, V. (2020). Territorial effects of shared-living heritage regeneration. Sustainability, 12(20), 8616. MDPI AG. Retrieved from https://doi.org/10.3390/su12208616

Escobar, A. (2018). *Designs for the pluriverse. radical interdependence, autonomy, and the making of worlds*. Duke University Press.

Fraga, D. (2020) The Manifesto of Ontological Design, Available online: https://medium.datadriveninvestor.com/the-manifesto-of-ontological-design-7fdb19169107. Accessed on September 2021.

Giddens, A. (1999). Risk and responsibility. *Modern Law Review., 62*(1), 1–10. https://doi.org/10.1111/1468-2230.00188

Giorgi, E. (2020). Technocene. In E. Giorgi (Ed.), *The co-housing phenomenon. environmental alliance in times of changes* (pp. 1–26). Springer, Urban Book Series. https://doi.org/10.1007/978-3-030-37097-8_1

Izaskun, C. (2011). Dare to Care. Design Brief from for Ecole Special Paris, 2011, Available online: http://www.esa-paris.fr/IMG/pdf/dare_to_care-_ecole_special.pdf. Accessed on June 2021.

Munari, B. (1971). *Codice ovvio [Obvious Code]*. Einaudi.

Papanek, V. (2005). *Design for the real world: human ecology and social change* (2nd ed.). Academy Chicago Publishers. (first published 1972).

Sim Van der Ryn and Stuart Cowan (2007) Ecological design, Tenth Anniversary Edition Annotated Edition. Island Press.

Willis, A.-M. (2006). Ontological designing. *Design Philosophy Papers, 4*(2), 69–92. https://doi.org/10.2752/144871306X13966268131514

'Vulnerable', item in Etymology dictionary. Available online: https://www.etymonline.com/word/vulnerable. Accessed on September 2021a.

'Vulnerable', item in Google's English dictionary provided by Oxford Languages. Available online: https://languages.oup.com/google-dictionary-en/. Accessed on September 2021b.

'Vulnerability', item in Google's English dictionary provided by Oxford Languages. Available online: https://languages.oup.com/google-dictionary-en/. Accessed on September 2021.

Winter, S. C. (2020). *Healthy communities: Exploring the multilevel and multidimensional nature of health and well-being in informal settlement communities*, Design for Vulnerable Workshop, Thursday, 7th of January 2021 (Unpublished).

Winter, S. C., Dreibelbis, R., Dzombo, M. N., & Barchi, F. (2019). A mixed-methods study of women's sanitation utilization in informal settlements in Kenya. *PLoS One, 14*(3), e0214114.

Part I
Sustainability and Vulnerabilities in Time of the Anthropocene

Chapter 2
How Technology Devices Can Help or Harm Vulnerable Communities in Technocene. Issues for Designers, Architects, and Policy Makers

Emanuele Giorgi

2.1 Introduction

Due to the effectiveness with which it indicates the contemporary relationship between humanity and the environment, the term Anthropocene has been at the center of the scientific debate for years. It is used to describe a new historical moment (a geological era) in which humanity finds itself (1) able to count on immense capacity to modify the global ecosystem and (2) at the same time, having to rethink many of the certainties that have characterized the social evolution up to this moment. In recent decades, in fact, humanity had to face new challenges, anxieties, and uncertainties that never had been placed on the path of human evolution and which directly depend on the unprecedented power that it has reached, with incredible repercussions on the responsibilities that this entails, in terms of ecological impact and social relations.

This relationship between technical power, human responsibility, and impact on the environment has been discussed extensively throughout the history of thought and religions. Technical capacity and responsibility, for example, were fundamental aspects in forging the basic concepts of the great Western religions and for Greek thinkers.

For Ancient Greek thought, in fact, in a world governed by the category of *ananke*, namely "compulsion" or "necessity," human responsibility was relegated exclusively to the city. Thus, in the outside world, the "power of man" could do nothing. Famous is the passage of Sophocles' *Antigone* (Sophocles, ca. 441 B.C./2001), which describes, in fact, the nature that is recomposed after the passage of human technology: both the earth that covers the furrow dug by a plow, and the surface of the sea that closes the wake left by ships. Even more famous is

E. Giorgi (✉)
Tecnologico de Monterrey, Escuela de Arquitectura, Arte y Diseño, Chihuahua, Mexico
e-mail: egiorgi@tec.mx

E. Giorgi et al. (eds.), *Design for Vulnerable Communities*, The Urban Book Series,
https://doi.org/10.1007/978-3-030-96866-3_2

the story of Prometheus who delivers fire to humanity, a symbol of technical knowledge. In the tragedy *Prometheus Bound* by Aeschylus (Aeschylus, ca. 460 B.C./1996): to the doubts raised by the choir whether the new technical knowledge does imply future damage to the harmony of nature, Prometheus replies that the environment, governed by *ananke*, is so perfect that it does not have to fear technical capacity of people. The limited technological power of that time dictated this conclusion and precisely for this reason it is very interesting to observe how the relationship between technical power (although limited) and the environment was already the subject of profound reflection for the Greeks.

Similarly, the Jewish-Christian conception of this relationship between humanity and nature is even more relevant, whereby God, once he created man and breathed the vital breath of *nishmat-hajjîm*, made him master and responsible for creation. This conception of ownership and responsibility for the environment has therefore characterized the relationship between man and the environment for the whole of Western society (Ravasi, 2009).

Up to a century ago, human actions in a more than millennial history certainly changed nature (creating barriers, artificial dams, or sea canals, for example) and led to the extinction of some species, but after all, the quality of the relationship between humanity and the environment did not change enormously. In fact, there was an insufficient power to transform human actions from an intervention internal to a substantially balanced system, to a threat that could break the equilibrium of the whole system. Today, instead, as the sixth report of the Intergovernmental Panel on Climate Change (IPCC) shows, the impact of human actions on the environment has been catastrophic, so that it is assumed that the effects of climate change cannot be eliminated anymore (IPCC_AR6_WGI_Full_Report.Pdf, n.d.).

In fact, as widely discussed by the scientific community, the last century human activities (from the economic-productive to the political level) had consequences not only on local subsystems but profoundly distorted the global ecosystem at different scales, from the nano-metric dimension of genetics to the mega-metric dimension of the atmosphere. If this worries a lot (at least the scientific community), it should worry even more the fact that nowadays humanity does not know how to regain control of the situation, putting into practice effective actions and truly regenerating a new balance for this relationship: a new balance, not a return to a *status quo ante*, which is already impossible to restore.

2.1.1 Technological Development and Society

In this complex scenario of the Anthropocene (Steffen, Grinevald, et al., 2011), it clearly emerges that it is the development of technological capabilities that closely accompanied humanity along in these changes and allowed it to have this impact on the planet. Without knowledge and technological means, humanity could never have changed the environment as much as it has done in recent decades (Steffen, Persson, et al., 2011). Moreover, the technological development and the scientific revolution, which underlies it deeply, contributed in changing the way humanity perceives the environment: "*not like a schoolboy who believes everything the teacher tells to him, but as a judge who forces the suspect to answer his questions,*" says Galimberti (Galimberti, 2009; Galimberti, 2012) to express this change.

If the scientific and technical vision of the world has changed the relationship between humanity and the environment, even more shocking is the transformation that is taking place in the management of this relationship. Before it was the human will (the political system) to indicate, with choices and decisions, which actions to undertake and which direction development should have, now the technological system imposes (to an extent never seen in previous historical phases) the direction towards which humanity must move. As Galimebrti affirms, today humanity is a co-historian of the technological system: "we have become co-historians: history is made by technical rationality and we take part in this history, waiting to become a-historians [...] History is no longer made by man but by technical-scientific rationality, with respect to which we are co-historians" (Boncinelli & Galimberti, 2000).

Technological development changed the production of goods and services, as well the social relations and the way people relate to their cities and territories. The firm position of numerous authors is that this unbridled technological development is the cause of an overall bewilderment of our society, the generator of the strong individualism with which we face every day, as well as the main force that promotes the constant increase of innovation and change. Technological development is, by its nature, a force driven by the need for constant change, also to resolve crises and changes in the environment. As pointed out by Marvin Harris in *Cannibals and Kings* (Harris, 1977), technical and technological evolution is often due to external crises and changes. Thus, for example, the transition from hunting with spears to hunting with bows is due to the reduction in preys' size, while the transition from hunting to agriculture would be due to important changes in the environment, which pushed populations to reduce migrations.

The speed of contemporary changes, however, entails enormous ethical and moral problems, which are widely debated (Brey, 2012) (Shrader-Frechette & Westra, 1997) (Belk, 2020) (Sharma & Chaturvedi, 2020) and highlights the urgent need to reformulate the role and responsibility of humanity. We need, therefore, to reflect on contemporaneity and on the new technological system that leads the development and aims at self-empowering, marginalizing the human will.

For this reason, to encourage a deep discussion on contemporary phenomena, it is useful to introduce the term *Technocene*, which quotes the famous term "Anthropocene"adding considerations on the role of technology (Giorgi, 2020d). In fact, alongside the indispensable considerations on the *Anthropocene* and its challenges for society, environment, and design, it is equally essential to reflect and understand the role that the technological system assumes in this global scenario. Hence, the term *Technocene* wants to highlight how all the events of the Anthropocene are taking place in a contemporaneity in which events and history are led mainly by the technological development and not totally controlled by human will. As it will be deeply discussed along this chapter, this idea has important consequences for vulnerable communities and for design strategies.

2.1.2 Market of Technological Development

In addition to these implications on an ethical and moral level, there is a more easily visible and understandable dimension: the market[1]. In fact, since the technological one is a system in constant change and renewal, it has an incredible impact on the market. In recent years, the size of the market for technology products has increased considerably and will increase even more soon. To represent this scenario, data presented by the United Nations "Technology and Innovation Report, 2021" are considered (Technology and Innovation Report 2021, 2021). The report refers to the 11 "digital frontiers," protagonists of the so-called *Industry 4.0*. They are technologies that are rapidly developing and emerging as the technologies that will change the world in the coming years, improving the quality of life and production: artificial intelligence (AI), Internet of things (IoT), big data, blockchain, 5G, 3D printing, robotics, drones, gene editing, nanotechnology, and solar photovoltaic (Solar PV).

Making predictions about the future of the technological system is difficult (Quinn, 1967), because predictions are difficult by themselves and because the technological system, by its very essence, does not respond to decisions or human needs, but is developed according to processes of continuous improvement that do not necessarily align with the needs of the market or of people.

Nevertheless, in the short term, we can define some directions in which the world of technology is heading. On one side there are macro trends, on the other side there are well-defined fields which, unless significant changes, can be considered as the next areas of development: precisely the "11 digital frontiers." These eleven technologies alone represented a $350-billion market in 2018, and this same market will grow to over $3.2 trillion by 2025. This means growth of nearly $3 trillion in just 7 years (Technology and Innovation Report 2021, 2021). Another interesting data about technological development, for example, concerns the diffusion of the Internet, which increased by 1331.9% between 2000 and 2021 (World Internet Users Statistics and 2021 World Population Stats, n.d.).

Nevertheless, when a phenomenon undergoes a significant quantitative change, as in this case, the same phenomenon undergoes also a significant qualitative change. In fact, based on this principle, the relationship people have with technology and how they use it to relate with the environment changed. Humanity entered

[1]Although technological development is still driven by a capitalist economic system, often oriented to the exclusive pursuit of profit, and the cause of many contemporary problems, this dimension is omitted in this work because the research aims to focus exclusively on the Technological System, which is self- develops independently of social control, representing an unprecedented phenomenon in history.

the fourth industrial revolution, in which the physical and digital dimensions are combined with the biological one (and this is the first time this happened). The transformation, therefore, is deep: it concerns both the context, but also the human being and the life itself. Elon Musk's company "Neuralink," with its goal to increase the potentialities of human intellect by installing neural implant in the brain, is perhaps the best example of these efforts that are being made in recent years to physically unite the digital and biological worlds (Gordon et al., 2018; Gupta et al., 2018; Kim et al., 2018).

Anyway, it is not only with digital grafts in the body that the human being is changing. Continuous exposure to the technological dimension profoundly changes the person so much that the terms "digital beings"(The Circle: A Novel by Dave Eggers) or "analog being in a digital environment" (Bailey, n.d.) start to be used, questioning the evolution of the species itself[2].

The environment has changed radically: the tools with which people live are all connected (or potentially all connected) in the IoT, allowing new information transmission that is ever faster, more effective, and richer in data. Furthermore, the IoT, combined with Big Data and AI, can easily allow effective predictions of the scenarios that can arise. The impact in this field is so great that today, the IoT and Big Data represent the largest business investment sector and the greatest demand for qualified personnel. In a virtuous circle, technological development improves the cost-effectiveness of the product, in turn allowing greater demand from the market and further technological improvements. The lowering of production costs and data retention should also allow for a massive democratization of the use of data, allowing more and more people to benefit from this phenomenon.

At the same time, the availability of a lot of data also changes their quality and the quality of the work that can be done with them. In this way, new IT opportunities that impact the lives of people, communities, and society are generated. There is talk of personalized and preventive medicine, of prediction of natural disasters (floods, droughts, etc.), or of social phenomena (poverty, terrorism, crime, etc.).

Data takes on an incredible value because they are more and more easily processed by the machines; thus, the "technology drives data, data catalyzes knowledge, and knowledge enables empowerment" process will assume unprecedented importance. This constantly strengthening technological system has significant consequences for the economic and political-administrative system: as the CEOs of a company make decisions based on the calculations and forecasts offered by a Big Data and AI system, the administrators of a city will be able to make decisions based on forecasts of the social phenomena that will affect a city. Data processing and forecasts can therefore be put at the service of society, for the well-being of vulnerable communities too. However, much remains to be done in the public and social sector, which has not yet fully exploited the potential offered by big data, IoT and AI and which could thus solve many of the problems that most afflict vulnerable

[2] As reported in Future Shocks, "the current pandemic has shown us is how important technology is for maintaining and facilitating communication - not simply for work purposes, but for building real emotional connections" (Future Shocks, n.d.).

communities. The World Economic Forum is convinced of this, for which the use of this technology will certainly bring many benefits in achieving the SDGs (Why Tech Will Be Key in Our Quest to Hit the SDGs, n.d.; How 5G Can Advance the SDGs, n.d.). Anyway, to meet these new scenarios without losing the "human dimension" of development, a cultural revolution is necessary, so to understand how the potential of these new technologies can be made available to society and in particular to the most vulnerable populations (Giorgi, 2020c).

2.2 Vulnerable Communities in Technocene

Undoubtedly, vulnerability has always been a very complex issue, whose characteristics could be defined in a variable way, on the basis of the geographical, cultural, and social context, and of the historical moment of reference as well (Brooks, 2003) (Schroeder & Gefenas, 2009). Although complex, the concept of vulnerability, however, has a very interesting feature, namely that of being able to highlight some aspects that go beyond a purely quantitative evaluation of a phenomenon. In fact, "vulnerability" differs from "risk" because of its ability to highlight aspects that are not only quantitative, which instead are very suitable for measuring "risk," and it is in this quantitative dimension that the success of risk concept lies (Martin, 1996). That is, alongside the more "easily" measurable evaluations of "risk," the concept of "vulnerability" allows to highlight the qualitative nuances that characterize contemporary territories, bringing to light those aspects of daily life, made up of dreams, hopes, and necessities that at times do not adapt to quantitative analytical structures.

In addition to this aspect of opportunity between qualitative or quantitative analyzes, it is appropriate to reflect how the concept of vulnerability is very up-to-date. In the past, in fact, repeating themselves almost identical from generation to generation, the challenges to society were clearly defined and, consequently, the aspects linked to the vulnerability of a population were much more easily defined. Today, on the contrary, the infinity of new challenges that contemporaneity launches to humanity requires a much deeper reflection on the issues that are involved in the definition of a vulnerable population, in order to be able to define, study, understand, and resolve them. The covid-19 pandemic has highlighted how some sectors of the population are more vulnerable than others, being able to react in a limited way to sudden crisis situations that will increasingly impact humanity (Abedi et al., 2021). Precisely the pandemic has shown how the most vulnerable populations have experienced more than others the impossibility of following the indications for social distancing, the impossibility of acquiring the necessary protection measures (masks, antibacterial gel, etc.), as well as the limitations to access health services in the appropriate times and ways. All these limitations have been added to the problems related to job insecurity and the restriction for not having access to information, or not to know how to react to information. Finally, the pandemic has shown how in the most vulnerable populations, the lack of technological means and knowledge has implied a decrease in access to education for more than a year for the youngest. This last example is a clear demonstration of how the different distribution of technological possibilities impacts a community, creating an even greater difference between those students from non-vulnerable communities, able to continue their lessons online, and students from vulnerable communities who, due to lack of opportunities, have had to stop school for more than a year (Masonbrink & Hurley, 2020).

2.2.1 Vulnerabilities in the Technocene?

Contemporary technological development is creating new and unprecedented challenges for society, contributing to enrich incredibly the complexity of the concept of vulnerability. Furthermore, technological development itself has always contributed creating important inequalities: from the first industrial revolution, and so for all the other revolutions driven by technical-scientific innovations. Whenever there was a significant change in technological development, there was also an increase in inequalities both within individual countries, as well as between countries and geographical macro-areas, and between the epicenter of technological revolutions (the western world) and the "periphery" (all other countries) (Technology and innovation report 2021, 2021).

However, despite this, in some way, humanity has always been able to face the new conditions, adapting to the changes imposed by technological evolution. In fact, albeit with difficulty, societies adapted to the new demands imposed by technology: skills changed, as production needs and, consequently, job opportunities did. The changes imposed by technological development had such a slowness to allow societies to react and ensure that the jobs that were lost could be recovered in some other way, thanks to the new technologies.

Today, on the contrary, according to the United Nations, this scenario is no longer plausible, since future technological revolutions will have such an impact and speed that the changes generated will be so strong and fast and it will be difficult to understand how society can react to these transformations (Technology and Innovation Report 2021, 2021). Consequently, it can be imagined that the next technological transformations will have an even greater impact on society and will create even stronger inequalities than the previous ones.

This prospect of ever more rapid and invasive future revolutions, with which humanity will not be able to keep up, is not limited only to the workplace but will have implications in many more areas of daily life. This will make even more complex and difficult being able to define the concept of vulnerability in the future: therefore limiting, as is often wrongly done, the new vulnerability to productive implications is dangerously deceptive. Vulnerability in the *Technocene* era thus takes on a much broader and more complex meaning. The aspect that summarizes this complexity could be recognized in the fact of not being able to access to technological tools (for shortage or disinterest). This barrier means the impossibility of being able to use technological platforms that will be increasingly indispensable to face everyday life and to get out from situations of vulnerabilities.

Often, when referring to the relationship between technology and vulnerable communities, attention is focused mainly on two cases:

- environmental emergency situations, in which a technological solution is adopted to respond to a situation of vulnerability suddenly created. This is very easily seen with the implementation of geo-spatial models that can help to organize responses in case of important events (Bazemore et al., 2011; Krstikj et al., 2021), in the use of drones as the first technological tool to understand the intensity of a disaster (Zwęgliński, 2020), in the use of big data for emergency forecast, management, and response (Song et al., 2020), or the more generic use of technological innovations and mathematical models to solve crisis or emergency situations (Chui & Ko, 2020; Srivastava, 2009; Sun & Yamori, 2018).
- attention to vulnerable groups (people with physical or psychological limitations, minorities, etc.) to whom technology can offer solutions to reduce vulnerability or facilitate their participation in social life. Examples are AR/VR projects designed to generate empathy towards the most vulnerable or applications that allow for solutions to facilitate access to goods and services by these groups (Merriam, 2009) (Ludert, 2018) (Chuah, 2019).

However, when referring to the impact of technology on the vulnerability of communities, the mention is to communities made up of people with different characteristics, dreams, and needs, who experience a state of vulnerability distributed over a period of time (a day, a week or a particularly cold season) and not just during an emergency event.

Although some authors highlight how the cohesion of a neighborhood can be associated with the "mechanism of cumulative advantage" (Méndez et al., 2021), community has always been an indispensable element for the well-being of society: a protective dimension for the individual, both in moments of calmness (Giorgi, 2020a) and in moments of crisis (Giorgi et al., 2021; Palgi & Getz, 2014), demonstrating that the communitarian dimension must be rediscovered in the contemporary world to allow society to overcome the challenges of the future (Giorgi, 2020b).

Therefore, the purpose of this research is precisely to open a reflection on how the ideas of vulnerabilities must change in the "Technocene," an era in which history is dictated by technological development that inserts powerful and unprecedented variables in society. The research aims to understand how technology can increase or decrease the vulnerability of populations and what the consideration urban-architectural designers should contemplate in these new environments: how the processes, methodologies, and objectives of design should eventually change.

2.2.2 *Scenarios*

Relating technological development and vulnerability mainly implies reflecting on two levels: the very essence of technology and the environment it creates (Martin, 1996). Leaving aside the implications that technological development can have for society as a whole, which would require analyzing the optimistic and pessimistic positions (in daily life respectively represented by the positions of Mark Zuckemberg and Elon Musk), there are two main scenarios that can be pre-figure and that interest the research at this level of discussion on vulnerable communities. On one hand, technological development is highlighting various possible practical applications, which, if well applied to vulnerable realities, could reduce the vulnerability of people and facilitate the transition to a much more equitable and emancipated society, capable of achieving in time the ambitious SDGs. On the other hand, a worrying scenario is envisaged, in which the inability to incorporate technological innovations leads vulnerable communities to be even more vulnerable and to fall into a state of immobility. This would make even more difficult (than it is already now) to reduce the gap that separates the most vulnerable communities from the whole society, thus creating an increasingly unjust, unequal, and unsustainable society.

2.3 Dangers and Negative Scenarios for Vulnerable Communities

2.3.1 Technology Illiteracy

Without a doubt, illiteracy is considered one of the main problems of contemporary society and the fight against it is among the priorities of both the United Nations (Goal 4|Department of Economic and Social Affairs, n.d.) and most individual countries. Even if the actions carried out in the last decades in this direction are certainly giving results that allow us to hope for a better future (SDG Indicators, n.d.), much remains to do, in particular due to the numerous crises humanity is going through, including the Covid pandemic (Freitag, 2021) and the technological development (Selfe, 1999) (Yienger, 2016) (Donovan, 2015).

Thus, how important the "traditional" literacy is, that is reading, understanding, and writing, just as important is and will increasingly be the "digital" literacy. It will be increasingly necessary to know how to move with critical skills in the immensity of the technological world and to be able to use (optimally) the technological resources that are becoming more and more prevalent in everyday life. Undoubtedly, just as "traditional" literacy brings with it numerous benefits (6 Benefits of Literacy in the Fight against Poverty|Con-cern Worldwide US, n.d.) (Post, 2016), so too digital literacy will bring positive effects (Mudra, 2020). According to the World Economic Forum, technology literacy could help in closing the skills gap: digital technologies applied to education can contribute to lower the cost and improve the quality (WEF—New Vision for Education_Report2015.Pdf, n.d.). In particular, education in the use of technology can allow improvement in the required skills for XXI century (Spires et al., 2019). We are used to assuming that our daily technological experience is common to all, but the reality is quite different. First, by cost or interest, access to technological means is limited in various parts of the world, as well as the infrastructure necessary to allow the effective use of digital media and the knowledge necessary to use technology too.

Currently (March 2021) Internet Usage Statistics estimates as 5,168,780,607 internet Users, which represent the 65.6% of global population, with a big disparity between North America users (93.9%) and Africa users (43.2%) (World Internet Users Statistics and 2021 World Population Stats, n.d.). The differences in the use of technology between countries, therefore, are very high, as well as within the same countries, and the level of digital literacy is very different, highlighting precisely the most vulnerable communities which, in terms of digitalization, also suffer for gender inequalities (Antonio & Tuffley, n.d.; Ong, n.d.). The danger is that "digital inequality" becomes a risk for "digital literacy". If this gap between genders, communities and countries were to continue to increase, it would become a problem both for the most vulnerable communities (limited in their ability to use technological resources) and for the whole local and global societies (O'Toole, n.d.). This would push away the possibility to reach the goals of sustainable and equitable development.

2.3.2 Data Production and its Density

Data are and will increasingly be a huge source of information and, consequently, of wealth. Technological development is moving in the direction of making the production, storage, and interpretation of data one of its priorities. As people who live in the duality of the physical and digital world, we are a continuous source of data that we—more or less consciously—transmit to companies so that they can process them and, in the best of cases, give us useful information. The more obvious example concerns traffic applications, in which users receive information on the circulation situation thanks to information shared in real time by other users.

This immense data management, called "Big data," opens the doors to make important reflections on the dimensions of the digital world that is becoming ever more immensely large[3] and the importance that data production is having. Those who produce data, which can be used by companies, receive something in return: services (as in the case of traffic apps), economic payments or, more generally, special attention. Therefore, the problem arises for those who do not have the economic, infrastructural, or numerical strength to generate large amounts of data. This means that an urban community will receive more attention from businesses than a rural community, where the connection capacities are more limited, the network slower and the quantity of devices capable of transmitting data is less. The density of data production will therefore be lower in vulnerable communities and consequently the interest of companies in them will be lower. An interesting project, in an area of a Biosphere Reserve in the Mexican Sierra Gorda, studies about design strategies to implement community participation, social sustainability, and local development, based on digital platforms (Cobreros et al., 2019).

[3] It was estimated that in 2020, 1.7 MB of data were created every second for every person on earth.

2.3.3 Exclusion

A further danger that, in particular as designers, we have to consider is the exclusion of the most vulnerable population from the generation of services and technological products. Technological platforms are now conceived, designed, and developed in a uniform way. While various platforms already think of vulnerable communities as primary users, it is very rare that people from these communities have participated in the design stages of the platform. This is a very big problem because culture is making the mistake of excluding the users from the design, of thinking that all users have the same needs and that an application can have the same effects for different contexts and users.

As in architecture, designers try to integrate residents in the design of buildings and urban spaces with participatory processes, so it is necessary to think about the design of these new digital spaces in which users are destined to entrust important aspects of their lives. The lack of shared planning, in addition to limiting the impacts of a digital resource, could also lead to unexpected processes of discrimination or segregation. For this reason, the development of new technologies will have to be increasingly contextualized, making the solutions tailored to communities' challenges.

2.4 Opportunities and Positive Scenarios for Vulnerable Communities

2.4.1 Reduction in Distances

To explain the importance of willpower, a story that is often used is the one of the Indian Dashrath Manjhi who lost his wife for not being able to reach the nearest hospital in a short time, from which only an impassable mountain separated them. Determined to change the situation, Manjhi dug a passage in this mountain for the next 22 years, reducing the distance between his village and the hospital from forty to a few kilometers. This story shows very clearly an aspect that characterizes the life of people in vulnerable communities: the impotence to face the lack of services or the impossibility of reaching them in adequate times.

Beyond to improve the infrastructures, as Manjhi did "artisanally," our society must absolutely take into consideration all those possibilities to offer better services in a homogeneous way throughout the territory. Technology undoubtedly offers this possibility and allows improvements in an innumerable number of fields. To name a few of them:

- *Telemedicine*: Certainly, a field that will develop considerably in the future is that of telemedicine, which makes it possible to carry out diagnostic tests and (potentially) interventions in vulnerable communities in remote areas. The pandemic has shown how, albeit still with important limitations, it has been possible to start offering online medical visits for virtual consultations and remote monitoring (Ajibade et al., 2020). It will be even more interesting to be able to count with surgical operations carried out by surgeons who operate miles away by means of a robot with results almost equal to or better than traditional surgery (Bakalar, 2021) (Morrell et al., 2021).

- *Online teaching*: As already discussed above, the pandemic has forced to experiment on a large scale those technologies supporting online teaching. This has highlighted the problems in terms of vulnerability and access to resources but, at the same time, it has allowed us to understand how much this technology can be relevant to standardize the educational offer in a territory in the near future, enabling to access information offers without the need to move. This means access to education without the consequences of commuting, renting, quitting a job, abandoning a family in difficulty, etc. If this is valid for a manager who from New York can follow a course given by the University of California, it is even more important and relevant for a Raramuri girl who, from her village in the Sierra Tarahumara, can follow the elementary class transmitted by her teacher 40 km away, without the necessity to commute every week along a dangerous route. Moreover, interesting studies, as the one by Krstikj, show how combination between technological devices and social innovation can turn students in agents of social change (Krstikj, 2021).
- *Counseling*: Being able to access consultancy from a remote community with an expert to solve practical problems, such as repairing an agricultural machine, improving welding techniques (Lai et al., 2020), or receiving advice on how to improve sales strategies. Benefits that residents of vulnerable communities can count on today thanks to the diffusion of technological products.

2.4.2 Self-production

Another aspect that characterizes the vulnerability of many communities, especially the isolated ones, is the impossibility of being able to obtain adequate supplies. Whether it is a need for everyday life or for emergencies, the difficulty in being reached by the production and supply chain can be a very serious problem. To these needs, technologies related to 3d printing are giving solid and effective answers. There are numerous cases of isolated communities that can print some necessary goods reducing thus their vulnerability dictated by the procurement of goods. Obydenkova et al., for example, report a case of 3d printing technology used to improve the work of a remote community of reindeer herders: printing of ear-tags, electric fence components, and lasso accessories (Obydenkova et al., 2018). Their research states that this technology can reduce production costs by up to 63% thanks to reduced costs of raw materials and transport, as well as being able to offer greater freedom in production. Another of the many examples about the benefits of 3D printing for vulnerable communities is described by King et al. that present two open-source mobile digital manufacturing facilities powered with solar photovoltaics capable of printing customizable Open Source Appropriate Technology (OSAT) in any community with access to sunlight (King et al., 2014). Therefore, 3d printers technologies can be used to produce customized goods in remote areas and to democratize production, leading to the rebirth of craftsmanship in a "digital" form.

2.4.3 Transportation

Furthermore, vulnerable communities can count, for all that cannot be produced locally by self-production techniques, on more effective and accessible transportation systems, which, both for goods and people, will be soon revolutionized thanks to technological innovations ('5 Advancements in Transportation Technology', 2018). Of course, the most vulnerable communities could benefit from them.

Relatively "simple" services such as Uber have already shown how much they can upset the transport market and how much they can affect users' changing habits. Important revolutions such as those that lie ahead in this field, especially in terms of autonomous cars, could have important positive effects. In fact, the efficiency of autonomous vehicles can reduce the costs of the service, thus making it more accessible and efficient (Automated Vehicles Comprehensive Plan | US Department of Transportation, n.d.).

Furthermore, the development of technologies related to drones could provide important benefits in terms of the transport of goods and people. Being able to deliver quickly, based on unplanned or unexpected needs, medicines or essential products is envisaged as something with a potentially very positive impact for vulnerable communities that today can hardly access emergency goods quickly. Examples are the projects of the World Food Programme implemented thanks to Unmanned Aircraft Systems (UAS) drone technologies (World Food Programme, 2020).

Certainly, all these technologies relating to means of transport depend very much on associated technologies such as the use of big data, artificial intelligence, GPS devices, research on sustainable and economic fuels, etc. but the impact that these changes will have on cities and territories will be impressive and the most vulnerable communities cannot stay out of them. A closing observation on this issue comes from a research by Sochor and Nikitas according to which vulnerable populations are not homogeneous when expressing attitudes towards transport technologies, but their assessment criteria tend to be "pro-social" as they usually consider that the social benefits outweigh the personal ones (Sochor & Nikitas, 2016). This highlights how improvements of transport means in vulnerable communities can have an even stronger impact on the social and community level.

2.4.4 Monitoring and Mitigation

We cannot forget how the concept of vulnerability is closely associated with the more quantitative concept of risk and how both of them become fundamental to highlight potentially dangerous situations or any changes that can suddenly impact a vulnerable community. For this reason, technology can help with the development of more efficient monitoring and mitigation systems: drones and earth observation technologies already demonstrate how equipment can be used to monitor all those situations of potential danger for vulnerable communities (Jayson-Quashigah et al., 2019; Brito et al., 2020, p. 19). As the case of Nanning, studied by Zhong et al., proves, technology and science improvements can also have higher relevance than social-economic or built-environment improvements in mitigating urban hazards (Zhong et al., 2020). Furthermore, as demonstrate by the UNICEF's experience in Uganda, Real Time Monitoring (RTM) can help in receiving data and transmitting information from and to vulnerable communities. Rapid SMS technology (mTrac and U-report) helps youths in vulnerable communities to participate in surveys and to receive information, allowing the agency to reduce costs, map in real-time, and provide better support (Cummins & Huddleston, 2013). Also, the research by Eyrich-Garg and Moss, about surveying homeless individuals, shows how technology devices can support in better understand social and urban phenomena involving vulnerable individuals (Eyrich-Garg & Moss, 2017). Technology therefore makes possible to map natural and social risks more easily than traditional systems, often limited to manual data collections in the most vulnerable areas (An et al., 2019). As Galer also states: "We have the intelligent data to help vulnerable communities" (Galer, n.d.).

2.4.5 Engagement (in Particular, Youth) with Global Issues

A last relevant aspect when talking about the benefits that technological development can bring to these communities regards the possibility of developing interesting forms of engagement between the community and global issues that, otherwise, can hardly enter the sensitivity of the most marginalized people. AR/VR, AI, new programs, and technological devices can bring important thoughts on global issues at the local level, in vulnerable communities, thus allowing residents to be aware of both the risks that their community could run in the coming years, and of the responsibility that everyone has in reducing the factors that increase the risks. As argued by Napawan, benefits in implementing technological devices are particularly visible in engaging youth with global and local problems (Napawan et al., 2017). This engagement can range from environmental issues, such as pollution or climate change, to social, health, or local production issues: various sectors can benefit from reaching these vulnerable communities by technologies that facilitate the commitment on local and global issues.

2.5 Policies, Strategies, Implementation, and Benefits

This contrast between potentially negative (Technology illiteracy, Data production and exclusion) and positive (distance reduction, self-production, transportation, monitoring and engagement) effects of technological development must be strongly considered at the moment to imagine the future of a vulnerable community (Fig. 2.1). Combined with institutional and political wise strategies, the development of technological devices can empower communitarian resilience everywhere, from urban "first-world" contexts to rural "third-world" (Shiferaw et al., 2014).

The empowerment of communitarian resilience can be particularly significant if it's addressed to face the crisis, related to climate change and forecasted environments (Yan & Roggema, 2019). There is no doubt that these considerations must be made by designers who are called to intervene in vulnerable communities because, as we have seen, the reality of Technocene and the dimension of the technological apparatus are increasingly gaining a role of protagonist in defining the destinies of a community. In these contexts, in addition to the skills and sensitivity of the designers, other factors are also important, such as public policies, education, infrastructures, and the market. The research presents some considerations about the importance of each of these factors.

2.5.1 *Public Policies and Engagement*

Even if, as the theoretical basis of this chapter, the research has taken the idea that technology is the historical actor of our contemporaneity and the system that defines historical processes, the role that public policies have in supporting the diffusion of technological tools is still fundamental. Indeed, bringing technological innovations

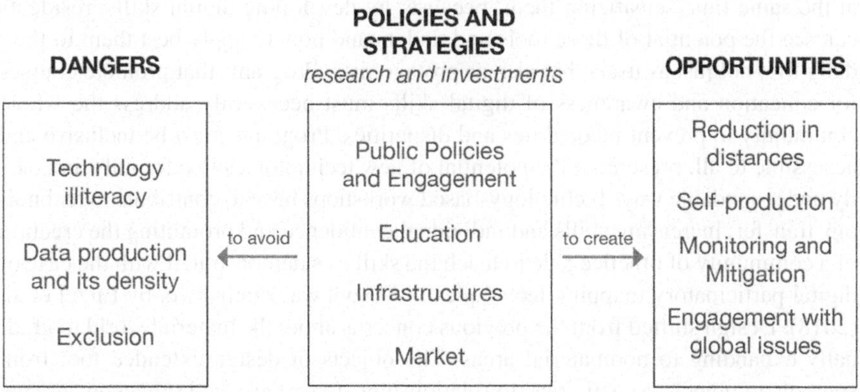

Fig. 2.1 Impacts of Policies and Strategies in Opportunities and Dangers of Technological Development in Vulnerable Communities (scheme by the Author)

to vulnerable communities is a political choice that requires severe efforts—on one hand, there is the need to plan interventions and investments, and on the other hand, the need to foster the participation of the population in these initiatives. The ability to develop public-private partnerships plays a fundamental role so that these efforts also do not end in providing technological resources that make vulnerable communities' residents mere users, but they have to be occasions to generate production and economic opportunities. As argued by Sianipar et al., bringing technological services to vulnerable communities requires a clearer understanding to empower people rather than only giving technology (Sianipar et al., 2014). Moreover, without any doubt, merging the digital environment with the social participation can bring important benefits in terms of social well-being, reducing conflicts and strengthening democratic participation. This is particularly true when referring to important urban transformation as those described by Falcón and Fuentes (2019). An important challenge for the next few years, in vulnerable communities, will be the ability to guarantee equitable access to the use and the management of resources, above all the technological ones that with their ability to provide fast, personalized, and efficient solutions are among the most relevant for adaptation to climate change.

Finally, with regard to public policies, a significant problem to consider is the lack of regulation of the ways in which technology can support vulnerable communities, with reference to moral values, norms, and commitments.

(Wang, 2019).

2.5.2 *Education*

Alongside public policies and community engagement, it is essential to encourage digital education in vulnerable communities in order that the most population can take advantage of the development of technological systems.

Bringing digital infrastructure and technology to vulnerable communities implies at the same time sensitizing them, because, by developing digital skills, residents can see the potential of these tools and understand how to apply best them to their daily life, not just as users, but also as protagonists. Programs that promote courses for education and awareness of digital skills must necessarily address the whole community to prevent inequalities and disparities. Programs must be inclusive and accessible to all, presenting the potential of new technologies in a friendly and easily understandable way. Technology-based workshops have to contribute to technology transfer: increasing skills and individual confidence and promoting the creation of a community of practice able to teach the skills, as demonstrated with the case of digital participatory mapping techniques in Mazvihwa, Zimbabwe, by Eitzel et al. (2018). Design shifted from the previous concerns about the material world to gradually expanding to nonmaterial areas. The objects of design extended too: from symbols to objects, to activities, to relationships, to services and processes, to systems, environments, and mechanisms. Real innovation is often learning by doing (Ni & Cattaneo, 2019).

2.5.3 *Infrastructures*

Even if the cost of technology and data is constantly decreasing, for a private individual entering the technological world can be an important problem from an economic point of view, especially for the most vulnerable populations. Furthermore, as we have seen, the most remote areas are unattractive for private enterprise. For this reason, it is necessary to develop joint efforts between private and public initiative to allow all citizens to be able to count with the basic technological resources to face the digital challenges of contemporaneity and take advantage of digital resources to reduce the state of vulnerability wherever they are. Public policies, educational and awareness-raising efforts will have limited impact if not followed by effective applications that are possible only if the population can effectively rely on technological resources in their daily lives. Furthermore, it is precisely in being able to provide a network of structures and infrastructures suitable for the massive use of technology that the inequalities of gender, generation, purchasing power and access to education or health can be reduced.

2.5.4 *Market*

We should consider that technological solutions do not reach everyone uniformly. Different accesses to technological devices, different levels of e-literacy, and overall capacities to access to technology limit the effectiveness of strategies based on the use of such devices (Van Winkle et al., 2017). Of course, as stated by Van Winkle with a focus on healthcare sector, several solutions can be implemented to facilitate the usefulness of these technologies: design for different levels of e-literacy, role of the health workers as liaisons with patients, analytics and customer relationship management tools to provide the appropriate interventions (Van Winkle et al., 2017). According to Nabben K. and Gardner-Stephen P., design of technology systems, in particular for crisis contexts, must be focused on "S4" (simple, secure and survivable systems) principles, trying to solve the problem and to support end user communities (Nabben & Gardner-Stephen, 2020).

2.6 Results, Benefits, and Conclusions

Several benefits, observed up to here, certainly satisfy many of the United Nations Sustainable Development Goals (SDGs): (3) good health and well-being, (4) quality education, (5) gender equality, (8) decent work and economic growth, (9) industry, innovation, and infrastructure, (10) reduced inequalities, (11) sustainable cities and communities, (12) responsible consumption and production, (16) peace, justice, and strong institutions, and (17) partnership for goals. These goals are undoubtedly the direct object of the benefits that technological development can bring. Furthermore, as we have seen, many other SDGs benefit indirectly from technological development.

All these goals, and in particular the improvement of living conditions and equality in vulnerable communities, however, could never be achieved if effective inclusion strategies are not developed, allowing to collaborate with individuals and groups who are the main users of these new technologies: vulnerable people must be the co-authors of technological development. They must be the protagonists and authors of technological progress too. We might say that technology along with a humanistic design is the starting point for complete sustainability both in social and environmental terms (Giorgi et al., 2020).

There is no doubt that the complexity so far represents a further challenge for the designer who provides sustainable solutions for vulnerable communities as the impacts at a local level cannot be overlooked. Moreover, these considerations are more relevant, as the role of the designer, especially of those who deal with vulnerable communities, is rapidly evolving to meet the contemporary needs. In fact, the designer involved in difficult urban or rural situations has to play the role of accompaniment in processes of self-production or participatory design.

Designing for a sustainable environment begins and ends with a deep knowledge of the particular place, complete with its unique characteristics of climate, type of soil, native species, and patterns of living, technology along with participation can enhance the comprehension of a place (Cattaneo et al., 2019).

In this scenario, proposing design choices that allow the community to adopt easily technological solutions is fundamental. For example, developing design solutions that allow women living in a vulnerable community, isolated and characterized by male emigration, to access the knowledge offered online or to receive practical training to solve domestic problems, can be a turning point element. Empowering the use of technology among young people means allowing practical or academic training in vulnerable communities, promoting an increase in social capital and in the territory development and a decrease in migration. Technology will change the way we live in our communities, our cities, and our territories. Being aware of this will allow us to take full advantage of its positive effects: (1) empower local communities, (2) initiate community-based entrepreneurship, (3) maintain the continuity of technological changes (Sianipar et al., 2014), and (4) reduce global inequalities.

This research aimed to highlight the effects that the development and diffusion of frontier technologies can generate in vulnerable communities. Clearly, a limitation lies in the fact that the focus and point of view is not from a technology field but from architecture and urban design. The technologies were therefore analyzed on the basis of their potential impact on communities, cities, and territories and with a mainly philosophical theoretical basis to have a general and holistic vision of the phenomenon. After this first look at the technological landscape, however, a further interdisciplinary study will be necessary. Furthermore, a more in-depth field research represents the subsequent development of this reflection to analyze the limitations and the potential in order to adopt technological tools in some vulnerable communities where the author and his colleagues are already working.

Acknowledgments The author thanks the external reviewers for providing critical and helpful comments on earlier versions of this chapter and on the book proposal. This work was funded by Fondo de Investigación 2020–2021 del Observatorio de Ciudades del Tecnologico de Monterrey (Mexico), through the research project "Design for vulnerables," project duration from October 2020 to December 2021.

References

5 Advancements in Transportation Technology. (2018, March 14). Ohio University. https://online-masters.ohio.edu/blog/5-advancements-in-transportation-technology/

6 Benefits of literacy in the fight against poverty | Concern Worldwide US. (n.d.). Concern Worldwide. Retrieved 5 September 2021, from https://www.concernusa.org/story/benefits-of-literacy-against-poverty/

Abedi, V., Olulana, O., Avula, V., Chaudhary, D., Khan, A., Shahjouei, S., Li, J., & Zand, R. (2021). Racial, economic, and health inequality and COVID-19 infection in the United States. *Journal of Racial and Ethnic Health Disparities, 8*(3), 732–742. https://doi.org/10.1007/s40615-020-00833-4. Scopus.

Aeschylus. (1996). *Prometheus Bound (Dover Thrift Editions)*. Dover Publications.

Ajibade, A., Younas, H., Pullan, M., & Harky, A. (2020). Telemedicine in cardiovascular surgery during COVID-19 pandemic: A systematic review and our experience. *Journal of Cardiac Surgery, 35*(10), 2773–2784. https://doi.org/10.1111/jocs.14933

An, A., Cao, L., Jia, Q., Wang, X., Zhu, Q., Zhang, J., Ye, X., & Gao, D. (2019). Changing abundance and distribution of the wintering swan goose anser cygnoides in the middle and lower yangtze river floodplain: An investigation combining a field survey with satellite telemetry. *Sustainability (Switzerland), 11*(5). https://doi.org/10.3390/su11051398. Scopus.

Antonio, A., & Tuffley, D. (n.d.). Digital literacy in the developing world: A gender gap. The Conversation. Retrieved 5 September 2021, from http://theconversation.com/digital-literacy-in-the-developing-world-a-gender-gap-28650

Automated Vehicles Comprehensive Plan | US Department of Transportation. (n.d.). Retrieved 6 September 2021, from https://www.transportation.gov/av/avcp

Bailey, D. D. E. (n.d.). The digital human: The cyber version of humanity's quest for immortality. The Conversation. Retrieved 16 June 2021, from http://theconversation.com/the-digital-human-the-cyber-version-of-humanitys-quest-for-immortality-108081

Bakalar, N. (2021, August 16). Are Robotic Surgeries Really Better? The New York Times. https://www.nytimes.com/2021/08/16/well/live/robotic-surgery-benefits.html

Bazemore, A., Phillips, R. L., & Miyoshi, T. (2011). Harnessing geographic information systems (GIS) to enable community-oriented primary care. *Journal of Map and Geography Libraries., 7*(1), 71–86. https://doi.org/10.1080/15420353.2011.534691. Scopus.

Belk, R. (2020). Ethical issues in service robotics and artificial intelligence. *The Service Industries Journal, 0*(0), 1–17. https://doi.org/10.1080/02642069.2020.1727892

Boncinelli E., Galimberti U. (2000). E ora? La dimensione umana e le sfide della scienza, con Giovanni Maria Pace. – Torino, Einaudi. p. 70.

Brey, P. A. E. (2012). Anticipating ethical issues in emerging IT. *Ethics and Information Technology, 14*(4), 305–317. https://doi.org/10.1007/s10676-012-9293-y

Brito, P. L., Kuffer, M., Koeva, M., Pedrassoli, J. C., Wang, J., Costa, F., & de Freitas, A. D. (2020). The spatial dimension of COVID-19: The potential of earth observation data in support of slum communities with evidence from Brazil. *ISPRS International Journal of Geo-Information, 9*(9). https://doi.org/10.3390/ijgi9090557. Scopus.

Brooks, N. (2003). Vulnerability, risk and adaptation: A conceptual framework. Tyndall Centre Working Paper No. 38, 21.

Cattaneo, T., Giorgi, E., & Ni, M. (2019). Landscape, architecture and environmental regeneration: A research by design approach for in-clusive Tourism in a Rural Village in China. *Sustainability, 11*(1), 128. https://doi.org/10.3390/su11010128

Chuah, S. H.-W. (2019). Wearable XR-technology: Literature review, conceptual framework and future research directions. *International Journal of Technology Marketing, 13*(3–4), 205–259. https://doi.org/10.1504/IJTMKT.2019.104586. Scopus.

Chui, C. H.-K., & Ko, A. (2020). Converging humanitarian technology and social work in a public health crisis: A social innovation response to COVID-19 in Hong Kong. *Asia Pacific Journal of Social Work and Development.*, 1–8. https://doi.org/10.1080/02185385.2020.1790412. Scopus.

Cobreros, C., Melón, M.E., Rosado, E., Lugo, N., & Maya, M. (2019). Rural community participation digital platform. In Cumulus Conference Proceedings Bogotá 2019. Sensing the city, sensing the rural.

Cummins, M., & Huddleston, B. (2013). Real time monitoring for the most vulnerable: UNICEF's experience in uganda. *IDS Bulletin, 44*(2), 57–68. https://doi.org/10.1111/1759-5436.12017. Scopus.

Donovan, S. (2015, September 16). School computer use 'may be affecting literacy and numeracy skills'. ABC News. https://www.abc.net.au/news/2015-09-16/computer-use-may-be-leading-to-literacy-numeracy-decline/6779986

Eitzel, M. V., Mhike Hove, E., Solera, J., Madzoro, S., Changarara, A., Ndlovu, D., Chirindira, A., Ndlovu, A., Gwatipedza, S., Mhizha, M., & Ndlovu, M. (2018). Sustainable development as successful technology transfer: Empowerment through teaching, learning, and using digital participatory mapping techniques in Mazvihwa, Zimbabwe. *Development Engineering, 3*, 196–208. https://doi.org/10.1016/j.deveng.2018.07.001. Scopus.

Eyrich-Garg, K. M., & Moss, S. L. (2017). How feasible is multiple time point web-based data collection with individuals experiencing street homelessness? *Journal of Urban Health, 94*(1), 64–74. https://doi.org/10.1007/s11524-016-0109-y. Scopus.

Falcón, C. E., & Fuentes, J. (2019). Improving social well-being through platform cooperativism. *CIRIEC-Espana Revista de Economia Publica, Social y Cooperativa., 95*, 161–190. https://doi.org/10.7203/CIRIEC-E.95.12655. Scopus.

Freitag, E. (2021, July 6). The Pandemic Will Worsen Our Reading Problem. Another Outcome Is Possible. Education Week. https://www.edweek.org/teaching-learning/opinion-the-pandemic-will-worsen-illiteracy-another-outcome-is-possible/2021/07

Future shocks: 17 technology predictions for 2025. (n.d.). World Economic Forum. Retrieved 22 June 2021, from https://www.weforum.org/agenda/2020/06/17-predictions-for-our-world-in-2025/

Galer, S. (n.d.). SAP BrandVoice: We have the intelligent data to protect society's most vulnerable people. Forbes. Retrieved 6 September 2021, from https://www.forbes.com/sites/sap/2021/07/13/we-have-the-intelligent-data-to-protect-societys-most-vulnerable-people/

Galimberti, U. (2009). Man in the age of technology. *Journal of Analytical Psychology, 54*(1), 3–17. https://doi.org/10.1111/j.1468-5922.2008.01753.x

Galimberti, U. (2012). Man in the age of technics. *Phainomena, 21*(82/83), 127–143.

Giorgi, E., Cattaneo, T., & Manzoni, G. D. (2020). Participation for sustainability: Why looking back to design the future? In K. Hadjri, L. Madrazo, & I. O. Durosaiye (Eds.), *Global dwelling: Approaches to sustainability, design and participation* (WIT transactions on the built environment) (Vol. 193, pp. 235–246). WIT Press. https://doi.org/10.2495/GD170191

Giorgi, E. (2020a). Co-housing. In E. Giorgi (Ed.), *The co-housing phenomenon. Environmental alliance in times of changes* (pp. 93–115). Springer, Urban Book Series. https://doi.org/10.1007/978-3-030-37097-8_4

Giorgi, E. (2020b). Hopes. In E. Giorgi (Ed.), *The co-housing phenomenon. Environmental alliance in times of changes* (pp. 259–263). Springer, Urban Book Series. https://doi.org/10.1007/978-3-030-37097-8_6

Giorgi, E. (2020c). In E. Giorgi (Eds.), *The co-housing phenomenon. Environmental alliance in times of changes* (pp. 27–58). Cham, Switzerland: Springer, Sharing as Cultural Revolution. Urban Book Series. https://doi.org/10.1007/978-3-030-37097-8_2.

Giorgi, E. (2020d). Technocene. In E. Giorgi (Ed.), *The co-housing phenomenon. Environmental alliance in times of changes* (pp. 1–26). Springer, Urban Book Series. https://doi.org/10.1007/978-3-030-37097-8_1

Giorgi, E., Martín López, L., Garnica-Monroy, R., Krstikj, A., Cobreros, C., & Montoya, M. A. (2021). Co-housing response to social isolation of covid-19 outbreak, with a focus on gender implications. *Sustainability (Switzerland), 13*(13). https://doi.org/10.3390/su13137203. Scopus.

Goal 4 | Department of Economic and Social Affairs. (n.d.). Retrieved 5 September 2021, from https://sdgs.un.org/goals/goal4

Gordon, C. R., Santiago, G. F., Huang, J., Bergey, G. K., Liu, S., Armand, M., Brem, H., & Anderson, W. S. (2018). First in-human experience with complete integration of neuromodulation devicewithin a customized cranial implant. *Operative Neurosurgery, 15*(1), 39–45. https://doi.org/10.1093/ons/opx210. Scopus.

Gupta, G., Pequito, S., & Bogdan, P. (2018). Re-thinking EEG-based non-invasive brain interfaces: Modeling and analysis. 275–286. https://doi.org/10.1109/ICCPS.2018.00034. Scopus

Harris, M. (1977). *Cannibals and kings* (1st ed.). Random House.

How 5G can advance the SDGs. (n.d.). World Economic Forum. Retrieved 17 June 2021, from https://www.weforum.org/agenda/2019/04/how-5g-can-advance-the-sdgs/

IPCC_AR6_WGI_Full_Report.pdf. (n.d.). Retrieved 30 August 2021, from https://www.ipcc.ch/report/ar6/wg1/downloads/report/IPCC_AR6_WGI_Full_Report.pdf

Jayson-Quashigah, P.-N., Appeaning Addo, K., Amisigo, B., & Wiafe, G. (2019). Assessment of short-term beach sediment change in the Volta Delta coast in Ghana using data from Unmanned Aerial Vehicles (Drone). *Ocean and Coastal Management*, 182. https://doi.org/10.1016/j.ocecoaman.2019.104952. Scopus.

Kim, Y., Meade, S. M., Chen, K., Feng, H., Rayyan, J., Hess-Dunning, A., & Ereifej, E. S. (2018). Nano-architectural approaches for improved intracortical interface technologies. *Frontiers in Neuroscience, 12*(JUL). https://doi.org/10.3389/fnins.2018.00456. Scopus.

King, D. L., Babasola, A., Rozario, J., & Pearce, J. M. (2014). Mobile open-source solar-powered 3-D printers for distributed manufacturing in off-grid communities. *Challenges in Sustainability, 2*(1), 18–27. https://doi.org/10.12924/cis2014.02010018

Krstikj, A. (2021). Social Innovation in the Undergraduate Architecture Studio. *Societies, 11*(1), 26. https://doi.org/10.3390/soc11010026

Krstikj, A., Contreras Ruiz Esparza, M. G., Vargas, J. M., Escobar, L. H., López de la Rosa, C., González Calderón, S. T., Giorgi, E., & Hinojosa, K. H. (2021). Decision-support tool for coordination of volunteers in large-scale lockdowns. *International Journal of Disaster Risk Reduction*, 102420. https://doi.org/10.1016/j.ijdrr.2021.102420

Lai, C.-H., Wu, T.-E., Huang, S.-H., & Huang, Y.-M. (2020). Developing a virtual learning tool for industrial high schools' welding course. *Procedia Computer Science., 172*, 696–700. https://doi.org/10.1016/j.procs.2020.05.091. Scopus.

Ludert, E. (2018). *"Diseñar para lo invisible, investigación-acción para hacer visible el autismo en la Ciudad de México*. Posgra-do UNAM.

Martin, B. (1996). Technological vulnerability. *Technology in Society, 18*(4), 511–523. https://doi.org/10.1016/S0160-791X(96)00029-2

Masonbrink, A. R., & Hurley, E. (2020). Advocating for children during the COVID-19 school closures. *Pediatrics, 146*(3) Scopus. https://doi.org/10.1542/PEDS.2020-1440

Méndez, M. L., Otero, G., Link, F., López Morales, E., & Gayo, M. (2021). Neighbourhood cohesion as a form of privilege. *Urban Studies, 58*(8), 1691–1711. https://doi.org/10.1177/0042098020914549. Scopus.

Merriam, S. (2009). *Qualitative research: A guide to design and implementation*. Jossey-Bass.

Morrell, A. L. G., Morrell-Junior, A. C., Morrell, A. G., Mendes, J. M. F., Tustumi, F., De-Oliveira-e-silva, L. G., & Morrell, A. (2021). The history of robotic surgery and its evolution: When illusion becomes reality. *Revista Do Colegio Brasileiro de Cirurgioes, 48*, 1–9. https://doi.org/10.1590/0100-6991e-20202798. Scopus.

Mudra, H. (2020). Digital literacy among young learners: How do EFL teachers and learners view its benefits and barriers? *Teaching English with Technology, 20*(3), 3–24.

Nabben, K., & Gardner-Stephen, P. (2020). S4: Simple, Secure, Survivable Systems Human-first crisis technology design principles. 2020 IEEE Global Humanitarian Technology Conference, GHTC 2020. Scopus. https://doi.org/10.1109/GHTC46280.2020.9342856.

Napawan, N. C., Simpson, S.-A., & Snyder, B. (2017). Engaging youth in climate resilience planning with social media: Lessons from #ourchangingclimate. *Urban Planning, 2*(4), 51–63. https://doi.org/10.17645/up.v2i4.1010. Scopus.

Ni, M., & Cattaneo, T. (2019). Design for Urban Resilience: A Case of Community-led Placemaking Approach in Shanghai China. Lecture Notes in Computer Science (Including Subseries Lecture Notes in Artifi-cial Intelligence and Lecture Notes in Bioinformatics), 11577 LNCS, 207–222. Scopus. https://doi.org/10.1007/978-3-030-22580-3_16.

Obydenkova, S., Anzalone, N. C., & Pearce, J. M. (2018). Prospects of applying 3-D printing to economics of remote communities: Reindeer herder case. *Journal of Enterprising Communities: People and Places in the Global Economy, 12*(4), 488–509. https://doi.org/10.1108/JEC-08-2016-0029

Ong, A. (n.d.). Excluded: How women suffer from digital poverty in the UK. Retrieved 5 September 2021, from https://www.aljazeera.com/features/2021/9/1/excluded-how-women-suffer-from-digital-poverty-in-the-uk

O'Toole, F. (n.d.). Fintan O'Toole: No one is safe when half of us are digitally illiterate. The Irish Times. Retrieved 5 September 2021, from https://www.irishtimes.com/opinion/fintan-o-toole-no-one-is-safe-when-half-of-us-are-digitally-illiterate-1.4575846

Palgi, M., & Getz, S. (2014). Varieties in developing sustainability: The case of the Israeli kibbutz. *International Review of Sociology, 24*(1), 38–47. https://doi.org/10.1080/03906701.2014.894344. Scopus.

Post, D. (2016). Adult literacy benefits? New opportunities for research into sustainable development. *International Review of Education, 62*(6), 751–770. https://doi.org/10.1007/s11159-016-9602-5

Quinn, J. B. (1967, March 1). Technological Forecasting. Harvard Business Review. https://hbr.org/1967/03/technological-forecasting

Ravasi G. (2009). Breve storia dell'anima (2003), Milano, Arnoldo Mondadori Oscar Saggi. p.80. Translation by the author.

Schroeder, D., & Gefenas, E. (2009). Vulnerability: Too Vague and Too Broad? *Cambridge Quarterly of Healthcare Ethics, 18*(2), 113–121. https://doi.org/10.1017/S0963180109090203

SDG Indicators. (n.d.). Retrieved 5 September 2021, from https://unstats.un.org/sdgs/report/2021/goal-04/

Selfe, C. L. (1999). Technology and literacy: A story about the perils of not paying attention. *College Composition and Communication, 50*(3), 411–436. https://doi.org/10.2307/358859

Sharma, S., & Chaturvedi, V. (2020). Ethical and legal issues of AI technology and its applications. *International Journal of Law and Legal Jurisprudence Studies, 6*(1), 24.

Shiferaw, B., Tesfaye, K., Kassie, M., Abate, T., Prasanna, B. M., & Menkir, A. (2014). Managing vulnerability to drought and enhancing livelihood resilience in sub-Saharan Africa: Technological, institutional and policy options. *Weather and Climate Extremes, 3*, 67–79. https://doi.org/10.1016/j.wace.2014.04.004. Scopus.

Shrader-Frechette, D. K., & Westra, L. (1997). *Technology and values*. Rowman & Littlefield Publishers.

Sianipar, C. P. M., Dowaki, K., & Yudoko, G. (2014). Technological solution for vulnerable communities: How does its approach matter? *IOP Conference Series: Materials Science and Engineering, 58*, 012022. https://doi.org/10.1088/1757-899X/58/1/012022

Sochor, J., & Nikitas, A. (2016). Vulnerable users' perceptions of transport technologies. *Proceedings of the Institution of Civil Engineers - Urban Design and Planning, 169*(3), 154–162. https://doi.org/10.1680/jurdp.14.00054

Song, X., Zhang, H., Akerkar, R. A., Huang, H., Guo, S., Zhong, L., Ji, Y., Opdahl, A. L., Purohit, H., Skupin, A., Pottathil, A., & Culotta, A. (2020). Big data and emergency management: concepts, methodologies, and applications. *IEEE Transactions on Big Data*, 1–1. https://doi.org/10.1109/TBDATA.2020.2972871

Sophocles. (2001). Antigone (Paul Woodruff, Ed.). Hackett Publishing Company.

Spires, H. A., Paul, C. M., & Kerkhoff, S. N. (2019). Digital literacy for the 21st century [chapter]. In *Advanced methodologies and technologies in library science, information management, and scholarly inquiry*. IGI Global. https://doi.org/10.4018/978-1-5225-7659-4.ch002

Srivastava, S. K. (2009). Making a technological choice for disaster management and poverty alleviation in India. *Disasters, 33*(1), 58–81. https://doi.org/10.1111/j.1467-7717.2008.01062.x. Scopus.

Steffen, W., Grinevald, J., Crutzen, P., & Mcneill, J. (2011). The anthropocene: Conceptual and historical perspectives. *Philosophical Transactions of the Royal Society A: Mathematical, Physical and Engineering Sciences, 369*(1938), 842–867. https://doi.org/10.1098/rsta.2010.0327. Scopus.

Steffen, W., Persson, Å., Deutsch, L., Zalasiewicz, J., Williams, M., Richardson, K., Crumley, C., Crutzen, P., Folke, C., Gordon, L., Molina, M., Ramanathan, V., Rockström, J., Scheffer, M., Schellnhuber, H. J., & Svedin, U. (2011). The anthropocene: From global change to planetary stewardship. *Ambio, 40*(7), 739–761. https://doi.org/10.1007/s13280-011-0185-x. Scopus.

Sun, Y., & Yamori, K. (2018). Risk management and technology: Case studies of Tsunami Evacuation Drills in Japan. *Sustainability (Switzerland), 10*(9). https://doi.org/10.3390/su10092982. Scopus.

Technology and Innovation Report 2021: Catching technological waves -. (2021). United Nations.

Van Winkle, B., Carpenter, N., & Moscucci, M. (2017). Why aren't our digital solutions working for everyone? *AMA Journal of Ethics, 19*(11), 1116–1124. https://doi.org/10.1001/journalofethics.2017.19.11.stas2-1711. Scopus.

World Food Programme (2020). WFP Drones Information Management Field Guide. https://drones.wfp.org/sites/default/files/node/resource/field_attached_document/2021-01/WFP%20Drones%20Information%20Management%20-%20FIELD%20GUIDE.pdf

Wang, N. (2019). 'A Success Story that Can Be Sold'?: A Case Study of Humanitarian Use of Drones. 2019-November. Scopus. https://doi.org/10.1109/ISTAS48451.2019.8938015.

WEF - New Vision for Education_Report2015.pdf. (n.d.). Retrieved 5 September 2021, from http://www3.weforum.org/docs/WEFUSA_NewVisionforEducation_Report2015.pdf

Why tech will be key in our quest to hit the SDGs. (n.d.). World Economic Forum. Retrieved 17 June 2021, from https://www.weforum.org/agenda/2019/09/technology-global-goals-sustainable-development-sdgs/

World Internet Users Statistics and 2021 World Population Stats. (n.d.). Retrieved 5 September 2021, from https://www.internetworldstats.com/stats.htm

Yan, W., & Roggema, R. (2019). Developing a design-led approach for the food-energy-water nexus in cities. *Urban Planning, 4*(1), 123–138. https://doi.org/10.17645/up.v4i1.1739. Scopus.

Yienger, M. E. (2016). Too much tech harms reading retention in young children. Inquiries Journal, 8(03). http://www.inquiriesjournal.com/articles/1374/too-much-tech-harms-reading-retention-in-young-children

Zhong, M., Lin, K., Tang, G., Zhang, Q., Hong, Y., & Chen, X. (2020). A framework to evaluate community resilience to urban floods: A case study in three communities. *Sustainability (Switzerland), 12*(4). https://doi.org/10.3390/su12041521. Scopus.

Zwęgliński, T. (2020). The use of drones in disaster aerial needs reconnaissance and damage assessment – three-dimensional modeling and orthophoto map study. *Sustainability, 12*(15), 6080. https://doi.org/10.3390/su12156080

Chapter 3
Poverty and Design, an Economics and Policy Perspective

Gustavo Merino

3.1 The Poverty Challenge

Poverty is one of the greatest global challenges. In 2017, close to 700 million people across the world faced extreme poverty according to the World Bank, living with less than $1.90 US dollars per day (World Bank, 2020a). The good news is that poverty fell very rapidly over the last three decades by historical standards. In 1990, 36% of the world's population was considered extreme poor but by 2017 it was less than 10%. Rapid economic growth in many regions resulting in greater incomes and employment, improved sanitation, health and education, as well as better and more widespread social programs and infrastructure, among other factors, contributed to the downward trend in poverty. In fact, the world was well positioned to reach Sustainable Development Goal 1 of the 2030 Agenda convened by the United Nations: End poverty in all its forms everywhere. The Covid pandemic and ensuing economic crisis, however, reversed this downward trend. The World Bank estimated that up to 150 million additional people could have fallen in poverty in 2020 (World Bank, 2020b). While the pandemic and the world economic crisis has been a tremendous setback for poverty reduction, the success of the previous decades shows that it is possible to face this challenge.

This chapter explores how design and innovation can contribute to the eradication of poverty and to improve the livelihoods and opportunities for those facing poverty. The rest of this section will discuss the various dimensions of poverty, as well as important global trends to keep in mind. The next section focuses on how design and innovation, of goods, services, and infrastructure, but also of policies and programs, can have a big impact on poverty alleviation and development, and

G. Merino (✉)
Escuela de Gobierno y Transformación Pública, Tecnológico de Monterrey, Mexico City, Mexico
e-mail: gustavo.merino@tec.mx

E. Giorgi et al. (eds.), *Design for Vulnerable Communities*, The Urban Book Series,
https://doi.org/10.1007/978-3-030-96866-3_3

important elements that need to be considered for greater impact. A multidisciplinary approach is followed that draws heavily on economics, policy, and business.

3.1.1 Dimensions of Poverty and Evolving Global Trends

Poverty is commonly measured in terms of income in relation to a given poverty line. An income of $1.90 US dollars per person is the line currently used for comparison purposed across countries by many international organizations, but there are other international and national lines including separate measurements for poverty and extreme poverty. Poverty, however, has several dimensions and manifestations beyond income. It is often associated with hunger or food insecurity—lacking regular access to enough safe and nutritious food for normal growth and development and for an active and healthy life—, according to the Food and Agriculture Organization of the United Nations (FAO). The FAO estimated that 690 million people went hungry or were undernourished in 2019 and that 9.7% of the world population—close to 750 million people—, faced severe levels of food insecurity (FAO, IFAD, UNICEF, WFP, and WHO, 2020). But poverty is more than that. It implies deprivations and lack of adequate access to essential goods and services essential for development and well-being, including health, education, social protection, and water and sewerage services, among others. People in poverty frequently face overcrowding in their homes and that leads to other problems. Deprivations can even persist, even if family income would be sufficient to pay for needed goods or services, if they are not available in their communities.

Some countries and organizations have sought to develop and use multidimensional measures of poverty and Mexico was one of the pioneers in this regard. Its poverty measurement includes the income dimension, distinguishing between general poverty and extreme poverty, but also six additional dimensions: nutrition, access to social security and pensions, education, health, quality, and space in the household, as well as basic services available in homes. The situation in each household for each of these dimensions is compared to an ideal, typically what the laws determine in terms of rights and obligations. People are considered poor and extreme poor depending on their income level and the extent to which they face deprivations for the various dimensions.[1] There are other multidimensional measures of poverty and development, such as the Human Development Index produced by the United Nations Development Program (UNDP), but a larger discussion exceeds the scope of this chapter.

Multidimensional measures of poverty can present a more accurate and informative picture of poverty conditions, and this serves as a better guide for policy

[1] People in poverty have an income below the poverty line (*línea de bienestar económico*) and face deprivation in at least one of the dimensions. They would be in extreme poverty if their income falls under the extreme poverty line and they face deprivation in three or more dimensions. For a detailed explanation on the methodology used in Mexico see Coneval (2019).

decisions and for organizations concerned with development, than a measure based solely on income. Governments do not have many tools at their disposal to directly affect the income of individuals other than through transfers and some elements of fiscal policy, although they have indirect means, such as creating an environment conducive to economic growth. They can, however, have a direct impact on the other dimensions through appropriate policies that improve access to infrastructure and services, promote better housing, raise educational attainment, or improve health outcomes, for instance. For the same reason, multidimensional measurements of poverty are also of greater value to foundations, NGOs and others concerned with reducing poverty and improving development.

Climate change, perhaps the greatest global challenge faced, can affect the poor disproportionally as they may have fewer means to protect themselves and adapt. While climate change will have impacts worldwide, several regions will be particularly hard hit by rising sea levels and atypical and severe weather patterns. In addition, the world is experiencing dramatic biodiversity loss with implications for food production, health and others, as well as overexploitation of many natural resources. Innovative design that helps combat climate change and simultaneously contributes to reduce poverty and vulnerability, should be a global priority.

It is also useful to consider evolving trends that are having significant impact on livelihoods for all groups but that can disproportionally affect the poor and marginalized. These trends imply risks and challenges but also opportunities. Rapid urbanization in all regions of the world is one of these trends. According to the United Nations, over half of the world's population currently lives in cities, while in 1950 it was just one third. It is expected that by 2050, two thirds of the world's population will live in urban areas (UN Population Division, 2021). This creates challenges and opportunities for sustainability, economic growth, and poverty reduction. It also represents new challenges for urban design and design of goods and services.

Demographic change is another largely irreversible trend, with population aging fast in many regions of the world. The provision of support and services including health for an aging population, is and will continue to be a significant challenge, requiring not only significant resources but also innovation. Further, while population growth overall has fallen in Europe, North and South America, and parts of Asia, it is still growing relatively fast in sub-Saharan Africa, where it is projected to grow from 1.3 billion today to 3 billion by 2060 (Guillén, 2020). This region is the poorest in the world and the growing population will burden existing infrastructure and services. However, having a younger population as the rest of the world is aging is an opportunity for Africa—that will benefit from a demographic bonus—, and for the rest of the world that will need the dynamism and productivity of the young.[2]

A third significant trend is rapid technological change, including significant progress in artificial intelligence, medicine, genetics, and robotics among others, as it is changing the way we live, consume, produce, and interact with each other. The

[2] A demographic bonus occurs when those of productive age represent a larger portion of the population than those who are dependents (the very young and the elderly).

rapidity with which the world was able to shift to remote work and education for millions during the pandemic, and with which vaccines against Covid-19 were developed, is a testimony of such rapid technological change. Opportunities abound, such as better treatments for health problems and capacity to deliver them faster and farther than before, greater educational opportunities and expanded markets that can be accessed even from remote areas, to cite a few. However, technological change also poses challenges and risks, such as the prospect of unemployment for those lacking the new skills required, or greater inequality for those in the wrong side of the digital divide. Here is a significant role to play for innovators and designers.

Large-scale migration as people move to seek a better life, escape violence or civil unrest and, more recently, avoid the effects of climate change and environmental loss, is a further trend that calls for innovation and design solutions, to both reduce the need to migrate and to help migrants and society in general to address the multiple challenges associated with migration. While in 1990 128 million people lived in a country other than where they were born, it is estimated that in 2020, 281 million did, almost 4% of the total population (IOM, 2019). This figure does not consider significant migration within countries, also part of the urbanization trend mentioned above.

3.2 Design, Innovation, and Poverty

There is no unique solution to the poverty challenge and it can well be characterized as a "wicked problem" as described by Rittel and Webber (1973) in their classic paper. International experience presents some necessary factors for poverty reduction, such as inclusive economic growth, improved infrastructure, and investment in basic capabilities such as education, health, and nutrition. There is evidence of programs and policies that have proven successful at addressing aspects of poverty and that can in some cases be adapted to specific national or regional contexts. However, the challenge persists.

The poor have underserved or unmet needs as discussed above. Design and innovation can go a long way in improving the livelihoods of those facing poverty, promote greater human development and create opportunities to raise incomes and for social mobility that reduces poverty permanently. Inclusive design, much like inclusive innovation as discussed in an OECD report, should be able to harness science, technology, and design know-how for low income, disadvantaged, and poor sectors of the population (OECD, 2013). By inclusive I mean design that has an impact on all groups of society and particularly the poor, vulnerable, and marginalized.

Raising incomes directly or indirectly is the main objective of many poverty alleviation strategies worldwide. Examples of strategies seeking a direct impact on income or consumption include cash or in-kind transfers as well as subsidies or tax credits. Among the policies seeking to raise family incomes indirectly are those aiming to enhance access to employment or economic opportunities, including job-training, provision of child-care services, or access to loans or other financial services. In some cases, poverty alleviation strategies combine direct income support with other types of assistance. An example of the latter are some conditional cash transfer (CCT) programs such as the former Progresa, later known as Oportunidades and Prospera program in Mexico, or the Bolsa Familia Program in Brazil. These programs present a good example of innovation and design in policy formulation and implementation which will be discussed below.

As poverty is a multidimensional phenomenon, increasing personal or family incomes may not be sufficient to overcome poverty. Goods and services demanded by people in poverty might not be available to them, or might be too costly and not meet their requirements, as they were not designed with them in mind. This might occur with those that directly impact developmental objectives such as food of sufficient variety and nutritional value; health services and health care products including personal hygiene; sewerage and water services as well as education services and materials. It might also affect access to other goods and services that, while not having a significant or direct effect on developmental goals, do have a large impact on their well-being and livelihood. Lack of access to appropriate goods and services can also impair performing economic activities that could increase family income and hence social mobility. Those in poverty, like everyone else, use goods and services not just for consumption but also for production. Note that, in some cases, essential goods and services might be available but not all the time or in sufficient quantity; hence, their impact on well-being or for income generation might be limited. For instance, electricity services might just be sufficient for a few light bulbs but not to power appliances or tools.

3.2.1 Improving Design and Delivery of Goods and Services

As Jagtap (2019) underlines in his literature review of research on design for marginalized communities, appropriately designed products can have a significant impact for social and human development. Jagtap notes that the concept of design to improve conditions and the life of marginalized societies can be traced to the early 70s. Since then, work on this issue has grown.

A starting point is to reflect on why the needs of the poor are often underserved in terms of the goods and services available to them. There are various reasons for this from the point of view of producers and sellers. Many companies might not see serving the poor as a profitable venture. Low income comes with low consumption levels and low demand for many goods or services; hence such societies might not be considered a target group. To aggravate the problem, many of the poor live in communities that may be far away from larger consumption centers or be relatively isolated, raising distribution and other costs and thus lowering profitability. An additional problem is that product design is often centered on more affluent groups of the population. Low spending capacity and high costs of serving people in poverty might deter some companies from engaging in the effort needed to better understand the needs and preferences of this group and invest in product development and changes in distribution chains to better serve them.

This situation is not limited to poor societies but can be seen in the strategies followed by western companies entering emerging markets in the 80s, as Prahalad and Lieberthal (2003) indicate. Companies targeted products and pitches to the relatively affluent consumers—who most resemble the typical western consumer—, foregoing the opportunity to reach a larger market. According to Prahalad and Lieberthal, these companies suffered from an "imperialist mind-set." The assumption was that emerging markets were ready for their existing products and that the middle class and the affluent in these markets were just like those in Europe or the United States and faced similar conditions. Therefore, not much was required in terms of modifying products and pitches. However, while this can be true of usually a small segment of the population, it was not the case for the majority, and companies were not able to improve sales according to their expectations. Companies started to wise up, realizing that emerging markets were not just extensions of their home markets. It was not just a question of reducing costs of making some cultural adaptations, it required strategic changes such as rethinking brand management, costs and, of course, product design, packaging and distribution channels. Design, for instance, had to consider local preferences, requiring in some cases, modification of existing products or even distribution channels. An example of the latter presented by the authors was the need of a western maker of frozen desserts entering India, to reformulate its products considering that most retail outlets did not have adequate refrigeration equipment to keep the product sufficiently cold and that frequent interruptions reduced reliability of available infrastructure. A further example is changes in packaging to enable consumers to purchase smaller quantities of a given good as the "normal" sized packages in western markets might be too costly for them given their income stream.

While it is undeniable that income and hence spending power among the poor is lower, this does not mean that there are no business opportunities. Prahalad (2004) further argues that there is a "fortune at the bottom of the pyramid" in the sense that it is possible, with the aid of good strategy, to provide goods and services to low-income groups of the population, meeting their needs, improving their livelihoods and even their incomes, and do so profitably. The lesson is that firms who shun this market are foregoing an important business opportunity that simultaneously contributes to the well-being of people in poverty. The author presents multiple examples in support of these arguments, of firms that designed or redesigned products and services successfully for this market and even created business opportunities for the poor.

Capacity and interest among the poor in acquiring goods and services can sometimes be greater than expected by companies. The poor and disadvantaged are sometimes forced to pay more for the same product than the noon poor, precisely because of insufficient availability or inadequacy of the good or service in question. Take the case of water. In many cities and regions water supply is limited or people lack access to running water in or near their homes forcing them to pay a higher price per unit of water by buying it from water-trucks or other forms of transportation. Price paid per liter is therefore more expensive for them than for more affluent families who obtain their water from the public system. This is an example of the deprivations faced by the poor, and of the opportunity to improve policy and to find better solutions through design. It also shows that the poor are willing to dedicate their limited resources to make purchases according to their needs and preferences. The fact that they do spend more per liter than in other parts of the city shows they would be willing to invest in better and more affordable solutions, thus creating investment opportunities to better meet their needs.

Through the analysis of the introduction of the mobile telecommunications in sub-Saharan Africa, Christensen et al. (2019) provide another interesting example on the possibility of meeting the needs of the poor through innovation. Cell phone technology is cheaper and faster to deploy than traditional fixed lines, but investors and telecom companies had shunned the sub-Saharan market given the rampant poverty afflicting a large proportion of the population, until an entrepreneur saw an opportunity where others did not. While many considered that telephone services would be too expensive for this group and hence not a viable business opportunity, he noticed that people were already spending important sums in transportation costs and use of their time, just to keep in touch with loved ones. Mobile phone service, although relatively expensive, was less onerous than the alternative, hence the business opportunity that simultaneously improved welfare for people in poverty. Despite the skepticism of many around him, including potential investors and financers, he was able to build a mobile telephone network in six countries in Africa (Celtel). This involved significant effort and ingenuity to overcome barriers of all kinds. It worked because mobile telecommunications technology can be installed at a lower cost than fixed phone lines and there is ample demand for such services, even among the poor. Today, Africa has a sophisticated mobile telephone industry with close to a billion phone lines. Falling prices of mobile telephony, as occurs with computers and internet service, have improved access to such services immensely.

These examples provide useful insights. First, there is ample demand among the poor for goods, services, and infrastructure that will allow them to have a better life. Second, through design, ingenuity, and the search for opportunities, it is possible to service the needs of the poor and marginalized, improving their well-being and even their capacity to generate better economic opportunities. The poor often spend a large part of their income pursuing what they need or want. In addition, technology can help overcome many obstacles that currently leave the poor and marginalized with unmet needs and create opportunities for increasing the income of the poor. Mobile telephone networks, for example, do not just facilitate communication over the phone but open a whole new world for the poor and vulnerable with the internet and all it has to offer including information, entertainment, and multiple services of all kinds, such as education, health, and financial services, which the poor might otherwise find difficult to access. Financial inclusion, for instance, is an important development tool.

Access to savings, credit, insurance, and other financial services can make a major difference for both improved consumption and as a means for engaging in economic activity. Where few have formal bank accounts, millions had mobile telephone services. Today we find there is an extensive and widespread use of mobile payment systems using mobile phone networks, among the poor in low-income countries, particularly in sub-Saharan Africa, such as the M-Pesa system that started in Kenya, now being replicated in various parts of Africa and Asia. According to an IMF document (Adrian & Mancini-Griffoli, 2021), there are a billion registered mobile accounts in 95 countries. Sub-Saharan Africa leads the way, with around half of the mobile money accounts worldwide, which is possible because of the widespread use of mobile phones. Indeed, as the IMF document states, such broad and relatively inexpensive access to digital money using mobile telephones can bring financial services to 1.7 billion people without a bank account. Offering such services requires not only careful design of the platform used for financial operations, considering users might have little experience with financial transactions, the use of mobile systems or even have low educational levels. As explained to me by an M-Pesa executive some years ago, they had to develop chips that allowed some operations for users that had traditional mobile phones (not smartphones). As Guillén (2020) notes, when very few have access to bank accounts or credit cards, they rely on mobile phone systems to make payments. This also allows access to multiple other financial services they would otherwise do without. Low financial inclusion persists in many regions, including some richer and better developed such as Latin America. Results from sub-Saharan Africa show that intelligent and inclusive design and innovation can go a long way.

Mobile technology can provide many additional benefits to people facing poverty beyond access to financial services. For those engaged in agriculture, as many of the poor in rural areas are, mobile technology can also be used to improve access to information relevant for producers, including prices and other market information as well as climate reports. It can also be used for extension services and to improve distribution networks and reduce the need of "middle-men" that can capture part of the profits. Additional applications can help improve health outcomes through preventive medicine and better treatments, as well as job training and education.

The eagerness with which people in poverty, just like most everyone else in society, embrace mobile technology and communications debunks the commonly held perception that technology is for those better-off and that the poor do not need, care for or are even capable of taking advantage of advanced technologies, having other more pressing concerns such as meeting basic needs. This perception has led to many well-intentioned efforts, by governments, NGOs, and civil society, to assist the poor shunning technology in the search for alleged simplicity, but which leave the intended beneficiaries short-changed because of this. Simplicity might be called for in some cases, such as reducing frills, but without "dumbing-down" or reducing the most relevant applications and uses of products for this segment of the population, as this does not help improve their welfare and opportunities in the long run.

There are many other examples where good design and the use of technology overcomes challenges in providing goods, services, or adequate infrastructure to the poor. Technology and design, for instance, can facilitate the generation of electricity using renewable sources (sunlight, waterpower, and others) through equipment that can be deployed relatively cheaply to small communities and even people's homes previously not connected to the grid, enabling them to have access to services fundamental for their well-being, but also for the accumulation of human capital and even income generation activities. Similarly, innovations in rainwater collection capacity, treatment, and desalination, as well as improvements in irrigation technology, can be used to provide water to poor and marginalized communities currently with adequate access to them.

While businesses and markets can better serve people in poverty as discussed above, not all their needs and preferences will be met this way. Business opportunities might not be profitable enough in all cases, given current technologies and local conditions. Lack of knowledge and information of markets where the poor live presents an additional problem, and the mechanisms firms frequently use to gather information might not work well for this market.[3] In addition, what economists call "market failures" might occur in some instances, leading to suboptimal provision of goods or services. They can be corrected by well formulated government or other intervention. Furthermore, the structure of production costs of many goods and services implies rising marginal and average costs after certain levels of production (they decrease at first). For this reason, some people are often left out, when serving them becomes too costly. As a result, there is a need also for non-market solutions and hence the need for groups concerned about improving livelihoods and development to become involved, such as governments, non-governmental organizations (NGOs), foundations, universities, think-tanks, as well as social entrepreneurs or firms seeking to enhance their social impact. Again, there is significant opportunity for design or redesign and the use of technology to address the challenge of unmet needs of people in poverty.

Whether it is a for-profit enterprise seeking to meet demand for goods or services of the low-income segment of the population, or a government entity or civil organization concerned with development, it needs to consider particular "demand conditions" of the targeted population. This means that products and services as well as distribution and delivery channels may need to be adjusted. Reducing production and delivery costs can be essential to make goods more affordable and available to the consumer. Such designs can be cheaper and simplified versions of existing products as are "no-frills" cars, low-cost laptops or tablets (Innovation Policy Platform, 2021). As mentioned, care must be taken to ensure functionality loss is minimal and other concerns such as safety are addressed. Demand conditions other than high sensitivity to price, should also be considered, such as the fact that the final users might lack access to services or complementary goods required for the use of the product being designed. For instance, electricity service may be irregular in many marginalized areas, hence, either products need to be adapted to function with other

[3]This is a common problem for the financial services industry seeking to evaluate credit risks of potential customers, as many poor people work in the informal sector and do not have a fixed income, have never bought anything using formal credit channels and do not have assets or use credit cards.

forms of energy, as is the case of pedal powered sewing or washing machines, or alternative means to provide electricity need to be found, such as the use of solar panels as discussed in OECD (2013).[4]

New and affordable products and services can also help face other social challenges, such as exclusion and inequality. The Covid-19 pandemic and global economic crisis disproportionally affected some groups, including women and those in the wrong side of the digital divide, reinforcing these and other inequalities and widening the gaps in society. Innovation and design can also contribute to address the global trends indicated in the previous section. New and better goods and services can be designed to address demographic change by helping the elderly stay healthier longer, live independently, and counter diminishing physical capabilities. It can also contribute to the creation of more sustainable cities, improve livelihoods and opportunities in rural areas, or reduce green-house gas emissions, overexploitation of the environment, and biodiversity loss.

Some design innovations, while relatively simple, can have a big impact on families. An example of this is the "water wheel" of which there are various designs (such as the Wello Water Wheel introduced in India).[5] Millions of people across the world do not have access to running water in their homes and must often walk some distance to a stream, other body of water, or public water pump. Rather than having to carry the water back home in a container on their backs or heads, users can fill the wheel with water and roll it. Not only is this easier on the body, but as more water can be carried in this way, fewer visits to the water source are needed, and their time can be used in more productive ways. The Bill and Melinda Gates foundation also promotes development of solutions to some problems faced by millions around the world as is the lack of adequate toilets and sanitation facilities in many places. To encourage innovation in this field, the foundation awarded grants to universities and others, to develop waterless toilets, microbacterial fuel cells to power sanitation of the waste and other innovations.[6] Universities are also including in their curricula on design, policy and other courses, development of innovations to address social challenges. The Reach Alliance formed by the University of Toronto who pioneered the Reach initiative, Tecnológico de Monterrey, Oxford University and University College London, promotes research on how to bring to all what is important to them, reaching the hardest to reach.[7] Stanford University developed the "Design for Extreme Affordability" multidisciplinary course, where students learn to design products and services seeking to improve conditions for the poor across the world. A result of this effort was the development of a lightweight, inexpensive, and

[4] An example is Mera Gao Power in Uttar Pradesh, India, that creates a microgrid using solar panels in roofs of houses that also provide power to neighboring homes in a cost-effective way, cited in OECD (2013, Box 3.2).

[5] https://www.engineeringforchange.org/solutions/product/wello-water-wheel/

[6] See https://www.gatesfoundation.org/ideas/media-center/press-releases/2013/12/gates-foundation-awards-grants-to-waterless-toilets (accessed June 2021)

[7] On this see https://reachalliance.org/ Accessed June 2021, with case studies from around the world.

easy-to-use incubator for premature and low-birth weight babies, helping many families who otherwise would lack access to the expensive incubators available only in hospitals (Stanford University, 2021).

Design innovations developed in poor communities or with significant participation of the poor or marginalized segments of the population can also contribute to developmental objectives and as means for increasing their incomes. These "grassroots designs" might not always be as technically sophisticated but can take good advantage of local knowledge including a better understanding of what members of the community are lacking. Supporting and fostering these design innovations is therefore called for.

3.2.2 Elements of Design for the Poor

A basic element of design is to clearly understand the needs, expectations, and other relevant aspects of the ultimate user of the design. While this seems straight-forward, it is not always followed. Here, I refer not just to goods and services, but also infrastructure and even policy and program design. As a result, many efforts fall flat or are not able to meet the objective of making an important difference in the livelihoods of the intended population. Even worse, they might undermine progress in this regard.

The problem is recognized in the literature on design and poverty. Jagtap (2019), quoting research in the field, mentions the need for a "bottom-up design approach" and "micro-level understanding" which aims to gain insights on their life, their social networks and interactions, their economic activities, how they produce and consume, as well as other cultural aspects and environmental conditions. Deep immersion into the lives of the poor can allow the designer to "move away from sympathy to informed empathy" and hence develop better solutions for the problem at hand. Ideally, the users should also be involved in the design process.

Even apparently well-designed products can sometimes not lead to the expected results. This could result from failure to consider all relevant elements in the design phase, but also from not paying attention to other factors such as the ease of adoption of the innovation in question. An example is the so-called "improved cooking stoves" or "ecological stoves" (also known by other names). Such stoves, of which there are various models, seek to address problems associated with indoor open-fires prevalent in millions of poor households across the world that, lacking electricity or gas, use firewood and sometimes other biomass as fuel, including agricultural residues and even dung. This creates a severe health hazard for the families, including damage to lungs and eyes; it requires significant use of the time of household members collecting firewood and other biomass, which could be used for more productive purposes, and presents a serious environmental risk including higher CO_2 emissions, increased deforestation, and risk of wildfires. Improved stoves mitigate some of these problems as they have chimneys to channel the smoke outside, and they require a fraction of the wood or other biomass to generate the same amount of heat. In addition, they can heat water for other household uses or contribute to heating homes in cold weather. Most designs are easy and inexpensive to build and install (can even be a do-it-yourself project). Consequently, such stoves are often hailed as an excellent solution where gas or electricity is not available, and government and NGOs across the world promote their use.

Some studies show improvements and significant uptake of the stoves by households while others present more disappointing results. Hanna et al. (2012), for instance, analyzed use and impact of improved stoves with a randomized control trial in rural Orissa, India. They found limited long-run impacts. For instance, while at first the stove reduced smoke exposure for the primary cook, long run exposure was not significantly lower and there were no evidently observable positive health effects. Apparently, use of the stoves declined in time for various reasons including low maintenance and breakdowns, but also preferences that did not lead to the expected behavioral changes. I was able to personally witness the problem of low adoption of these types of stoves after visiting families who had one installed as part of a national government program in Mexico. Some stoves had never been used or were used very infrequently, and families continued cooking on an open flame, frequently right next to the stove. When asked why, families would often mention that they were used to and liked the traditional way and some just did not believe in the supposed benefits of the product.

Research results, much like the anecdotal evidence presented for Mexico, raise questions to reflect on. First, as has been mentioned, product design will be more effective when it is able to capture the needs and preferences of the user, but attention should also be given to how it will be used and other relevant aspects. Design is rarely final. It should adapt continuously to changing conditions and preferences. If it is part of a strategy to change behavior, it should facilitate or even induce such change.

Second, adoption is not automatic, even when the benefits or usefulness of a product appear evident and its design seems appropriate. While in a "normal" consumer good this is important, it is particularly relevant when there are objectives regarding its use that exceed the benefit to the final user. In the case of the ecological stoves, the public or social benefits of this product are greater than the immediate benefits to the families who possess and use them. Among the social benefits are those related to public health (reduction in health damages from exposure to smoke and open fires), environmental sustainability (reduced deforestation and emissions), and poverty alleviation (more productive use of time for families allowing for greater incomes or for human capital accumulation). In this case, merely helping families build or install such stoves is not sufficient, benefits for the family and for the population at large only come about if the stoves are used continuously. Part of the solution is related to the design of the product itself. It must be sufficiently appealing to the intended users for them to be willing to use it (and purchase it, if applicable). However, it is not enough that the product meets this criterion in a laboratory-like setting but also when applied in the real world, where many factors outside the design of the product might be in play: tradition, culture, expectations, perception of results, and others. In the public program to install the ecological stoves in Mexico, for instance, the government agency rapidly understood that if families were merely asked if they would like a stove installed free of charge, almost all responded in the affirmative yet, as discussed, having the stove did not mean it would be used. Carefully explaining the benefits of the stove for the family during the installation process helped but did not guarantee adoption. Further experimentation and conversations with an NGO experienced with this type of equipment suggested adoption was more likely if those responsible for the stoves accompanied families after installation for a few days, to answer questions and encourage use of the stoves, a lengthy and costly process, difficult to perform by an overburdened agency.

These results show that adoption of a new product is not automatic and cannot be taken for granted. Lack of adequate or sufficient adoption over the long run will probably not lead to the expected results. Behaviors evolve over time as do perceptions on the benefits of the innovation in question, which may or may not correspond to reality. As the study of Hanna and colleagues quoted above show, testing over a long-term horizon is often needed to understand the effectiveness and impact of an innovation.

Other innovations seeking to improve conditions for the poor present similar problems and other issues to reflect on. Hobbes (2014) presents an example of a product targeting poor communities that received wide attention and funding. Known as the "Playpump," its design appeared to simultaneously solve many problems in the community in a relatively affordable package. The Playpump was a merry-go-round for children, connected to a pump that drew underground water for use by the community. By playing and enjoying themselves in the Playpump, the children would be filling the water tank. Paid advertisements in the tank helped cover the cost of maintenance. Price was low, it was easy to install and could provide water for over 2000 people. Funding for the project poured in and many were installed. However, a few years later many were not in use and in need of repair and there were accusations of child labor being used to pump the water. Again, a seemingly good and carefully designed product might deliver less than expected including for reasons beyond design of the product.

Anti-malaria nets provide an additional example. The World Health Organization estimates 229 million cases of malaria worldwide in 2019. Malaria is a serious vector-transmitted disease that causes great hardship to those afflicted and their families (World Health Organization, 2021). Beyond the effects on health, it causes a significant loss in productivity and hence family income and even has an impact on local and national development in affected regions of the world.[8] A relatively inexpensive and effective way to protect families is the use of specially designed nets treated with insecticide that people can sleep under and thereby avoid mosquito bites. The benefits of such nets protect not just those who use them but also reduce the risk to others as there is less contagion. That is, use of safety nets creates what economists call a positive externality to others. By using the net and therefore

[8] See for instance the work of McCord and Sachs (2013).

significantly lowering the risk of becoming infected, that person is not just protecting him or herself, but also reducing the risk of contagion to others. However, when making decisions about the use of the net, people are considering the benefits to themselves and not to others. Consequently, less nets are used than what is optimal from a social perspective, generating a market failure. It makes sense, therefore, for governments, donors, NGOs to encourage the use of the anti-malaria nets, a well-designed product. The issue here is how to go about it. The path chosen will likely affect decisions taken by the targeted population by changing their incentives, as Banerjee and Duflo (2011) point out. If the nets are handed out for free, will people value them less and not use them? Should they be handed out to all or target only the poor, and if the latter, are the targeting mechanisms functioning well? If they are just subsidized or delivered at cost, will people buy enough of them to protect all the household? The delivery mechanism can also have impacts beyond decisions taken by the users. For instance, free provision of nets might inhibit the development of a local market for nets. As people will not expect to have to pay or pay full price, even if they need a replacement or an additional net, fewer nets will be produced and available for purchase. This also hurts net producers and sellers, some of whom might also be poor. Answering these questions requires empirical research on decisions taken by people when faced by various delivery modes.[9]

As with the case of the ecological stoves discussed earlier, the malaria net example illustrates again that success hinges not merely on the functionality and characteristics of the product in question to meet its immediate objective, but also on its adoption and proper use, which in turn is affected by cultural and environmental factors but also by the incentives faced and behavioral changes.

3.2.3 Design in Poverty Alleviation Programs and Policies

As with products and services, good design is vital for any policy or program seeking to reach poor and marginalized societies and improve their present and future conditions. More effective, less expensive, and faster-working solutions for a problem can be found, that often counter conventional wisdom or appear counter intuitive. Research undertaken by economists Michael Kremer and Edward Miguel shows that school performance and even future earnings could be promoted by deworming children using already existing school infrastructure (J-PAL, 2021). Deworming is relatively cheap and certainly less so than improvements in educational materials such as textbooks. It reduces sick-days and improves school attendance and concentration as well as overall nutrition, improving cognitive development and productivity. This does not mean that educational strategy should

[9] Abhijit Banerjee, Esther Duflo, and Michael Kramer have used and helped develop randomized control trials to answer these and other questions, helping build into economics research and methodologies from the health sciences. This helped analyze the impact on incentives and decisions, which has also led to a better understanding of poverty.

be based on public health interventions such as deworming children, nor that improving infrastructure, educational materials, teacher training, or others are not important. It shows, however, that appropriate research to understand what leads to bad outcomes—poor school performance with lifetime implications in this example—, can lead to improved policy or program design.

Failure to properly account for existing conditions, incentives, and behavior in the design of policies and programs is relatively common. For instance, because poverty is often associated with hunger, numerous poverty alleviation initiatives across the world provide some form of food assistance such as food delivery, school breakfasts, community soup-kitchens or food vouchers and subsidies for certain foods. However, while many people in poverty face hunger, not all have a problem accessing sufficient food. Many of the food support programs did not have a prior diagnosis of the conditions of the population being targeted or did not consider that the type of intervention created incentives that could affect behavior and therefore its impact. It is puzzling for many, for instance, when a secondary market of food that is delivered emerges. How can it be possible, they ask, that poor people not eat the food provided and instead give it away, use it to feed livestock or sell it for less than what it's worth. With these questions, they are failing to consider that delivered food might not correspond to local preferences, that some might be unable to eat the delivered product due to an allergy or other health condition, that they might crave other types of food (such as fresh produce), or even that, despite being poor, they might not suffer from hunger and instead face other deprivations. The money received from selling foodstuffs in the secondary market could well go into other essential goods for the family such as medicines, products for personal hygiene or others, including those needed to make a harsh life more pleasant and bearable, such as entertainment options like a TV.

Consider also the case of school-breakfast programs. Their aim is to improve nutrition and educational outcomes and many rely on the assumption that school breakfasts supplement food eaten at home. However, for many families with already over-stretched budgets, school breakfasts become an opportunity to save meager resources on food and use them instead for other basic needs, a reasonable and understandable reaction. Many children thus go to school without a bite to eat at home for breakfast. Thus, the program substitutes rather than supplements food eaten at home and nutritional outcomes, rather than improve, might suffer. Does this mean school breakfasts are a bad idea? No, it means that the design of this program should take this potential behavioral change into account. The program design could assume that children will not have breakfast at home and improve the contents of school breakfasts or, if this is not possible for budgetary or logistical reasons, find ways to encourage families not to skip breakfast for their children.

Banerjee and Duflo (2011) provide additional examples where resulting behavior reflects incentives and conditions and might lead to results different than those intended by the policy or program. They cite a program in China that subsidized basic staples in two regions, seeking to expand their consumption. This makes sense at first glance. For what economists call "normal" goods, a lower price leads to a higher quantity demanded. In this case, however, the opposite occurred, people were spending less on the subsidized commodity, choosing instead to spend more on better tasting but more expensive foodstuffs they would normally eat less of. They were not obtaining better nutrition, but their lives became more pleasant. The subsidy has apparently made them feel a bit richer as the staple in question represented a large part of their consumption.

Policy design must go beyond preconceived notions, it requires analysis, research and should be evidence-based. Its focus should not be limited to the needs that should be satisfied, but also into the factors that influence people's decisions and behavior, as they will have a large bearing on the success of the policy. In general, people make rational decisions, the poor and non-poor alike. These decisions and ultimately their behavior, is shaped by the underlying conditions, expectations and incentives people face. Just as product design improves by understanding the user, policy design can be much more effective when understanding how people make decisions.[10] Doing so can dramatically improve results and make a better contribution to poverty reduction and improvements in the well-being of these families.

Generating the appropriate incentives in policy design can even help overcome long-standing behaviors and traditions that hinder development or overcoming poverty. Mexico's conditional cash transfer program, Progresa, later renamed Oportunidades and eventually Prospera, implemented between 1997 and 2018, is a case at point. This program sought to combat the intergenerational transmission of poverty by focusing investments in building basic capabilities, particularly nutrition, education, and health. This program included many policy innovations and built into its benefits structure, a series of incentives to encourage behaviors conducive to achievement of the program goals. Eligible families received a cash transfer to purchase food and other basic goods, conditioned on all members of the family visiting the local health clinic at determined intervals according to health risks, for evaluation and prevention, thereby reinforcing the nutrition and health components of the program. Scholarships were provided for students enrolled in school conditioned on their continued attendance. Failure to show up at the clinic or attending school would mean losing the relevant transfer.[11]

Additional design innovations sought to address persistent problems with important developmental implications. Girls were more prone than boys to abandon school starting in middle school, reflecting long held conceptions that investments in girl's education was not necessary, and it was preferable for them to stay at home and help the mother with household chores and taking care of the family or livestock. The program designers recognized the complexity of changing beliefs and traditions, a venture that, if successful, would probably take many years. Rather

[10] The work of Banerjee and Duflo (2007) on the economic lives of the poor presents valuable insights and evidence on decisions taken by people on poverty in several countries regarding investments, education, nutrition, work, and others and how behaviors are shaped by incentives and conditions faced

[11] The Progresa-Oportunidades-Prospera programs were subject to numerous independent evaluations and scholarly research. See, for example, Parker and Todd (2017). For more information on the program see Dávila Lárraga (2016)

than attempting to do so, they chose a different approach, significantly raising the cost to families of having their daughters stay at home instead of attending school. This was achieved by making scholarships for girls substantially higher than for boys. This incentive proved to be more powerful than the traditional way of thinking that had left many girls without education. Boy's education was not affected by this. Scholarships for both genders increased by grade to counter the higher opportunity costs of students faced by remaining in school instead of working full time, so the appropriate incentives were already built into the program to encourage schooling. As a result of this design, enrollment increased for both girls and boys.

The program design innovated in other ways too. Program designers preferred to provide cash rather than in-kind transfers. The latter were a logistical nightmare and the program designers believed the families themselves knew better than the government what they were lacking. Critics of this idea expressed concern that the beneficiaries of the program would use the resources to pay for goods and services that did not contribute to the program goals. Building on prior research and conjectures, the designers proposed to transfer the resources directly to the mother of the household or, if lacking, another female in the household such as the grandmother or daughter, under the assumption that women would prioritize spending the resources for the benefit of the family, more so than the male members. This hypothesis, like other elements in the design of the program, was put to the test during the pilot phases of the program and was proven largely accurate and incorporated in the final design of the program.

The example above reinforces the point that careful policy and program design that considers incentives and behavioral changes will be more conducive to achieve the intended objectives. Much can be gained if in the initial phase of the program pilots can be conducted to test hypotheses and evaluate initial responses to finalize design. Further evaluations, including those focusing on impact, are desirable to make any adjustments or redesign if needed or even substitute for other policies and programs. Much like underlying conditions, technologies and environmental factors can change over time, as do preferences, behaviors, knowledge, as well as policy priorities, so evaluation and redesign are needed on a continuous basis.

3.2.4 Raising Incomes and Economic Opportunities for the Poor

Improved design of products, services and infrastructure can also lead to better economic opportunities for poor families, raising incomes and helping them leave poverty.

This can occur in several ways. First, by providing people in poverty with better tools and products to engage in economic activities, whether as employees or while engaged in an entrepreneurial activity such as small business. Second, by freeing time for economic activities which was previously taken up by other activities, such as collecting firewood, water, or transportation. Third, by facilitating access to savings, credit, and other financial services or by making it easier to participate in and increase earnings from their activities, through better market information using mobile applications and the internet. Fourth, through the creation of markets. "Market creating innovations" according to Christensen et al. (2019), "transform complex and expensive products and services into simple and more affordable products." In this way, they become more accessible for segments of the population. Such innovations foster economic activity: they create jobs—thereby raising incomes—, they can increase profits in entrepreneurial activities (including the microenterprises many of the poor engage in), and they also raise government revenue that can fund social programs and infrastructure.

Some of the research in design and poverty has focused on initiatives to raise family incomes by, for instance, helping families use local designs to produce goods including crafts or otherwise use their designs for commercial products, although there is some controversy regarding this issue including the problem of cultural appropriation by multinational corporations. In this regard, organizations promoting fair trade seek to ensure that local originators of such crafts and designs benefit.[12] While benefits can come from this, the need of the poor and vulnerable for all kinds of goods and services strengthening income generating activities for this segment of the population requires looking beyond traditional products and activities, hence the importance of innovative design in general.

3.3 Conclusion

Great strides have occurred in terms of poverty reduction in the last three decades. The COVID-19 pandemic and the ensuing global economic crisis reversed the downward trend in poverty reduction for the first time in over two decades, also creating new or deepening some existing inequalities and vulnerabilities. Much must be done to better face this challenge and meet Sustainable Development Goal 1: End poverty in all its forms, everywhere. Whether it is through firms seeking a profit or though efforts from social entrepreneurs, governments, NGOs, universities, and others concerned with development, inclusive design and innovation are essential in these efforts by improving or creating new products, services, and processes that make it possible for the poor and vulnerable to meet their needs, improve human development, and create new markets or economic opportunities for them to abandon poverty.

The poor do not just lack adequate access to many goods, services, and infrastructure. They also live in a world where many institutions, laws, and programs were not made for them or even with them in mind. As Banerjee and Duflo (2011) indicate, for the poor "It means making decisions about things that come with a lot

[12] See Thomas (2006) for a discussion on this issue.

of small print when you cannot even properly read the large print." To be successful, design for people in poverty should seek to understand their needs and preferences and why they have not been met, the underlying conditions they face as well as how they go about in their lives, including their "economic" lives: how they consume, how they make decisions, and how they engage in economic activities given the challenges they face. It is not a question of merely adapting existing designs to make them more affordable. It means going beyond developing designs that work well in laboratory-like settings or under living conditions different than those faced by the poor. It means continuing to perfect designs that work well for them despite their challenges and limitations, so that they can be adopted. Better design should also be integral in policies and programs seeking to address the needs of the poor. This requires building in the right type of incentives to induce behaviors conducive to the achievement of the development goals sought. Finally, design for the poor should harness advances in science and technology to overcome obstacles currently faced that impede development and that can reduce existing inequalities between groups.

References

Adrian, T., & Mancini-Griffoli, T. A New Era of Digital Money, (IMF, 2021), https://www.imf.org/external/pubs/ft/fandd/2021/06/online/digital-money-new-era-adrian-mancini-griffoli.htm. Accessed June 15, 2021.

Banerjee, A., & Duflo, E. (2007). The economic lives of the poor. *Journal of Economic Perspectives, 21*(1—Winter 2007). pp. 141–168.

Banerjee, A., & Duflo, E. (2011), Poor economics. A Radical Rethinking of the Way to Fight Global Poverty, Public Affairs.

Christensen, C., Ojomo, E., & Dillon K. (2019). The Prosperity Paradox. How Innovation Can Lift Nations Out of Poverty, Harper Business.

Coneval. (2019). Metodología para la medición multidimensional de la pobreza https://www.coneval.org.mx/InformesPublicaciones/InformesPublicaciones/Documents/Metodologia-medicion-multidimensional-3er-edicion.pdf

FAO, IFAD, UNICEF, WFP, WHO. (2020), The State of Food Security in the World.

Guillén, M. (2020). *2030: How Today's Biggest Trends Will Collide and Reshape the Future of Everything*. St. Martin's Press.

Hanna, R., Duflo, E., & Greenstone, M. (2012). Up in Smoke: The Influence of Household Behavior on the Long-Run Impact of Improved Cooking Stoves, Harvard Center for International Development (CID) Working Paper 241, April. https://www.hks.harvard.edu/sites/default/files/centers/cid/files/publications/faculty-working-papers/241_HannaDufloGreenstone.pdf

Hobbes, M. (2014). Stop Trying to Save the World. Big ideas are destroying international development. New Republic, November 18, 2014. https://newrepublic.com/article/120178/problem-international-development-and-plan-fix-it

Innovation Policy Platform. (2021). Addressing Social Challenges, http://www.innovationpolicyplatform.org/www.innovationpolicyplatform.org/content/addressing-social-challenges/index.html. Accessed June 2021.

International Organization for Migration of the United Nations (IOM). (2019). World Migration Report 2020. https://publications.iom.int/system/files/pdf/wmr_2020.pdf

Jagtap, S. (2019). Design and poverty: a review of contexts, roles of poor people, and methods. *Research in Engineering Design, 30*, 41–62. https://doi.org/10.1007/s00163-018-0294-7

J-PAL. (2021). Deworming to increase school attendance, Abudul Latif Jameel Poverty Action Lab, https://www.povertyactionlab.org/case-study/deworming-increase-school-attendance, accessed June 2021.

Lárraga, D. L. (2016). How Does Prospera Work? Best Practices in the Implementation of Conditional Cash Transfer Programs in Latin America and the Caribbean, Interamerican Development Bank Technical Note 971. https://publications.iadb.org/publications/english/document/How-does-Prospera-Work-Best-Practices-in-the-Implementation-of-Conditional-Cash-Transfer-Programs-in-Latin-America-and-the-Caribbean.pdf

McCord, G., & Sachs, J. (2013). Development, Structure and Transformation: Some evidence on comparative economic growth. NBER Working Paper 19512. http://www.nber.org/papers/w19512

OECD. (2013). Innovation and inclusive development report. Conference discussion report. February 2013 version. https://www.oecd.org/sti/inno/oecd-inclusive-innovation.pdf. Accessed June 2021.

Parker, S., & Todd, P. (2017). Conditional cash transfers: The case of Progresa/Oportunidades. *Journal of Economic Literature, 55*(3), 866–915.

Prahalad, C. K. (2004). *The fortune at the bottom of the pyramid: Eradicating poverty through profits*. Wharton School Publishing.

Prahalad C. K., & Lieberthal, L. (2003). The end of corporate imperialism, Harvard Business Review, August.

Rittel, H., & Webber, M. (1973). Dilemmas in a general theory of planning. *Policy Sciences 4*, 155–169.

Stanford University. (2021). Design for extreme affordability. https://extreme.stanford.edu/projects/embrace/. Accessed June 15, 2021.

Thomas, A. (2006), Design, poverty, and sustainable development. *Design Issues, 22*(4). MIT Press.

United Nations Population Division. https://www.un.org/development/desa/pd/themes/urbanization, accessed June 2021.

World Bank. (2020a). Global Action Urgently Needed to Halt Historic Threats to Poverty Reduction, October 7, 2020. https://www.worldbank.org/en/news/feature/2020/10/07/global-action-urgently-needed-to-halt-historic-threats-to-poverty-reduction. Accessed June 2021.

World Bank. (2020b). Covid-19 to add as many as 150 million extreme poor by 2021, https://www.worldbank.org/en/news/press-release/2020/10/07/covid-19-to-add-as-many-as-150-million-extreme-poor-by-2021

World Health Organization. (2021). Fact-sheet Malaria, https://www.who.int/news-room/fact-sheets/detail/malaria. Accessed June 2021.

Chapter 4
A Pandemic of Vulnerability: Is Design a Painkiller or a Vaccine?

Paolo Ceccarelli and Pilar M. Guerrieri

4.1 Vulnerability Status

Vulnerability is a negative aspect of the human condition that is permanent and recurrent throughout the millennia in all places in the world. Some of its characteristics have remained unchanged because they are linked to the difficult relationship between mankind and nature, while others have changed as a result of the relationship between human beings themselves. Vulnerability sometimes is linked to earthquakes, floods and profound climatic variations; at other times it depends directly on the behaviour of human beings, such as in the case of wars, certain forms of a pandemic, famine and the effect of economic and technological changes in societies. Of course, over the course of time, various responses to conditions of vulnerability have always been sought, obviously resorting not only to geographical choices and political, technological and social responses, but also to the design of ways of living, working and relating to others. Very clear examples of the importance attached to the design of new forms of settlement and ways of life are those found between the end of the eighteenth and the beginning of the nineteenth century, linked to the urbanisation processes generated by the rise of the industrial revolution.

Immigrants leaving the countryside of England, Belgium and Germany to settle on the fringes of historic cities lived in extremely vulnerable social, economic and health conditions. This gave rise to a complex system of technical solutions to improve housing conditions, make the places where people lived and worked hygienic, reduce the risk of fire, etc. The aim here was to reduce vulnerability in

P. Ceccarelli
International Laboratory of Architecture & Urban Design (ILAUD), Italy
e-mail: cec@unife.it

P. M. Guerrieri (✉)
Politecnico di Milano, DASTU, Milan, Italy
e-mail: pilar.guerrieri@polimi.it

E. Giorgi et al. (eds.), *Design for Vulnerable Communities*, The Urban Book Series,
https://doi.org/10.1007/978-3-030-96866-3_4

order to ensure a workforce capable of producing in a manner appropriate to the organisational needs of the emerging capitalist industry. In essence, then, design served to reduce the negative elements that could undermine a desire for new exploitation, which in turn would create new vulnerabilities over time. On the other hand, it is undeniable that the first English workers' settlements, imagined by Owen[1] and then developed by programmes carried out by the various corporations or public bodies, have deep roots in the idea of design as a positive factor in resolving physical fragilities and social weaknesses. The same applies to the projects of the Mietskasernen in Berlin or the large-scale sanitary works in Paris.

This reference to a vulnerability created by a process of economic transformation that needs to be resolved in order to avoid its more general negative effects, is one of the many aspects of vulnerability and the response to it. It is therefore worth examining a number of cases of vulnerability that contemporary society has to deal with in order to see which are the permanent and common elements of the responses to it and which are the innovative solutions that are produced for particular circumstances. This will help us to understand the extent to which, in the face of natural disasters, the effects of wars, environmental crises, forms of economic exploitation and technological changes, the architectural and urban planning project is used and can have more or less positive results.

4.2 Vulnerability and Natural Phenomena

However, the problem of vulnerability to natural phenomena is not only to be seen in terms of responding to disasters, but also in anticipating and solving a priori problems that may arise in a given location. It should not be forgotten that human settlement often took place in areas that were already structurally vulnerable and that had been chosen because they were able to protect against aggression or somehow manage to survive in very difficult climatic conditions.[2]

From the point of view of the choice of a vulnerable environment precisely because it was capable of protecting against enemy incursions and destruction, the best known and most obvious example is that of Venice. The choice of the first nucleus of absolutely marginal islands made it possible, on the one hand, to prevent invaders from the Roman mainland territories from killing more of the population; but on the other hand, it forced the invention of ways of managing the very unstable natural environment in order to avoid the damage it could produce.

Venice has thus become a city which, having emerged in an extremely vulnerable environment, has taken this characteristic into account in the way it has been built and managed over the centuries, gradually introducing new solutions and

[1] Prentoulis, Marina, *Owen and Cooperative Communities: Towards Revolutionary Trade-Unionism or Political Reformism*, Draft ECPR Conference 2004, https://ecpr.eu/Filestore/PaperProposal/fea0429f-edee-4648-a7c4-aa37ef2106ab.pdf (accessed: 20/04/21).

[2] Pielke, Roger A., *Climate Vulnerability* (Academic Press, 2013) ISBN: 9780123847041.

adaptations according to changing technological, socio-economic and, of course, environmental situations. We should not forget, for example, the "floating" solution adopted to construct buildings on platforms made of vertical piles sunk into the mud and horizontal embankments to form the base of the buildings.

In terms of similar examples, we are spoilt for choice, with a huge variety of different situations. One thinks, for example, of the complex system of dwellings in the ice of the Inuit populations of North America and Greenland, and the problems they faced in ensuring adequate protection in a very difficult environment, which is now in crisis, becoming a source of different vulnerability due to the rise in temperature and the melting of the ice in many areas.

In contrast to this example, different types of settlements in the desert, in oases, or in caves have had situations that were already very fragile, made extremely vulnerable due to climate change, especially in recent years. In these settlements, the role of design and careful planning in the use of the various natural and artificial components of the settlement has always been a determining element in the life of the communities that lived there.

Another very particular case to reflect on is that of large communities aggregated in settlements on boats, especially in the deltas of important Asian rivers, such as those of the Mekong in Vietnam and the Ganges in eastern India. In this case, the choice of floating dwellings is due to the economic activities of the inhabitants, but also to the possibility of avoiding the disastrous effects of flooding.

In all these cases, it is clear that design played a role, often with very sophisticated solutions in terms of concept and implementation. In fact, it can be said that the vulnerability arising from a site has been controlled and greatly reduced precisely because of the complexity of the designs adopted. It is worth reflecting on this ambivalence of the relationship with vulnerability, which on the one hand is an issue that we have to suffer because of events beyond our control, while on the other hand it is seen as a challenge; sometimes in a very risky and subtle game between life and death. Needless to say, cultivating the slopes of volcanoes can often lead to abundant and high-quality harvests, but it can also mean the risk of total destruction.

In cases such as the ski slopes designed in the shopping centre in the desert area of Dubai, design has been pushed to the extreme, dangerously distorting the original nature of the territory.

4.3 Cases of Humanitarian-Economic Vulnerability

We will now consider cases involving particularly serious situations of vulnerability to which attempts have been made to find solutions, perhaps temporary, that would avoid the worsening of humanitarian crises and social and economic disasters. The first group of such situations is exemplified by the reconstruction of the region destroyed by the Fukushima tsunami[3] in Japan in 2011, but similar situations can also be found in the USA as a result of the Katrina tornado, and in other parts of the world following major earthquakes or violent weather disasters.

Obviously, in the case of the reconstruction of the settlements destroyed or abandoned due to the reactor explosions in Japan, it was a question of creating completely new settlements, even in areas far removed from the contaminated areas, for a relatively homogeneous population that had lost all reference to the physical context in which it had developed. This clearly gave design a considerable role, which, however, was not forced to include particularly significant innovative elements, but rather had to recompose contexts that had been destroyed and subsequently abandoned, and had in any case become inaccessible for a very long time. In many cases, the construction of these new environments had to reproduce established situations and normal traditional ways of life, as well as reduce the traumas that the event had inflicted on many thousands of inhabitants. In this respect, the role played by participatory planning was very important, as in the *machizukuri* in the 1960s, as a way for local communities to actively engage in the improvement of their living environments.

"Japan commemorates the ten-year anniversary of the Fukushima triple disaster: the magnitude 9 earthquake, subsequent tsunami and radiation leakage from the nuclear power plant; a chain of events that killed more than 15,000 people. Ten years later, the northern part of the region, about 10 km from the Fukushima Daiichi nuclear power plant from which the radioactive material that compromised the area leaked, is still uninhabited, and access will be completely prohibited in more than 12% of the entire territory until at least 2023".[4]

Ishinomaki was the city that suffered the greatest damage in the Great East Japan Earthquake, 2011. Immediately after the disaster, the process of reconstruction began, and a huge number of volunteers came to the town to work on cleaning up the damaged buildings. Support organisations and bases of operations were set up in the central urban area. In the aftermath of the earthquake, it was soon clear that in the process of recovery, "the urban areas affected by the disaster cannot be maintained as a living environment merely by maintaining and preserving the original

[3] *Il triplice disastro di Fukushima, 10 anni dopo. Una tragedia non ancora finita* in Huffpost Journal, 11/03/2021 https://www.huffingtonpost.it/entry/il-giappone-commemora-il-decennale--del-triplice-disastro-di-fukushima_it_6049bbb4c5b636ed33782615 (accessed 20/04/21).

[4] "Il Giappone commemora l'anniversario dei dieci anni dalla triplice catastrofe di Fukushima: il terremoto di magnitudo 9, il successivo tsunami e la dispersione delle radiazioni dalla centrale nucleare, una serie concatenata di eventi che hanno causato la morte di oltre 15.000 persone. Dieci anni dopo, la zona nord della regione, a circa 10 km dalla centrale nucleare di Fukushima Daiichi, da cui è fuoriuscito il materiale radioattivo che ha compromesso la zona, è ancora inabitabile e l'accesso sarà vietato completamente in oltre il 12% dell'intero territorio almeno fino al 2023", see articles: n. 15: "Post disaster Reconstruction of central Ishinomaki through the formation of local initiatives" and n. 17 "Community Empowerment after the Chichi Earthquake: The Case of Taomi Ecovillage, Puli" in Satoh, Shigeru, *The First generation. Japanese Machizukuri and Community Engagement: History, Method and Practice* (London: Routledge, 2020).

buildings and architectural environment",[5] but by taking up new technologies, programmes, businesses and services. As well as the diverse, functional and participatory *machizukuri* forms of design, it was necessary to engender the active involvement of the community in decision making, including internal and external organisations, and community empowerment strategies.

A historical case that has some similarities with Fukushima, is the Messina earthquake[6] and tidal wave in Italy in 1908. The event was very serious, the city was largely destroyed and almost 100,000 people died. In this case too, the reconstruction project played a very important role in the social and economic recovery of the city. However, the events in Messina also highlight the risk of producing new forms of vulnerability by reconstruction processes that focused on certain aspects. In Messina, the richest and most representative part of the city was adequately reconstructed with buildings that rebuilt its environmental quality, but at the same time, the settlement quality of the most popular and dense areas of the city was neglected; in fact, the population were housed in shacks designed for a short duration, but in reality, they still existed and were used after more than a century. In fact, the reconstruction work was limited to solving only some of the vulnerability problems of the Messina area. This type of approach is found almost systematically in all the post-earthquake reconstruction plans of recent decades in Italy, which has obviously led to situations of serious local discomfort and often processes of abandonment of the areas affected by the earthquake. One exception is Friuli: after the 1976 earthquake, many of the buildings where refugees from the destroyed areas were temporarily housed were replaced by well-designed urban developments, complete with all the necessary functional elements.

The problems posed by the Katrina tornado, with the flooding and destruction it caused, were certainly also of a planning nature: reconstruction of safer contexts, demolished buildings to be rebuilt, etc., but also, and above all, related to protection by means of engineering infrastructures against the risk of flooding caused by the worsening climate. The fact remains, however, that in areas at high risk from natural phenomena, such as regions subject to strong earthquakes, tsunamis, tornadoes, etc., measures to reduce vulnerability are always relative. Japan's history, which is the longest and most complex in the fight against disasters caused by earthquakes, often resulting in terrible fires, still displays some very fragile situations. All the engineering measures introduced through very careful building design and very conscientious town planning, are often undermined, as in the case of Fukushima, by events that are extremely difficult to anticipate with certainty and to control

---

[5] Chapter n. 15: "Post disaster Reconstruction of central Ishinomaki through the formation of local initiatives" in Satoh, Shigeru, *The First generation. Japanese Machizukuri and Community Engagement: History, Method and Practice* (London: Routledge, 2020), p. 183.

[6] *The 1908 Messina Earthquake: 100-year Retrospective*, Report by Risk Management Solutions, 2008 https://forms2.rms.com/rs/729-DJX-565/images/eq_1908_messina_eq.pdf (accessed: 20/04/21).

completely. In many cases, therefore, the design of difficult and risky areas has only the role of containing and partly reducing vulnerability and its worst consequences.[7]

4.4 Conflict Zones

Another group of interesting vulnerable situations is that caused by wars and the destruction of entire cities, with thousands of people fleeing to safer areas. This has led to attempts at partial reconstruction once the war has ended, and the construction of new settlements consisting mainly of tents and temporary structures to house tens of thousands of refugees. Similar results have been produced by effects such as water shortages and widespread droughts which, often combined with situations of political instability, have driven millions of people from their places of residence. Wars have led to extremely serious situations in the Middle East and Yemen, and climate change has displaced large populations in parts of Africa and Asia. Many of these situations, such as scarcity of water and food resources, changing relationships with the use of territories where people have lived for millennia, have been produced by economic policies linked to the construction of large infrastructures—think of the system of dams on the Yangtze or the capture of water upstream of agricultural regions by countries crossed by major rivers in Africa. In this case, vulnerability could not be combated by action at a local level, but only by the forced displacement of a territory's inhabitants.

One type of vulnerable community, now home to millions of people worldwide, is that of temporary camps for refugees due to war, drought and destruction of natural resources, and migration in search of better living conditions. These cities, which often house tens or hundreds of thousands of people, are made up of tents which are not set up spontaneously but are the product of international assistance programmes, and bring together communities that are ethnically, socially and culturally totally different. These are undoubtedly vulnerable communities, officially intended to live temporarily in these conditions but, in reality, they have often been living in these camps for many decades. In these cases, the role of design and physical planning in trying to achieve a minimum of acceptable living conditions is clear, but so is the inadequacy of design to provide an acceptable response to the problems they pose.

However, two specific characteristics of these places, which differentiate them significantly from other vulnerable communities, should be borne in mind. The first is the fact that they are made up of inhabitants from very different cultural, linguistic and social backgrounds. Refugees often come from quite distant regions and are

---

[7] Deryugina, Tatyana, *The Economic impact of Hurricane Katrina on its Victims: Evidence from Individual Tax Returns*, October 2014: American Economic Journal Applied Economics 10(2), DOI: https://doi.org/10.1257/app.20160307, Ewing, Bradley T.; Jamie Brown Kruse; Daniel Sutter, *An overview of Hurricane Katrina and Economic Loss,* January 2009: Journal of Business Valuation and Economic Loss Analysis 4(2):1–1, DOI: https://doi.org/10.2202/1932-9156.1075.

often of different ethnicities, suddenly forced to live together without a common history. Even in less fragmented situations, mixing between refugees from large, half-destroyed cities and small, isolated villages in the same region is common. This implies the need to find solutions that meet often very different and unbalanced needs.

The design of settlements, which tends to be, as mentioned above, the result of standard models of international aid agencies, actually shows more or less relevant adjustments of various kinds to take into account these different needs and expectations.

A second characteristic is instead represented by the evolution of the initial models over time, as the temporary settlement inevitably becomes consolidated with its own history of responses to problems arising over time and its own culture. Certainly, refugee camps such as Dadaab in Kenya, which have existed for several decades now, do not resemble normal cities, but they have also developed original functional and spatial organisation solutions that are in any case, quite adequate for the normal functioning of the communities they host.

As is well known, Dadaab, which was launched in 1991 as a temporary haven for some refugees, has reached the dimensions of a large city with its 400,000 inhabitants, divided into three main settlements with service centres, schools, and meeting places that perform the functions of a large urban system, even though they have very unusual spatial and architectural characteristics. In this sense, Dagahaley, Ifo and Hagadera, and the most recent additions, Ifo II and Kambioos, are particularly noteworthy because they offer some practical examples of how camp-based communities can play a positive role in the management of their own affairs. All are in practical terms run by democratically elected community volunteers.[8] The camp has become a cluster of facilities and services: "healthcare and hospitals, schools and vocational training centers, communication and transport facilities, sports and entertainment, food and markets, social protection and empowerment programs, and so on, can be understood as public services, which may even exceed the local equivalent in their host regions".[9] These camps transcend from emergency measures and temporary shelters into much more ambiguous spaces, developing as humanitarian urbanism.

Another very interesting situation with respect to the physical organisation of the settlement and the different solutions given to some of its functions, is that of the Zaatari camp of Syrian refugees in Jordan, itself of very considerable size, both in terms of space and population gathered. In this case, there was also a lot of conceptual work behind the camp project, linked to the experience of Kilian Kleinschmidt,

[8] IRIN News: The refugee camp that became a city, https://www.un.org/africarenewal/news/refugee-camp-became-city (accessed: 20/04/2021); Perouse De Montclos, Marc-Antoine; Peter Mwangi Kagwanja, Refugee Camps or Cities? The Socio-economic Dynamics of the Dadaab and Kakuma Camps in Northern Kenya, *Journal of Refugee Studies*, Volume 13, Issue 2, June 2000, Pages 205–222, https://doi.org/10.1093/jrs/13.2.205

[9] Jansen, Bram J., *Cities in the making: Contours of the Urbanizing Refugee Camp*, 20/04/2020, https://www.bpb.de/gesellschaft/migration/kurzdossiers/307980/urbanizing (accessed: 20/04/20).

who later became responsible for running the camp. Kilian Kleinschmidt has a lot of experience in working in this area and has succeeded in stimulating many groups and institutions to also study types of accommodation and temporary service facilities to be used to house refugees.[10]

Here too, however, temporary solutions have inevitably become permanent outcomes, often shaped by the habits and cultural patterns of different immigrant groups. It is undeniable that in these situations, more or less formal design solutions have a considerable weight in reducing the precariousness of living conditions and contributing positively to alleviating vulnerability.

A particular case, interesting in many ways in terms of how it works, is that of the Sarawi camps[11] between Algeria and Morocco. Here, the refugees correspond to a homogeneous cultural group of landless people who have nevertheless reproduced ecosystems that function relatively well despite being in extremely difficult environmental and economic conditions. 173,600 Sahrawi refugees are living in five camps located in the Tindouf province. Most refugees arrived after the conflict in Western Sahara in 1975 or were born in the camps. 90,000 of them are considered particularly vulnerable due to their heavy reliance on humanitarian assistance to access food, water, education, and other necessities.[12]

Nevertheless, these camps are particularly interesting from a design point of view: "The Sahrawi camps are spaces in which inhabitants are in charge of their own lives—at least to the extent possible with the continuing occupation of their home country. It is a space that has given rise to a novel system of administration and new social structures, where nomadic traditions have transformed into modern concepts of family structures and new identities have been created. In stark contrast to the common conception that these camps are usually spaces where politics are prohibited, within the Sahrawi camps, politics are both facilitated and promoted. Not only is the Sahrawi population encouraged to engage themselves in political matters, but the camps themselves are seen and used as political projects in their anticipation of the Sahrawi nation state of the Western Sahara. The Sahrawi camps, therefore, give us proof of "the camp" as a form of urban space. At a time when places of control and surveillance are multiplying in our cities, where gated communities and corporate compounds withdraw ever more space from public and political interaction, the opposition of the urban condition to camp spaces becomes less and less valid. Maybe the Sahrawi camps represent a spatial quality that is in fact more urban than many of our cities".[13]

[10] See IPA Innovation and Planning Agency: Global Networking and Humanitarian Expertise, founded by Kleinschmidt, which helps us to connect the world's capacity with the world's needs through an advanced database specialized in connecting global technological, social ecological and economic solutions around the globe (see https://switxboard.org/).

[11] Dawn Chatty, *Deterritorialized Youth: Sahrawi and Afghan Refugees at the Margins of the Middle East* (New York-Oxford: Berghahn, 2010).

[12] https://www.acaps.org/country/algeria/crisis/sahrawi-refugees (accessed: 20/04/2021).

[13] See Herz, Manuel, *From Camp to City: Permanent is not the opposite of temporary*, in "Architectural Review", 16 May 2017, https://www.architectural-review.com/essays/from-camp-

4.5 Workers Camps

Similar characteristics to the temporary settlements associated with war-related migration are those hosting migrants in areas where they hope to find work and hope to reside permanently. The migrant camps along the Mediterranean coast indicate a situation that tends to fester instead of being resolved positively; often places designed as temporary reception solutions have become permanent, veritable cities with entire families who already have two generations of inhabitants behind them.

The temporary camps for agricultural workers are a different matter however, these are settlements of only young males from Africa and Asia, of different ethnicities, seeking entry into Europe and who partly work in agriculture and often partly have links to illegal drug-related activities. In several cases, these settlements have also become permanent settlements that are very vulnerable from both a social and a health point of view, but which do not have the possibility to evolve into forms of community typical of urban life. These settlements are almost always made by the landowners where the labourers work, and they often have links to organised crime that runs drug trafficking, prostitution, etc., in these areas. They are found in various southern European countries, such as Italy, Spain and Greece.

Many migrant workers gain rural employment through informal networks headed by *caporali*, which they join through various migration patterns. In the province of Foggia, for example, the vast majority of migrant agricultural workers are North African, sub-Saharan African and Eastern European, between 20 and 40 years of age.

The exploitation of migrant workers in Italian tomato production is a particularly crucial issue.[14] In fact, foreign agricultural labourers live in isolated rural ruins or shanty towns in Sicily and Calabria after being selected by agents who recruit seasonal workers to work as much as possible for a lower salary. This occurs not just for tomato pickers, but also for pepper, orange and kiwi cultivations. Shanti towns like San Ferdinando or Borgo Mezzanone, with leaking mobile homes, black walls, old freezers, tents, shacks and mud paths, are just two of dozens more slums in southern Italy.[15] It's a situation that has been described by NGO's as a "modern form of slavery", where vulnerability is both mental and physical. Similar issues can be found in rural Andalusia, which is well explained in the "Research on the Vulnerability of Human Rights in Andalusian Rural Areas: Migration, Labour

---

to-city-permanent-is-not-the-opposite-of-temporary (accessed: 20/04/21). In addition, here, design experiments in renewable energy have improved living conditions, such as in Taten Lehbib, where they used recycled plastic bottles filled with sand to make local homes able to resist floods and reduce the camp's overall recycling issues.

[14] Praticò, Giulia and Ole Henning Sommerfelt, *Counteracting exploitation of migrant workers in Italian Tomato Production*, a report by Ethical Trading initiative Norway, 2015, https://www.ethicaltrade.org/sites/default/files/shared_resources/italian_tomato_production_report.pdf (accessed: 20/03/21).

[15] Jones, Tobias, Ayo Awokoya, *Are your tinned tomatoes picked by slave labour?* In "The Guardian", 20/04/2021, https://www.theguardian.com/world/2019/jun/20/tomatoes-italy-mafia-migrant-labour-modern-slavery (accessed: 20/04/21).

Treatment, and other Forms of Exploitation for the Strengthening of Andalusian Development Agents" (Waldimeiry Correa da Silva e Carla Cingolani, 2019).[16]

Similar forms of settlements with a very specialised demographic and social structure, which makes them very different from the spontaneous slum formations on the fringes of large cities, are the communities of young immigrants that have developed over the last decades in various countries of the world: settlements of male immigrants in "urban villages" in China, communities of male immigrants from rural areas in Indian metropolises, different forms of temporary labour settlements in various countries of South-East Asia, as well as some African and Latin American regions. These communities are sometimes on the fringes of cities, as they are linked to agriculture, in others they are in the fabric of the metropolis as temporary labour camps, isolated from the social context of the cities in which they are located.

4.6 Urban Migrants

In the Chinese case, the legal situation that makes a considerable part of migration to the big cities illegal has favoured the formation of very high-density settlements with very poor housing conditions in areas of rural villages that have been incorporated into the rapidly expanding urban fabric. The relative autonomy of the government in these villages, often located in areas that have become central to the cities, has prompted the creation of housing that can accommodate a marginal and semilegal workforce operating in the service sector or construction industry, etc. For these inevitably single immigrants, the conditions of very high vulnerability were balanced by the proximity of the place of residence to the places of work, with an undoubted economic advantage for a workforce with very low wages. In these settlements, services are obviously reduced to a minimum and very often different forms of crime and social unrest flourish, creating security problems.

This Chinese model, which has obviously taken on different forms over time, highlights the exchange between the risks and benefits of vulnerability, an exchange that should not be underestimated and which to a large extent explains the continual recurrence of new situations of vulnerability even in contexts where it could be largely resolved through more articulate and careful planning policies.

Similar problems are found in India, where younger generations migrate towards the bigger megacities to find better job opportunities. The move towards cities has many push and pull factors away from rural life; people think that the city will provide lots of opportunities, such as social services (better housing, healthcare and education), economic opportunities (more jobs and higher wages), and enhanced

[16] Correa da Silva, Waldimeiry and Claudia Cingolani, *Labour Trafficking and Exploitation in Rural Andalusia*, (Intechopen, 2020, https://doi.org/10.5772/intechopen.90090, https://www.intechopen.com/books/modern-slavery-and-human-trafficking/labour-trafficking-and-exploitation-in-rural-andalusia (accessed: 20/03/21).

environmental conditions (better living conditions with a safer environment, e.g. drinking water and sanitation), but these perceptions are not always correct. "Every minute 25–30 people are migrating to Indian cities from rural areas in search for a better livelihood and lifestyle."[17] The Indian urban population is likely to reach 600 million by 2030. Such an influx presents a huge challenge for the cities and, at the same time, is creating a series of issues related to the abandonment of rural areas and villages. These migrants are becoming a vulnerable part of urban suburbs and generating vulnerable rural areas without any skilled young artisans to preserve lively ancient Indian art and craft traditions.

4.7 Tech Vulnerability

The expansion of vulnerability is also linked to technological development and the transformations it brings about in territories and urban settlements. This is the case for many areas that were originally mining areas or first industrial settlements. Not only is vulnerability evident from the changing economic geography and the role that certain places have played in the different phases of development, but it is the inevitable result of a process of continuous productive, distributive and organisational transformation. As has happened in the past with regions such as the Ruhr in Germany or the mining and industrial areas of Poland in Europe, the Rust Belt in the USA, or the crisis of the diamond mining areas in South Africa, it is inevitably that this will happen with the realisation of profound transformations in economic geography, such as those generated by the construction of the two main routes of the Chinese Silk Road. It is already clear that seaside cities that lived on a continuous flow of trade will be in crisis as a result of profound changes in the routes of goods, such as the cutting off of the Malacca peninsula and the relative marginalisation of Singapore. New landmarks will emerge and places that have long lived in prosperity will inevitably go into crisis.

A case that raised a lot of debate in this regard was, for example, the Three Gorges Dam project on the Yangtze River, which led to the displacement of 1.3 million people from some agricultural areas and the flooding of almost 650 km^2 to create the new hydropower reservoirs. People who were not vulnerable suddenly became so; they had to leave the territories where they had lived for centuries and settle in new places, often with great economic and social difficulties, thus becoming vulnerable.

[17] Pranav, Divay, *India preparing for the biggest human migration on the planet*, in "Invest India", January 02, 2018, https://www.investindia.gov.in/team-india-blogs/india-preparing-biggest-human-migration-planet (accessed: 20/04/21), see also Mehrotra, Santosh, *Mega Challenge of rural-urban migration*, in "Business Line", October 03, 2019, https://www.thehindubusinessline.com/opinion/mega-challenges-of-rural-urban-migration/article29577159.ece, (accessed: 20/04/21).

Other areas where there was industrial and mining development in the early stages of the last century were subsequently marginalised by profound technological changes, depletion of ore deposits, and the relocation of production chains to other regions of the world. Phenomena of this kind have produced pockets of social marginality in places that initially, and for fairly long periods, had enjoyed prosperity and significant social dynamics. Marginalisation and decadence have gradually taken root and it has been extremely difficult to overcome them, despite their proximity to areas and regions that are still developing.

Examples of this situation are numerous in industrial centres in Europe and the United States, which have gradually lost their population, partly becoming abandoned places or ghost towns.

A case in point is Detroit in the United States. Until the 1950s, Detroit was the undisputed world capital of automobile production. The Ford Motor Plant, the famous auto industry, opened in 1927, drawing a million new residents to the city. With the entry of Japanese production onto the world market, a progressive crisis began which, over the course of a few decades, marginalised this important city, a symbol of modernity, and transformed it into an urban settlement that was increasingly vulnerable and incapable of emerging from a demographic, economic and social crisis. The population of the city has fallen from a high of 1,850,000 in 1950 to 680,000 in 2015; vast areas of the city are in a state of severe urban decay and vulnerability.

A similar decline process happened in the Rust Belt area, once a powerful industrial area, formerly known as the "Breadbasket of America". Since the mid-twentieth century, heavy industry has declined in the region, bringing economic distress and poverty, and resulting in a decline in the population.[18]

Urban shrinkage is currently a problem of urbanisation processes in Poland, and also in the rest of Europe. The reasons behind the shrinking of cities are various and numerous. It is assumed that the basic issue is the crisis of the local economy, which began in 1990. When Poland opened to global processes, economic backwardness in many cities was disclosed and, consequently, it was necessary to restructure their economies. Adjusting to new market conditions was particularly difficult in cities with monofunctional production. The collapse of the previous economic base of the city was the reason behind increased unemployment, the impoverishment of society, social segregation and marginalisation, which in turn has led to the depopulation of cities and a significant process of urban shrinking.[19] The consequences of shrinking cities are many, such as demographic, social, economic, spatial, infrastructural and political. Urban shrinkage has a strong impact in many areas in the city.

The role of design in relation to the vulnerability generated by technological change and the consequent economic and social crises is undoubtedly important, but its characteristics vary profoundly from case to case. The Ruhr has been

---

[18] LeDuff, Charlie, *Detroit: An American Autopsy* (New York: Penguin, 2013).

[19] Musial-Malago, Monika, *The Process of Urban Shrinking in Poland*, in STUDIA MIEJSKIE tom 24 (2016) http://www.studiamiejskie.uni.opole.pl/wp-content/uploads/2017/03/S_Miejskie_24_2016-Musial.pdf (accessed: 20/03/21).

profoundly transformed from a mining and industrial region in serious crisis into an area of tourist and cultural attraction, as well as a new location for advanced industrial activities. This has allowed a profound regeneration of the area compared to the previous situation, and the creation of a new role and a new social and cultural base.

Some of the interventions carried out have become a reference model for other regions in the world, also a highly innovative key. In the case of industrial and mining cities in crisis in Poland, however, the interventions have often been aimed at partially adjusting situations containing the most negative effects and reducing situations of abandonment and depression. There have been interesting examples of small-scale interventions to contain the breakdowns, but not so as to guarantee real reversals. The situation in the Rust Belt in the USA is fragmented into different types of local responses that have not in fact contributed significantly to countering the crisis situation. Depressed cities and territories have often become even more vulnerable, as demonstrated by the economic and social crisis that continues to affect states such as Pennsylvania and Ohio, and which are confirmed by the voting patterns in the last presidential elections. An extreme case of an attempt to redesign the role of a city, such as that attempted in Detroit, making it an alternative to a major industrial centre emptied of its functions, a settlement based partly on urban agriculture and partly on the development of artistic activities requiring the use of empty spaces or the reuse of abandoned building structures, has not produced particularly significant results. In this case, the role of design, which has sometimes been very original, has had very little effect in reducing the vulnerability of the urban settlement.

4.8 Which Role Does Design Play?

What can be drawn from the cases we have examined? Design has certainly played a positive role in the recovery and reorganisation of certain parts of urban settlements, but without doubt, the possibility of recovery and the reduction of vulnerability lies first and foremost in the capacity for integrated strategic planning between the economy and the organisation of the territory.

What we have seen provides a series of images in different sectors and regions of the world; cases that could, however, be further enriched with many other references. For example, we did not want to touch on the problems of the very large-scale vulnerability of informal settlements, especially in Africa, Asia, and Latin America, because they have already been considered. Similarly, another area of design that will play an extremely important role with respect to vulnerable communities has been avoided, that of climate change with all the problems that will bring.

Despite the choice of particular cases in this study, it is clear that the problem of vulnerability, with its infinite different forms, is a structural and foundational element of the condition of mankind on the planet. A condition that has become increasingly complex and fragile as human domination of the earth has progressed;

a domination that has been able to positively control certain aspects, creating situations of relative stability and equilibrium, but which, in other areas, has instead increased the fragility of the human habitat, clearly entailing a continuous search for answers to the natural or man-made crisis processes that continue to occur. In this sense, it is evident that the project, the specific technical solution and local planning intervention, are indispensable and contribute in some way to avoiding the excessive worsening of crises, but it is also evident that in many cases, punctual and specific actions only partially and temporarily resolve the crisis situations.

Basically, what has happened and is happening once again points to the need to redefine the very concept of design, its role and its methods. As with many other aspects of culture, design only plays a positive role in some respects and is constantly changing. It probably only has the role of a painkiller rather than a vaccine that cures permanently, but this is already an absolutely extraordinary role in the present context.

References

Chatty, D. (2010). *Deterritorialized Youth: Sahrawi and Afghan Refugees at the Margins of the Middle East*. Berghahn.

da Silva, C. (2020). Waldimeiry and Claudia Cingolani, *Labour Trafficking and Exploitation in Rural Andalusia*, (Intechopen). https://doi.org/10.5772/intechopen.90090., https://www.intechopen.com/books/modern-slavery-and-human-trafficking/labour-trafficking-and-exploitation-in-rural-andalusia (accessed: 20/03/21).

De Montclos, P. (June 2000). Marc-Antoine; Peter Mwangi Kagwanja, Refugee Camps or Cities? The socio-economic dynamics of the Dadaab and Kakuma Camps in Northern Kenya. *Journal of Refugee Studies, 13*(2), 205–222. https://doi.org/10.1093/jrs/13.2.205

Deryugina, T. (October 2014). The economic impact of hurricane Katrina on its victims: evidence from individual tax returns. *American Economic Journal Applied Economics, 10*(2). https://doi.org/10.1257/app.20160307

Ewing, B. T., Kruse, J. B., & Sutter, D. (January 2009). An overview of hurricane Katrina and economic loss. *Journal of Business Valuation and Economic Loss Analysis, 4*(2), 1–1. https://doi.org/10.2202/1932-9156.1075

Herz, M., *From Camp to City: Permanent is not the opposite of temporary*, in "Architectural Review", 16 May 2017., https://www.architectural-review.com/essays/from-camp-to-city-permanent-is-not-the-opposite-of-temporary (accessed: 20/04/21).

Il triplice disastro di Fukushima, 10 anni dopo. Una tragedia non ancora finita in Huffpost Journal, 11/03/21 https://www.huffingtonpost.it/entry/il-giappone-commemora-il-decennale-del-triplice-disastro-di-fukushima_it_6049bbb4c5b636ed33782615 (accessed 20/04/2021).

IRIN News, *The Refugee Camp that became a City*, https://www.un.org/africarenewal/news/refugee-camp-became-city (accessed: 20/04/2021).

Jansen, B. J., *Cities in the making: Contours of the Urbanizing Refugee Camp*, 20/04/20, https://www.bpb.de/gesellschaft/migration/kurzdossiers/307980/urbanizing (accessed: 20/04/2020).

Jones, T., Awokoya, A. *Are your tinned tomatoes picked by slave labour?*, in "The Guardian", 20/04/21, https://www.theguardian.com/world/2019/jun/20/tomatoes-italy-mafia-migrant-labour-modern-slavery (accessed: 20/04/2021).

LeDuff, C. (2013). *Detroit: An American autopsy*. Penguin.

Mehrotra, S. *Mega Challenge of rural-urban migration*, in “Business Line”, October 03, 2019., https://www.thehindubusinessline.com/opinion/mega-challenges-of-rural-urban-migration/article29577159.ece, (accessed: 20/04/2021).

Musial-Malago, M. (2016). *The Process of Urban Shrinking in Poland*, in STUDIA MIEJSKIE tom 24. http://www.studiamiejskie.uni.opole.pl/wp-content/uploads/2017/03/S_Miejskie_24_2016-Musial.pdf (accessed: 20/03/2021).

Pielke, R. A. (2013). *Climate vulnerability*. Academic Press. ISBN: 9780123847041.

Praticò, G., & Sommerfelt, O. H.. *Counteracting exploitation of migrant workers in Italian Tomato Production*, a report by Ethical Trading initiative Norway, 2015., https://www.ethicaltrade.org/sites/default/files/shared_resources/italian_tomato_production_report.pdf (accessed: 20/03/2021).

Prentoulis, M. *Owen and Cooperative Communities: Towards Revolutionary Trade-Unionism or Political Reformism*, Draft ECPR Conference 2004., https://ecpr.eu/Filestore/PaperProposal/fea0429f-edee-4648-a7c4-aa37ef2106ab.pdf (accessed: 20/04/2021).

Satoh, S. (2020). *The first generation. Japanese Machizukuri and community engagement: History, method and practice*. Routledge.

The 1908 Messina Earthquake: 100-year Retrospective, Report by Risk Management Solutions, 2008. https://forms2.rms.com/rs/729-DJX 565/images/eq_1908_messina_eq.pdf (accessed: 20/04/2021).

Part II
Approaches, Principles and Paradigms to Contemporary Research and Practice for Vulnerable Communities

Chapter 5
Thinking Socially: Anthropological Approaches to Contemporary Research

Annie Malcolm

5.1 Introduction

> *"Go to Wutong Shan, I hear that's where the real artists live."*
> *"Wutong Shan is a looney bin."*
> *"I'm moving to Wutong Shan." "Why? Have you fallen ill?"*
> *"Everyone in Wutong Shan is a master."*
> *"Everyone in Wutong Shan is crazy."*
> *"Wutong Shan is special... the best place in Shenzhen."*
> *"Wutong Shan is not like the rest of Shenzhen and Shenzhen is not like the rest of China."*
> *"This is a mini Hong Kong."*
> *"Dafen is business. Wutong Shan is art."*
> *"If you want to study anything about the countryside, don't study here. This is the city."*
> *"Oh I don't rent out here, it's just farmer's houses."*
> *"Wutong Shan is not normal life. This is not real."*

All of these quotations are from interlocutors in an art village on the outskirts of Shenzhen, China, where I conducted research. They come from people with whom I ended up in situations of encounter in which they trusted me enough, or felt compelled, or indifferent, or annoyed enough, to state these opinions. I was doing what anthropologists call participant observation; I was participating in social life and observing its contours and events. Each of these utterances came bathed in a rich context, of which living in the environment afforded understanding. The question I explore here is, **how do we know what we know?**

This article is about the method of ethnography as a way of doing anthropology to produce knowledge that engages and produces something called the social, some provocations, new ways of thinking about research, human interaction and place. I bring up several lines of thought to promote a conversation about ethnographic

A. Malcolm (✉)
University of California, Berkeley, CA, USA
e-mail: a.malcolm@berkeley.edu

E. Giorgi et al. (eds.), *Design for Vulnerable Communities*, The Urban Book Series,
https://doi.org/10.1007/978-3-030-96866-3_5

engagement and meaning-making. I examine creative work on the outskirts of cities in China, primarily in one field site outside of Shenzhen called Wutong Shan. That means going to and living in art villages in order to understand how they relate to processes of urbanization.

In response to the neighboring Dafen Village, which churns out five million paintings a year, mostly copies of European art, Wutong was set up as the site for the production of original art. Instead of becoming a booming production site for original art objects, it has become an area where people produce the place itself as a hub of alternative lifestyle activities, spiritual seeking, and national arts revival. Ten years ago, a place where artists could rent cheap space, Wutong became an informal center of art practice. The government's intervention to make it an art village came in 2011, *after* private citizens had already begun forming the art village context. Government intervention to support creative industry was ineffective (nominal/neglectful), and in its wake, art itself is not the main commodity sold in the village because there isn't a market for original art. Rather, as the more successful artists sell their work in more urban and international spaces, art practice and the artistic atmosphere it produces are instrumentalized to sell products used in tea ceremonies, life cultivation (*yangsheng*), and spiritual practices. People in Wutong mix Daoist, Buddhist, and Confucian forms of thought and practices as part of their life cultivation, economic activity, and sociality. Life cultivation includes meditating, waking early, exercising in the morning, and exchanging health advice—life cultivation and its sociality are co-constitutive. These trends tend to be the way of older citizens, whereas the people in this site are young and middle aged, uniquely reflecting an ambivalence about navigating urban life.

5.2 Thick Description

Understanding people's cultural values in order to work *with* them instead of merely *on* their landscapes requires what Clifford Geertz calls thick description. Thick description frames anthropology as an interpretative science; the way a literary scholar interprets a text, the ethnographer interprets a reality. This method is fundamentally semiotic. In language, we only understand a word because we understand the language of which it is a part; Saussure famously called these two aspects *langue* and *parole*, roughly translated to mean language and speech (de Saussure, 1998). In anthropology, understanding a cultural rite, object or situation is made possible by studying the code of which it is a part. Culture may seem un-codified, but it is actually highly systematic, somewhat like language. Culture is a web of meaning in which man is suspended. Geertz writes, "Believing, with Max Weber, that man is an animal suspended in webs of significance he himself has spun, I take culture to be those webs, and the analysis of it therefore not an experimental science in search of law but an interpretive one in search of meaning" (Geertz, 1973).

5.3 Interpretative Work: Studying Culture in Wutong Shan

In Wutong Shan, migration, spiritual and art practices are illuminated by broader processes. For example, I only understand *rural* because I'm intimate with a categorical system that binarizes rural and urban. I only find meaning in the figure of the Buddha because I know something about Buddhism. I only know it is significant for people in contemporary China to move to a space caught between rural and urban and to decorate it with rural aesthetics and dreams because I know that migration in China is seen as a uni-directional process, from rural to urban. And I know that these people taking a renewed interest in Buddhism is exceptional because I know about the erasure of religion from China in recent history, during the Mao years. And I was taught these things—the rural, the Buddha, the need for spiritual nourishment—through talking to people, and the rest—the context that makes those things matter—through scholarship. One has to know what's important to people in general to analyze any one utterance they make. **Context creates meaning.**

5.4 Aesthetics

In my study, the formation of an art village is a specter of urbanization. Art villages are sometimes magical, sometimes rural-feeling, enclaves that are the results of processes of urban change. I attend to how artists reflect on and produce these peri-urban places in their life choices, like that to migrate to an art village, and also in their work. I combine direct observation, through the five senses, with accounts from interlocutors to construct a narrative of the village. Jacques Rancière defines aesthetics as, "the system of a priori forms determining what presents itself to sense experience" (Rancière, 2004). The concept of aesthetics speaks to both my object and my method, my entanglement in my object. Ethnography involves attending to the observable aspects of a situation, made possible by the senses. Sarah Pink, who writes on sensory ethnography, explains the method thus:

> ...ethnography is a way of creating and representing knowledge (about society, culture and individuals) that is based on ethnographers' own experiences. It does not claim to produce an objective or truthful account of reality, but should aim to offer versions of ethnographers' experiences of reality that are as loyal as possible to the context, negotiations and intersubjectivities through which the knowledge was produced. (Pink, 2015: 8).

Pink stresses **loyalty to the experience as well as its context**. In my work I prop the background needed to pose relevant questions equally with the sensorium that develops as a result of actually being there. "Being there" is big in anthropology.

5.5 Environmental Anthropology

Wutong Shan is defined largely as a physical environment and its inhabitants are deeply influenced by the natural world. Environmental anthropology takes a special look at the relationship between people and their environments. One way to think about this is through multi-species anthropology. Anna Tsing is an anthropologist whose book *The Mushroom at the End of the World: On the Possibility of Life in Capitalist Ruins* (Tsing, 2015) narrates the story of the matsutake mushroom to explore the question, how can we live amidst the ruins we have created? Tsing attends to the Anthropocene in her studies of how human and non-human agents come together in what she calls relations of corrupted diversity. We imagine nature as something pristine and human beings as corrupting it; Tsing argues that the combinations and variations of human and non-human species over time, or diversity, is a form that is always already corrupted.

5.6 Actor Network Theory

There's a lineage of this kind of thinking. Bruno Latour's Actor Network Theory imagines humans and non-humans as actors in networks, and the job of the social scientist as that of studying the traces those networks leave (Latour, 2005). While Latour's ANT is quite heavily theoretical, Latour's first major project, for example, was a study of the pasteurization of France, how the very chemicals used to pasteurize dairy changed French society. Scholarship in hydro-social studies thinks of water as in a network *with* humans to make meaning. Natural resources are also actors in a network with humans in my field site. In Wutong, water is very important. Wutong Art Village, at the foothills of Wutong Mountain, sits atop the Shenzhen reservoir (Hong Kong's water source); it is therefore not developing in the raze-and-build fashion characteristic of many of China's urban fringe areas. Rather, inhabitants change the built environment by converting factories and original village houses into art studios, live/work spaces, and shops, exhibiting a unique form of deindustrialization in Chinese cities (see Figs. 5.1 and 5.2).

5.7 Ethnographic Encounters

Anna Tsing recently mused, "one can do archival things on a landscape but a landscape is not just an archive." There is a special or specialized archive when it comes to the art village. But who are the gatekeepers? **We need** narrators to give us the discourse to rifle through these archives, these intimate knowledges, **people native to the problem we're observing who help us create meaning in context.** To that end, below are excerpts from my ethnography, at an acute level of specificity, and

Fig. 5.1 Wutong Art Village and Wutong Mountain (photo by author)

their analyses, which explore questions of urbanization, culture, globalization, industrialization, deindustrialization, and climate change.[1] The analyses show how anthropologists perform a *tacking* between exquisitely concrete moments and the space of knowledge production.

1. I met Zheng Xin on a gray March day in his studio, a converted factory in Kengbei Village, one of the eight villages that comprise the Wutong Mountain Scenic Area. Zheng is 30 years old from Guangdong's northern neighbor Jiangxi province, with striking cheekbones and lanky limbs. "At the beginning it was like the interior and the countryside… and there weren't a lot of artists here," said Zheng. Western music—his CD collection includes Tracy Chapman, Sinead O'Conner, and John Lennon—blared from large speakers. As his adorable 3-year-old daughter played for our attention, her mother came in and out of the room and his work—painting and sculpture—filled the space. We peeled through stacks of paintings of a figure—"always me" he said—and of horses, dark themes, and a sense of drowning. After graduating from the Guangxi Fine Art Academy in 2009, Zheng came to Wutong, having heard about it from friends,

[1] All names of the interviewees have been changed with pseudonyms.

Fig. 5.2 Wutong Shan factory building, repurposed for art studios (photo by author)

and finding it a fitting environment, away from excess noise and people. "Whether I'm inspired or not, I do art, everyday," said Zheng. "You can't use words to express paintings. It's feeling." Zheng sells his paintings in galleries in the city center of Shenzhen, an hour from the mountain, or friends come and buy his work. "Now the rents are rising, and it's getting more commercial," he said.

5.7.1 Analysis

China's interior is historically less developed than its coastal cities. It is less cosmopolitan, in the sense that it is less urban and also less connected to the rest of the world. It is variously referred to as hinterland, a step back in time and countryside. The interior is actually home to China's most polluted cities, the result of coal that fuels coastal urbanization, and dust from desertification. The interior has been largely depopulated since market reform, and many who remain can't find work. Among my interlocutors, stances toward the space of the interior varied along two paths. As Shenzheners, coastal city dwelling and seeking or taking advantage of work opportunities, many set themselves apart from the interior while some, as Wutong Shan "mountain dwellers," aligned themselves with the interior, talking about Wutong as a return to a less developed, anti-urban space. In some sense, these two viewpoints are not at odds; the urban is constituted through its negative relationship to the hinterland; i.e., one can be urban by not being rural. Zheng's interest in Western music speaks to the notion that male artists in China are catching up to Western contemporary art (Welland, 2018). Zheng's wife taking care of their child indexes the gendered dynamics at work. Zheng can *be* a painter in his context because he was trained as such, because he doesn't have to care for his child full time, and because he's made a choice to live in a place with low rents.

2. At the enzyme shop that night, I met a person who calls himself a culture teacher. "During the Cultural Revolution our culture was destroyed," he said. "Now we are picking it back up. Because of Hong fasi, the temple nearby, this is a place where many people gather to revive traditional Chinese cultural forms."

Analysis: Dispossessed of the cultural forms of the past, people in Wutong strive to revive some of them, to create a place in which to enact, through practice, a life of alternatives to the urban, materialistic, and secular (In Chinese, *meiyou xinyang* or "without beliefs"). The Mao period left meaning fractured, leaving signifiers dangling apart from their original signifieds in its wake. Dispossession serves as a motivator to migrate to a less urban place and re-possess signifiers from the past, disarticulated from their original signifieds.

I'll conclude with a conversation with an interlocutor, which took place in a basically abandoned art villa. As part of the government's creative industry designation in Wutong Shan, an art villa building was constructed but the government did not follow through with its plan and fill the space with art. Kong, a young man from Shandong, came to bring the culture of gǔqín, a traditional Chinese instrument, to Wutong, fulfilling what he saw as his duty.

> Kong: There's a responsibility to spread gǔqín culture, so I came. Whether the government had asked me to come or not, I would have come.
>
> Me: And they…
>
> Kong: They support me.
>
> Me: Did they have this dream, to make this museum to support traditional Chinese arts?
>
> Kong: Yes, right, *qin qi shuhua*—instruments, board games, books, and paintings—ancient Chinese arts hobbies—it's the basic things ancient Chinese did.

I was quite sure Kong was exaggerating when he said the government supported him, as I knew that the government had done very little in the area to actually live up to the creative industry designation they'd made. Kong's own words even make that clear; "whether they had asked me to or not, I would have come," he said emphatically. On another level, he's quite right; whether the government follows through with it or not, it has a dream—literally written into their China Dream policy—to reignite excitement about traditional Chinese culture in a nationalism push. Knowing that the villa had been empty for months, it was clear that Kong had actually come somewhat of his own volition. His enthusiasm for the ancient arts hobbies is evidence for a return to tradition in Wutong Shan and across China.

I see this conversation as parole—speech or utterance. Knowing the context of what was going on, the langue, the larger themes at play, helps me understand some of what happened in the interaction. This moment congeals as a result of both the context around it and the conversation itself. Interpretative work demands both a broad view of the conditions affecting the social and an attention to exquisitely fine detail. Ultimately what is gained from the encounter is a snapshot of cultural phenomena, an understanding of a network of elements that come together to express a mini-landscape, much like a Chinese landscape painting, at the foothills of a mountain, set to the plucked seven-string instrument associated with Confucius.

References

de Saussure, F. (1998). *Course in General Linguistics*. Open Court Classics. (Originally 1916).
Geertz, C. (1973). *The interpretation of cultures*. Basic Books.
Latour, B. (2005). *Reassembling the social - An introduction to Actor-Network-Theory*. Oxford University Press.
Pink, S. (2015). *Doing Sensory Ethnography*. SAGE Publications Ltd.. (Originally 2009).
Rancière, J. (2004). *The politics of aesthetics: The distribution of the sensible*. Translated by Gabriel Rockhill. New York: Continuum International Publishing Group.
Tsing, A. L. (2015). *The mushroom at the end of the world: On the possibility of life in capitalist ruins*. Princeton University Press.
Welland, S. S.-L. (2018). *Experimental Beijing: Gender and globalization in Chinese contemporary art*. Duke University Press.

Chapter 6
Thinking in Systems: Sustainability Cognition for Design Communities

Xiaocun Zhu and Pius Leuba dit Galland

6.1 Sustainability: System

6.1.1 Sustainability Dilemma

Life has been on earth for about 3.8 billion years and has profoundly shaped the planet through endless rounds of evolution. Many species became extinct over time, including the extremely successful Dinosaurs, who had been the dominant terrestrial vertebrates for more than a hundred million years. In contrast, modern humans (homo sapiens) have only been around for a few hundred thousand years. Yet we must be concerned with the doom of our own species already.

"In the long term, the biosphere will handle pretty much anything we throw at it, including climate change," claims astrophysicist Adam Frank (2018) in his recent article "Earth Will Survive. We May Not." Nevertheless, "if we don't take the right kind of action soon the biosphere will simply move on without us, creating new versions of itself in the changing climate we're generating now" (para. 7–8). "Hence, when we talk about sustainability, our challenge is not to address an ethic of values, but rather an ethic of survival …The goal of sustainable development is to ensure that the human species [will survive]" (Lou, 2018, p. 346).

This chapter was developed upon the authors' paper "Building up a Mindset of Design for Sustainability" in the Cumulus Conference Proceedings Wuxi 2018 Diffused Transition & Design Opportunities.

X. Zhu (✉)
College of Design and Innovation, Tongji University, Shanghai, China
e-mail: zhu_xiaocun@tongji.edu.cn

P. Leuba dit Galland
Grolimund+Partner AG, Bern, Switzerland

E. Giorgi et al. (eds.), *Design for Vulnerable Communities*, The Urban Book Series,
https://doi.org/10.1007/978-3-030-96866-3_6

In recent decades, human society has made remarkable progress in sustainability: through unprecedented levels of scientific findings, we far better understand the mechanisms of how our species affects planet Earth. "Sustainable" methods, technologies, and practices are increasingly being invented and put to application. Ever more "green" products become available in the market. Progressively, international consensus is reached among governments: in 2015, 195 countries adopted the first-ever universal, legally binding global climate deal at the Paris Climate Conference (COP21). The deal "brings all nations into a common cause to undertake ambitious efforts to combat climate change and adapt to its effects" (United Nations Framework Convention on Climate Change [UNFCCC], n.d.).

Yet, reality of human sustainability is scarcely satisfying. Our environmental impact keeps intensifying at exponential rates. New challenges appear behind relentless development and well-intentioned solutions. Every year, the Global Footprint Network computes how fast we exhaust Earth's natural resources "by comparing humanity's total yearly consumption (ecological footprint) with Earth's

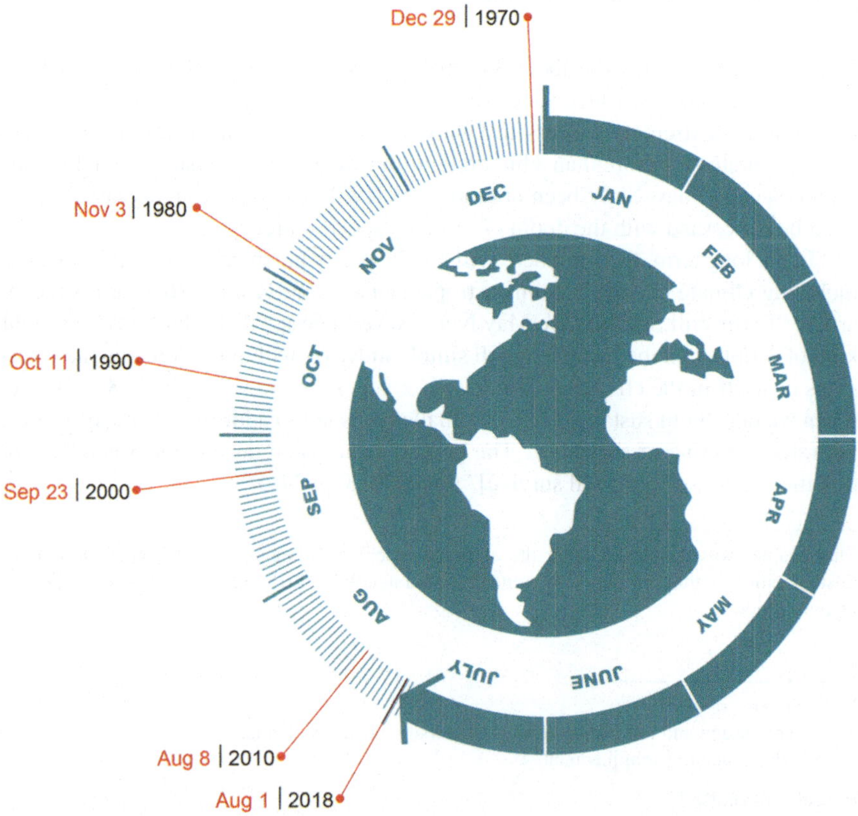

Fig. 6.1 Earth Overshoot days in 1970, 1980, 1990, 2000, 2010, 2018. Graphic by R. Tang, based on data from the Global Footprint Network, 2018

capacity to regenerate renewable natural resources in that year (biocapacity)" (Global Footprint Network, n.d., para. 1). Last year, August 1, 2018, was "the earliest date since ecological overshoot started in the early 1970s" (see Fig. 6.1). "Overusing Earth's ecosystems is one of the largest challenges facing humanity today, with climate change being a big portion of that challenge" (Global Footprint Network, 2018, para. 2 & 12).

In 2009, the concept of planetary boundaries (PB) was proposed by a group of scientists to define the environmental limits of our planet Earth, within which humanity can safely operate (Rockström et al., 2009). The findings of the following years revealed that human-induced changes are continuously and dramatically approaching these limits—even appearing beyond, already in the high-risk zone in

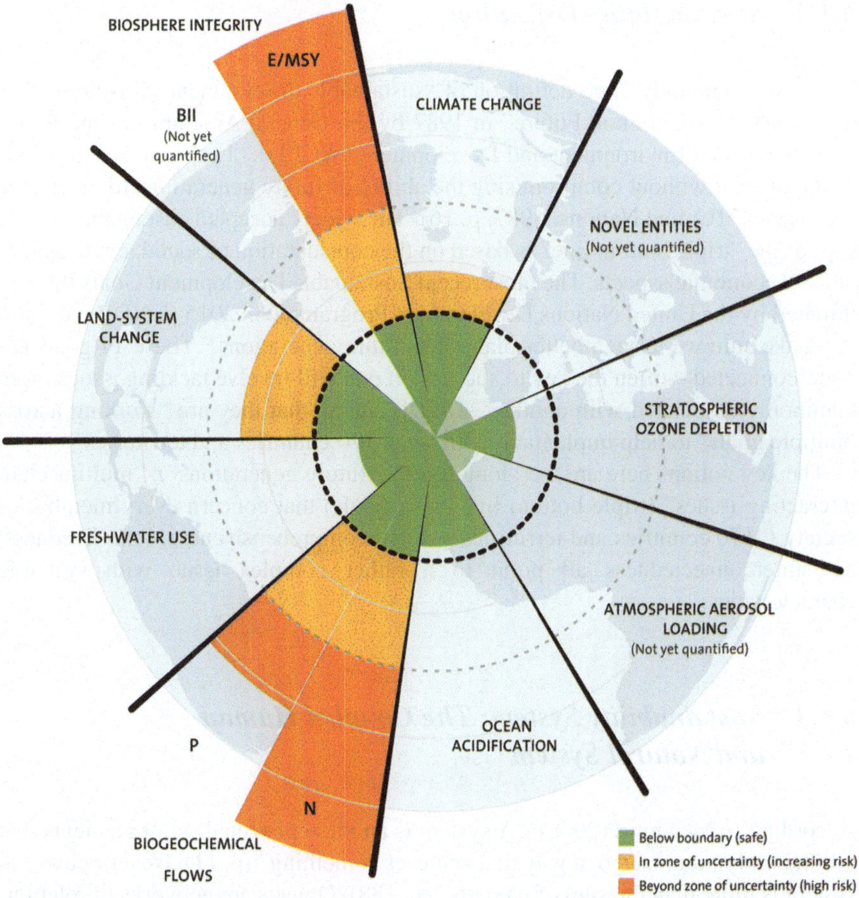

Fig. 6.2 The study on planetary boundaries in 2015. The green zone is the safe operating space, the yellow represents the zone of uncertainty (increasing risk), and the red is a high-risk zone. The planetary boundary itself lies at the intersection of the green and yellow zones. Credit: J. Lokrantz/Azote Images based on Steffen et al., 2015. Courtesy: https://www.stockholmresilience.org/research/planetary-boundaries.html

sectors such as biochemical flows and generic diversity (Steffen et al., 2015) (see Fig. 6.2).

The consequences of devouring our ecological budget become gravely evident worldwide. We are flooded with messages about threats of global warming, toxins in everyday items, extinction of species, habitats and ecosystems, in addition to appeals from the scientific community that we must change before it's too late (Goleman, 2010). Human sustainability appears increasingly urgent, yet society's response is utterly inadequate.

If one looks closer at this dilemma, one can find that it is mainly due to the unique characteristics of true sustainability.

6.1.2 Sustainability Definition

The most commonly cited definition of "sustainable development" was defined in the report "Our Common Future" in 1987 by the General Assembly of the World Commission on Environment and Development (WCED): "[…] it meets the needs of the present without compromising the ability of future generations to meet their own needs." (United Nations, 1987, p. 16). The widely accepted sustainability concept of the "triple bottom line" is based on the consideration of social, environmental, and economic aspects. The most recent Sustainable Development Goals (SDGs) initiated by the United Nations Development Programme in 2015 (UNDP, n.d.) aim "to make sure we leave a better planet for future generations." These 17 goals are "interconnected—often the key to success on one will involve tackling issues more commonly associated with another." UNDP claims that they are "working across multiple goals" to help implantation "in some 170 countries and territories."

The key notions here are the "long-term" ("future generations"), "multifaceted, interacting issues" (triple bottom line or 17 goals) that concern every member of society ("170 countries and territories"). Such comprehensiveness, long-livedness, and interconnectedness all point to a rather complex issue with systemic characteristics.

6.1.3 Sustainability System: The Coupled Human and Natural System

According to Meadows (2008), "A system is an interconnected set of elements that is coherently organized in a way that achieves something" (p. 11). Nevertheless, "A system is more than the sum of its parts" (p. 188). Objects are networks of relationships embedded in larger systems, and many relationships in systems are nonlinear. "There are no separate systems. The world is a continuum" (p. 190).

Recently, "one major advance has been recognizing Earth as a large, coupled human and natural system, consisting of many smaller coupled systems linked through flows of information, matter, and energy and evolving through time as a set of interconnected complex adaptive systems" (Liu et al., 2015, para. 2). This "coupled human and natural system" (CHANS) represents countless integrated systems in which people interact with natural components (Liu et al., 2007).

Proof for this interconnectedness is becoming increasingly prominent in the details of our daily life. While already getting tired of the endless, depressing news on how creatures globally suffer from our inventions with plastics, we had to become familiar with the reports that tiny pieces of plastic were discovered in our table salt. Yet we did not realize how serious the problem was until researchers recently showed that 90% of commonly home-used salt brands around the world are contaminated with microplastics (Conley, 2018). This research, published in the journal *Environmental Science & Technology* in October 2018, suggests that the average adult ingests about 2000 microplastics per year due to its presence in the world's water bodies and thus in our food production processes. In the same month, microplastics were confirmed to have reached human gut as they were found in human stool in a small pilot study across 8 western countries (Wüstneck, 2018).

These findings should not come as a surprise, due to the complete interconnectedness and oneness of Earth's planetary system. We, too, live in this system. As our microplastics are now ubiquitous, permeating even remote places and species around the globe, mankind will not be spared. We tend to ignore this "inconvenient truth" (Al Gore). Yet our countless sustainability issues are interwoven with our lives in so many ways that we cannot really afford to disregard them anymore. Everything exists in a web of relationships in one way or another.

This "big picture" becomes clearer as more of the jig-saw puzzle pieces are being noticed and assembled. Overwhelming scientific evidence shows that "human activity has become the main driver of this burgeoning crisis, one that gravely threatens, well, you and me. We are collectively enmeshed in activities that inexorably endanger the ecological niche that houses human life" (Goleman, 2010, p. 9). Therefore, our every action at any moment, scale, and place, including every trade, industry, and profession and every single person on Earth, affects human sustainability or survival, respectively.

6.1.4 Sustainability Complexity

"Many key global sustainability challenges are closely intertwined (…). These challenges include air pollution, biodiversity loss, climate change, energy and food security, disease spread, species invasion, and water shortages and pollution. They are interconnected across three dimensions (organizational levels, space, and time) …" (Liu et al., 2015, para. 1). "Because they are embedded in larger systems, however, some of our "solutions" have created further problems. And some problems, those most rooted in the internal structure of complex systems, the real messes? Have refused to go away." "No one deliberately creates those problems, no one wants them to persist, but they persist nonetheless" (Meadows, 2008, p. 4).

Thus, without holistic, systemic approaches, Earth's most urgent environmental crises will continue to persist or worsen. The systemic nature of sustainability calls for systems-based solutions which must yield "fundamental discoveries and sustainability actions that are not possible by using conventional disciplinary, reductionist, and compartmentalized approaches" (Liu et al., 2015, para. 2).

Unfortunately, starting with the Greek philosophers, humanity has developed its analytical skills much more than its synthetic ones; we are much better at breaking down an issue into its minute parts than to see the unity or connections among a multitude of aspects. Those which focus on a narrow set of features, or even a tiny fraction of sustainability, invariably create new problems elsewhere, no matter how good their intentions are. Most of the current green products or projects in the market "too often lull us to more readily ignore the way that what we now think of as 'green' is a bare beginning, a narrow slice of goodness among the myriad unfortunate impacts of all manufactured objects." "Everything made has innumerable consequences; to focus on one problem in isolation leaves all the other consequences unchanged" (Goleman, 2010, p. 26).

To date, barely any of our solutions are comprehensive enough. Despite this, however, the complexity of sustainability is only one half of the problem. The other half is the challenge of recognizing this issue at a society-wide scale. "Understanding is the first step toward finding solutions that will lead to change and redemption" (Tavris & Aronson, 2015, p. 12).

6.2 Cognition: System Thinking

6.2.1 Evolution of Thinking

According to evolutionary psychology theory, our thinking occurs in two patterns that operate differently: the reflexive and the deliberative systems (Marcus, 2008). The reflexive system perceives/assesses the world and reacts/behaves in response to it. It acts rapidly, effortlessly and essentially in the subconscious realm. It's about making snap judgments—it's what kicks in to save us (i.e., instinctive, reacting). The deliberative system, however, consciously reflects the logic of our goals and choices. It is slow, effortful, and judicious. It is about considering, planning, and making decision for the long term (i.e., devising, actually thinking).

"The two streams of thinking rely on fairly different neural substrates" in our brains (Marcus, 2008, p. 51), as they represent different products of evolution. The reflexive system depends on the evolutionary old brain systems (hindbrain), dating back as early as half a billion years ago. It controls basic functions that were as critical to a dinosaur as they are to a human. "The deliberative system, meanwhile, seems to be based primarily in the forebrain, in the prefrontal cortex," which "is a lot newer, found in only a handful of species, perhaps only humans" (p. 51).

"Evolution built the ancestral reflexive system first and evolved systems for rational deliberation second" as the progressive overlay of technologies (Marcus, 2008, p. 86). Such an evolved and organically grown product tends to be good enough but not perfect. Therefore "Even though the deliberative system is more sophisticated, the latest in evolutionary technology, it has never really gained proper control" (p. 52).

6.2.2 *Hidden Impacts and Blind Spots*

Our "human mind is so consistently vulnerable to error" (Marcus, 2008, p. 15) that we rarely make the best decisions. This has been shown across various fields—most famously financial markets, among others. "Over hundreds of millions of years, evolution selected strongly for creatures that lived largely in the moment ... The fact that organisms tend to value the present far more than the future" is called "future discounting" (Marcus, 2008, pp. 84–85). To act promptly rather than thinking properly was more crucial "for our ancestors, who lived almost entirely in the here and now (as virtually all non-human life forms still do)" (Marcus, 2008, p. 37). It was the obvious and indispensable natural adaptation to a threatening environment.

Today, humans live in completely different circumstances, which would require us to reflect beyond the immediate, sensible present. Our complex environments and lifestyles are supplying ever more information and require ever more input, which in turn necessitate far more attention, understanding, evaluation and planning than was required from the "hunter-gatherer." Unfortunately, "the routines of our daily lives go on completely disconnected from their adverse impacts on the world around us" (Goleman, 2010, p. 42). "We have no sensors, nor any innate brain system designed to warn us of the innumerable ways that human activity corrodes our planetary niche" (Goleman, 2010, p. 46). The reflexive system determines many of our everyday actions, although we may not be aware of it. Our so-called "fight-or-flight" (survival) mode explains why we respond primarily to immediate and biologically sensible threats and make shortsighted decisions to satisfy our direct and immediate natural needs. Future discounting is still deeply rooted in our genes, even though it seems no longer suited or necessary in our drastically modified environments.

Why we humans still can't get over the inherited tendency of future discounting can be explained by evolutionary inertia: human society has developed much faster ever since acquiring the mind, than what evolution can catch up with. Since the industrial revolution, our environments have completely been turned upside down. "By the standards of evolution, that's not a lot of time for debugging, and a long time for the accumulation of prior evolutionary inertia" (Marcus, 2008, p. 14).

This cognitive dilemma is what we humans are currently facing on our quest to avert our own extinction: we are suffering from a blind spot that is masking distant/indirect/invisible interconnections. As this is undoubtedly fatal for our existence as a species, we can no longer afford to leave these issues be hidden from us.

6.2.3 New Sensitivity: Systems Thinking

From the above discussion we find that to reach human sustainability we must enable our deliberative thinking to somehow compensate for our cognitive blind spot and overpower our reflective behavior where necessary.

"We have to acquire a new sensitivity to an unfamiliar range of threats, beyond those our nervous system's alarm radar picks up—and learn what to do about them" (Goleman, 2010, pp. 46–47). Even more so, as we live "in a world that rapidly needs to shift behaviors arising from very complex systems" (Meadows, 2008, p. Xiii). We need to reflect in a new way, a way of seeing wholes, "to see the hidden patterns that connect human activity to the larger flow of nature, to understand our true impact on it, and to learn how to do better" (Goleman, 2010, p. 44). This requires a change of mind: from seeing ourselves as different from the world, to connected to the world; from seeing problems as caused "somehow," to seeing how our own actions create these problems (Senge, 2006).

Systems thinking, which pays more attention to interdependence than independence, can contribute critically to our new sensitivity to the compensation of our cognitive blind spots. For instance, systems thinking can help us emphasize on hidden/indirect/distant impacts and relations, rather than the visible/direct/immediate ones. "It is a framework for seeing interrelationships rather than things, for seeing patterns of change rather than static snapshots" (Senge, 2006, pp. 68). "In mastering systems thinking, we give up the assumption that there must be an … individual agent responsible. … Everyone shares responsibility for problems generated by a system" (p. 78).

This new sensitivity is required both at the scale of individuals as well as society in order to produce lasting changes via many types of actors, on multiple levels and on manifold attitudes. Up until now, the environmental movement has mostly only involved groups of disconnected individuals on diverse and seemingly disconnected issues. That is not enough and is the reason why the many endeavors have not yet been successful after more than half a century of struggles. A shared sustainability awareness is a crucial missing piece in our collective efforts. Only such a shared cognition among all humans (involving every single individual of the species) will allow us to understand the complex interconnections and to become aware of the myriad hidden, indirect impacts on sustainability ubiquitous in our daily lives. This collective human culture of sustainability will be a major step towards true human sustainability.

"As our world continues to change rapidly and becoming more complex, systems thinking will help us manage, adapt, and see the wide range of choice we have before us. It is a way of thinking that gives us the freedom to identify root cause of problems and see new opportunities" (Meadows, 2008, p. 2). System thinking will help us to overcome the sustainability dilemma (that the situation is still getting worse) and potentially our cognitive dilemma (that we let it happen). Our thinking processes might be flawed by nature, but "we have a chance to make the most of the noble but imperfect minds we did evolve" (Marcus, 2008, p. 17). Education, effective policies, big data, and artificial intelligence might be useful tools in achieving this, too, of course.

6.3 Design for Sustainability: System-based Approaches

6.3.1 Design: Basic to All Human Activity

We all live in a designed world. Everything we make, use and discard has been designed by someone, somewhere. “Design is basic to all human activity. The planning and patterning of any act towards a desired, foreseeable end constitutes the design process.” (Papanek, 1991, p. 3). “In this age of mass production, when everything must be planned and designed, design has become the most powerful tool with which man shapes his tools and environment (and, by extension, society and himself)” (Papanek, 1991, p. ix). Today, design is confronted with “far greater challenges than merely satisfying individuals’ needs and is operating within an ever-expanding context” (Lou, 2018, p. 349).

Unfortunately, Papanek’s words from a generation ago still fully apply today: “…by creating whole new species of permanent garbage to clutter up the landscape, and by choosing materials and processes that pollute the air we breathe, designers have become a dangerous breed” (Papanek, 1991, p. ix). As society can only conveniently employ these designs as fittingly, healthily, or positively as they have been designed for, designers, as both members of society and creators of means for society, are doubly responsible.

“In many ways, the environmental crisis is a design crisis. It is a consequence of how things are made, buildings are constructed, and landscapes are used…Our present forms of agriculture, architecture, engineering, and industry are derived from design epistemologies incompatible with nature’s own. It is clear that we have not given design a rich enough context. We have used design cleverly in the service of narrowly defined human interests…Such myopic design cannot fail to degrade the living world, and, by extension, our own health” (Van der Ryn & Cowan, 2007, p. 9).

Meanwhile, as every transition is a decision-making process that can be nudged by design (Lou, 2018), design appears to be predisposed to attain human sustainability.

6.3.2 Make Design Positive: Integrate Human and Natural Systems

The sustainability-conscious architects and founders of GIGA, Raefer Wallis and Ryan Dick, proposed that we ought to make design positive to the environment, and that everything we make should serve to leave the environment in a better state than when we found it. Otherwise, the world would have been better without us. Indeed, "there is no such thing as a 'neutral' design," as Thaler & Sunstein stated in their renowned book Nudge (2009, p. 3). Every design defines interactions between Human and Natural systems, either positive or negative ones. We can make design positive to the environment through adjusting the relationship between human and nature (Lou, 2018). This creates an opportunity, where design for sustainability will drive a demand for a wave of design innovations.

Hence design must imperatively be pursued within a context of sustainability. This requires fundamental understanding of how things and nature work, recognizing the countless ways human systems interact with natural ones (Goleman, 2010). Designers must learn to think in systems so as to uncover hidden interrelationships and to thus achieve better design.

Densely populated urban scenarios of human society represent the extreme of the human system. It is where design actions are intensively ongoing and where Earth's nature has been almost completely altered. But natural processes, physical laws and chemistry cannot be muted. Multiple interactions between the coupled systems frequently take place on various scales and in various forms and affect the environment far beyond the city's boundaries. Rendering design positive to the environment in the urban context will make a significant and far-reaching contribution to saving our species and civilization.

6.3.3 Sustainability Cognition Required for the Future Designers

Nevertheless, in current design practice, it seems that dealing with sustainability aspects is often seen as tedious or the responsibility of some external expert who consults the designer. In other words, it is the work of "others," not necessarily the designer's, as designers tend to regard the sustainability topics as simply too technical, boring, incomprehensible or frustrating. Some designers might pick it up as an alternative source of inspiration, to liven up the design or to respond the trend. Very few are giving their best to render their design as sustainable as possible.

Human sustainability is an increasingly urgent challenge at global scale and with a huge impact. Training some more specialists equipped with better knowledge and newer tools is nowhere enough to achieve this sustainability. We need a different breed of designers—designers who possess a sustainability cognition, think systemically and therefore can work towards system-based approaches to this epochal problem.

For humanity's sake, sustainability must become the core of design, not merely a supplement. It must be the foundation, the vocabulary, or the guiding principle throughout any design program. Hence, it is much more beneficial, meaningful, and enduring to present and perceive sustainability as an inevitable (but benefitting) requirement of design and human life for the entire design community. We need new design communities more than new design experts (Lou, 2018).

It is indeed a great challenge to transform our current design education and practice. This challenge asks for a shared notion and a collective effort from the entire design community, as well as the overall human society as a whole—each and every one of us. Design can be a tool to inspire the whole society to extend the notion of human sustainability, thereby enabling our society to achieve sustainable societal changes on a larger scale, making it more profound and effective. Humanity needs a joint culture of sustainability to make the next step in evolution.

6.4 Conclusion

It is about "different way of seeing and thinking" and thus taking different actions (Meadows, 2008, p. 4). Obtaining a sustainability cognition will enable the entire design community to define and accept the sustainability of the human future as a starting point of its work. Learning to think in systems will empower the designers to work on multiple scales and reach a deeper level. Systems-based design approaches which tackle the coupled human and natural systems across all dimensions are necessary to address the complexity of sustainability challenges and identify effective solutions.

There is a call for systemic views on our sustainability, in design and in everyday life. Moving from analytical (breaking down) thinking towards synthetical (bringing together) thinking, human sustainability is less a topic for focused specialists than for a design community based on networked thinking. Large communities of designers as well as ordinary persons could create designs for sustainability based on democratically organized data (big data, science), technology (artificial and human intelligence), and values. Chances are, we may find such design relevant and rewarding.

References

Conley, J. (2018, October 18). 90% of Table Salt Is Contaminated With Microplastics. Retrieved March 25, 2019, from https://www.ecowatch.com/table-salt-microplastics-2613395969.html

Frank, A. (2018, June 12). Earth Will Survive. We May Not. *The New York Times*. Retrieved March 24, 2019, from https://www.nytimes.com/2018/06/12/opinion/earth-will-survive-we-may-not.html

Global Footprint Network. (2018, June 13). *Earth Overshoot Day 2018 is August 1, the earliest date since ecological overshoot started in the early 1970s*. Retrieved from https://www.footprintnetwork.org/2018/06/13/earth-overshoot-day-2018-is-august-1-the-earliest-date-since-ecological-overshoot-started-in-the-early-1970s/

Global Footprint Network. (n.d.). Past Earth Overshoot Days. *Earth Overshoot Day*. Retrieved on September 24, 2018, from https://www.overshootday.org/newsroom/past-earth-overshoot-days/

Goleman, D. (2010). *Ecological intelligence: The hidden impacts of what we buy*. Broadway Books.

Liu, J., Dietz, T., Carpenter, S. R., Alberti, M., Folke, C., Moran, E., Pelli, A. N., Deadman, P., Kratz, T., Lubchenco, J., Ostrom, E., Ouyang, Z., Provencher, W., & Charles l. Redman,

Stephen H. Schneider Taylor, W. W. (2007). Complexity of coupled human and natural systems. *Science, 317*(5844), 1513–1516. https://doi.org/10.1126/science.1144004

Liu, J., Mooney, H., Hull, V., Davis, S. J., Gaskell, J., Hertel, T., & Li, S. (2015). Systems integration for global sustainability [Structured Abstract and Abstract]. *Science, 347*(6225), 1258832. https://doi.org/10.1126/science.1258832

Lou, Y. (2018). Designing interactions to counter threats to human survival. *She Ji: The Journal of Design, Economics, and Innovation, 4*(4), 342–354. https://doi.org/10.1016/j.sheji.2018.10.001

Marcus, G. F. (2008). *Kluge: The haphazard construction of the human mind.* Houghton Mifflin.

Meadows, D. H. (2008). In D. Wright (Ed.), *Thinking in systems: A primer.* Chelsea Green Publishing.

Papanek, V. (1991). *Design for the real world: Human ecology and social change.* Thames and Hudson.

Rockström, J., Steffen, W., Noone, K., Persson, Å., Chapin, F. S., III, Lambin, E., Lenton, T. M., Scheffer, M., Folke, C., Schellnhuber, H., Nykvist, B., De Wit, C. A., Hughes, T., van der Leeuw, S., Rodhe, H., Sörlin, S., Snyder, P. K., Costanza, R., Svedin, U., … Foley, J. (2009). Planetary boundaries: Exploring the safe operating space for humanity. *Ecology and Society, 14*(2), 32. https://doi.org/10.5751/es-03180-140232

Senge, P. M. (2006). *The fifth discipline: The art and practice of the learning organization.* Random House Business Books.

Steffen, W., Richardson, K., Rockstrom, J., Cornell, S. E., Fetzer, I., Bennett, E. M., & Sorlin, S. (2015). Planetary boundaries: Guiding human development on a changing planet. *Science, 347*(6223), 1259855–1259855. https://doi.org/10.1126/science.1259855

Tavris, C., & Aronson, E. (2015). *Mistakes were made (but not by me): Why we justify foolish beliefs, bad decisions, and hurtful acts.* Houghton Mifflin Harcourt.

Thaler, R. H., & Sunstein, C. R. (2009). *Nudge: Improving decisions about health, wealth, and happiness.* Penguin.

United Nations Framework Convention on Climate Change (UNFCCC). (n.d.). *The Paris Agreement.* Retrieved July 28, 2018, from https://unfccc.int/process-and-meetings/the-paris-agreement/the-paris-agreement.

United Nations, Our Common Future: Report of the World Commission on Environment and Development (WCED). (1987). Retrieved March 24, 2019, from http://www.un-documents.net/our-common-future.pdf

United Nations Development Programme (UNDP), Sustainable Development Goals. (n.d.). Retrieved March 22, 2022, from https://www.undp.org/sustainable-development-goals

Van der Ryn, S., & Cowan, S. (2007). *Ecological design.* Island Press.

Wüstneck, B. (2018, October 23). In a first, microplastics found in human poop. Retrieved March 25, 2019, from https://www.nationalgeographic.com/environment/2018/10/news-plastics-microplastics-human-feces/

Chapter 7
Towards Conscious Design and Urban Planning: Inspiration from Consciousness in Business

Christiane Molina

7.1 Introduction: An Urgent Need of Consciousness

Various social and environmental problems prevailed in the world before 2020. In that year, five of the most likely risks for the globe were related to the environment (World Economic Forum, 2020). Natural disasters, such as Hurricane Maria, which left Puerto Rico in ruins in 2017 (Scott, 2018), Hurricane Dorian, which devastated the Bahamas in 2019 (Law, 2019) and the never-ending 2019 and 2020 wildfires in Australia (BBC News, 2020), were more frequent. Income inequality still represents a global challenge, as wealth remains concentrated in a small percentage of the population (United Nations, 2020b). Other types of inequality, e.g., that of access to broadband internet services, not only continued but also, as they gained relevance in the knowledge economy, accounted for the inability of certain segments of the population to access opportunities for social mobility (Conceição, 2020).

Then, the COVID-19 pandemic struck and with it came a setback in the advances made towards the solution of these and other wicked problems. Starting with people's well-being, the pandemic represented a pause in healthcare programs such as child vaccination and prevented individuals from undergoing health screening or receiving treatment for ailments other than COVID-19 (The Lancet Public Health, 2020). Moreover, by the beginning of August 2021, the virus had caused the loss of more than 4 million human lives (World Health Organization, 2021). Concerning economic growth, in 2020, the world experienced a contraction larger than the Great Depression and other recessions that had followed it, such as the financial crisis of 2008 (Gopinath, 2020; Ping Chan, 2020; World Bank, 2020). Naturally, such recession was accompanied by unemployment and a reduction in working hours. The International Labor Organization estimated that from 2019 to 2020, more than a

C. Molina (✉)
EGADE Business School, Tecnologico de Monterrey, Mexico City, Mexico
e-mail: christiane.molina@tec.mx

E. Giorgi et al. (eds.), *Design for Vulnerable Communities*, The Urban Book Series,
https://doi.org/10.1007/978-3-030-96866-3_7

hundred million jobs were lost (International Labour Organization, 2021). Furthermore, job losses directly impacted poverty and hunger. Millions of people will fall into poverty for the first time in more than two decades of consistent decline in the poverty headcount. These negative impacts of the pandemic are even more acute for already vulnerable groups, such as children and women.

Multiple challenges of environmental, social, and economic nature also affected the fields of urban planning and architectural designs before 2020. The urban population has continuously increased in past decades; on some continents, the urban population has quadrupled, and this trend is expected to continue, expanding from approximately 55% of the population living in urban areas in 2018 to almost 70% by 2050 (Buchholz, 2020; Satterthwaite, 2020; United Nations, 2018). At the same time, pressure is mounting on the provision of public services such as transportation and housing alternatives in urban areas. Furthermore, the number of people living in slums also rose to 24% in 2018 (United Nations, 2021). To ensure that vulnerable groups benefit from urban living, new policies for urban growth are crucial (United Nations, 2018). In addition to the social issues accompanying urbanization, numerous environmental problems are also emerging due to the concentration of increasing population proportions in urban areas. For instance, the need to control greenhouse emissions is due not only to its effect on climate change (Satterthwaite, 2020) but also to its linkage to air pollution, which causes millions of premature deaths (United Nations, 2021).

During the COVID-19 pandemic, other challenges associated with the increasing concentration of the population in urban areas emerged or worsened. Higher percentages of contagion were observed in urban zones (United Nations, 2021). Slum-dweller numbers increased, along with the number of people who fell into slum-like living conditions. Most importantly, the number of people who could be considered vulnerable surged (The Lancet, 2020). A disproportionate exposure to risk is what defines vulnerability (The Lancet, 2020). Thus, amid the uncertainty and volatility of COVID times, many groups that were not previously considered vulnerable can become so. Financial troubles, mental health and other issues can force diverse people into vulnerable groups. Seemingly easy measures to avoid contagion, such as handwashing, can prove problematic for people who live in crowded areas and lack optimal access to clean water, among other services.

While calls for resets in areas such as business were prominent in years prior to the pandemic, the level of emergency that we face today appears to leave us with the need to rethink how we live and approach different facets of human activity, including that of design, which, as explained above, is not without challenges. In the business field, the concept of consciousness is proposed by movements such as conscious capitalism (Mackey & Sisodia, 2013) and conscious business (Kofman, 2006) as the key element of an alternative management model that will enhance the role and positive impact of enterprise in society and the planet for the potential solution of the great challenges described herein. In this chapter, we will explore possible implications of central aspects of conscious business perspectives for the field of design and urban planning. The following section and second part of this chapter presents the definition of consciousness and offers and outline its application to business. This section also offers an overview of three of the elements of conscious business: purpose, stakeholder engagement, and sustainability.

7.2 Consciousness: Definition and Application to Business

Consciousness can be defined as an awareness state in which a person is cognizant of the inner self and of its surroundings (Merriam-Webster Dictionary, 2021a). It can also be viewed as being entirely awake and present in the moment (Kofman, 2006; Mackey & Sisodia, 2013) or the opposite of sleepwalking (Marques, 2020). Consciousness as the recognition of inner and external conditions does not happen overnight either for individuals or for organizations. Thus, consciousness may be conceived as a process or a journey (Mackey & Sisodia, 2013; Sisodia et al., 2018). Similarly, consciousness is neither simple nor a quick fix. Consequently, developing consciousness demands dedication and care (Kofman, 2006).

Recently, consciousness has profoundly affected the field of business, primarily because, although capitalism has undoubtedly generated much progress, it has also been accompanied by many social and environmental problems (Freeman, Martin, & Parmar, 2007;Kofman, 2006; Mackey & Sisodia, 2013). In response, an alternative business paradigm enriched with the notion of consciousness has emerged (e.g., Kofman, 2006; Mackey & Sisodia, 2013). This proposal is supported by principles such as purpose, stakeholder engagement, conscious culture (Mackey & Sisodia, 2013) and leadership (Kofman, 2006; Mackey & Sisodia, 2013) and manifests through conscious practices that comprise those related to sustainability. In this chapter, the author draws from three of these elements, purpose, stakeholder engagement, and sustainability, to discuss potential implications for both architecture and urbanism. The following section describes the concept of purpose and how it is applied in business while expanding on its role in these professions.

7.2.1 Purpose: Beyond Profit

There is a widespread idea that businesses exist solely to generate wealth for their owners or investors (Freeman, Martin, & Parmar, 2007). Moreover, a myth has been built around the proposal, credited to Milton Friedman, that the goal of business is that of profit maximization (Mackey & Sisodia, 2013). Such idea has created many problems, such as short-term orientation and little attention to working conditions (Mackey & Sisodia, 2013). However, this myth can be debunked if we view maximizing profits as a result rather than businesses' central goal (Freeman, Martin, & Parmar, 2007; Mackey & Sisodia, 2013). In this vein, different streams of research in the field of business, such as stakeholder theory and conscious capitalism, propose that lucrative business has a purpose beyond the generation of returns (Freeman, Martin, & Parmar, 2007; Mackey & Sisodia, 2013).

Purpose stems from answers to questions about the reasons behind a company's existence and the impact it aims to exert on the world (Mackey & Sisodia, 2013). One's ability to articulate a purpose has many benefits. Purpose inspires organizations and their members (Mackey & Sisodia, 2013), allowing them to approach stakeholder value creation more easily (Freeman et al., 2010). Purpose may also create a context of reduced tensions and enhanced harmony in the organization (Mackey & Sisodia, 2013).

Just as the purpose of business extends beyond return generation, the existence of professions is tied to a higher purpose that responds to societal needs (Mackey & Sisodia, 2013). Such is the case for architecture (Mackey & Sisodia, 2013), which involves more than merely designing buildings and can play a major role in supporting the development of communities and promoting humanitarian assistance aims (Makstutis, 2010). Furthermore, the products of architecture reflect cultural identity, provoke various emotions, and serve as expressions of individuals' beliefs and values (Makstutis, 2010). Urbanism can also be regarded as more than studying the specifications for city spaces (Merriam-Webster Dictionary, 2021b). Urban development and renewal projects, for example, can enhance people's quality of life (Jianwei et al., 2021; Zhuang et al., 2019). Nonetheless, the intrinsic purpose of professions can also be lost when financial gains are the only or the primary objective for professionals (Mackey & Sisodia, 2013).

Great purposes are unique to each individual or organization, but there are certain general aspects to consider when articulating them. First, purpose holds great power (Mackey & Sisodia, 2013). When it reflects stakeholders' commitment, purpose can naturally motivate their satisfaction (Freeman, Harrison, & Wicks, 2007) and can also be a source of inspiration and energy that transcends and touches the stakeholders themselves (Mackey & Sisodia, 2013). Second, purpose can evolve following individual and organizational growth (Mackey & Sisodia, 2013).

In general, the tenet of purpose that is central to conscious business seems effortlessly applicable to architecture and urbanism. Nevertheless, such application requires care and attention to benefit from its full potential as a guiding light in conscious practice. Next, we explore the principle of stakeholder engagement, enriching the proposal of conscious business literature with contributions from stakeholder theory.

7.3 Managing for Stakeholders: Creating Different Forms of Value

The stakeholder perspective is one of the pillars of conscious business philosophies. This approach broadens the view of businesspeople because it draws attention to "any group or individual who can affect or is affected by the achievement of the organization's objectives" (Freeman, 1984, p. 46). Stakeholder theory has varied applications and is not restricted to the field of business. In fact, while its origins in the field of business naturally lead to major impacts in its related areas, the theory has also influenced other disciplines, such as healthcare and public administration (Freeman et al., 2010).

The pertinence of stakeholder theory to the fields of architecture and urbanism stems from their very nature. For instance, architecture refers to "the art or practice of designing and building structures and especially habitable ones" (Merriam-Webster, 2021), and as such, its exercise may involve several stakeholders, including the various clients to whom architects' work is tied (Makstutis, 2010). Urbanism is a field that studies the requirements of societies in urban areas (Merriam-Webster Dictionary, 2021b). Thus, its practice affects and may be affected by numerous groups of people or by individuals alone. For example, stakeholder participation appears crucial for the success of the growing number of urban renewal projects taking place worldwide (Zhuang et al., 2019).

To achieve stakeholder involvement, architects and urbanists have to address central questions of stakeholder theory, starting with who the stakeholders are (Laplume et al., 2008). The answer may vary according to the specific characteristics of architectural and urban planning projects. However, there are two major categories to consider. Primary stakeholders include those who can exert direct influence upon the project or will be directly impacted by the implementation of the project and, as a result, are essential for its success (Freeman, Harrison, & Wicks, 2007). For instance, clients who hire the architect (Makstutis, 2010) and residents or users who live in the spaces designed (Bredenoord, 2015; Estaji, 2017) may always fit in this category, even if the project is financed by others. In addition, governments can be major stakeholders of projects such as those of social housing (Oyebanji et al., 2017). Secondary stakeholders are those whose influence is exercised indirectly through primary stakeholder groups (Freeman, Harrison, & Wicks, 2007). As sustainability becomes a major topic in architecture and urbanism (Grant, 2009; Kim & Kwon, 2018), environmental groups or regulatory agencies could be considered secondary stakeholders.

Once stakeholders are identified, how to engage them becomes a central matter. While we can assign categories such as primary versus secondary to stakeholders, interaction with all stakeholders is essential (Freeman, Harrison, & Wicks, 2007). Stakeholder engagement refers to "the process of effectively eliciting stakeholder views on their relationship with the organization" (Freeman, Harrison, & Wicks, 2007, p. 152). The depth of involvement or quality of the process may vary, depending on both the process itself and its outcomes as well as on the degree of responsiveness to the demands or needs of the stakeholders (Freeman, Harrison, & Wicks, 2007). Consequently, stakeholder engagement can range from nonparticipatory types of communication, in which stakeholders receive information about decisions made by the organization, to active participation, in which stakeholders exert majoritarian control over decision-making.

In architecture and urbanism, as in business, stakeholder engagement is essential, and its lack can cause many problems, including cultural damage, confrontations, social disparities, and social injustice (Jianwei et al., 2021; Zhuang et al., 2019). Furthermore, in social housing projects, overlooking residents' needs could lead to degrading, unsafe and gloomy housing solutions (Makstutis, 2010). In other cases, an ill-developed relationship with the client can reduce the success of a project and even, under certain circumstances, damage the relationship to the point where the client goes elsewhere (Makstutis, 2010). In contrast, stakeholder interaction and inclusion in decision-making may improve project outcomes, for instance, leading to enhanced quality of life and prosperity for urban dwellers (Jianwei et al., 2021; Liang et al., 2021). Moreover, stakeholder participation can work to the advantage of architects and urbanists, easing processes such as the implementation of urbanism's nature-based solutions (Ferreira et al., 2020).

Despite the indisputable relevance of stakeholder engagement, local residents and even vulnerable groups that are supposed to be the beneficiaries of urban and architectural projects are frequently overlooked (Ferreira et al., 2020; Jianwei et al., 2021). This oversight may be due to the perception that stakeholder participation is time consuming because of their divergent interests (Ferreira et al., 2020). However, although stakeholders differ, their needs and wants may often converge (Freeman, 1984), enabling their simultaneous satisfaction and avoiding tradeoffs (Freeman, Harrison, & Wicks, 2007). Thus, how to engage stakeholders and how to better understand them are two relevant questions that demand further attention.

Current evidence suggests that participation in project development phases and decision-making is among the critical success factors of sustainable social housing projects (Oyebanji et al., 2017). For example, Ugolini et al. (2018) propose the constant conversation and exchange of creative ideas as an essential element in urban park management. Similarly, Mayer et al. (2005) highlight the potential usefulness of tools that support decision-making that, combined with simulations, can generate a shared understanding of several issues that emerge in sustainable urban development projects. Participatory design methods that comprise interviews and multiple site visits to users have also proven effective in enabling architects and residents to form strong bonds and promoting the delivery of customized interior design solutions in social-interest housing projects (Salome et al., 2017).

The best-functioning stakeholder relationship types may also be a topic of additional study. Depending on the types of projects, the kind of relationship that architects establish with their clients may vary (Makstutis, 2010). For example, a close, friendly relationship may work well for residential design projects. In urbanist projects such as the design and management of public parks, networking at a personal level appears important for establishing new cooperative relationships (Ugolini et al., 2018). Research in stakeholder theory also shows that for vulnerable groups that may be "silent," stakeholder relationships charged with moral commitment to the advancement of stakeholders could exert a more lasting impact (Davila & Molina, 2017). Furthermore, when the relationship is also established with the aim of satisfying stakeholder needs, otherwise vulnerable or silent groups can gain salience. Overall, in terms of relationship building, we must not forget that stakeholders are people "with names and faces," and, as such, effective relationships could benefit from a human touch (Freeman, Harrison, & Wicks, 2007, p. 57).

In addition to building ties with stakeholders and engaging them, relationships need continuous management. Therefore, getting to know stakeholders and obtaining a deep understanding of them are crucial activities. To accomplish this goal, Freeman and colleagues (Freeman, Harrison, & Wicks, 2007) propose the use of seven strategies, among which is stakeholder assessment. This topic has been extensively studied in the project management literature, offering useful guidelines for professionals in any discipline. For instance, Eskerod and Lund (2013) suggest that an ideal stakeholder assessment includes both the contributions and expectations of each stakeholder. Contributions refer to what stakeholders offer towards the achievement of goals or the implementation of the project. In turn, expectations comprise their needs, concerns, requirements, wishes, etc., and information on how all of these are prioritized by the stakeholders themselves. In architectural projects, some basic questions to direct stakeholder assessment may address how residents or beneficiaries of a project live and their aspirations (Makstutis, 2010). For urban planning projects, a few factors that can also be explored relate to stakeholder needs of mobilization, social cohesion, and safety (Alessandria, 2016).

All the elements of stakeholder management explored in this section converge to enable value generation. For this reason, stakeholder theory and its approach to management can enrich how architects and urbanists approach their professional arenas. According to the stakeholder perspective, value is created through satisfying the needs of multiple stakeholders at the same time (Freeman et al., 2010; Freeman, Harrison, & Wicks, 2007). Architecture and urbanism naturally influence people's quality of life, and embracing the stakeholders' management perspective might truly fulfill their potential, even if harmonizing stakeholder expectations can be a complex task. In the following section, the author lays out some thoughts on sustainability, a manifestation of the recognition of the external environment that is consistent with the concept of consciousness and proposals of conscious business.

7.3.1 Sustainability, a Matter of Survival

Sustainability is not a new concept; however, it continues to gain attention from society and to influence an increasing number of dimensions of human life because of the urgency for action (United Nations, 2020a). Sustainable development can be defined as "development that meets the needs of the present without compromising the ability of future generations to meet their own needs" (World Commission on Environment and Development, 1987). Since 2015, the agenda for sustainable development has been guided by 17 goals defined by United Nations member states. While advances have been made worldwide in all Sustainable Development Goals (SDGs), the COVID-19 pandemic that began in late December 2019 (World Health Organization, 2020) has jeopardized the whole agenda (Sachs et al., 2020; The Lancet Public Health, 2020). Business as well as urbanism and architecture early embraced sustainability as integral part of their practice. This subject has actually been present in planning records since the introduction of the sustainable development definition and quickly merged with novel urbanism, among other trends, particularly in North America and Europe (Grant, 2009).

Sustainability is found at the intersection of three dimensions: economic, social, and environmental. As cities continue to expand, sustainability plays a more prominent role in design and planning. In finding the needed balance among the three dimensions of sustainability, architecture and urbanism seem closely intertwined. For instance, architectural design may foster urban sustainability (Kim & Kwon, 2018). In this vein, an understanding of the local context, including the characteristics of the natural environment, and the subsequent adaptation of architecture and urbanism practices seem vital (Kim & Kwon, 2018). Such practices range from local sourcing of building materials (Bredenoord, 2015) to ensuring social inclusion through attention to the use of space (Liang et al., 2021).

Although there may be possible negative impacts associated with the implementation of architectural and urbanism projects, particularly in construction activities due to carbon emissions, for example (Costa et al., 2020), contributions of urbanism and architecture to sustainable development abound. Public architecture can contribute to sustainability in different manners, preserving history, for example, or enabling cost reduction through recycling (Kim & Kwon, 2018). Urban parks can enhance people's quality of life (Ugolini et al., 2018). Furthermore, low-cost housing can also be sustainable (Bredenoord, 2015). However, among sustainability dimensions, the social dimension has received less attention (Boström, 2012). Possibly contrary to expectations, this appears to be a constant in various subjects related to architecture and urban planning projects, such as social housing (Edwards, 2019; Jensen et al., 2012) and urban renewal (Jianwei et al., 2021; Zhuang et al., 2019).

Achieving sustainability through architecture and urbanism is not an easy task. Evidence has revealed multiple factors for the potential success of projects in architecture and urban planning. Concerning social housing, research shows (Oyebanji et al., 2017) that these types of projects are affected by several aspects, from contextual conditions that comprise favorable legal frameworks to considerations of what materials to use. Solutions to the challenges of sustainability include prioritization of practical issues rather than dimensions (Brorström, 2015) and the participation of all stakeholders (Oyebanji et al., 2017), particularly local communities and organizations (Bredenoord, 2015). Additionally, an embeddedness view of sustainability dimensions, that is, considering how the social is intertwined with the economic and environmental dimensions (Edwards, 2019), can help ongoing efforts to reach sustainability in both design and urbanism.

Overall, finding a balance between sustainability dimensions appears easier in theory than in practice. Despite its complex nature, sustainability is an urgent matter, and different disciplines may contribute to its achievement, including both architecture and urbanism. Next, the author presents some conclusions on how the notion of consciousness and learning from its application to conscious business can also contribute to a more conscious design practice in architecture and urbanism.

7.4 Conclusions

Despite the great economic and social progress experienced in past decades, various complex and pressing problems remain today. Many of these challenges were exacerbated by the COVID-19 pandemic that struck the world in 2020 (The Lancet Public Health, 2020) and has already taken millions of lives (World Health Organization, 2021). Voices worldwide are calling for urgent action to address this major test of humanity. In the field of business, consciousness is thought to be the missing piece of an alternative model that holds potential solutions to many of the issues that afflict society.

This chapter explored the consciousness concept and learnings from its application to business with the aim of advancing consciousness in the practice of architecture and urbanism. As the author explored three elements of conscious business, i.e., purpose, stakeholder engagement, and sustainability, it became clear that all three are pertinent to these disciplines. An overview of the insights from conscious business and its potential application to design is presented in Table 7.1. First, all professions, including architecture, urbanism, and business, have a purpose tied to societal

Table 7.1 Conscious business and conscious design

Principles	Conscious business	Conscious design
Purpose	• Lies beyond profit generation. • Answers why a company exists. • Refers to the impact sought by the business.	• Associated with the profession. • Architecture's purpose lies beyond design. • Urbanism's purpose lies beyond city spaces' study.
Stakeholder engagement	• Managerial attention focused on "any group or individual who can affect or is affected by the achievement of the organization's objectives" (Freeman, 1984, p. 46). • Co-creation of value.	• Attention of architecture and urbanism enriched by attention to stakeholders. • Mechanism of stakeholder engagement: Participatory design methods; stakeholder assessment.
Sustainability practice	• Sustainability as the intersection of social, economic, and environmental dimensions.	• Environmental dimension: Local sourcing of materials. • Social dimension: Design for social inclusion; preservation of history. • Economic dimension: Cost reduction from recycling.
Challenges	• Engrained profit maximization paradigm. • Complex balance between sustainability dimensions.	• Potential loss of purpose when profits are prioritized. • Stakeholders, particularly vulnerable groups, often overlooked. • Lack of favorable legal frameworks for sustainability practices.

needs (Mackey & Sisodia, 2013), and while practitioners can profit from them, their ultimate end lies beyond financial gains or the mere activities performed. Second, stakeholder engagement is naturally embedded in all the phases of projects in architecture and urbanism, yet, more commonly than expected, several stakeholder groups, including vulnerable groups, are neglected because of time and economic considerations. Third, sustainability, a topic that demands attention and action, may be at risk in urban design and architecture because of the predominance of environmental and economic dimensions, to the detriment of the social dimension.

Drawing from the experience of conscious business and embracing its principles may not be simple. However, there seem to be areas of convergence that leave room for opportunity. For instance, the limited attention to the social dimension of sustainability and the inadequate engagement of stakeholders may be solved by adopting strategies from stakeholder management, such as formal processes of stakeholder assessment (Freeman, Harrison, & Wicks, 2007), as core elements of architectural and urbanism projects. The cooperation of multiple actors (Bredenoord, 2015) and perhaps a multidisciplinary collaborative approach could also help urbanists and architects address many of the challenges, such as unfavorable regulatory frameworks or existing political commitments, that they may face when designing and implementing sustainable and conscious solutions to the problems of residents and cities, including challenges.

Humanity faces various complex and deeply rooted challenges exacerbated by a global pandemic. Time is running out, and without the actions of all societal actors, the solutions could grow even more distant. The author hopes that this chapter provides some ideas of how consciousness can be applied to disciplines other than business, particularly urbanism and architecture. Future research on this integration is encouraged.

References

Alessandria, F. (2016). Inclusive city, strategies, experiences and guidelines. *Procedia - Social and Behavioral Sciences, 223*, 6–10. https://doi.org/10.1016/j.sbspro.2016.05.274

BBC News. (2020). *Australia fires: Evacuation call stepped up as crisis worsens*. Retrieved July 28, 2021, from https://www.bbc.com/news/world-australia-50986363

Boström, M. (2012). A missing pillar? Challenges in theorizing and practicing social sustainability: Introduction to the special issue. *Sustainability: Science, Practice, and Policy, 8*(1), 3–14. https://doi.org/10.1080/15487733.2012.11908080

Bredenoord, J. (2015). Sustainable housing and building materials for low-income households. *Journal of Architectural Engineering Technology, 05*(01), 1–9. https://doi.org/10.4172/2168-9717.1000158

Brorström, S. (2015). Strategizing sustainability: The case of River City, Gothenburg. *Cities, 42*(PA), 25–30. https://doi.org/10.1016/j.cities.2014.08.005

Buchholz, K. (2020, November 04). *How has the world's urban population changed from 1950 to today?* Retrieved July 28, 2021, from https://www.weforum.org/agenda/2020/11/global-continent-urban-population-urbanisation-percent

Conceição, P. (2020). *Human Development Report 2020: The next frontier, human development and the Anthropocene*. Retrived July 28, 2021, from http://hdr.undp.org/sites/default/files/hdr2020.pdf

Costa, E., Shepherd, P., Velasco, R., & Hudson, R. (2020). Automating concrete construction: Sustainable social housing in Colombia. *24th International Conference of the Iberoamerican Society of Digital Graphics*, (pp. 254–259). https://doi.org/10.5151/sigradi2020-35.

Davila, A., & Molina, C. (2017). From silent to salient stakeholders: A study of a coffee cooperative and the dynamic of social relationships. *Business and Society, 56*(8), 1195–1224. https://doi.org/10.1177/0007650315619626

Edwards, M. (2019). *Social sustainability and housing for vulnerable groups in Sweden: An integrated literature review*. Retrived, July 28, 2021, from https://www.tracentrum.se/media/2304/social-sustainability-and-housing-for-vulnerable-groups-in-sweden.pdf

Eskerod, P., & Lund, A. (2013). *Project stakeholder management*. Routledge.

Estaji, H. (2017). A review of flexibility and adaptability in housing design. *International Journal of Contemporary Architecture, 4*(2), 37–49. https://doi.org/10.14621/tna.20170204

Ferreira, V., Barreira, A. P., Loures, L., Antunes, D., & Panagopoulos, T. (2020). Stakeholders' engagement on nature-based solutions: A systematic literature review. *Sustainability (Switzerland), 12*(2), 1–27. https://doi.org/10.3390/su12020640

Freeman, R. E. (1984). *Strategic management: A stakeholder approach*. Pitman.

Freeman, R. E., Harrison, J. S., & Wicks, A. C. (2007). *Managing for stakeholders: Survival, reputation, and success*. Yale University Press.

Freeman, R. E., Harrison, J. S., Wicks, A. C., Parmar, B. L., & De Colle, S. (2010). Stakeholder theory: The state of the art.

Freeman, R. E., Martin, K., & Parmar, B. (2007). Stakeholder capitalism. *Journal of Business Ethics, 74*(4), 303–314. https://doi.org/10.1007/s10551-007-9517-y

Gopinath, G. (2020, April 14). *The great lockdown: Worst economic downturn since the Great Depression. IMF Blog*. Global Economy. Retrieved July 28, 2021, from https://blogs.imf.org/2020/04/14/the-great-lockdown-worst-economic-downturn-since-the-great-depression/

Grant, J. L. (2009). Theory and practice in planning the suburbs: Challenges to implementing new urbanism, smart growth, and sustainability principles. *Planning Theory and Practice, 10*(1), 11–33. https://doi.org/10.1080/14649350802661683

International Labour Organization. (2021). *ILO Monitor: COVID-19 and the world of work. Seventh edition Updated estimates and analysis*. Retrieved July 28, from https://www.ilo.org/wcmsp5/groups/public/@dgreports/@dcomm/documents/briefingnote/wcms_740877.pdf

Jensen, J. O., Jørgensen, M. S., Elle, M., & Lauridsen, E. H. (2012). Has social sustainability left the building? The recent conceptualization of "sustainability" in Danish buildings. *Sustainability: Science, Practice, and Policy, 8*(1), 94–105. https://doi.org/10.1080/15487733.2012.11908088

Jianwei, L., Shengju, S., & Jingang, L. (2021). The dawn of vulnerable groups: The inclusive reconstruction mode and strategies for urban villages in China. *Habitat International, 110*(March), 102347. https://doi.org/10.1016/j.habitatint.2021.102347

Kim, S., & Kwon, H.-A. (2018). Urban sustainability through public architecture. *Sustainability, 10*(4), 1–21. https://doi.org/10.3390/su10041249

Kofman, F. (2006). *Conscious business: How to build value through values*. Sounds True.

Laplume, A. O., Sonpar, K., & Litz, R. A. (2008). Stakeholder theory: Reviewing a theory that moves us. *Journal of Management, 34*(6), 1152–1189. https://doi.org/10.1177/0149206308324322

Law, T. (2019, Septermber 7). *Here's what it's like in the Bahamas after it was devastated by hurricane Dorian*. Retrieved January 10, from https://time.com/5669725/bahamas-after-hurricane-dorian/

Liang, D., De Jong, M., Schraven, D., & Wang, L. (2021). Mapping key features and dimensions of the inclusive city: A systematic bibliometric analysis and literature study. *International Journal of Sustainable Development and World Ecology.*, 1–20. https://doi.org/10.1080/13504509.2021.1911873

Mackey, J., & Sisodia, R. (2013). *Conscious capitalism: Liberating the heroic spirit of business*. Harvard University Press.
Makstutis, G. (2010). *Architecture: An introduction*. Laurence King Publishing.
Marques, J. (2020). Sleepwalking versus mindfulness: A conscious leadership choice. In K. Dhiman & Satinder (Eds.), *The Routledge companion to mindfulness at work* (1st ed.). Taylor & Francis.
Mayer, I. S., van Bueren, E. M., Bots, P. W. G., van der Voort, H., & Seijdel, R. (2005). Collaborative decisionmaking for sustainable urban renewal projects: A simulation - Gaming approach. *Environment and Planning B: Planning and Design, 32*(3), 403–423. https://doi.org/10.1068/b31149
Merriam-Webster. (2021). Architecture. In *Merriam-Webster Dictionary*. Merriam-Webster.com dictionary. https://www.merriam-webster.com/dictionary/architecture
Merriam-Webster Dictionary. (2021a). Consciousness. In *Merriam-Webster Dictionary*. https://www.merriam-webster.com/dictionary/consciousness
Merriam-Webster Dictionary. (2021b). Urbanism. In *Merriam-Webster Dictionary*. https://www.merriam-webster.com/dictionary/urbanism
Oyebanji, A. O., Liyanage, C., & Akintoye, A. (2017). Critical success factors (CSFs) for achieving sustainable social housing (SSH). *International Journal of Sustainable Built Environment, 6*(1), 216–227. https://doi.org/10.1016/j.ijsbe.2017.03.006
Ping Chan, S. (2020, April 14). *Coronavirus: "World faces worst recession since Great Depression."* Retrieved July 5, 2021, from https://www.bbc.com/news/business-52273988
Sachs, J., Schmidt-Traub, G., Kroll, C., Lafortune, G., Fuller, G., & Woelm, F. (2020). *The sustainable development goals and COVID-19. Sustainable development report 2020*. Cambridge University Press.
Salome, A., Benavides, J., & Dávila, E. (2017). Arquitectura interior en viviendas de interes social: una ruptura de esquemas [Interior architecture in social interest housing: Breaking schemes]. *Revista Centifica de Arquitectura y Urbanismo [Scientific Journal of Architecture and Urbanism], 38*(1), 113–125.
Satterthwaite, D. (2020, April 9). *An urbanising world*. International Institute for Environment and Development. Retrieved July 12, 2021, from https://www.iied.org/urbanising-world
Scott, M. (2018, August 1). *Hurricane Maria's devastation of Puerto Rico*. National Oceanic and Atmospheric Administration, Climate.Gov. Retrieved January 10, 2020, from https://www.climate.gov/news-features/understanding-climate/hurricane-marias-devastation-puerto-rico
Sisodia, R. S., Henry, T., & Eckschmidt, T. (2018). *Conscious capitalism field guide: Tools for transforming your organization*. Harvard Business Review Press.
The Lancet. (2020). Redefining vulnerability in the era of COVID-19. *The Lancet Global Health, 395*, 1089. https://doi.org/10.1016/S2214-109X(20)30116-9
The Lancet Public Health. (2020). Will the COVID-19 pandemic threaten the SDGs? *The Lancet Public Health, 5*(9), e460. https://doi.org/10.1016/S2468-2667(20)30189-4
Ugolini, F., Sanesi, G., Steidle, A., & Pearlmutter, D. (2018). Speaking "Green": A worldwide survey on collaboration among stakeholders in urban park design and management. *Forests, 9*(8), 1–20. https://doi.org/10.3390/f9080458
United Nations. (2018). *68% of the world population projected to live in urban areas by 2050, says UN*. Department of Economic and Social Affairs News. Retrieved July 28, 2021, from https://www.un.org/development/desa/en/news/population/2018-revision-of-world-urbanization-prospects.html
United Nations. (2020a). *'Urgency to act' for sustainable development, greater than ever as coronavirus pandemic continues*. UN News. Retrieved July 28, 2021, from https://news.un.org/en/story/2020/05/1063742
United Nations. (2020b). *World social report 2020: Inequality in a rapidly changing world*. Retrieved July 28, 2021, from https://www.un.org/development/desa/dspd/wp-content/uploads/sites/22/2020/02/World-Social-Report2020-FullReport.pdf

United Nations. (2021). *Goal 11, make cities and human settlements inclusive, safe, resilient and sustainable*. Retrieved July 28, 2021, from https://sdgs.un.org/goals/goal11

World Bank. (2020, June 8). *The global economic outlook during the COVID-19 pandemic: A changed world*. Retrieved August 29, 2020, from https://www.worldbank.org/en/news/feature/2020/06/08/the-global-economic-outlook-during-the-covid-19-pandemic-a-changed-world

World Commission on Environment and Development. (1987). *Report of the world commission on environment and development: Our common future*. https://doi.org/10.1080/07488008808408783.

World Economic Forum. (2020). *The global risks report 2020*. Retrieved July 28, 2021, from http://www3.weforum.org/docs/WEF_Global_Risk_Report_2020.pdf

World Health Organization. (2020, March 23). *Rolling updates on coronavirus disease (COVID-19), Summary*. Retrieved March 25, 2020, from https://www.who.int/emergencies/diseases/novel-coronavirus-2019/events-as-they-happen

World Health Organization. (2021). *Who Coronavirus (COVID-19) Dashboard*. https://covid19.who.int/

Zhuang, T., Qian, Q. K., Visscher, H. J., Elsinga, M. G., & Wu, W. (2019). The role of stakeholders and their participation network in decision-making of urban renewal in China: The case of Chongqing. *Cities, 92*, 47–58. https://doi.org/10.1016/j.cities.2019.03.014

United Nations. (2021). Goal 11: Make cities and human settlements inclusive, safe, resilient and sustainable. Retrieved Nov 15, 2021, from https://sdgs.un.org/goals/goal11
World Bank. (2021). [illegible] COVID-19 pandemic. [illegible] https://www.worldbank.org/en/news/[illegible] [illegible] during the [illegible]

Chapter 8
From Storytelling to Numbers: A Discussion on Vulnerability in the Global and Local Context

Roberto De Lotto

8.1 Introduction

UN 2030 Sustainable Development Goals define a worldwide framework that embraces all the human activities. Goals, targets, and indicators are basic elements of the pragmatic scheme that UN identified to update well-known concepts included in the word "sustainability" since 1989. Earlier, the awareness about opportunities and limits to growth (Meadows et al., 1972) was clearly stated in the 26 principles of the "United Nations Conference on the Human Environment" that was held in Stockholm, Sweden, from June 5–16 in 1972. Since almost 50 years, the United Nations emphasize the necessity to overcome singular and specific needs towards a global and long-term vision. In the last half of a century, this is the most relevant worldwide story that tries to address the individual and the general behaviors towards more responsible actions.

Apart from the practical outcomes that demonstrate the difficulties to reach aimed results, in this chapter author wants to stress the necessity to reconcile some dichotomies that emerge in sustainable planning and design for vulnerable communities:

1. Qualitative VS quantitative approach.
2. Subjective VS objective point of view.
3. General VS specific remark.

These three issues are diffusely treated trying to remark common objectives and similar viewpoints of human settlements and territory.

R. De Lotto (✉)
Department of Civil Engineering and Architecture, Urban Project Laboratory – UPLab, University of Pavia, Pavia, Italy
e-mail: roberto.delotto@unipv.it

E. Giorgi et al. (eds.), *Design for Vulnerable Communities*, The Urban Book Series,
https://doi.org/10.1007/978-3-030-96866-3_8

Sustainability concept and applications are a clear example of how to mix all these components. The specific case study presented in this book also allows to face the interrelation among concepts like vulnerability, risk, exposition, sensitivity, pressure, impact, damage, and resilience, which are becoming more and more part of our diffuse understanding. Moderate cultured persons easily comprehend these cited elements even without a deep knowledge of related scientific implications.

The tension between "numbers" and "storytelling" is usually resolved by planners and designers with a combination process: the representation of results is always a mixed composition of qualitative and quantitative elements that are specifically addressed to different users.

These activities need the capacity to synthesize solutions for a certain set of problems; when they are cooperative actions, they start with the building of a direct dialogue with communities to acquire data. The ability of the professional is to create a shared set of evidence from vague information and, at the end of the process, to communicate technical knowledge in a non-technical way.

8.2 Goals, Targets, and Indicators: Narration and Measure

On the top of the website of the 17 SDG (https://sdgs.un.org/goals, visited in June 2021) there is a list of numbers: 17 goals, 169 targets, 3082 events, 1303 publications, 5458 actions. This quantitative indication aims to demonstrate the strong commitment of UN.

For each Goal there are: a general overview based on specific dedicated numbers, a list of events described throughout a narrative system (texts and video), and a list of publications with short description. At the Goal level, both measure and storytelling are used to catch the widest interests by technical and non-technical readers.

Targets are defined in a very synthetic way (e.g., for Goal 11 targets are described with less than 50 words, only target 11.b is 53 words) in order to be clear and precise. Indicators are textually presented, and their value consist in precise measures that, combined each other, can give a quantitative response to relevant entities.

8.3 Storytelling and Numbers: Shared Goals

In the 1960s of the last century with postmodernism, in architecture and urban planning social sciences and the "qualitative" world become fundamental parts of knowledge building and dissemination. The diffuse social engagement together with the rise of "planning theory" carried architecture and urban planning to have a wider sphere of action than during the modern movement. Nowadays it is a half of a century that the so-called "soft sciences" were recognized, together with the historically robust "hard sciences," as foundation of the territorial and local analysis for any planning and design process.

Even if the theoretical passages, the practical applications, and the worldwide recognition were neither easy nor continuous, it is quite difficult nowadays to declare that one component prevails on the other.

In the third millennium, both storytelling and numbers try to reach a common aim: to build a shared vision of the world.

Storytelling was born to create a social identity and a common understanding of the world since the first homo sapiens groups appeared in the hearth (Harari, 2018). The relevance of storytelling is so high that it can overcome some of the strongest key issues like "truth" and "certainty." In his essays about the development of human culture, the historian Yuval Noah Harari (Harari, 2018, 2019a, b) underlines the power of storytelling to generate the sense of community: subjects who have the same convictions and objectives cooperate each other. And cooperation is the key for success (as biological species, as communities, as economical groups, etc.). Powerful narrative examples cited by Harari are: religions, economic organizations, and human rights. Nowadays the most diffuse discourse is about the social liberal theory, or liberalism.

On the one hand Harari states the superiority of storytelling as a glue for human communities, on the other hand he declares that science and rigorous methods are indispensable to state unmistakable truths.

Many other authors coming from social sciences strengthen the discourse value in planning and design activities, moreover when they face dialogic and bottom-up processes (Ameel, 2017; Sandercock, 2010).

In architecture field, after the technocratic modern movement axioms, participatory processes and social engagement were proved as experimental newness in Giancarlo De Carlo and Guido Canella works since the 1960s of the nineteenth century (De Carlo, 2008; Katuyuki, 1983).

In urban planning it was the turning point of planning theory that clarified the relevance of narration, discourse, and argumentative features (Fisher & Forester, 1993; Hajer, 1993; Throgmorton, 1993).

At the didactical level Given, in "The SAGE Encyclopedia of qualitative research methods" (Given, 2008), organized the comprehensive field of "soft sciences." The author underlines some methodologies that have been deeply developed by scholars, such as phenomenology, grounded theory, and discourse analysis.

Moreover, new modes to build territorial and community knowledge look to nonconventional narrations like people's histories, literature novels, and movies (Leone, 2010; Longo, 2012; Tzatzadaki, 2021). So, the space of qualitative research is continuously expanding.

In sustainability field, the necessity to bond individual behaviors with governments decisions carried to a global storytelling able to capture the interests of scholars (both scientists and sociologists) and non-cultured people; so, storytelling and numbers walk together. Sustainability narration is based on scientific data, and it is reinforced by the mathematical projections about, in example, global warming and increase of temperature, glaciers and sea ice melting, contradictions in richness distribution in the world.

As it is well known, scientific data are not sufficient to have worldwide consensus. In example, in the last decades many negationists or skeptical points of view about global warming emerged and had strong influence in some governments' decisions. It is probable that, more than the capacity of data to demonstrate reality, the biggest driving force of such mismatches was the pressure by strong economic stakeholders (Angeles & López, 2021). Generally speaking, the attitude to deny scientific truth in favor of more preferrable discursive scenarios is a fact that must be considered (Dice, 2017).

From what expressed above, when approaching sustainability issues in planning and design activities it emerges a common guideline in the different approaches: the need to imagine and organize a certain reading of the territory, of communities, of the social and economic environment that is connected to general and worldwide shareable values.

8.4 Storytelling and Data Society

Even though the power of narration to build shared knowledge is recognized, it is a fact that we are living in a "data society." Data and numbers are the basis of our ICT-oriented life.

In "Data Action—Using Data for Public Good" the author, Sarah Williams, exposes clearly how to use data to help planners and urban managers to increase knowledge and to solve different kind of problems (Williams, 2020). In her approach two aspects are very interesting: at the beginning she declares that data were used in the past to reinforce unjust policies and this fact raised skepticism about their use. Subsequently, she represents many opportunities to make data be a part of common and shareable look at the city and at the environment.

The optimism in data benefits is strictly connected with the public use of their elaboration. About the urban settlements, it is an automatic passage because the city is, by definition, a public fact (Ratti & Claudel, 2016).

About using numbers for a storytelling, a very easy experiment reader can try is connecting at the web page: https://www.worldometers.info and combine some numbers to create a predetermined discourse. In example, we can demonstrate how much rich countries abuse of food while poor countries suffer for starvation informing that nowadays (July 31, 2021) there are 788,384,323 obese people in the world and that 6,492,646 people died this year for hunger. Or, to create a misperception about the Covid-19 pandemic tragedy, we can gather together data as follows: in 7 months of this year 2021, 2,902,131 people died by smoking, 1,451,991 died by alcohol abuse, 622,545 committed suicides; in 7 months the people who died because of their intentional decisions are 4,976,667 while in 14 months 4,230,535 died by Covid-19. Covid-19 killing capacity is less than a half of human will.

Following Williams and the previous two short experiments, nowadays it is extremely important to verify not only the source of information but also the way data and numbers are treated.

Without any doubt data management is one of the key issues about the future furthermore considering the fast development of Artificial Intelligence that is stimulating questions and worries. The AI prefigures the possibility to create a digital model of qualitative and quantitative, subjective and objective, general and specific issues and topics. All of these models will have a binary structure and various interpretation algorithms. Psychological manners, biological processes, and creative skills will be unified in mathematical numbers.

In this credible future scenario, which will be the future of human beings, of their jobs and social roles, when automation will be able to substitute all human actions?

Which will be the role of qualitative experience when algorithms will be able to calculate and forecast every human biological process?

These are not only practical questions but, above all, ethical and moral ones. There is the hope that AI will carry to an "Augmented Intelligence" able to support human beings and not to substitute them (Pasquale, 2021).

The possibility to measure qualitative objects is a very stimulating research issue: it is the case of "resilience," probably the trendiest word recently. Brunetta, Faggian, and Caldarice edited a book entitled "Bridging the Gap. The Measure of Urban Resilience" (Brunetta, Faggian and Caldarice, 2021) in which they inserted in a unique publication modeling, measuring throughout indicators, assessing, visualizing, and mapping many case studies of resilience. It is remarkable that in scientific literature these attempts maintain a high level of relevance. In fact, the coexistence of qualitative and quantitative aspects is extremely stimulating from the intellectual and systematic point of view. Nowadays methodologies like fuzzy logic, multicriteria analysis, pairwise comparison, and Analytic Hierarchical Process are diffused in many fields of human activities probably because they are not entirely "black boxes" (users can keep control of passages and results).

Vulnerability is another inspiring example. It is a very frequented field of investigation because it covers multi-disciplinary approaches: the most intuitive is the relation between a specific context and general problems at which the vulnerability is related. As examples: quantitative descriptive indicators can be related to the GDP per capita (valid at the local and global scale); qualitative descriptive discourse can be related to the feeling of desolation that is a psychological facet due to the distance between desires and reality (valid at the very general and at the individual scale). Moreover, the scientific Risk analysis studies describes Vulnerability as a component of an event, together with Exposition and Danger, that describes the level of criticism in a mathematical way ($R = V \times D \times E$).

In vulnerable communities the local situation, the specific conditions, and the way these elements are communicated to people involved and to decision makers (using conventional and non-conventional sources) have greater relevance than pure numbers. At the same time, numbers are essential for local governments to improve strategies and policies to renew these communities towards healthier conditions.

So, a big research question emerges: how to accurately measure something (like vulnerability of communities) that can be better defined in a qualitative way?

If it is not possible, then: how to address relevant national policies basing on intuitive aspects?

It is obvious that both approaches are necessary, no one alone is sufficient.

8.5 Theoretical Framework Towards Flexible Approaches

Globalization and postmodernity signed a significant trace in many fields starting from anthropology, political life, and society ethics (Sennett, 1998; Sassen, 1999). The passage from the local scale to the global one carried to a profound twisting of all the previous certainties. Bauman's definitions regarding the loss of long-lasting values towards a transient reality (Palese, 2013) influenced both hard and soft sciences. To justify theories, to demonstrate them and also to define rational criteria for legitimating theories, the "anything goes" principle conferred to methodology a key role (Russell, 1983).

It is not possible to synthesize such a strong and tormented transition in a few words, but it is noteworthy to resemble the passage from a comprehensive technocracy to the loss of strong beliefs within the development of the complex system theory.

The word "complex" is commonly used to describe something complicated, difficult to be understood or demonstrated. As technical word, complexity involves fields of science that come from physics like chaos theory and complexity theory (Warren, 2013). Human settlements (urban, rural-urban, rural) and the environment interact each other, and the non-linearity of the interactions measures the complexity of the system.

A set of elements that have interactions is called system. Generally, the more they are the elements and the more the topology of interactions is articulate, the more the system is complex. There are different kinds of complexity, but in planning and design fields their differences are not so relevant. A simplified view of such systems could compare complexity with the number of elements, with the heterogeneity of interactions among them, with the level of uncertainty in defining structure and interactions among parts. A system can be represented in suboptimal models: deterministic models are appropriate in low complex systems (with a few components or with simple relations, i.e., tree structure); statistical models are appropriate in big and simple systems (i.e., a system with a lot of elements with simple relations); heuristic models are appropriate in system in which phenomena always follow similar rules (i.e., experimental observation and direct experience).

In natural science and in socioeconomic ones it is proofed that complexity depends on the number of elements, on the interactions among them but mostly on the "nature" of these interactions.

Nonlinear interactions (the principle of effects overlapping is not valid) and close-ring topology (retroaction) generate systems that are difficult to be described even if they have a small number of elements (i.e., human being behavior in a circle of friends). Nonintuitive phenomena occur, such as bifurcation, self-organization, disaster, and newness generation. Casti considered that the capacity to generate surprise and to have not predictable behavior are essential characters of complex systems. Moreover, complex systems cannot be divided in parts without losing their essential aspects. Consequently, complex systems' behavior is not predictable starting from the single elements behaviors. Among all, self-organization is the capacity to define a space-temporal order without an external action nor internal code. The elements of the system interact following the principle of minimum resistance (Casti, 1994). Organization is the instrument a complex system uses to manage its complexity. From the different possible configurations, the systems assume the one that minimizes resistance. Self-organization is the way the system responds to the environment, opposing to destructive forces and safeguarding its integrity (Engelen, 1988).

Given this, it is obvious that all human settlements are complex systems. So, it is possible to describe them choosing among the multitude of methods and approaches, but it is impossible to figure out precise forecasts of the future. In this sense, the rigorous analysis and the liquid behavior may find a common field of cooperation.

Human settlements are generally considered as result of a certain planning and self-organization. The less the settlement is well structured, the more we can recognize vulnerable attributes.

So, as complex systems, vulnerable communities are extremely interesting because the self-organization attributes are prevalent over the planned or designed characteristics.

This means that, in vulnerable situations, pre-constituted models or abstractions are not valid in favor of a case-by-case approach.

8.6 The Paradigm of Flexibility in Vulnerable Contexts

City and territorial planning, as much as urban and architectural design, focus on a certain picture of the future that planners or designer may figure. The more the object of the study is the result of self-organization, the less the application of general rules and case studies is effective. So, vulnerable settlements require a high level of analysis of the specific context and a low level of generalization.

The definition of flexibility is a possible guideline in such fuzzy frameworks (De Lotto & Morelli di Popolo, 2015).

Stated that the city and human settlements have been explained as a complex system (above all, Batty, 2012), "Which is the meaning of complexity in our daily life?" is a question that may have not a unique answer. Rouelle (1991) and with a different approach Meyers (2011) underlined those complex systems are problematic to be scientifically validated. Surprise that complex systems create, and self-organization are two of the main distinguishing elements of such systems. Portugali analyzed different authors (i.e., Allen, Batty, Brillouin, Proigogine, and Schrodinger) and stated that cities exhibit all the proprieties of natural complex systems: they are open, complex, bottom-up, and often chaotic. In the last decades, when the calculation power of computers emerged fast, some mathematical models were developed to study complex systems applied to cities (Portugali, 2000; 2013). These attempts aimed to formalize complexity: the results were significant in descriptive phase but were not satisfactory at the foreseeability stage.

The fuzziness in forecasting, carries planners and designers to pay attention to subsystems like "physical elements" and "actors of settlements' development process."

So, following Portugali, a question emerges: "Which is the relationship between the physical (artificial) part of a city and a complex system"?

The artificial city is a set of objects that can be described by a simple system. In fact, it is possible to define all the elements that compose the set, and to describe the mutual functional relationships among them (Gargiulo & Papa, 1993). The urban agents are elements with interaction ability that are capable to transform the artificial city in a complex system. The behavior of "urban agents" is the hint that characterizes and identifies the city with its differences and quality. Moreover, the city appears as result of relations among the different urban elements and, being an open system, as consequence of the interaction between urban elements and the external environment (McLoughlin, 1969). Urban elements can be approached at very different scales (from simple buildings units to districts) using various disciplines like planning or design, and for each scale there is a possible external environment.

Urban planning is the science that studies the development of city; for what have been told so far this cannot be a deterministic-based science (Batty, 2012).

Furthermore, architectural design and co-design have a strong relation with social sciences and so they cannot be compared to hard scientific processes. This means that both quantitative and qualitative elements must be considered in the two processes.

A system is deterministic if the knowledge of initial state allows predicting the future. The behavior of a human settlement seems closer to a deterministic chaos than to a perfectly predictable scenario. It is, more precisely, a dynamically complex system, but this definition does not carry to any tangible advance in practice. It can only exclude the certainty of some rigid procedures.

Nowadays the impact of regulation on modern life has influenced every planning and design action on the city. Regulations define: what can be built and transformed and where; what features should have and what standards must respect the buildings; what activities may occur in certain buildings or places and under what conditions; what community services must be guaranteed by private real estate developments; what forms of inclusion and exclusion to apply in public spaces and private spaces. (Moroni, 2013).

A significant number of additional features are based on the interactions among the different elements (artificial and human) that belong to the city. In particular, the ability to organize becomes one of the fundamental properties of a complex system and it can be expressed as the evolution of relational interactions. Organization thus becomes the constituent property of a system (Gargiulo & Papa, 1993).

Once the deterministic certainty is no more valid, and even when the comprehensive knowledge of the social and economic context becomes more and more obscure, the role of urban planning and architectural design should focus on adaptive scenario. The adaptation is achieved through the constant redefinition of the relationship between the system and its environment (co-evolution). The organisms evolve and co-evolve with other bodies and with the surrounding environment.

The concept of co-evolution offers good perspectives to co-planning and co-design activities.

With the modifications of the environment, the system assumes different forms, transforming the scheme of interactions among the elements. So, inside the system, the organization structure is not fixed: it has a dynamical configuration. Self-organization is the way the system adapts to the environment, opposing the destructive actions and safeguarding its integrity (Allen & Sanglier, 1981; Bertuglia & Staricco, 2000).

The self-organization (De Roo, 2016) can be proposed as a sort of evolution ability. Through this characteristic, the system can preserve adaptation to the environment. The idea of a system that naturally progresses is linked to evolution theories and interpretations.

Starting from different scales and different approaches to urban studies, the city can be considered as an organism, if the socioeconomic contest is the environment-study. In another approach, the city can be considered the environment, if the organisms are the citizens.

The changing of role is strictly related to the observation scale. Inside a globalized world, many roles tend to be elastic and the separation between individual and group, such as between subject and object is fuzzy.

In vulnerable communities, social roles are often vague, and modifications of behaviors are frequent.

Geddes identified the cities as organisms. City is the place where men can evolve themselves, and in which they can address all the knowledge, such as the heritage from the past. This heritage is the principle for progress towards future. In Geddes' urban theory, cities' shapes and functions are not static because of their evolution (Batty & Marshall, 2009; Welter, 2002). Moreover, the organic point of view is perfectly coherent with the idea of liquid society as a non-static entity.

Being the result of self-organization, vulnerable communities are a perfect example of organic human settlements. So, where the settlement is unstable and where it has with a weak structure, planners and designers can only define scenarios and general principles.

In contemporary urban debate, the evolutionist approach expects that there are not fixed and constant functions for urban spaces, and that operative destinations of the buildings can vary also in short time (De Lotto, 2011; Morelli di Popolo, 2013). It is important to underline that Geddes' evolution theory does not mean to transfer the laws of nature to the field of society or to the physical elements of the city. Evolution is a story that waves together innovation and memory, transformation techniques, collective ideals and the preservation of traditions and most remote institutions, always differently (Geddes, 1915).

So, Geddes' evolution is more a storytelling than a data combination.

The city is a human fact, and for this reason the quick modifications of the society need to be reflected in the settlement. The organic urban theory modifies the bases of technocracy, to create more adaptable plans and programs, with margins of flexibility.

Bauman's liquid society is "easy-fitting" with the temporary context (socioeconomic and urban) (Bauman, 2000). At the same time, the static nature of the physical part of the city presents many conflicts with such a dynamic society.

So that flexibility has been expressed in different dimensions, considering diverse characteristics of the city as a complex system.

In example, the flexibility in relational systems carried to the so-called "variable geometry" that overcomes the jurisdictional boundaries of cities thanks to the communication means. Considering the government systems, many authors in different periods (i.e., H.J. Gans 1993, Poli 2010, R.C. Wood 1958 and 1959) supported adaptive short-term policies connected to their capacity to be effective. In connection with the environmental system and the related ecological planning, the concept of resilience has been deepened. The acceptance of the human limited predictive capacity moved urban simulation from the definition of a static and ideal picture to the assessment among possible scenarios.

For vulnerable communities, I recommend strengthening flexibility.

A flexible context does not need a programmed shape; only the structural elements and the relation nets could be defined considering its changeability. The flexible settlement is characterized by the following basic principles (De Lotto, 2011):

- Temporal dimension: Flexibility makes sense only if modifications are accepted in different temporal periods. Modifications of the physical part of the settlement should consider the changing of their anthropological meaning. Obviously, the temporal dimension involves not only the settlement and its studies but also its regulation (when existing) and the way planners and designers understand their activity.
- Variable geography: The dimension and shape of the settlement are not predetermined. They vary following the socioeconomic context. Rem Koolhaas proposed interesting suggestions: any type of intervention on the city does not change its main structure. This approach entailed a functional open-endedness (Koolhaas, 2001). These variations (in density, in functions and in the intensity of use) need a strong backbone of infrastructures able to sustain modifications without compromise their effectiveness. That means a significant public investment, or a public-driven private investment that allows the establishment of the settlement.
- Reversibility: Towards a sustainable world, expansion and contraction phases should be controlled by a comprehensive life cycle assessment that regards the whole settlement and not only the single buildings. In vulnerable settlements the contraction phase implies the disuse of places. Planners and designers of flexible settlements must be aware of this phenomenon and should consider that a neighborhood might also regress its dimension.
- Functional un-differentiation: according to the first principle, the settlement should adapt its functions to the contingent needs without a complete twisting of the entire structure. This means that all the "functional containers" must consider hosting different functions in different times. Each urban function carries its urban requirements in terms of facilities and in terms of infrastructures. The compatibility among different functions (and their possible twisting) involves

also the forecast of all the urban equipment for every possible functional configuration.
- Ethero-organization: A balance between the necessity to address the behavior of urban agents and to let them freely evolve together with the whole system is needed.

Flexibility is not a synonym of resilience. In fact, resilience assumes to come back to an equilibrium phase after a certain impact, while flexibility means continuous adaptation to the modifications that singularly might have a small magnitude but that, on the whole, modify the environment.

8.7 Ethero-organization for Co-design Activities

Once the main knowledge fields have been sketched, it is necessary to identify appropriate approaches that a designer/planner needs when facing specific communities' contexts. Matters of scale (global and local), of points of view (subjective and objective) and of communication codes must be inserted in a clear organizational model. The organization aspect is very important to define the role of scholars in relation to the community they work on. Moreover, it is fundamental to clarify an order in the organization to define the typology of information and knowledge that can be used and treated and for pre-defined goals.

In the urban context it is difficult to find some evident forms of self-organization unless when the normative system is weak or lacking. In a general (and qualitative) overview of city management top-down government is still predominant in comparison to bottom-up one. It is difficult to pursue a kind of city management that involves citizens in all the governance steps because of the implicit lack of supervision in the conflicts. On the other hand, it is obvious that increasing the level of sharing information and knowledge means awareness of the limits and possibilities of a community. The relation between technicians and citizen needs particular skills in storytelling more than in numbers and data management. In fact, planners and designers have to adapt their knowledge building process acquiring information from non-conventional sources (Sandercock, 2003). Moreover, planners and designers must find the most appropriate way to communicate the results of their work to be accepted by the community. And this is possible throughout a storytelling that shows a specific and shared horizon for the whole community. In local policies practices, planners and designers become more facilitators than future builders.

The vulnerability of self-organized systems is evident looking at the poor conditions in which people live in self-built settlements such as the Mexican or Brazilian favelas or Paso del Norte establishment.

The "urban agents" have a natural capacity to become planners and designers, and this capacity permeates the urban life: it is the capacity that can transform the elements into artifacts. In a flexible and adaptive approach to urban studies and urban actions, it emerges how it is desirable an ethero-organization system that balances the top-down planning with the self-organization (bottom-up planning). In this way the planning process becomes a real adaptive instrument for the development of the city. Results are not physically fixed, but the knowledge to reach a result (whatever it is) is clear to everyone. Basic elements that guide life of an organism (individual or collective) are coherence (self-organization), relationship with the environment (ethero-organization), and evolution.

8.8 Conclusions

In this paper author described some of the infinite application of the two most diffuse way to create a shareable knowledge: storytelling and numbers. Here knowledge is not considered as a unique description of reality, but it is a way to connect objects (ideas, numbers) to communicate a certain scenario. Both storytelling and numbers, such as soft and hard sciences, do not have the supremacy in knowledge building. This epistemological issue has been discussed and treated for centuries and, in urban studies, it passed from a complete faith in numbers and algorithms, to the trust in discursive knowledge, to a return of data management and simulation together with a suspicion upon them. Nowadays, the world of big data is built with billions of information that, combined each other, may describe a certain reality with the same precision of a heavy heuristic method.

My personal point of view derives directly from the complex system theory. What can help planners and designers to have a comprehensive description of the settlement or environment they are studying, is the system of relations among objects, among stories, among numbers, among individuals. So, both approaches are equally necessary.

Once accepted the definition and description of complex systems, and avoiding to define the supremacy of one discipline over the others, a comprehensive approach that integrates both rigorous and liquid pictures of the world is surely the way to reach the best results at least regarding human settlements.

In vulnerable communities and settlements, that have a very low character of stability, a flexible approach can be helpful for planners and designers because it can be intended as a resolutive way of thinking to combine numbers and storytelling.

In fact, every simplified model (deterministic, heuristic, statistic) has specific implications where it has great performance. In example, statistics are extremely useful to describe global phenomena and to compare specific contexts to a general trajectory. The weak point is the number of considered objects, that is the statistical meaning of data.

Flexibility is a midpoint between hard quantitative approach and a soft qualitative one.

Human settlement complexity is maybe one of the most interesting issues for planner and designers. It must be noted that very often planners tend to apply deterministic or statistic methods, while designers are mainly oriented to heuristic methodologies. In both fields, storytelling and numbers are indispensable aspects of the same experience.

Vulnerable communities study involves the need to compare very local characteristics and needs, to the global and shared vision of community that is part of worldwide narration. So inter-scalar networks are fundamental parts of the whole study process.

Global and local necessities emerge strongly in data description of vulnerable communities in example in the measure of GDP per capita that is far from the poverty threshold. In weak contexts, flexibility can be helpful in providing suitable opportunities.

Planning and design activities aim to synthesize both qualitative and quantitative aspects and, at the moment, a comprehensive shared methodology needs more on-site research. The theoretical state of the art is defined, some practical actions must verify the credibility of such hypothesis.

References

Allen, P., & Sanglier, M. (1981). Urban evolution, self-organization and decision making. *Environment and Planning, 13*, 169–183.

Ameel, L. (2017). Towards a narrative typology of urban planning narratives for, in and of planning in Jätkäsaari, Helsinki. *Urban Design International., 22*(4), 318–330. https://doi.org/10.1057/s41289-016-0030-8. ISSN: 1468–4519 (electronic).

Angeles, M., & López, A. (2021). *Climate Change: Negationism, Skepticism, and Misinformation*. Tabula Rasa [online], n.37, pp.283-301, Epub Mar 02, 2021. ISSN: 1794–2489. https://doi.org/10.25058/20112742.n37.13.

Batty, M. (2012). Urban regeneration as self-organization, in Architectural design, n.° 215.

Batty, M., & Marshall, S. (2009). The evolution of cities: Geddes, Abercrombie and the New physicalism. *Town Planning Review, 80*(6), 551–574. Liverpool University Press.

Bauman, Z. (2000). *Liquid modernity*. Polity Press.

Bertuglia, C. S., & Staricco, L. (2000). *Complessità, autorganizzazione, città*. FrancoAngeli Editore, Roma, EAN: 9788846419040.

Brunetta, G., Faggian, A., & Caldarice, O. (2021). *Bridging the gap. The measure of urban resilience*, MDPI, Basel, ISBN 978-3-0365-0766-8 (Hbk), ISBN: 978-3-0365-0767-5 (PDF).

Casti, J. (1994). *Complexification: Explaining a paradoxical world through the science of surprise*. HarperCollins. ISBN: 9780060168889.

De Carlo, G. (2008). *Questioni di architettura e urbanistica*, Maggioli Editore, Santarcangelo di Romagna, ISBN-10: 8838742391, ISBN-13: 9788838742392.

De Lotto, R. (2011). Flexibility principles for contemporary cities, in *Changing Shanghai – from Expo's after use to new green towns*, by Zheng Shiling e Angelo Bugatti, Officina Edizioni, Rome.

De Lotto, R., & Morelli di Popolo, C. (2015). Complex, adaptive and hetero-organized urban development: the paradigm of flexible city, Proceedings International Multi-Conference on Complexity, Informatics and Cybernetics, IMCIC 2015, Orlando, FL, March 10–14, 2015, http://www.iiis.org/CDs2015/CD2015IMC/IMCIC_2015/PapersPdf/ZA737TX.pdf

De Roo, G. (2016). Self-organization and Spatial Planning - Foundations, challenges, constraints and consequences, In G. De Roo, and L. Boelens (Eds.), *Spatial Planning in a Complex Unpredictable World of Change – Towards a proactive co-evolutionary type of planning with the*

Eurodelta, Edition: 1, Chapter: 3, InPlanning, Groningen, Germany, ISBN 978-94-91937-26-2 (print) 978-94-91937-27-9 (e-book).

Dice, M. (2017). The True Story of Fake News: How Mainstream Media Manipulates Millions, The Resistance Manifesto (1 novembre 2017), ISBN-10: 1943591024, ISBN-13: 978-1943591022.

Engelen, G. (1988). The theory of self-organization and modelling complex urban systems. *European Journal of Operational Research, 37*(1), 42–57., ISSN 0377-2217. https://doi.org/10.1016/0377-2217(88)90279-2

Fisher, F., & Forester, J. (1993). *The argumentative turn in policy analysis and planning*. Duke University Press. ISBN: 978-0-8223-1372-4.

Gargiulo, C., & Papa, R. (1993). *Caos e caos: la città come fenomeno complesso in Per il XXI secolo: una enciclopedia e un progetto*, Università degli Studi di Napoli, Federico II, – in Italian.

Geddes P (1915), Cities in Evolution, 1915,

Given, L. M. (2008). *The SAGE Encyclopedia of qualitative research methods*. SAGE Publications. ISBN: 9781412941631.

Hajer, M. A. (1993). Discourse coalitions and the institutionalization of practice: the case of acid rain in Great Britain. In F. Fischer & J. Forester (Eds.), *The argumentative turn in policy analysis and planning* (pp. 233–253). Duke University Press. ISBN: 978-0-8223-1372-4.

Harari, Y. N. (2018). Homo deus. Breve storia del future, Giunti/Bompiani, Firenze, ISBN: 9788845298752, English version, (2015) Homo deus. A brief history of tomorrow.

Harari, Y. N. (2019a). Sapiens. Da animali a dei. Breve storia dell'umanità, Giunti/Bompiani, Firenze, ISBN: 9788845296499, English version, (2011) From animals into gods. A brief history of humankind.

Harari, Y. N. (2019b). *21 lezioni per il XXI secolo, Giunti/Bonpiani, Firenze*, ISBN: 9788830100824, English version (2018) 21 lessons for the 21st century.

Katuyuki, S. (Eds.). (1983). Guido Canella, Zanichelli Editore, Bologna, ISBN-13: 978-8808011763

Koolhaas, R. (2001). Junckspace, 2001, Quodlibet, Macerata, ISBN 9788874621125, Italian version.

Leone, D. (2010). *Sequenze di città. Gli audiovisivi come strumento di studio e interpretazione della città*, Francoangeli, Milano, ISBN: 9788856835717, In italian.

Longo, M. (2012). *Il sociologo e i racconti. Tra letteratura e narrazioni quotidiane*, Carocci, Roma, ISBN: 9788843067664.

McLoughlin, J. B. (1969). *Urban & regional planning. A system approach*. Faber and Faber.

Meadows, D. H. et al. (1972), *The Limits to Growth: A report for the Club of Rome's project on the predicament of mankind, Potomac Associates - Universe Books*, ISBN: 0876631650, digital version: https://collections.dartmouth.edu/teitexts/meadows/diplomatic/meadows_ltg-diplomatic.html (visited on May 2nd 2021).

Meyers, R. (2011). *Encyclopedia of complexity and systems science*. Springer.

Morelli di Popolo, C. (2013). La rilettura della città come sistema organico verso nuove interpretazioni, in *Il governo della città nella contemporaneità*. La città come motore di sviluppo, a cura di Sbetti F., Rossi F., Talia M., Trillo C., Dossier n.°4, INU Edizioni, Rome.

Moroni, S. (2013). *La città responsabile. Rinnovamento istituzionale e rinascita civica*. Carossi Editore, Rome, 2013 – in Italian.

Palese, E. (2013). Zygmunt Bauman. Individual and society in the liquid modernity. *SpringerPlus, 2*, 191. https://doi.org/10.1186/2193-1801-2-191

Pasquale, F. (2021). *Le nuove leggi della robotica. Difendere la competenza umana nell'era dell'intelligenza artificiale*. Luiss University Press, Rome, ISBN: 9788861055995. original version: *New laws of robotics: defending human expertise in the age of AI*, Harvard University Press, 2020.

Portugali, J. (2000). *Self-Organization and the City*. Springer.

Portugali, J. (2013). What makes city complex?, Paper published on spatialcomplexity.info, by M. Batty 2013 – accessed on 2014 July 20.

Ratti, C., & Claudel, M. (2016). *The city of tomorrow: sensors, networks, hackers, and the future of urban life*. Yale University Press. ISBN: 9780300204803.

Rouelle, D. (1991). *Chance and Chaos*. Princeton University Press. ISBN: 9780691021003.
Russell, D. (1983). Anything goes. *Social Studies of Science*, *13*(3) (Aug., 1983), pp. 437–464, Sage Publications, Inc, available at: https://www.jstor.org/stable/284799?seq=1#metadata_info_tab_contents
Sandercock, L. (2003). Out of the closet: the importance of stories and storytelling in Planning Practice. *Planning Theory and Practice, 4*, 11–28. https://doi.org/10.1080/1464935032000057209
Sandercock, L. (2010). From the campfire to the computer: an epistemology of multiplicity and the story turn in planning. In L. Sandercock & G. Attili (Eds.), *Multimedia explorations in urban policy and planning: beyond the flatlands* (pp. 17–37). Springer. ISBN 978-90-481-3209-6.
Sassen, S. (1999). *The global city: New York, London, Tokyo*. Princeton University Press.
Sennett, R. (1998). *The corrosion of character: The personal consequences of work in the new capitalism*. Norton.
Throgmorton, J. A. (1993). Survey research as rhetorical trope: electric power planning arguments in Chicago. In F. Fischer & J. Forester (Eds.), *The argumentative turn in policy analysis and planning* (pp. 117–144). Duke University Press. ISBN: 978-0-8223-1372-4.
Tzatzadaki, O. (2021). La narrazione non convenzionale come innovazione nel processo analitico socio-spaziale. Il caso della città diffusa del Nordest, XXIII SIU Conference Proceedings, Downscaling, Turin June 17–18, 2021, Planum Publisher, ISBN: 9788899237288, DOI: https://doi.org/10.53143/PLM.C.121, Vol 1, pp.147–153. https://www.worldometers.info, accessed from June 2018 until today.
Warren, K. (2013). *Chaos Theory and Complexity Theory, Encyclopedia of Social Works*. Oxford Press. https://doi.org/10.1093/acrefore/9780199975839.013.45
Welter, W. M. (2002). *Biopolis: Patrick Geddes and the city of life*. MIT Press.
Williams, S. (2020). *Data action. Using data for pubic good*. The MIT Press. ISBN: 9780262044196.

Chapter 9
Urban Design in the Age of Climate Change: Paradigms and Directions

Ali Cheshmehzangi, Chris Butters, Amir Cheshmehzangi, Ayotunde Dawodu, Ruoyu Jin, and Eugenio Mangi

9.1 Introduction

In recent years, the frequency of climate change impacts on cities has been increasing (Cheshmehzangi & Dawodu, 2018; World Bank, 2010). To combat them, cities need to minimise their adaptation strategies and move towards mitigation patterns that are more durable and forward-thinking. Therefore, tangible actions are needed

A. Cheshmehzangi (✉)
Department of Architecture and Built Environment, University of Nottingham Ningbo China, Ningbo, China

Urban Innovation Lab, University of Nottingham Ningbo China, Ningbo, China

Network for Education and Research for Peace and Sustainability (NERPS), Hiroshima University, Hiroshima, Japan
e-mail: ali.Cheshmehzangi@nottingham.edu.cn

C. Butters
GAIA Architects, Oslo, Norway

A. Cheshmehzangi
School of Engineering, The University of Warwick, Warwick, UK

A. Dawodu
Department of Architecture and Built Environment, University of Nottingham Ningbo China, Ningbo, China

Urban Innovation Lab, University of Nottingham Ningbo China, Ningbo, China

E. Mangi
Department of Architecture and Built Environment, University of Nottingham Ningbo China, Ningbo, China

R. Jin
Built Environment and Architecture/Construction, Property and Surveying, London South Bank University, London, UK

E. Giorgi et al. (eds.), *Design for Vulnerable Communities*, The Urban Book Series,
https://doi.org/10.1007/978-3-030-96866-3_9

to reverse some of the existing trends in design and planning (Cheshmehzangi & Dawodu, 2018) and towards a comprehensive decarbonisation era. Much of the most needed actions are policy-driven or based on the development or restructuring of our institutional arrangements, or are purely based on our actions that are held with multiple stakeholders. As such, Cheshmehzangi and Dawodu (2018), while arguing the importance of institutional attributes, remind the role of people who are recognised as key actors for sustainable urban development. Now, in this chapter, the focus is mainly on the role of urban design as a discipline, and particularly the role of urban designers and planners, as the ones believed to be able to make new pathways for better paradigms (Cheshmehzangi & Butters, 2017a; Cheshmehzangi & Butters, 2017b). These, if planned and implemented rightly, ought to lead us to make our cities more sustainable, viable, and effective to face the increasing climate change impacts. These ideals should not only appear to make cities more resilient but should also equip us with the possibilities of reversing some of our somewhat unsustainable planning and design trends.

In line with our central argument, new paradigms are essential to finding sustainable pathways that are transitional and hopefully innovative in the end. Cities need to identify context-specific solutions that are then transformational in multiple ways. As highlighted by Dodson (2010, p. xix), cities are in urgent need of transformation:

> We are faced with the need to develop new knowledge, a better understanding of how biophysical and socio-economic systems interact, develop new strategies and actions, and new international cooperative arrangements in order to solve what is the defining challenges of the twenty-first century.

Therefore, the need for such transformation is ever more important in an era of climate change impacts on cities.

In this chapter, we first emphasise the importance of urban design in the age of climate change. This is narrated based on existing patterns of urban development and opportunities that await us for new practice and policy directions. The shifts are, undoubtedly, essential to what we can regard as sustainable urban design models. The majority of focus is on the environmental dimension and environmental tools (Ameen et al., 2015; Cheshmehzangi, 2016), while progress on developing a multidimensional sustainability agenda is perceptible. In the later parts of the chapter, we delve into a set of design principles proposed to highlight how we can achieve sustainable urban design instead of high-rise design topologies that are widespread in many parts of the world. These design principles are discussed in more detail with some examples before we elaborate more on the role of urban designers in the age of climate change and in facing the climate impacts on cities and urban life. These will provide us with enough materials to conclude the chapter with some general suggestions that require careful adaptation or redevelopment strategies in the contexts in need of immediate strategies in planning and design.

9.2 Urban Design in the Age of Climate Change

It is already known that cities are commonly the main source of energy use and climate emissions. Much of this comes from planning and design issues that changed our built environment into less sustainable living habitats over the years. In many cases, the rapid growth and development impacted vast city restructuring and lower quality redevelopment projects. These weakened the role of urban design and empowered the position for planned and unplanned city layouts and configurations. In the field of sustainable urbanism, urban design models play a significant role in achieving new paradigms of change that are environmentally sound and beneficial for multiple stakeholders. Of our concern amongst all stakeholders is the stakeholder that defines people as the group of end-users and the ones residing in housing compounds that occupy most of the built environment structures in any contemporary city. Therefore, it is essential to address: what urban design models can provide enhanced energy efficiency and better quality of life at the same time? And how these can provide us with better ideals of design to combat climate change impacts on cities?

Urbanisation is happening at a rapid pace, and more than fifty percent of the world population now live in cities around the globe. Currently, this is estimated at 4.2 billion people worldwide. The pace will continue as it has in the last few decades. It is estimated that three out of five people will live in cities by 2030. With these estimates, it is also expected that two-thirds of the world's population will live in urban areas by 2050. Most of them, especially in the rapidly urbanising contexts, faced today's greatest challenge and threats, particularly 'climate change'. The effects are immense and increasing. Urban environments and municipalities are vulnerable and exposed at different levels/scales to the range of hazardous stresses and associated changes such as extreme weather, heatwaves, rising sea levels, frequent storms, and flooding (Moraci et al., 2018). The threat of climate change is so compelling, and long-term in nature that necessary constitutional changes (Cheshmehzangi & Dawodu, 2018) are required in the way cities are planned and managed.

As a major threat, it affects and unsettles the public health, local economies, infrastructure, and quality of life for the urban inhabitants. Cities are important economic powerhouses and the prominent source of generating wealth; therefore, it is paramount to minimise these impacts. In the face of the many effects of climate change, cities must be redefined and transformed as a whole to respond more effectively to the current situations of heat and climatic complexities. We also anticipate such vulnerabilities with adequate and resilient infrastructure at the urban scale. Hence, urban designers have a wide range of roles in promoting sustainable urbanism and development in the form of adequate land-use zoning (Cheshmehzangi & Butters, 2017b; Zhang et al., 2019), transportation planning (Kulinska & Kulinska, 2019; Wey, 2019), and green infrastructure planning and design (Cheshmehzangi & Butters, 2015a, 2015b). As a result, it is crucial to radically transform urban design strategies, redesign cities and make them more sustainable, resilient, resource-efficient, and zealous to future changes. This has to be done through comprehensible and coordinated policies, considering how to utilise and manage the land use. This ultimately leads to creating energy-efficient built environments, with multiple methods of integrating agriculture and green infrastructure into the urban settings. These should enable us to rethink few factors, such as how can we better manage water supplies in these areas? How can we mitigate low emission and greener transportation? And how can we generate and distribute energy to the larger scale of communities and beyond?

In the light of the above questions, we need to urgently develop and plan methods or systems to address rapid urbanisation, and build more sustainable communities. By doing so, we have to encourage low emission development strategies and facilitate connectivity and economic development to enable ways of reducing carbon footprint and greenhouse gas emissions. Hence, sustainable urban development should provide an opportunity for renewable energy supplies (Luo et al., 2019; Worighi et al., 2019) on an urban scale (e.g. renewable energy-powered micro-grids) and, at the same time, promote social cohesion and a sense of community (Cheshmehzangi & Butters, 2017b). By doing so, we are enabling new pathways to make our cities more resilient and adaptive at first and later to equip them to recover from the challenges linked to climate change. These will ultimately lead to a shift towards mitigation strategies that are more effective in the long run.

According to the UN-Habitat's principles of sustainable neighbourhood planning (see Dawodu et al., 2017), urban design should be in favour of compaction with specialised limited land use (Burgess, 2000), high-density green buildings (Cheshmehzangi & Butters, 2017a; Huo et al., 2019), and narrower streets. Such typology allows for more integrated green space and better connectivity that allows for adequate space for mobility between the urban environments. In such ways, the proximity to transit and improved public transport, such as light rail and rapid bus transit schemes, enables new paradigms that will eventually adopt new policies to promote walkability and cycling at the urban design and even the planning levels. It is already evidenced that scattered low-density settlements (UN Habitat, 2014; IPCC, 2015; UN Habitat, 2015) consume more resources and produce higher levels of greenhouse gas (GHG) emissions than the compact ones (i.e. encouraging car-dependency and energy consumption). It also occupies more land, forest, and open spaces needed to alleviate heat and reduce the risk of flooding. This is evidenced in research conducted by the Royal Town Planning Institute (RTPI Website, 2015) in the UK, which provides a briefing on urban form and sustainability (see Butters, 2012; Jabareen, 2006; Cheshmehzangi & Butters, 2017b). In addition, the three C's of 'compactness', 'connected', and 'coordinated' are essential to make cities more productive, resilient, and cleaner, with significant reductions in the cost of providing services and urban infrastructure requirements.

Most cities characterised by hot and dry as well as hot and humid climates are facing excruciating conditions of urban heat island effect (UHIL) and intense warming in populated and dense areas (Cheshmehzangi & Butters, 2015a; Cheshmehzangi & Butters, 2015b). This major issue, and the increase in the frequency of heatwave periods and hot days, add to rising energy demands and prices. The practice of making cities and urban environments more sustainable always lags behind the simple and inexpensive solutions that can make the change that is needed. Hence, the larger shift towards adaptation rather than mitigation strategies is an alarming matter. Consequently, Mi et al. (2019, p. 582) refer to cities as '*the core of climate change mitigation*', meaning that our strategies need to be inclusive and comprehensive enough to reduce carbon emissions and energy consumption.

Finally, we would like to investigate methods of achieving sustainable urban design and opposing high-rise layouts, mainly from a pragmatic perspective. This enables us to suggest principles that are thoughtful, inexpensive, and hopefully game-changers of contemporary practice and business-as-usual.

9.2.1 Methodology

The methods used for this study are twofold, literature review and the use of case study examples. The study first highlights 12 principles of sustainable urban design (SUD) based on the existing literature review. The selection of these principles is based on four key areas of (1) planning, (2) energy, (3) services and materials, and (4) design. These are suggested as the backbone of understanding and evaluating sustainable urban design directions. Each principle is then discussed independently, backed up by a relevant literature review. The selection and overall narrative is based on the findings of a research programme, titled Energy and Low-Income Tropical Housing (ELITH), which was successfully completed in 2017. The analytical studies and observations made during the project timeline provided us with the rich literature on sustainable urban design pathways, some of which are published as journal papers, policy guidelines, and concept papers/notes. Each principle is then supported with case study examples and not case study analysis. Discussions for each principle allow us to highlight key factors under each category and summarise the debates from the urban design perspective. The combined approach provides enough evidence to have key findings from a non-scientific research narrative but from a literature-based perspective.

The following sections follow this proposed narrative by first introducing 12 principles for sustainable urban design and delving into each principle through literature review and observation of case study examples. Afterward, the chapter ends with discussions and concluding remarks on the topic, highlighting the role of the urban designer in the age of climate change, and summarise paradigms and directions to achieve the sustainable urban design. Discussions with reflect on key practical and policy directions and navigating the viewpoint on the importance of sustainable urban design in achieving the overarching goal of urban sustainability. Figure 9.1 summarises the methodological approach of the chapter.

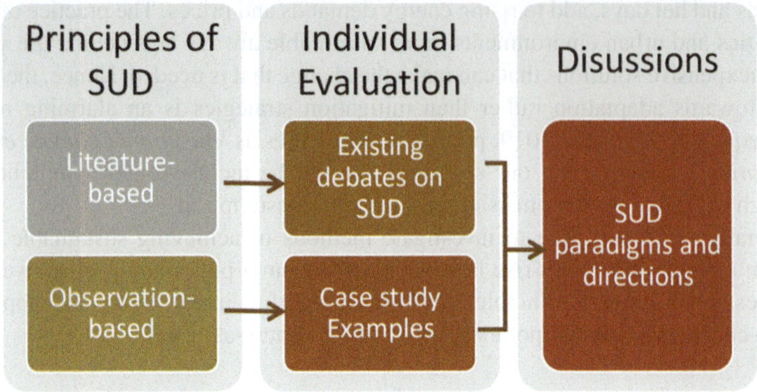

Fig. 9.1 Summary of methodological framework for the chapter

9.3 Introduction to 12 Principles for Sustainable Urban Design (SUD)

Here, we highlight some key viewpoints on sustainable urban design principles and explain briefly how we can implement them to mitigate climate change impacts on cities through better planning. In this respect, the following 12 principles, mostly related to design factors, are introduced and discussed in line with the aims of this research. These 12 principles for sustainable urban design are recognised as key strategies (Fig. 9.2) opposing the existing and non-sustainable trends of housing development that often occur as models of high-rise towers and in gated communities, where accessibility is minimised, and density is increased via the means of floor area ratio (FAR) and building heights. These 12 principles comprise the following:

1. Density.
2. Energy efficiency.
3. Building services.
4. Urban microclimate.
5. Solar/heat.
6. Embodied energy.
7. Operational energy.
8. Embodied maintenance.
9. Ventilation.
10. Materials transportation and construction.
11. Post-use.
12. Design.

In the following part, we provide a brief explanation for each proposed design principle. The study's narrative is mostly focused on paradigms against the contemporary high-rise typology and in support of the low-rise one. The arguments are,

Fig. 9.2 Summary of 12 sustainable urban design principles discussed in this chapter

therefore, associated with specific contexts and patterns. The suggested principles are important for design consideration and practice re-considerations that should essentially lead to policy reforms and new paradigms.

9.3.1 Density

It is investigated that density has a large impact on sustainable development (Cheshmehzangi & Butters, 2017a; Cheshmehzangi & Dawodu, 2021). Based on urban modelling studies of multiple urban residential areas in various cities in China and the European contexts, the impact of floor area ratio (FAR) is recognised as one of the key elements to optimise the energy efficiency of the housing units. The FAR calculations often involve multiple parameters, one of which is the relationship between the overall floor area space and the building heights. While in some contexts FAR of 1 is often accepted, equivalent to 100% of the site area in multiple floors, the higher FAR figures between 4 and 8 (and even higher) are also recorded (ibid). In some contexts, like in China, this is compromised with lower surface coverage areas. It follows the current regulations that allow for more outdoor amenities, green spaces/services, and car parking. Therefore, the impact is on the development of higher buildings that cover a smaller part of the overall site. This typology, in most occasions designated as residential units, is regarded here as less sustainable and with a heavier impact on the urban landscape, inclusive of both the cityscape and the actual green infrastructure planning (see example of Figs. 9.3 and 9.4).

However, our studies indicate a mixed-use layout or a low-to-midrise typology that can accommodate more opportunities for green infrastructure utilisation, renewable energy integration, and better cooling (or heating) strategies. Some of these are essential from the perspectives of city planning and at a larger scale (Cheshmehzangi & Butters, 2017b), while some are effective at the building design dimension (Dawodu & Cheshmehzangi, 2017). At all scales of the built environment, the main aspect would remain at the design level, which is discussed throughout the following principles.

Fig. 9.3 Cityscape: the urban landscapes are changing drastically with mid-to-high rise residential towers often in a gated community layout in Ningbo, China (Source: Ali Cheshmehzangi)

Fig. 9.4 Urban density: the new high-rise buildings increase both density and FAR of the central parts of Sydney, Australia, impacting the overall population density in some of the older residential and mixed-use areas (Source: Ali Cheshmehzangi)

9.3.2 Energy Efficiency

Hsieh et al. (2017) suggest methods of defining density and land uses under the energy performance targets. One of these targets is energy efficiency, which is a significant part of combatting climate change in cities. When it comes to studies in the built environment and in association with climate change issues, a major aspect is the building's energy efficiency. It is an important part of multiple sectors, and from both urban design and building design perspectives, it also is a major point of design consideration. This is even more substantial for growing economies like China, as the growth in energy demand would lead to higher prices and more energy consumption. In a typical sub-tropical location, where we conducted research between 2013 and 2017, there are at least 3–4 months of heat and extreme heat. Many contemporary buildings are designed in a way that is not resilient or adaptive but dependent on the use of multiple air conditioning units in a singular household or building. For instance, a contemporary 3-bedroom house in sub-tropical China will require four to five air conditioning units to maintain adequate human comfort. This is also affected by the high-rise building typology, where we expect to have more surface exposure to a higher level of solar gain. Such typology allows for higher chances of retaining the heat, and at the same time, creates more shadow for the surrounding buildings. A proper solution for height adjustment can lead to better units that can accommodate a more balanced heat gain and loss, and as a result, be less dependent on air conditioning devices and for a shorter period of time.

9.3.3 Building Services

Typically, a high-rise structure requires more materials and usage of building services than a low-rise one (Cheshmehzangi & Butters, 2017a). This viewpoint also includes larger structural beams, columns, and building services shaft and capacity. And this means higher use of construction materials and a larger space allocated for lifts, circulation routes, stairs, and fire emergency escapes. Hence, a minimum of 15% of each floor area space is often the requirement for circulation routes and lifts, as well as more space for the building services is needed in between each floor. These factors raise the overall cost of the building, and they increase the overall management responsibilities and maintenance cost. Therefore, once again, we argue against the high-rise typology and advocate for less usage of building services that is feasible in the lower rise typology (see example of Figs. 9.5 and 9.6).

Fig. 9.5 Building services: air conditioning units as one of the main building services visible on the building façade in Changsha, China (top), and on the roofs in Osaka, Japan (bottom) (Source: Ali Cheshmehzangi)

Fig. 9.6 Building services: air conditioning units as one of the main building services visible on the building façade in Changsha, China (top), and on the roofs in Osaka, Japan (bottom) (Source: Ali Cheshmehzangi)

9.3.4 *Urban Microclimate*

A primary principle is the applicability of urban microclimate for larger-scale projects. It is a less studied area of research for building design but a very important topic at the urban design level. An example is the utilisation of green infrastructure and how it can be integrated with urban design (Santamouris, 2006). This particular aspect is a recognised method of improving building performance (Vuckovic et al., 2016) and utilising specific indicators (Schinasi et al., 2018) for sustainability measures. These are specifically associated with cooling strategies (Cheshmehzangi & Butters, 2015a; Cheshmehzangi & Butters, 2015b; Cheshmehzangi & Butters, 2017b) and energy reduction from primary cooling demand (Meggers et al., 2016). From our earlier studies (Cheshmehzangi & Butters, 2017a), we identified that the low-rise typology in China fosters a favourable microclimate, which was common practice almost 20 years ago. Most projects of the time had a simple form/layout before the housing policy reforms of 1998, shifting the move from low-rise to mid-to-high rise typologies across the entire country. The dramatic change in FAR figures caused by this reform meant taller buildings and less genuine green infrastructures (and mostly conceived as green spaces only). A consideration of microclimate can only be provided through an integrated design and consideration of the multiplicity of the urban environment. This requires careful design and integration of green infrastructures for various means and uses.

9.3.5 Solar/Heat

In a typical high-rise building, solar and heat gain are generally higher than the low-rise typology. As expressed earlier, much of this is linked to the larger surface exposure to solar radiation. It is also partly due to different material use, where we identified widespread employment of glazing and windows for towers. The heat gain is a crucial factor that must be moderated through better material use and less façade exposure to direct solar radiation (see example of Fig. 9.7). Design possibilities for more shading devices (both horizontal and vertical), louvres, and even balconies can help to reduce the heat gain in various ways. Figure 9.8 demonstrates differences in degree day analysis of different FARs, showing respective heat gain and heat losses for each variation.

9.3.6 Embodied Energy

A less studied but crucial design principle is the consideration of embodied energy in buildings (Butters & Thomas, 2016; Chang et al., 2012) and material selection. Typically, we anticipate having a higher rate of material used in a high-rise building. Therefore, higher buildings often have higher embodied energy. For instance, we often employ more steel or reinforced concrete in towers than low-rise structures. The energy used for such material production is higher, also meaning a higher embodied energy (see example of Fig. 9.9).

Fig. 9.7 Façade design: large ratio of glazed facades for a cluster of residential towers in Hong Kong, China, with extreme hot and humid climatic conditions and with a high cooling demand (Source: Ali Cheshmehzangi)

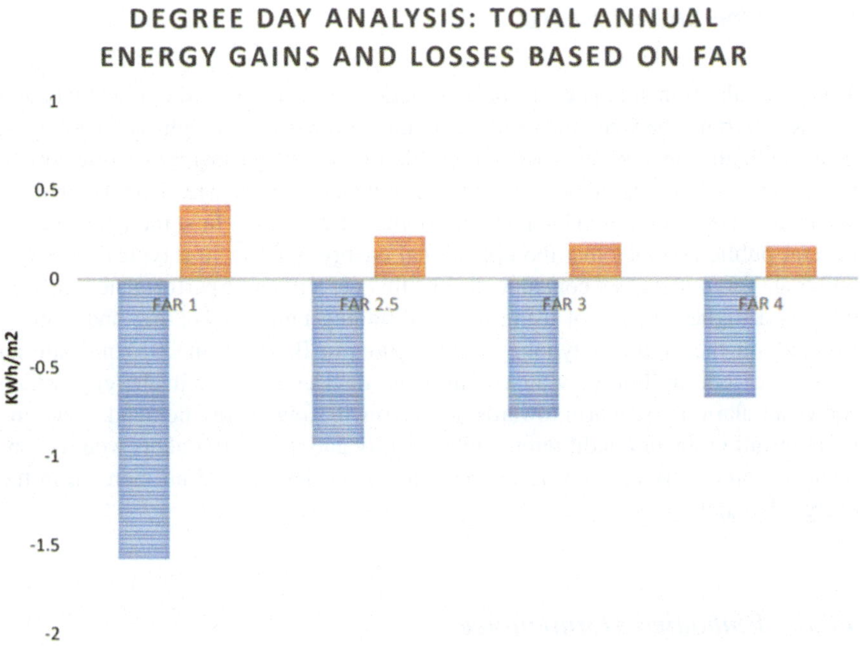

Fig. 9.8 Degree day analysis demonstrating total energy gains and losses based on FAR variations (Adapted from Cheshmehzangi & Dawodu, 2021)

Fig. 9.9 Material use: more steel and reinforced concrete for construction of new high-rise buildings in Ningbo, China (Source: Ali Cheshmehzangi)

9.3.7 Operational Energy

This particular principle is well studied globally. However, it is identified that there is little difference between the residential units of low-rise and high-rise typologies. In general, the operational energy depends on associated factors of orientation, façade design, building layout, and other design factors. Some are also related to the occupants' consumption behaviour for cooling and heating. In some contexts, the general public assumes that the operational energy is lower in a typical high-rise building. Yet, our studies conclude against this perception of performance. In fact, there is a higher chance of using air conditioning units for heating and cooling demands in the high-rise typology. The figures in Bangkok in 2015 indicate an increase of one million new air conditioning devices per year in the city, which shows an alarming scenario towards an ever-increasing energy demand. Once the users install more air conditioning units, the city and its environments would eventually become warmer, causing the need for more devices and an increase in the energy demand.

9.3.8 Embodied Maintenance

In association with the issues of embodied energy, the cost of maintenance is higher, and the overall maintenance process is more complicated for high-rise buildings. As part of the building's life cycle assessment (LCA) and its recurrent embodied energy (Dixit, 2019), maintenance adds to building materials' overall longevity and usability. Hence, material specifications and the quantity used for construction are both fundamental factors that should be fully considered for LCA and maintenance planning.

9.3.9 Ventilation

In general, ventilation is often considered a crucial problem in the construction of high-rise buildings. Energy load reduction for both cooling and heating needs is associated with the overall building design. Examples of large residential towers, both in height and in width, indicate a possibility of having around 25 apartments per floor. This figure increases the chances of having a large number of single-side ventilated units and some inner units (facing the courtyard or the inner court) with no natural ventilation. In addition, such a layout allows a minimal opportunity for cross-ventilated or double-sided ventilated units. In particular, the inner ones would become heavily dependent on mechanical ventilation such as air conditioning. Furthermore, natural lighting becomes a significant issue, and again the artificial illumination will become essential. Hence, ventilation strategies must be fully integrated into the overall design, including orientation, space layouts and openings, the direction of openings, etc.

9.3.10 Materials Transportation and Construction

Materials transportation and construction energy are indeed higher in high-rise buildings. In fact, it is often not easy to verify their source and the kind of transportation adopted for their delivery. This is associated with embodied energy considerations and should consider the overall energy required for transportation and material shipping. There is a higher chance of using local materials for low-rise buildings due to the lower technological requirements needed for their realisation. However, there is less likelihood of local material selection and usage for high-rise buildings during the construction period. Hence, the embodied energy is higher from two aspects of material quantity and transportation.

9.3.11 Post-use

This is believed to be a major aspect with minimal attention in research. The post-use consideration is not only about the usability of the building after completion, but it also considers the overall life cycle of the building after its construction and completion phase (Sartori & Hestnes, 2007; Zhang & Wang, 2015). This should include the demolition phase and the re-use of materials, both fundamental factors involved in the overall life cycle assessment of a typical building. In this regard, the post-use of high-rise buildings is generally more complex. This is relevant to all contexts that require demolition, renovation, and redevelopment strategies. All these three types of post-use strategies, although at the end of the life cycle, are very important to do planning for material use and recycling.

9.3.12 Design

The design itself is an important criterion for the development of sustainable buildings and urban environments. Design should take into consideration the following aspects: location, related to context-specific and climatic conditions as well as the location of the site and its surroundings; urban layout, comprising building orientation, heights, shape, and the overall layout of the buildings; and user, comprising user orientation, use, and requirements that may be related to design of the building. A typical high-rise building offers little flexibility in design and hence a minimal chance for extensions, installations, and modifications over time. Instead, these can occur for the low-rise buildings with little complication (see example of Fig. 9.10).

Fig. 9.10 Design and planning: the number of high-rise compounds are increasing significantly in rapidly developing cities like Shanghai, China (Source: Ali Cheshmehzangi)

In the high-rise typology, one can only modify the ground floor layout and perhaps implement minor cosmetic changes with minimal effect. Hence, as a precursor to energy solutions, design is the ultimate and the most careful consideration of all the 12 principles.

Following from exploration of these 12 principles, the study highlights how they are then taken into consideration from the building to the urban design level. Yet again, the importance is on what paradigm may be suggested to allow transitions and new directions in the field of the built environment. This, we regard as a major turning point (Cheshmehzangi & Dawodu, 2018), and an opportunity to reverse some of the unsustainable trends in urban development.

9.4 The Role of Urban Designer

Urban designers should ideally play a major role in developing innovative and sustainable paradigms to combat climate change impacts on cities. From earlier studies, it is argued that inappropriate dense urban environments/districts can produce and contribute to poor microclimates. Hence, the construction material selection is a crucial factor at the building and urban design levels. For instance, impermeable and heat-absorbing materials like asphalt and concrete will make the neighbourhood hotter and increase the urban heat island effect. This fact leads to more energy use and, therefore, more GHG emissions. The impervious surfaces also contribute to a higher risk of flooding as they cannot absorb rain and stormwater. These effects can be mitigated with a proper approach and design strategies (Wayne Lee & Kohm, 2014). For these reasons, the building level design may not be sufficient for such integrated thinking in design. In addition, building heat exchange performance at the surface could be improved by using heat-resistant materials and reflective surface coating. Other measures include increased vegetative covering, lower outdoor temperature, enhanced air quality, and reduced pollution. They also minimise rain and floodwater runoff, which helps to better utilisation and management of water retention and flood management (Sun et al., 2020). Such methods enable us to simply mitigate the urban heat island (UHI) effect and simultaneously lower the carbon footprint (Carter et al., 2015). In more contemporary research, the inclusion and integration of nature-based solutions (NBSs) are important to contemporary urban planning and design research. Some examples include green roof design, parks, and green walkways, and other green infrastructure facilities, which can support city environments by reducing the temperature, mitigating the UHI effect, and improving the air quality in the long run.

In line with improving the design quality of buildings at the micro-level and making them more energy-efficient, we also need to reduce temperatures at the urban district level, at the meso and macro levels. For instance, improvements in the form and layout of buildings and neighbourhoods by planned shading and orienting them according to the sun path can significantly impact the buildings' energy efficiency and the larger-scale urban environments. As such, the overall design of the buildings should allow connectivity with the healthy outdoor microclimates (Cheshmehzangi & Butters, 2017b; Piselli et al., 2018) and encourage natural urban ventilation that reduces the need for energy use, and as such, help to minimise the need for heating and cooling. While reducing these energy demands to zero is not possible at a low cost, achieving large reductions in energy use and GHG emissions is feasible.

Another factor to take into full consideration is the people/citizens dimension. The citizen role is also discussed thoroughly by Cheshmehzangi and Dawodu (2018) as a critical part of achieving sustainable urban development in the face of climate change. As expressed in an article by Zenghelis and Stern (2015), 'the physical shape of cities will determine the behaviours of its citizens and the responsiveness of its institutions'. For instance, pedestrian and cycling corridors will encourage walkability and reduce greenhouse gas (GHG) emissions, especially when incorporated with green-space infrastructure planning in cities. Hence, by putting pressure on governmental bodies (e.g. climate change governance, see Nagel et al., 2019) and policymakers, the push is to improve and provide better infrastructure fitting the citizens' needs. By doing so, we go through step-by-step processes to promote health and well-being and also reduce carbon emissions. This can be recognised as a win-win solution, which is inexpensive, should the institutions provide enough support.

The overall understanding is that climate change is a critical challenge, and its impacts are already experienced worldwide. Cities are more vulnerable (Moraci et al., 2018) due to multiple reasons of higher population, less natural elements, more physical infrastructure, and less resilience. Yet, it is everyone's interest to combat the problem before it is beyond our control. Are we late? Perhaps, yes. But are we too late? We believe not. In response to the changing climate today, these considerations coupled with progressive thinking should provide us with enough tools to stop exploiting natural systems and maximise the urban environments' efficiency with appropriate policymaking that is in line with sustainable development ideals. By doing so, we still have many opportunities for innovation and extraordinary impacts, which can ultimately future-proof the built environment and well-being of our cities and their citizens.

9.5 Conclusions

This study provides a holistic overview at the design level and particularly at the urban design level. Integrating the investigated principles in design strategies offers inexpensive and straightforward solutions that significantly impact the built environment. Our study is essentially the starting point for future research that will hopefully set new grounds for transformational city paradigms and new planning/design directions. These principles at a larger scale of urban design enable us to take a broader approach to climate change impacts. The solutions would therefore become more holistic in comparison to the smaller scale of building design where the pace of sustainability transitions is believed to be rather slow.

Figures 9.11 and 9.12 above illustrate one hot and humid and one hot and dry conditions, both of which have a high cooling demand. We see two very different extremes of hot climates, with partial integration of some cooling strategies and façade design, but often with major flaws that are yet to be resolved in contemporary practice. The two compounds show how only a few design principles are considered, proving that a comprehensive approach is needed should we wish to develop a sound building design.

Finally, we argue in favour of urban transitions and decarbonisation, essential for our cities. The complexity of urban issues combined with climate change impacts is evident in every part of the world. To be resilient and adapt to the ongoing and forthcoming changes will make us stronger, yet with little progress in reversing any of the unsustainable trends. To become innovative and progressive in the right direction, we need to develop and implement mitigation strategies at multiple levels. In this chapter, we highlighted some of the fundamental principles that are transitory between two scales of building and urban design.

Acknowledgements Ali Cheshmehzangi would like to thank NSFC for project number 71850410544, and Ningbo Soft Science for project number 2017A10072, for the provided funding between 2016 and 2019. We also acknowledge the Energy and Low-Income Tropical Housing (ELITH) project, which was completed in 2016–2017. The findings from the ELITH project helped to develop several set of guidelines, policy reports, and planning paradigms. We also acknowledge the support and funding from the Ministry of Education, Culture, Sports, Science and Technology (MEXT), and the support provided to Network for Education and Research on Peace and Sustainability (NERPS), Hiroshima University, Japan.

Fig. 9.11 Design and planning paradigms: a new high-end and high-rise residential compound in hot and humid Bangkok, Thailand, incorporating some key design features but still with a large ratio of glazed facades (top) and a low-to-midrise residential compound in hot and dry Shiraz, Iran, incorporating scale and façade configuration, but still with little consideration of spatial strategies for green infrastructure integration (bottom) (Source: Ali Cheshmehzangi)

Fig. 9.12 Design and planning paradigms: a new high-end and high-rise residential compound in hot and humid Bangkok, Thailand, incorporating some key design features but still with a large ratio of glazed facades (top) and a low-to-midrise residential compound in hot and dry Shiraz, Iran, incorporating scale and façade configuration, but still with little consideration of spatial strategies for green infrastructure integration (bottom) (Source: Ali Cheshmehzangi)

References

Ameen, R. F. M., Mourshed, M., & Li, H. (2015). A critical review of environmental assessment tools for sustainable urban design. *Environmental Impact Assessment Review, 55*, 110–125.

Burgess, R. (2000). The compact city debate: A global perspective. In M. Jenks & R. Burgess (Eds.), *Compact cities, sustainable urban form for developing countries*. Spon Press.

Butters, C. (2012). A holistic method of evaluating sustainability. In T. Haas (Ed.), *Sustainable urbanism and beyond*. Rizzoli.

Butters, C., & Thomas, T. H. (2016). *Comparing the Carbon Impact of Buildings to that of the Site Works*, research publication from the ELITH project, Warwick University, UK, Available at: https://warwick.ac.uk/fac/sci/eng/elith/publications/all_publications/elith-w07.pdf (Accessed on: 22 September 2019).

Carter, J. G., Cavan, G., Connelly, A., Guy, S., Handley, J., & Kazmierczak, A. (2015). Climate change and the city: Building capacity for urban adaptation. *Progress in Planning, 95*, 1–66.

Chang, Y., Ries, R. J., & Lei, S. (2012). The embodied energy and emissions of a high-rise education building: A quantification using process-based hybrid life cycle inventory model. *Energy and Buildings, 55*, 790–798.

Cheshmehzangi, A. (2016). Multi-spatial environmental performance evaluation towards integrated urban design: A procedural approach with computational simulations. *Journal of Cleaner Production, 139*, 1085–1093.

Cheshmehzangi, A., & Butters, C. (2015a). *Refining the Complex Urban: The Study of Urban Residential Typologies for reduced Future Energy and Climate Impacts*, in proceedings for the international conference on 'True Smart and Green City?, The 8th Conference of the International Forum on Urbanism (IFoU), held in Incheon, South Kora, June 2015.

Cheshmehzangi, A., & Butters, C. (2015b). *Urban Green Infrastructure for Cities of Developing Countries*, Policy Guideline of ELITH project, published in Ningbo.

Cheshmehzangi, A., & Butters, C. (2017a). Chinese urban residential blocks: Towards improved environmental and living qualities. *Journal of Urban Design International, 22*(3), 219–235.

Cheshmehzangi, A., & Butters, C. (Eds.). (2017b). *Designing cooler cities: Energy, cooling and urban form: the asian perspective*. Palgrave Macmillan Publishers.

Cheshmehzangi, A., & Dawodu, A. (2018). *Sustainable urban development in the age of climate change – people: The cure or curse*. Palgrave Macmillan Publishers.

Cheshmehzangi, A., & Dawodu, A. (2021). Towards a sustainable energy planning Strategy: The Utilisation of Floor Area Ratio for Residential Community Development Planning and Design in China, *Frontiers in Sustainable Cities*, Urban Energy End-Use, https://doi.org/10.3389/frsc.2021.687895.

Dawodu, A., Akinwolemiwa, B., & Cheshmehzangi, A. (2017). A conceptual re-visualization of the adoption and utilization of the Pillars of Sustainability in the development of Neighbourhood Sustainability Assessment Tools. *Sustainable Cities and Society, 28*, 398–410.

Dawodu, A., & Cheshmehzangi, A. (2017). Passive cooling energy systems SWOT analyses for energy-use reductions at three spatial levels. *Energy Procedia, 105*, 3411–3418.

Dixit, M. (2019). Life cycle recurrent embodied energy calculation of buildings: A review. *Journal of Cleaner Production, 209*, 731–754.

Dodson, J. (Ed.). (2010). *Changing climates, earth systems and society*. Springer.

Hsieh, A., Schuler, N., Shi, Z., Fonseca, J. A., Marechal, F., & Schlueter, A. (2017). Defining density and land uses under energy performance targets at the early stage of urban planning processes. *Energy Procedia, 122*, 301–306.

Huo, X., Yu, A. T. W., Darko, A., & Wu, Z. (2019). Critical factors in site planning and design of green buildings: A case of China. *Journal of Cleaner Production, 222*, 685–694.

Intergovernmental Panel on Climate Change (IPCC). (2015). *Human Settlements, Infrastructure, and Spatial Planning*. Climate Change 2014 Mitigating Climate Change. pp. 923–1000. https://doi.org/10.1017/cbo9781107415416.018

Jabareen, Y. R. (2006). Sustainable urban forms, their typologies, models, and concepts. *Journal of Planning Education and Research, 26*, 38–52.

Kulinska, E., & Kulinska, K. (2019). Development of ride-sourcing services and sustainable city logistics. *Transportation Research Procedia, 39*, 252–259.

Luo, X., Peng, P., Shao, Y., Li, J., & Yi, G. (2019). A coordinated power control strategy for urban micro-grid. *Energy Procedia, 158*, 6626–6631.

Meggers, F., Aschwanden, G., Teitelbaum, E., Guo, H., Salazar, L., & Bruelisauer, M. (2016). Urban cooling primary energy reduction potential: System losses caused by microclimates. *Sustainable Cities and Society, 27*, 315–323.

Mi, Z., Guan, D., Liu, Z., Liu, J., Viguie, V., Fromer, N., & Wang, Y. (2019). Cities: The core of climate change mitigation. *Journal of Cleaner Production, 207*, 582–589.

Moraci, F., Errigo, M. F., Fazia, C., Burgio, G., & Foresta, S. (2018). Making less vulnerable cities: Resilience as a new paradigm of smart planning. *Sustainability, 10*(3), 1–18.

Nagel, M., Stark, M., Satoh, K., Schmitt, M., & Kaip, E. (2019). Diversity in collaboration: Networks in urban climate change governance. *Urban Climate, 29*, 100502.

Piselli, C., Castaldo, V. L., Pigliautile, I., Pisello, A. L., & Cotana, F. (2018). Outdoor comfort conditions in urban areas: On citizens' perspective about microclimate mitigation of urban transit areas. *Sustainable Cities and Society, 39*, 16–36.

Royal Town Planning Institute (RTPI). (2015). *Preventing climate change: the crucial role of planning*, Available at: https://www.rtpi.org.uk/briefing-room/rtpi-blog/preventing-climate-change-the-crucial-role-of-planning/, Accessed on: 15 September 2019.

Santamouris, M. (Ed.). (2006). *Environmental design of urban buildings: An integrated approach*. Earthscan Publications.

Sartori, I., & Hestnes, A. G. (2007). Energy use in the life cycle of conventional and low energy buildings: a Review Article. *Energy and Buildings, 39*(3), 249–257.

Schinasi, L. H., Benmarhnia, T., & De Roos, A. J. (2018). Modification of the association between high ambient temperature and health by urban microclimate indicators: A systematic review and meta-analysis. *Environmental Research, 161*, 168–180.

Sun, J., Cheshmehzangi, A., & Wang, S. (2020). Green infrastructure practice and a sustainability key performance indicators framework for neighbourhood-level construction of sponge city programme. *Journal of Environmental Protection, 11*(2), 82–109.

UN Habitat. (2014). Planning for Climate change: Guide – a strategic, value-based approach for urban planners, part of 'Planning for Climate Change' document, Available at: http://mirror.unhabitat.org/downloads/docs/PFCC-14-03-11.pdf. (Accessed on 19 August 2019).

UN Habitat. (2015). *A new strategy of sustainable neighbourhood planning: Five Principles*. 2015:8.

Vuckovic, M., Kiesel, K., & Mahdavi, A. (2016). Toward advanced representations of the urban microclimate in building performance simulation. *Sustainable Cities and Society, 27*, 356–366.

Wayne Lee, K. and Kohm, S. (2014). *Cool pavements*. Green Energy Technologies (Vol. 204, pp. 439–453). https://doi.org/10.1007/978-3-662-44719-2_16.

Wey, W. M. (2019). Constructing urban dynamic transportation planning strategies for improving quality of life and urban sustainability under emerging growth management principles. *Sustainable Cities and Society, 44*, 275–290.

Worighi, I., Maach, A., Hafid, A., Hegazy, O., & Van Mierlo, J. (2019). Integrating renewable energy in smart grid system: Architecture, virtualization and analysis. *Sustainable Energy, Grids and Networks, 18*, 100226.

World Bank. (2010). Cities and Climate Change: An Urgent Agenda (Vol. 10). Available at: https://openknowledge.worldbank.org/handle/10986/17381 (Accessed on: 22 August 2019).

Zenghelis, D., & Stern, N. (2015). Climate change and cities: a prime source of problems, yet key to a solution, The Guardian, article on 19 November 2015, available at: https://www.theguardian.com/cities/2015/nov/17/cities-climate-change-problems-solution (Accessed on 31 Oct 2019).

Zhang, Y., Long, H., Tu, S., Ge, D., Ma, L., & Wang, L. (2019). Spatial identification of land use functions and their tradeoffs/synergies in China: Implications for sustainable land management. *Ecological Indicators, 107*, 105550.

Zhang, X., & Wang, F. (2015). Life-cycle assessment and control measures for carbon emissions of typical buildings in China. *Building and Environment, 86*, 89–97.

Part III
Designing for Vulnerabilities. Applications and Actions

Chapter 10
Social Housing in Historic Centers: Contemporary Experiences in Latin America

Nivaldo Vieira de Andrade Junior and Gabriela Amado Chetto Coutinho

10.1 Social Housing and Experimentation in Latin America

The architectural critic Justin McGuirk, in a book focused on analyzing some of the most important architectural experiences in social housing held in Latin America in the last decades, asks himself: “Why is Latin America special? And what can we learn from it?”. He answers:

> […] the countries of South and Central America were home to some of the greatest experiments in urban living of the twentieth century. [...] It was in an attempt to deal with the scale of urban migration that Latin American architects picked up the gauntlet thrown down by the European modernists. If standardised, industrialised housing was the future, then it would be adapted to the scale of the New World.” (McGuirk, 2014, pp. 07–08).

These “great experiments in urban living in Latin America” are the massive social housing complexes built in the major cities of the region between the 1940s and 1960s, such as the Unidad Vecinal n. 3 in Lima (1946–1949), designed by Fernando Belaúnde Terry and his team, housing 1112 apartments; the Pedregulho Housing Complex in Rio de Janeiro (1947–1952), designed by Affonso Eduardo Reidy, comprising 384 apartments; the 23 de Enero Housing Complex in Caracas (1955–1957), designed by Carlos Raúl Villanueva, comprising more than 9000

This chapter is dedicated to the late professor and architect Esterzilda Berenstein de Azevedo (1949–2020), who, in the wake of Lina Bo Bardi, bet on social housing in the historic center of Salvador and succeeded —at least for some time—to convince public managers and politicians of this goal. She involved dozens of young architects and architecture students in this proposal, among them one of the authors of this text. Her intelligence, aggregating character and enthusiasm continue to serve as a reference for new generations of architects, and even some who have not had the opportunity to work directly with her.

N. V. de Andrade Junior (✉) · G. A. C. Coutinho
Faculty of Architecture, Federal University of Bahia, Salvador, Brazil

E. Giorgi et al. (eds.), *Design for Vulnerable Communities*, The Urban Book Series,
https://doi.org/10.1007/978-3-030-96866-3_10

apartments; the Unidad Vecinal Portales in Santiago, Chile (1954–1966), designed by the Studio Bresciani, Valdés, Castillo and Huidobro, housing 1860 apartments in 19 blocks; or the massive Nonoalco-Tlatelolco housing estate in Mexico City (1960–1964), designed by Mario Pani, with more than 100 buildings and 15,000 apartments, housing nearly 100,000 people—the equivalent of a medium city.[1]

Since the late 1960s social housing experiments in Latin America took other paths, leaving the utopia of gigantic complexes and focusing on other issues such as incremental housing, as in the Experimental Housing Project – PREVI (1968–73), promoted in Lima by the United Nations and the Government of Peru,[2] and in the projects designed by Elemental, as the Quinta Monroy complex (2004), in Iquique, Chile, which was mostly responsible for the 2015 Pritzker Prize awarded to the group leader, the Chilean architect Alejandro Aravena.

Although Latin America houses several listed urban sites, 35 of which inscribed in UNESCO's World Heritage List,[3] the experiences of social housing in historic centers in the region are exceptions and began much more recently, different from Europe, where initiatives such as the Piano per l'Edilizia Economica e Popolare (PEEP) of Bologna's historic center, conceived from 1963 and whose first housing units were delivered between 1970 and 1971, already highlighted the importance of social housing as a central element in historic centers' public policies (Cervellati et al., 1981; Cervellati & Scannavini, 1976). Other significant experiences on a smaller scale were carried out in other European countries from the 1970s onwards, such as the São Victor housing complex, designed by Álvaro Siza Vieira under the program SAAL (Serviço de Apoio Ambulatório Local), created after the Carnation Revolution in 1974, and built in the historic center of Porto in the mid-1970s, with 52 units.

In general, the scarcity of social housing experiences in Latin America stems from the fact that public policies implemented in these areas have privileged, in the last 50 years, tourism-related activities. This prioritization of tourism is associated with a prejudiced view that social housing is incompatible with tourism. With this,

---

[1] The Nonoalco-Tlatelolco housing estate was built in a historic site, where the Aztecs where finally defeated by the Spanish, and where there is also the sixteenth century church of Santiago Tlatelolco.

[2] The PREVI complex consists of 24 housing types in clusters of approximately 20 units each. The types were designed by 13 foreign architects or studios, directly invited by the promoters for their previous experience in social housing, and 13 Peruvian architects or studios, selected by an open competition. Among the foreign architects, world famous professionals such as the British-American Christopher Alexander; the Parisian office Candilis, Josic and Woods; the Indian Charles Correa; the Dutch Aldo van Eyck; the Japanese Kiyonori Kikutaki, Kisho Kurokawa and Fumihiko Maki; the British James Stirling; and the Colombian Germán Samper. Among the Peruvian architects, important names such as José García Bryce, Jacques Crousse and Luis Miró Quesada (Land, 2015)

[3] Antigua Guatemala, Arequipa, Camagüey, Campeche, Cartagena de Indias, Casco Viejo de Panamá, Cienfuegos, Colonia del Sacramento, Cuenca, Cuzco, Diamantina, Goiás, Guanajuato, Lima, Mexico City, Morelia, Oaxaca, Old Havana, Olinda, Ouro Preto, Paraty, Potosí, Puebla, Querétaro, Quito, Salvador de Bahia, San Miguel de Allende, Santa Cruz de Mompox, Santo Domingo, São Luís, Tlacotalpan, Trinidad de Cuba, Sucre, Valparaiso, and Zacatecas.

in recent decades, many Latin American countries continue to produce massive housing complexes, with hundreds or even thousands of units in the outskirts of large cities and without the same quality of design that characterized the experiences held between the 1940s and the 1960s. In some countries, such as Brazil and Venezuela, to cite just two examples, programs like Minha Casa Minha Vida (2009–2020) and Gran Misión Vivienda (from 2011) continue to be responsible for the construction of millions of housing units of poor architectural quality in the urban peripheries, far from the basic infrastructure and services, corresponding to most of the social housing production in these countries in the last years.

10.2 The Uruguayan Experience of Housing Cooperativism

One exception seems to be Uruguay, with more than half a century of experiences with housing cooperativism, which emerged in the second half of the 1960s, in a scenario of social and economic crisis that allowed the emergence of new political organization models, capable of articulating and legitimizing alternative strategies to hegemonic modes of production. Given the context of complete inactivity of the construction industry and the fragility of public policies to promote housing (Cecilio, 2015), the emergence of Uruguayan housing cooperativism have reflected the social movements action in the fight for decent housing, which led to the development of "an authentic system of production and social management of habitat" (Vallés, 2015, p. 16).[4]

The Uruguayan National Housing Law, established in 1968, brought significant advances to the development of housing cooperatives, guaranteeing the right to decent housing, and incorporating cooperativism as a regulated housing production regime (Vallés, 2015). Through the creation of the National Housing Fund, the law provides the financing organization for state, private, and cooperative projects; it institutionalizes the Technical Assistance Institutes (IATs), multidisciplinary nonprofit teams responsible for providing all necessary support to cooperatives; and "recognizes the various forms of ownership and cooperative modalities for access to housing and services" (Cecilio, 2015, p. 31).

The modalities of cooperatives are classified according to two criteria, which refer to the model of the ownership regime and the type of initial contribution to the financing. According to the first criterion, housing units can be owned individually or collectively—in this case, it is a mode of use, in which residents have the right to use a unit for an indefinite period, being able to pass it on for generations. The second criterion recognizes and classifies cooperatives according to two financing modalities: ahorro previo (mutual savings), which mainly serves the portion of the population with average income, and ayuda mutua (mutual aid), in which

[4]The first experiences of housing cooperatives emerged from three initiatives carried out by the Uruguayan Cooperative Center, in the regions of Florida, Salto and Rio Negro, in 1966.

low-income families have the possibility to contribute with working hours in the construction of their own houses (Vallés et al., 2011).

It should be noted that this policy has an enormous range capability, due to the different types of cooperatives offered, being able to adapt to different urban contexts, including diverse social profiles of residents, different locations, and various types of financing, which makes it a system of enormous potential for expansion and effectiveness.

The success and longevity of Uruguayan cooperativism are reflected in the execution of about 30,000 units over the 50 years of existence of the system (Vallés, 2015). Even during the period of the military dictatorship (between 1973 and 1985), when the government closed the IATs and tried to extinguish collective property, to disarticulate the system, Uruguayan society found mechanisms of resistance and demonstrated the depth of the system's roots, by resuming cooperative practices in the post-dictatorial democratic scenario (Vallés et al., 2011).

In general, the Mutual Aid Housing Cooperatives (CVAM) are structured on principles of democratic management and social participation promoted through the mutual aid undertaken by members of the cooperative and the state: the government facilitates access to the soil, to loans and grants; users provide the workforce and are responsible for the management of the cooperative; and the IATs technically advise the working groups. The CVAM "refers and transforms a unique tradition of popular sectors—self-construction—into a practice through which a family with limited and occasional external support raises its own home" (Iglesias, 2008, p. 46).

This type of construction offers a very pertinent response to the social, economic, and spatial demands of the residents, who demonstrate the capacity and autonomy to build houses with spatial and constructive quality at low cost. The residents, associated in community organizations, become protagonists of the integral process, from the moment of choosing the site to the management of the post-occupation communities, in a collective effort that builds meanings. About this aspect Vallés et al. (2011, s.p.) consider that:

> From the sociological and anthropological dimension, it is possible to affirm that the CVAM is constituted in an area of production of a differentiated cultural identity, understanding culture from the scope of its production and not from its consumption. This implies that, through the reelaboration or symbolic representation of material structures, one understands, reelaborates and transforms the social system and the struggle for hegemony.

In Brazil, the Uruguayan cooperativism has influenced several relevant initiatives since the 1980s, especially in São Paulo, such as the self-managed housing projects of Vila Nova Cachoeirinha, started in 1982 (Baravelli, 2006), and the work of Usina, technical assistance office in architecture and urbanism created at the end of that decade in São Paulo and operating in several cities over the country. João Marcos de Almeida Lopes (2018, p. 243), one of the Usina founders, recorded that:

> For us technicians, the references to an associative political culture were and are, to a large extent, Eurocentric [...]. Other exogenous references, such as Uruguayan cooperatives, producing houses through mutual aid and self-management, should be also included.

10.3 Latin American Historic Centers: Tourism as Salvation and Social Housing as Taboo

The event that established the defense of tourism as a priority activity to be implemented in Latin American historic centers is the Meeting on the Preservation and Utilization of Monuments and Sites of Artistic and Historical Value, held in 1967 in Quito, Ecuador, by the Organization of American States (OAS) whose final report became known as the "Norms of Quito." Based in the European experience, the Norms of Quito defend, in the section "VII - Monuments as Touristic Attractions," the use of monuments and sites as touristic attractions as the only way to preserve them:

> 1. Intrinsic cultural values are neither weakened nor compromised by association with tourist interests; on the contrary, the increased attraction of the cultural properties and the growing number of outside admirers confirm awareness of their importance and national significance. [...] Europe owes to tourism, directly or indirectly, the salvation of much of its cultural heritage condemned to complete and irreparable destruction, and modern man, more visually than literarily sensitive, finds increasing opportunities for self-enrichment through viewing examples of western civilization, scientifically rescued because of the powerful incentive of tourism.
>
> [...]
>
> 7. The economic and social advantages of tourist travel vis-a-vis monuments are evident in most modern statistics, particularly in those European countries that owe their present prosperity to international tourism and include among their major sources of wealth the inventory of their cultural properties. (Icomos, 1967)

In this period, the first missions to Latin American countries by international consultants in heritage preservation took place, based on requests to UNESCO by national governments. Among the missions effectuated in the period, we highlight those carried out in 1966 and 1967 by Michel Parent in 35 Brazilian cities within the scope of the "Cultural Tourism" program of UNESCO. Parent was at the time Inspector General of French Historical Monuments and the theme of touristic exploitation of monuments and historic centers was the focus of his report:

> Tourism can certainly be one of the sources of the future development of national income and provide an economic alibi for the considerable efforts that must be made if we are to safeguard the vast cultural heritage that has long been in danger, but whose ruin will soon be irreversible (Parent, 2008, p. 46).

Despite Parent's defense of tourism as a strategy for transforming cultural heritage into a vector for economic and social development, in line with the thinking that was then prevalent at UNESCO—and which would also be defended in the "Norms of Quito"—Parent (2008, pp. 46–51) himself had reservations about this strategy:

> But tourism cannot do everything. Leaving the heritage solely to the criteria of immediate tourist profitability would be making contestable choices, aggravating certain imbalances, confronting "facade effects" with the increase of internal degradation [...]. On the contrary, it is important that tourism does not constitute an end in itself, nor even a means of simultaneously satisfying the curiosity and comfort of non-Brazilians or a few Brazilians disconnected from the national reality, but that the technical model of infrastructure associate the

way of knowing the Brazilian culture with the way of living it and, in this way, can integrate the tradition, science and the safeguarding of the values of ancient Brazil to the development of the future Brazil.

Paulo Ormindo de Azevedo (2001) remembers that this model based on tourism was introduced in the region through projects sponsored by UNESCO or OAS and consisted in creating touristic development poles in the most important squares in the historic centers. Azevedo also remarks that Parent's report would serve as the main guideline to the first interventions held in the 1970s in the Historic Center of Salvador. Also in the early 1970s, Carlos Flores Marini, a Mexican architect who was one of the two Latin Americans to directly contribute to the Venice Charter in 1964 and who was among the signatories of the Norms of Quito, developed technical missions as an OAS consultant in the historic centers of Panamá and Cartagena de Indias, proposing the same tourism-based model.

However, the most ambitious among those projects developed by international agencies in Latin America was, undoubtedly, the Plan Copesco (*Proyecto Nacional de Conservación y Puesta en Valor del Patrimonio Cultural en el Perú*), created in 1969 and that focused on the tourist exploration of cultural heritage in a large area of 84 thousand kilometers that included Cuzco and the margins of the Titicaca Lake. The Plan Copesco comprised building urban and touristic infrastructure, as well as the restoration of several monuments.

Azevedo (2001, p. 303) notes that the adoption of cultural tourism as a lifeline for the built heritage and for national development was a solution that, in Latin America,

> charmed the authorities and national elites, who, in addition to the possible external benefits, saw in it the opportunity to reappropriate the historic centers abandoned to the excluded. With this inspiration, and the illusion of solving all the complex social and physical-environmental problems with a magic word, many resources have been spent with very modest results.

Azevedo (2001, p. 315) also records that the mistakes and successes of these experiences based on cultural tourism:

> should not be credited exclusively to national or local authorities, but also to international organizations and agencies that, without knowing the local problems, have led to the adoption of experiences developed in very diverse contexts. Cultural tourism as a panacea for all the ills of the historic center is the most obvious example.
>
> However, little by little, it is understood that the solution for our historic centers inevitably involves issues that directly affect the local population, such as the slum and informal work. Although these problems are structural in Latin American society, there are always creative ways to mitigate them. It is necessary to overcome prejudices and cut privileges that do not contribute to the solution and tend to perpetuate the problem. We must learn to work with street vendors and to turn tenements and slums into decent homes for their inhabitants.

For Azevedo (2001, p. 315), the cost-benefit of experiences aimed at the local population of Latin American historic centers "is greater than the building of expensive tourist infrastructures that cannot be profitable without the previous surpassing of the existing social framework in these areas":

> Those who choose to live in the center are subjected to subhuman conditions of habitability and wholesomeness. The acquisition of the buildings by these residents is practically impossible both for their low income and lack of real guarantees, as for the indivisibility of mansions and townhouses, leading to a progressive subdivision and ruining of the buildings, operated as a sublet. (Azevedo, 2001, p. 299).

Thus, according to Azevedo (2001, p. 308), there is no escape from the conclusion that:

> The matter of housing [...] is one of the key issues in the rehabilitation of historic centers. Firstly, because the historic center has always been a polyfunctional area and in this lies its great dynamism and vitality. Second, because housing creates affective links with the building and with the neighborhood, which facilitate the preservation of the historic center. Third, because the presence of a fixed population creates a neighborhood typical economy, represented by artisan production, shops of first need and personal services. Fourth, because only housing can guarantee life and security 24 h a day, dispensing expensive security and animation systems. The architectural typologies of the historic center are basically habitational or mixed, and the progressive elimination of this function means not only its functional adulteration, but also of its architectural and heritage aspects.

Despite this, Azevedo (2001, pp. 308–309) recognizes that:

> However, housing has always been a taboo theme in the restoration programs of Latin American historic centers. On the one hand, due to lack of familiarity with the subject, housing agencies allege that their residents do not have the capacity to participate in social housing programs and that there is no demand from other social groups to live in the historic center.
>
> There is also a widespread prejudice that the recognition of the right of these populations to continue living in the historic center would render their rescue for tourism unfeasible. [...]
>
> But the bottom line is another. About 80–90% of the owners of the historic center properties have abandoned them and have no interest or capacity to recover them, but do not detach them because of their potential value. Added to this is the fact that the mansions and townhouses of aristocratic rural families do not adapt to the possibilities and needs of today's nuclear families, demanding their transformation into multi-household condominiums. This situation of impasse demands that the State intervene in these areas to carry out such transformations, which forces a courageous political decision.

This article will focus on this "courageous political decision," through the analysis of some social housing projects in Latin American historic centers carried out since the 1980s and specially in the last 20 years.

10.4 The Pioneering Experiences of Social Housing in the 1980s

In the 1980s, several technical meetings were held in Latin America dedicated to discussing the importance of housing in the historic centers' conservation. Estela Cañellas, Viviana Colella, and Natalia Da Representacao (Cañellas et al., 2008, p. 4) observed that:

> Following the movement that is in development in Europe but adapted to the context of Latin American cultural and socio-economic reality, numerous instances are developed that promote a Latin American vision of the problem, [...]. Among the most relevant are: the ICOMOS meetings in 1982, 1984 and 1985; the Second Inter-American Symposium on the Conservation of Monumental Heritage in Morelia (1981) and the Third in 1983, which ratify the social aspects of rehabilitation; the Seventh International Symposium on Conservation of the Monumental Heritage of Puebla (1986), where emphasis is placed on integral rehabilitation; the Symposium "Housing today in yesterday's architecture" in Mexico (1987), about the reconstruction of the historic center after the earthquake, and the Seminar on Housing Rehabilitation in Historic Areas held in Mexico and Havana (1987) whose final declaration had much significance in our environment since "While reiterating the points made in previous documents, the document is an important step in the process of raising awareness of the values that the housing heritage rehabilitation assumes for our American circumstance".

This period coincides with the end of the military dictatorships installed in the previous decades and with the consequent processes of redemocratization in many countries of the region, such as Peru, Bolivia, Argentina, Uruguay, Brazil, Paraguay, and Chile,[5] which have contributed to strengthening the social movements and to broadening the debates on citizens' participation in public policies.

10.4.1 La Boca Urban Recovery and Development Program in Buenos Aires

The 1980s also correspond to the time some pioneering social housing projects were executed in Latin American historic centers. The La Boca Urban Recovery and Development Program in Buenos Aires, better known as RECUP-Boca, was conceived in 1985, in the context of the Argentine redemocratization. According to Guevara (2011), RECUP-Boca "was an ambitious program that aimed to promote the integral development of the [La Boca] district, both on the urban and social levels. It incorporates, in turn, the participatory dimension as a fundamental component of design." Guevara (2011) remembers that:

[5] In 1980, the architect Fernando Belaúnde Terry was elected for the second time president of Peru, ending 22 years of military dictatorship, which began with the coup of Juan Velasco Alvarado in 1968; in 1982, Hernán Siles Zuazo returned to the presidency of Bolivia, elected for the second time, re-establishing democracy after 18 years of political instability and military governments; in 1983, the election of Raúl Alfonsín to the presidency in Argentina marks the end of one of the continent's bloodiest dictatorships, started in 1976, with the coup that overthrew President María Estela Perón and installed the "National Reorganization Process"; in 1984, the election of the civilian Julio María Sanguinetti for the presidency of Uruguay ends the civic-military dictatorship started in 1973; the following year, the election—albeit indirect—of Tancredo Neves for the presidency of Brazil ends a period of 21 years of military dictatorship, started in 1964; in 1989, the end of the General Alfredo Stroessner's government, in Paraguay, closes one of the continent's longest dictatorships in the twentieth century, which lasted 34 years; in 1988, in Chile, a plebiscite approves the end of another long military dictatorship, General Augusto Pinochet's government, which began with the military coup against President Salvador Allende, in 1973.

The RECUP recognised two main action lines. On the one hand, the immediate actions, aimed at alleviating the situations of housing emergency, to guarantee minimum conditions of habitability in the buildings, while the definitive works were carried out. On the other hand, actions aimed at the general rehabilitation of the neighborhood, through the incorporation, rehabilitation and progressive transfer of properties to beneficiary families, allowing to restart the investment cycle with the recovery of credits. A key element of the rehabilitation component was the existence of a significant reserve of fiscal land on the estate known as the "Casa Amarilla". RECUP intended to use this reserve as a land bank to relocate families of tenement houses and dwellings that were considered irrecoverable or were in critical overcrowding, while progress was being made in the rehabilitation of properties that were considered recoverable.

Unfortunately, the implementation of RECUP-Boca had no follow-up, and its performance was largely limited to the purchase, by court auction, of 21 tenement houses between 1989 and 1990, and the management of the rehabilitation works for four of them. Guevara (2011) notes that:

This program, which "have put in tension the municipal logic" for its novel formulation in terms of rehabilitation, failed in breaking down the "prevailing order": the *laissez faire* of the State in the face of "occasional displacements" of the low-income population and their expulsion from the central city.

10.4.2 The Special Program for the Recovery of Historic Sites of the City of Salvador

Between 1986 and 1989, the historic center of Salvador—which, in 1985, had been inscribed on the UNESCO World Heritage List—is the subject of another innovative but equally failed experiment of social housing in historic buildings. At the invitation of the mayor Mário Kertész, the Italian born architect Lina Bo Bardi elaborated the "Special Program for the Recovery of Historic Sites of the City of Salvador" (PERSH), in which she positions herself frontally against a gentrification based on the tourist activity, advocating the maintenance of the resident population in the place:

It is not a tourist job, done with the intention of turning the Pelourinho into an ice cream city.

The bases are for us old acquaintances: it is the fight against folklore. [...]

The case of the Historic Center of Bahia is not the preservation of important architectures, [...] but the preservation of the Popular Soul of the City. In a nutshell: the plan must be socio-economic to avoid repeating the mistakes of known interventions in illustrious cities such as Rome, Bologna, Venezia, and countless wonderful corners of the Old World that have changed the social base of entire Regions, with the residents of years and years thrown away and middle-middle-class taking over.

To avoid this, we are looking to associate work and housing [...] and small shops: a kind of underground economy.

[...]

In general, people are taken out of their homes, other shelters are provided on the outskirts of the cities. In the restored buildings are installed boutiques for tourists, exhibitions, crafts made in São Paulo, etc. Now, the main idea of recovery in Salvador is precisely to maintain the population who lives in the houses that need to be restored, recovered.

> The general idea is to do, on the ground floor, small shops [...], production of foods, small works, recovery, restoration, repair of things. And, on the upper floors, housing. That what currently exists [1987] in Bahia is just this kind of thing.
>
> This is the idea that accompanies the architectural restoration of the Historic Center of Salvador. (Bardi, 2008, p. 270, p. 295).

The PERSH, conceived by Lina Bardi, included a several cultural facilities, many of them, such as the Casa do Benin in Bahia and the Casa do Olodum, aimed at valuing the African contribution to Bahian culture, but its main axis was the refurbishment of hundreds of tenements houses in ruin process, through the adoption of a building system in reinforced mortar developed by the architect João Filgueiras Lima, the Lelé. During Kertész's tenure, Lina could execute only the pilot project on Ladeira da Misericórdia, which should serve as a model for the interventions in the rest of the historic center. The pilot project corresponded to five adjoining buildings in one of the oldest slopes that connected the Upper City and the Lower City of Salvador, then quite degraded. The three remaining old houses, which almost entirely preserved the original facades, would house five commercial establishments on the ground floor and seven apartments on the upper floors. An existing ruin between the houses would house the "bar of the 3 arches," for selling "liquor + crabs," while in the empty land located at the north direction would be built the most striking of Lina's works in the complex, housing the "Coatí Restaurant."

Ladeira da Misericórdia buildings were never formally occupied as social housing, except for a short period in which, with the intervention of Lina already executed, the properties were invaded by homeless families living in the historic center. This project failure had several reasons, among which, no doubt, the main was what Azevedo (2001, p. 309) identified as being "the widespread prejudice that the recognition of these populations the right to continue living in the historic center would make it impossible to rescue them for tourism." Today, the five buildings restored by Lina Bardi in the second half of the 1980s, although belonging to the Municipality of Salvador, are empty and again in the process of ruin, except for one of them that is used by the Military Police.

10.4.3 The Manzana de San Francisco in Buenos Aires

In the late 1980s, Buenos Aires have witnessed the rehabilitation of the Manzana de San Francisco, in the heart of the Argentine capital urban center. This intervention, carried out through a partnership between the Municipality of Buenos Aires and the Regional Government of Andalusia, consisted in the recycling of seven buildings belonging to the Banco de la Ciudad, located in the same block of the San Francisco Church, in the Monserrat neighborhood, 100 meters from the Plaza de Mayo. The buildings were tenements and housed 114 families who, organized in a "neighborhood commission," actively participated in the whole planning and project process, initiated in 1989, following the realization of the San Telmo Open Congresses, events to establish jointly, between neighbors and municipal public managers, the rehabilitation strategies of the Bonairense center. The architectural project of the Manzana de San Francisco rehabilitation was designed by architects Armando Otero, Jorge Ponce, Ricardo Tiraboschi, and Judith Vinocur, under the coordination of Cristina B. Fernández and Salvador Moreno Peralta, and the works were completed in 1995 (Cañellas et al., 2008; Municipality of Buenos Aires, 1991).

According to Cañellas et al. (2008), the Manzana de San Francisco operation takes place "As a response to the occupation of buildings and in a context in which social organizations claim the right to living and having as antecedent a form of intervention that had started with the RECUP-Boca project."

10.5 Social housing in Latin American Historic Centers in the 1990s

Despite those experiences in the historic center of Salvador and in La Boca neighborhood and *Manzana de San Francisco*, public policies adopted in Latin America since the 1990s frequently has been favoring tourism over residents.

10.5.1 The Recovery Plan for the Historic Center of Salvador

In Salvador, the State of Bahia Government implements, from 1992, the "Recovery Plan for the Historic Center of Salvador," focused on tourism and involving US$ 100 million investment approximately in infrastructure works (such as peripheral parking lots), refurbishment of public spaces and restoration of buildings. 531 buildings were recovered; the State Government acquired 432 buildings and obtained the use of other 133, becoming the largest owner of the Historic Center of Salvador (Gordilho Souza, 2010, p. 92).

The "Historic Center of Salvador Recovery Program" has transformed in a remarkable way the landscape and the social frame of the old *Maciel-Pelourinho* neighborhood, provoking great controversy by the social sanitation and gentrification processes that it promoted, with the forced transfer of 2909 families from the 470 houses recovered in the first six stages of the project implementation (1992–1997), transformed into cultural spaces, shops, bars, and restaurants aimed at tourists.

In Roberto Marinho de Azevedo words (1994, p. 131), "Pelourinho has become a setting. That neighborhood where one felt the old Salvador appears today as a theater where one represents Salvador for tourists." In addition, the expulsion of the population from the Historic Center and the reduced amounts paid as compensation to these families resulted in the immediate occupation of neighboring areas:

> The massive withdrawal of old residents, the attraction to the informal market, the existing social ties, are some factors that have contributed, in recent years, to promote the intensive

occupation of the slopes of Pilar, Lapinha, Santo Antônio and Taboão, in addition to the emergence of new tenements in Baixa dos Sapateiros and Saúde, followed by the systematic occupation of idle buildings and ruins [...] (Gordilho Souza, 2010, p. 93).

10.5.2 La Boca Art District in Buenos Aires

In Buenos Aires, in turn, since 2008, several "creative districts" have been invented through municipal laws, such as the Design District, in the Barracas neighborhood, in 2013; and the Arts District, in La Boca and in sectors of San Telmo and Barracas neighborhoods, in 2012. According to Ana Gretel Thomasz (2016), this model of "creative city" adopted in Buenos Aires is indebted to liberal urbanism and "is based on a peculiar mode of public intervention in which the State acts as a market facilitator, providing economic benefits to inverters interested in settling in the New Districts."

Ethnographic research conducted by Thomasz (2016) "indicates that the intervention in La Boca violates various rights enshrined in the World Charter for the Right to the City." Among the testimonies collected by Thomasz on the District of the Arts creation, there are those who argue that the action, under the guise of bringing artists to a neighborhood that already has them, without consulting them, "It's a cover for a real estate project" and that the law that created the Arts District "has nothing to do with the promotion of arts, culture. It is real estate business. It is totally expulsive to small artists," stimulating the installation in La Boca of two "large conglomerates" linked to the cultural and entertainment industries, giving "entrance to the large chains of bookstores, of tango for tourists" and preparing ground.

for a process of privatization, mercantilization and spectacularization of local artistic-cultural expressions, like what have been done in Caminito, where old tenement houses were acquired by investors, displaced and converted into workshops, *tanguerias* [tango houses], restaurants or "thematic" shops for tourists [...]

"Caminito is an antecedent of the Arts District. It was privatized and transformed into a place where three or four capitalists took all. They have not created jobs, or put the art in representation of all", remarked Jimena, member of a La Boca social organization [...]. (Thomasz, 2016).

10.5.3 Casa de los Siete Patios and La Victoria Brewery in Quito

At the same period, the historic center of Quito was the scene of interesting experiences, even if punctual. The best known is the Casa de los Siete Patios, executed in an agreement between the Municipality of Quito, through FONSAL[6], and the

[6] Quito was the first site to be inscribed on the UNESCO World Heritage List in 1978, along with the historic center of Krakow, Poland. After the serious earthquake of March 5, 1987 that mobilized the entire Ecuadorian society and resulted in the declaration of state of emergency, the

Regional Government of Andalusia between 1989 and 1993, with a project designed by architects Jorge Carvajal, Patricia Fondello, José Roman Ruiz, and Emilio Yanez. After the intervention, the Casa de los Siete Patios could house 39 families, of which 23 already lived in the property, occupied by low-income families since the 1970s. Another relevant experience was the conversion of the old La Victoria Brewery, with more than 14,000 square meters of built area, carried out by FONSAL between 1989 and 1998 and designed by architects José Ordoñez, Marcelo Bravo, Enrique Vivanco, and Diego Salazar. The experiences of social housing in the historic center of Quito continue throughout the 1990s, although always punctually, with the rehabilitation of 508 Caldas street, performed between 1995 and 1997 by FONSAL, this time through an agreement with the French NGO Pact Arim (Quito 1992, 2000).

10.5.4 COVICIVI I in Montevideo

In Uruguay, through the 1990s, in the scope of the experiences of the mutual aid cooperatives, a new modality was created called recycling, which consists in reusing the existing physical structure in the city, generally found in the central area, with the potential to implement housing use through the application of social technology based on self-construction and self-management. This modality seeks to reconcile the resolution of the quality housing problem with the right of its inhabitants to remain in their traditional neighborhoods, through the insertion of housing use in rehabilitated buildings with historic value (Vallés et al., 2011).

As in other large and medium-sized Latin American cities, the historic center of Montevideo has undergone a process of population loss and reduction of quality of urban space, reflected in the progressive degradation of its physical structure and in the emptying of the buildings. On the other hand, the historic center of Montevideo remains a very well-served sector in terms of urban infrastructure and services compatible with appropriation for housing use. This context favors the appropriation of the underused or empty properties existing in the central area in actions of mutual aid housing cooperatives, in an initiative that articulates rehabilitation of the built

National Government promulgated the Law of the "Cultural Heritage Rescue Fund," delegating to the Municipality the guardianship of the real estate of Quito's historic center; the National Congress enacted, in December of the same year, a law creating the Cultural Heritage Rescue Fund (FONSAL), funded initially with 10% of the resources of the National Emergencies Fund (FONEN)—equivalent, in the first year, to six million dollars—and with the transfer of 3% of the price of all tickets sold for public shows held in the city, as well as national and foreign donations. Subsequently, the transfer to FONSAL was defined as a percentage of the Income Tax of all taxpayers living in the city of Quito. FONSAL existed until 2010, when it was renamed the Metropolitan Heritage Institute (IMP), a name it still retains today. Due to its administrative autonomy, which allowed the continuity of actions and projects even with the changes in municipal management, and also due to the exceptional budget compared to other Latin American cities, FONSAL and the historic center of Quito quickly became a paradigm of urban historical sites management in the region.

heritage with the housing demand, in addition to enable to the reestablishment of the property social function.

In the 1990s, the Municipality of Montevideo has developed four pilot programs[7] to demonstrate the feasibility of mutual aid cooperatives in architectural recycling experiments. One of these experiences was the "Las Bóvedas" program, started in 1993, under the shared responsibility of the Old City Housing Cooperative (COVICIVI) and the Technical Assistance Institute Hacer-Desur (Vallés, 2008). Within the scope of technical assistance, the COVICIVI has had to overcome some challenges to formulate a working methodology compatible with the introduction of this system in the rehabilitation of pre-existing properties. It was necessary to make adaptations in relation to the model guided by the construction of new buildings. In relation to these processes:

> [...] was a challenge for all actors. For the inhabitants it meant a change in the collective imaginary of traditional private housing of the cooperatives: single family in open set, with shared free spaces; with front and/or back courtyard, as well as the construction materials and technologies typicals of this system. Recycling existing buildings was equivalent to modifying the typology of housing and therefore the common spaces obtained. For the technicians, the challenge of designing and managing a work in which the builders do not have previous experience, is even greater, because they are pre-existing buildings with heritage value (Vallés et al., 2011).

The construction, inaugurated in 1998, combined the recycling of existing buildings and new constructions in an emblematic sector of the historic center, totaling 34 housing units and recomposing the urban morphology of the block where it was deployed, in addition to creating collective spaces for the complex inside and on the terrace (Vallés, 2008).

10.5.5 The New Alternatives Program in Rio de Janeiro

In Rio de Janeiro, during the first administration of Mayor César Maia (1993–1997), the Municipality created a series of social urbanism programs that became a reference, the most famous of them being the *Favela Bairro*, created in 1994 and focused on the implementation of urban infrastructure, services, public equipment, and social policies in the favelas, aiming to integrate them into the formal city. The same creators of this program—architects Luiz Paulo Conde and Sérgio Magalhães, then at the head of the city's urban planning and housing departments—created, in 1996, the New Alternatives Program (PNA), intended for the refurbishment of residential buildings, initially throughout the city and, in a second moment, concentrated in the central area neighborhoods. Among the PNA's lines of actions were the

---

[7] The "Mujefa" Program, a cooperative made up of female heads of households; the "Pretyl" Program, recycling by the Civil Association "Pretyl"; the "Goes" Program, carried out by the Cooperative COVIGOES I; and the "Las Bóvedas" Program, which will be discussed further in this article (See Vallés, 2008, p. 156)

rehabilitation of old proletarian villages, the renovation of tenements and the restoration of ruined buildings. The PNA continued during the administration of the mayor-architect Luiz Paulo Conde (1997–2001) and in César Maia's second tenure (2001–2005), and between 1998 and 2005 delivered 119 housing units distributed in rehabilitated or new buildings in the central area of Rio de Janeiro. According to João Carlos Monteiro (2015, p. 447),

> Despite the insignificant number of houses produced, the PNA is frequently cited as a successful experience of incorporating the housing component into urban "revitalization" projects, having served as an inspiration for such actions in other Brazilian cities.

The first projects carried out under the PNA consisted of the recovery of tenements, especially the oldest in Brazil, located at Rua Senador Pompeu, n° 34. The building, about 120 years old and declared cultural heritage by the Municipality, was recovered in 2003, with 23 one-bedroom housing units.

10.6 Social housing in Latin American Historic Centers (2000–2020)

Intervention experiments aimed at adapting old buildings located in Latin American historic centers to social housing have intensified over the last 20 years; although they remain exceptions in quantitative terms, represent interesting and diverse examples of project and management challenges.

The projects that will be analyzed have been implemented in contexts as distinct as the historic centers of Mexico City, João Pessoa, Lima, Montevideo, Quito and Salvador, as well as São Paulo's central area (see Fig. 10.1). The last one, although not being a "historic center" in the traditional sense, concentrates a high amount of individually listed buildings, including many office towers built between the 1940s and the 1970s that pose other design challenges in the face of its conversion into housing.

10.6.1 Quito: From Esquina de San Blas to the Old Hotel Colonial

In Quito, the Historic Center Company[8] carried out, between 2000 and 2008, 10 rehabilitation or new buildings for housing purposes, totaling 390 apartments. The financing was carried out through public funds and loans offered to buyers. The largest projects are "*Esquina de San Blas,*" with 103 units, and "*Camino Real,*"

[8] The Historic Center Company was created by the Municipality of Quito with mixed capitals to run the historic center rehabilitation program

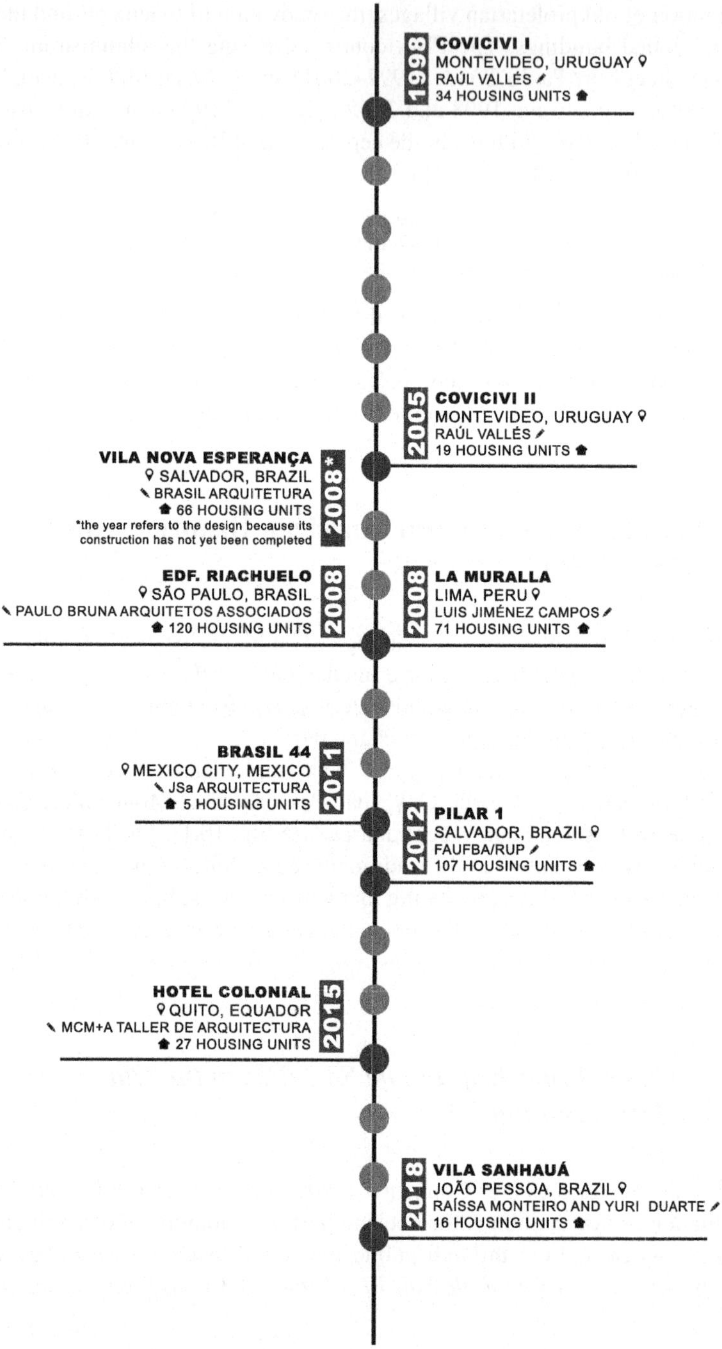

Fig. 10.1 Social housing projects analyzed in this chapter

Fig. 10.2 Panoramic view of the *Hotel Colonial*. Source: *MCM+A Taller de Arquitectura* Collection. Credits: Sebastián Crespo (Photographer)

with 95 units, both located at the northern end of the Ecuadorian capital's historic center (Del Pino, 2010, p. 38).

More recently, between 2012 and 2015, the Metropolitan Heritage Institute of Quito, in cooperation with the Regional Government of Andalusia, restored and adapted in social housing the old *Hotel Colonial*, converted into 27 apartments, 11 of them in the block built in the 1930s and 16 others in the block built in the 1960s and in its extension (see Fig. 10.2). The architectural design by the office *MCM+A Taller de Arquitectura* was chosen through public competition and was recognized with the National Award in the category Rehabilitation and Recycling in the Pan American Architecture Biennale of Quito, in 2016. Despite this, for various reasons, the rehabilitated architectural complex has not yet been occupied, 6 years after the completion of the works.

10.6.2 The Housing Regeneration Plan and Brasil 44 in Mexico City

In Mexico City, the Federal District Government, through the Federal District Housing Institute (Invi) and the Office for Urban Development and Housing (Seduvi), in partnership with the Regional Government of Andalusia, has developed

Fig. 10.3 *Brasil 44*'s facade after the intervention. Source: *JSa Arquitectura* Collection

a housing regeneration plan for the historic center, with the aim of reversing the population loss process, resulting from changes in land use, the physical degradation of buildings, and the serious earthquake of 1985, that destroyed several buildings. This process resulted in the loss of 118,609 inhabitants in the historic center between 1970 and 1995. Add to this that the area daily receives a floating population of approximately 1.2 million people, generating conflicts about interest for the occupation and use of public and private land. The historic center vitality makes it an attractive location for the vulnerable population of the city, leading them to the occupation of abandoned or ruined buildings (Suárez Pareyón, 2004).

Among the social housing projects carried out as part of the housing regeneration plan of the historic center of Mexico City, it is worth highlighting the building located at number 44 of República de Brasil Street, less than 600 meters from the Metropolitan Cathedral and Zócalo, which used to house five low-income families and a wedding dress shop (see Figs. 10.3 and 10.4). The project was carried out between 2006 and 2011 after these old tenants had acquired the property through a government credit. According to the architect Javier Sánchez, from JS[a] Architectural Studio,

> Using minimal resources, the program sought the flexibility to allow the users to help adapt a 65 m^2 space according to their individual needs. The ceiling heights were used to con-

Fig. 10.4 Detail of a *Brasil 44*'s apartment. Source: *JSa Arquitectura* Collection

struct mezzanine levels in both public and private areas, together with the use of common areas such as the roof terrace, passageways and central courtyard, elements that recovered their original dimensions to provide improved illumination. (apud Montaner & Adriá, 2013, p. 140).

10.6.3 COVICIVI II and Other Cooperative Experiences in the Old City of Montevideo

In the last 20 years, the historic center of Montevideo, known as the Old City, has hosted new interventions of preexisting buildings recycling in social housing, promoted by cooperatives. In the same block of COVICIVI I (see Figs. 10.5, 10.6, and 10.7) and designed by the same architect Raúl Vallés, the Old City Housing Cooperative carried out, between 2001 and 2005, a second recycling project, COVICIVI II (see Figs. 10.8 and 10.9), which stands out for having been the first intervention in a mutual aid system in a protected building: the house of Brigadier-General Bernardo Lecocq, built in 1795 and listed in 1975 (Vallés, 2015).

Lecocq house is an example of the civil architecture of the colonial period and integrates one of the most important historical complexes of the city, located in front of the port area. The building was in an advanced state of degradation, but still retained its stylistic, typological, and spatial characteristics, which allowed the

Fig. 10.5 COVICIVI 1's facade after the intervention. Source: Architect Raúl Vallés Collection

Fig. 10.6 Detail of COVICIVI 1's courtyard. Source: Architect Raúl Vallés Collection

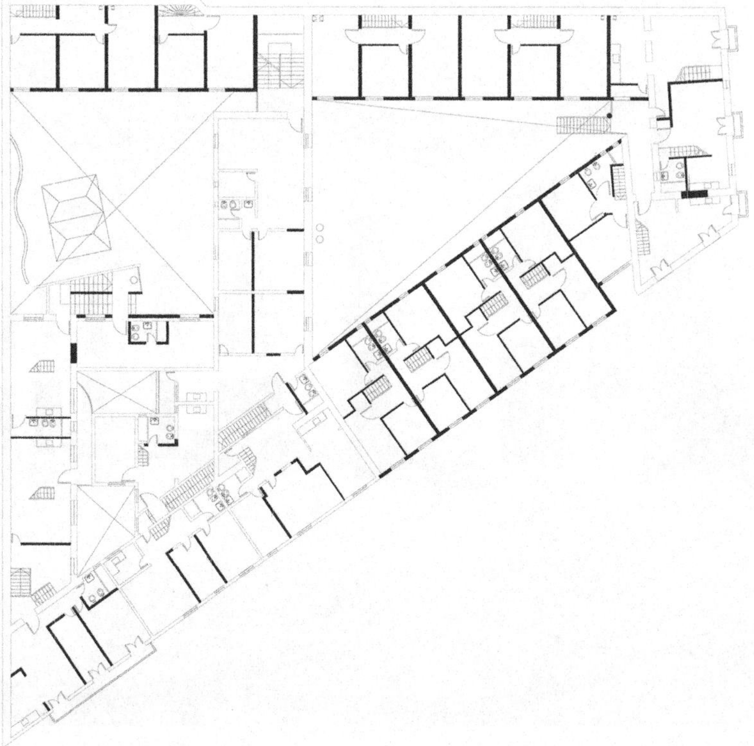

Fig. 10.7 COVICIVI 1's ground floor plan. Source: Architect Raúl Vallés Collection

restoration of the facades and the maintenance of the central courtyard. Internally, the floors were redivided for better use in the distribution of the units. The complex also includes collective spaces on the ground floor that connect the two blocks, as well as commercial points with direct and independent access to the street, which facilitate integration with the external environment and the square (Vallés et al., 2011). A volume inside the complex was demolished due to the critical state of its structure, which enabled the construction of a new block, with four floors. In total, the COVICIVI II houses 19 apartments, with an average of 67 m^2 each, 11 in the existing building and 8 in the new building (Vallés & Castillo, 2015, p. 119).

The construction was executed in a mixed technique, associating the pre-existing wall structure in self-supporting masonry with the insertion of prefabricated parts and restoration procedures.

> We must highlight a constant search for constructive rationality calling for light prefabrication, almost on an industrial scale, that the cooperative movement adopted from an adequate technical assistance and that potentiates the supply of unskilled labor, organized by the mutual aid system. (Vallés, 2015, p. 18).

In the following years, less than a kilometer from COVICIVI II and in front of the Porto Market, the emblematic Jaureguiberry Building, built in 1911 and characterized by its mansardas and enclosed galleries along the facades was rehabilitated

Fig. 10.8 COVICIVI 2's facade after the intervention. Source: Architect Raúl Vallés Collection

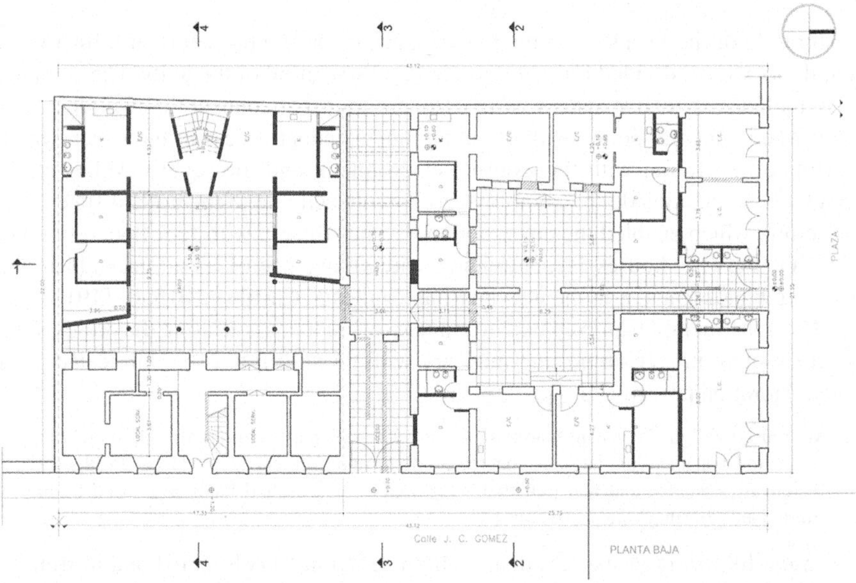

Fig. 10.9 COVICIVI 2's ground floor plan. Source: Architect Raúl Vallés Collection

by the cooperative *El Resorte*, formed predominantly by architects and architecture students.

The rehabilitation of the Jaureguiberry Building, executed between 2004 and 2009 and designed by the architects R. Béhèran, A. Mazzini, E. Mazzini, and O. Otero, was financed by the Municipality of Montevideo and the Provincial Deputation of Barcelona, with the support of the Ministry of Housing, Territorial Planning and Environment. The building now houses 16 units, with duplex apartments on the third and fourth floors. Taking advantage of the location in one of the main tourist attractions of historic center, the ground floor is occupied exclusively by commercial activities. Although it is an example of housing rehabilitation for people of a slightly higher economic level than the other projects, it demonstrates the vitality of the architectural recycling experiences conducted by cooperatives in the historic center of Montevideo.

Southeast of the historic center, in Barrio Sur neighborhood, where the Afro-Uruguayan community traditionally lives, an old abandoned industrial building was rehabilitated between 2000 and 2010 by an Afro-descendent female family heads belonging to the *Mundo Afro* organization. The *Ufama al Sur* housing complex was designed by architects Gonzalo Morel and Guillermo Rey with 36 housing units of a single room but with the possibility of extension: 22 may become two-bedroom apartments and 6 may be extended up to three bedrooms. The complex also includes a civic center and sports facilities.

The Irupé Building (2004–2005), in turn, corresponds to a new housing complex, built in a cooperative regime on a small site (220 m^2) in Montevideo historic center, in front of Solis Theatre, one of the most important monuments of the city. The building, designed by the architects Gonzalo Guevara and Rodolfo Schwedt with assistance from the institute ECO, houses 17 units, being 12 of one room and 15 of two rooms, besides two communal halls, one on the ground floor and another in the basement, open to a courtyard where the remains of the historical wall of the citadel, discovered during the excavation, can be appreciated (Vallés & Castillo, 2015).

10.6.4 La Muralla Housing Complex in Lima

In the historic center of Lima, listed as a UNESCO World Heritage Site in 1991, the La Muralla housing complex was built between 2005 and 2008, next to the Government Palace and 150 meters from the Plaza de Armas (see Fig. 10.10). The name refers to a historical wall discovered on the site during archaeological inspections and denotes the site importance within Lima's urban context (Jiménez Campos, 2021). Despite the privileged location, the existing buildings of the block chosen for the complex implementation were in a very degraded condition, inserted in a context of economic decay and social vulnerability of the residents.

Fig. 10.10 Insertion of the La Muralla Housing Complex in the pre-existing urban context. Source: Chapter's author Collection

As in other Latin American metropolises, the Historic Center of Lima also underwent a process of physical degradation and functional emptying, starting in the 1950s and 1960s (Jiménez Campos, 2021). Faced with the critical situation in which it was, in the early 2000s, the Municipality launched the "First Program for the Urban Renewal of the Historic Center of Lima" with the intention of identifying the areas that needed improvement and carrying out urban rehabilitation projects at the central area. The initiative to conceive the La Muralla Housing Complex arised from this programme (Jiménez Campos, 2020).

To enable the implementation of the Program, in 2003, the municipal government sanctioned a new legal framework based on the Law of Promotion of Private Investment in Urban Renewal Actions, on the creation of the Metropolitan Fund for Urban Renewal and Development (FOMUR), as their financing agent, and on the foundation of Municipal Real Estate Company of Lima (Emilima), agency responsible for managing the Programme. The organization of this new legal structure demonstrates the Municipality's intention to carry out urban renewal actions, which made it possible to finance and promote large-scale and important initiatives in this way.

Inside the block, there were seven old buildings listed by the National Institute of Culture (INC), whose structures were in critical condition, and which had been transformed into tenements (Jiménez Campos, 2020). The buildings typology—organized around internal courtyards (see Fig. 10.11)—and the existence of an

Fig. 10.11 Panoramic view of the La Muralla Housing Complex's new housing blocks. Source: Chapter's author Collection

empty lot—a residual space left by a demolition—had enabled the installation of another informal occupation model inside the block: houses built in an improvised way in mudbrick and wattle-and-daub, in extremely precarious conditions of habitability and health and devoided of urban infrastructure systems and basic sanitation (Jiménez Campos, 2021).

In parallel to the demolition and restoration activities, started in August 2004, the Municipality developed a great work of social assistance, integration with the community and training of residents, building a relationship of trust with the beneficiaries of the project. Social work has integrated the central objective of the renovation plan: to improve people's quality of life through access to decent housing, skills, and employment, as well as to rehabilitate the deteriorated urban heritage, through the recovery of the block and the restoration of examples of the built cultural heritage in Lima.

The participatory model was implemented through the creation of a Management Committee, composed by representatives of the residents and the governing agency (Emilima), and the legal organization of the local community around the "*Rastro de San Francisco*" Association. The strategy implemented by Emilima's management was highly successful with the incorporation of residents and their cultural practices in the development of the Program, bringing it credibility and sustainability and building a relationship of trust with them, particularly during the period in which they were transferred to a temporary shelter for the works execution.

This model reflects the understanding that public policies aimed at central historic areas need to overcome the paradigm of urban rehabilitation based only on the material dimension of the space, to the detriment of the social and cultural dimensions. Such policies should combine initiatives of morphological and landscape restoration and rehabilitation of buildings to mechanisms that favor the permanence of people in their traditional territories and the valorization of cultural expressions and meanings practiced in places, to thus promote the preservation of the neighborhood's memory and affirmation of the population's identity along with quality of life and social transformation.

In this sense, a training program to residents for services related to civil construction, restoration, carpentry, etc., with the intention of taking advantage of them to work in the construction of their own houses and in the restoration works of the old buildings. Informal workers, many in extreme poverty, the residents had the opportunity to qualify themselves and achieve conditions to gain space in the work market and enable a way to pay for the property financing installments. The author of the project, the architect Luis Jiménez Campos (2020, p. 252), highlights the importance given to the figure of women in social inclusion and empowerment program:

> On the other hand, the inclusion of women as a fundamental part of the program was emphasized, and the training processes were incorporated for their inclusion in the work related to the project. The protagonism assumed has empowered the women of La Muralla in its organization and administration.

The architectural project, chose by a public competition, seeks to establish relations of unity and identity between the pre-existing building fragments, the surrounding historical houses, and the new architecture. The project establishes a harmonious dialogue with the historical surroundings and reinterprets striking elements of the pre-existing houses—as the scale and the high of the buildings, the rhythm of the facades, the striking colors, the eclectic language of preexistence, the balconies, etc.—in contemporary architectural language, which is affirmed by the absence of decorative elements, the apparently random distribution order of the openings in the facades—but which strictly corresponds to a pre-established pattern—and in the use of concrete walkways that connect the spaces of circulation and conviviality created around the residential pavilions.

The La Muralla Complex benefits 71 families with housing units distributed in duplex apartments, with three bedrooms and area of 82 m^2 (see Fig. 10.12), and triplex apartments, with five bedrooms and area of approximately 90 m^2 (see Fig. 10.13). Each unit had a maximum estimated cost of around $10,000 and were subsidized through the *Techo Propio* program, through a credit offered by Lima Metropolitan Bank. The project also includes a multipurpose room, four commercial points, a place for tourist use, a municipal guard post and an amphitheater, besides to the housing blocks (Jiménez Campos, 2020).

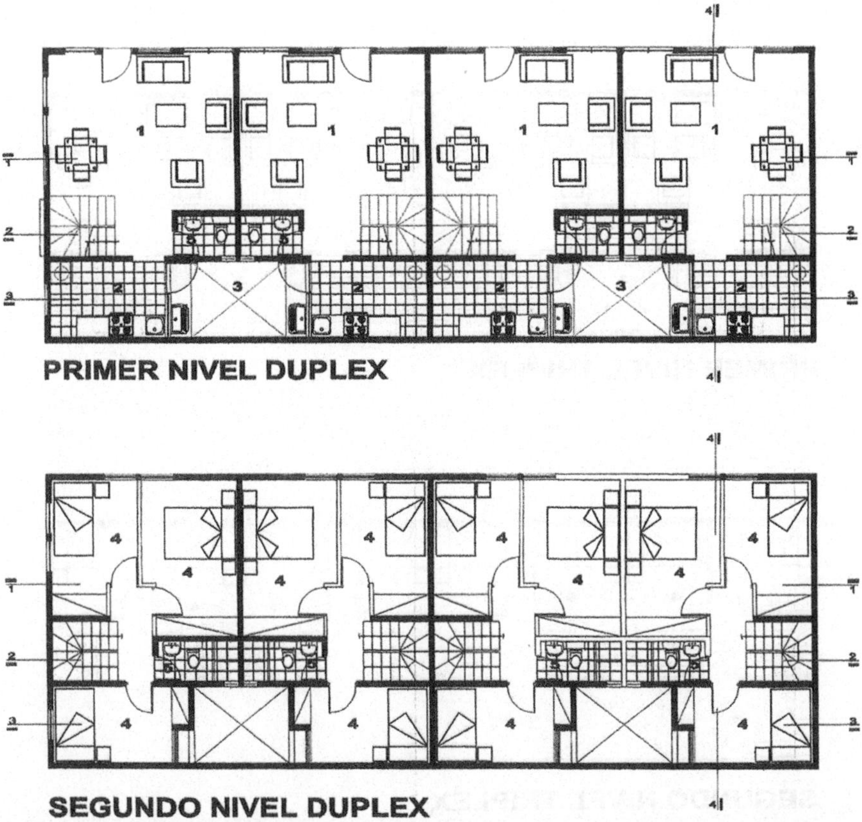

Fig. 10.12 Duplex apartments floor plans of La Muralla Housing Complex. Source: Architect Luis Jiménez Campos Collection

10.6.5 Social Conflicts and Abandoned Projects in Salvador

After the implementation of the first six stages of the "Recovery Plan for the Historic Center of Salvador" (1992–1997), focusing on cultural and leisure activities aimed at tourists and which resulted in the expulsion of almost three thousand families, the program was forced to incorporate the low-income population resident in the area: in the seventh stage of the Plan, 103 families demanded the right to remain in the area, through a Direct Action of Unconstitutionality (Adin) filed by the Association of Residents and Friends of the Historic Center of Salvador (Amach). Faced with this demand, the Public Prosecutor's Office filed a Public Civil Action that, through a Term of Conduct Adjustment (TAC), signed with the Government of the State of Bahia in 2005, managed to guarantee the permanence of those families that wished to remain in the historic center, through the Social Housing Program (PHIS), with resources from the Ministry of Cities initially estimated at almost R$ six million

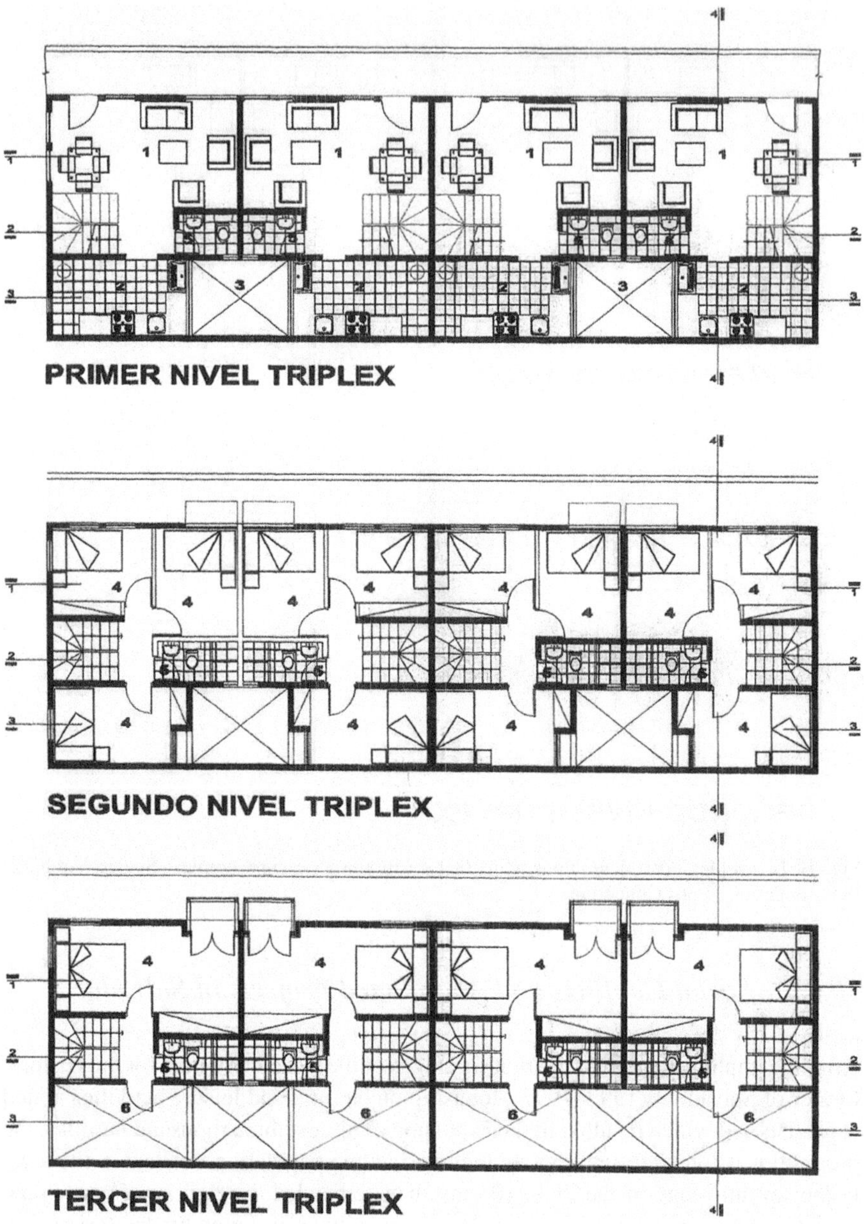

Fig. 10.13 Triplex apartments floor plans of La Muralla Housing Complex. Source: Architect Luis Jiménez Campos Collection

(Bahia, 2010). Fifteen years after the signing of the TAC, almost a third of the 103 families benefiting from PHIS have not yet received their permanent residences and continue to live in precarious "temporary houses."[9]

While the seventh Stage interventions began to be carried out in the historic center of Salvador, two other State Government initiatives were conceived aimed at recovering historic townhouses and adapting them to social housing. In the Santo Antônio Além do Carmo neighborhood, on the north end of the historic center, the government of the State of Bahia, within the scope of the Caixa Econômica Federal Residential Rental Program (PAR), with the support of UNESCO and the Inter-American Development Bank (IDB) resources, recovered some properties in ruins that were transformed into residential complexes, preserving the façades and reconstructing the external volumetric reading of the buildings, based on a project by the architect Demetre Anastassakis. Of the various properties contemplated in the project, only a few were restored, such as the ruins of an old townhouse covered with azulejos at Rua Direita do Santo Antonio, n° 19.

Another contemporary initiative to the seventh Stage is the Pilar Urban Rehabilitation Plan (RUP), which results from an agreement between the Government of the State of Bahia and the Faculty of Architecture of the Federal University of Bahia (FAUFBA), signed in 2003. It included some empty lots and, mainly, dozens of townhouses in an advanced state of degradation located on the streets of Pilar, Julião and Caminho Novo do Taboão, at the foot of the slope that separates the Lower and Upper Cities, on the northern limit of the historic center of Salvador.[10]

Many of the townhouses had been occupied and transformed into tenements, and the main objective identified by the Plan, which guided the architectural projects prepared between 2003 and 2008, was to promote housing for the population already residing in the area, qualifying the properties in terms of habitability. The first project developed within the scope of the RUP corresponded to the adaptation and expansion of the former State Ice Factory, a robust-reinforced concrete structure erected in the 1940s between Jequitaia avenue, the hillside and the access bridge to the tunnel connecting to the Upper City. The old structure was supposed to be adapted to house 107 families residing in irregular buildings erected a few meters away, at the foot of the Ladeira do Pilar. The majority (73%) of these families had an income of less than half the minimum wage and their transfer was urgent, either because of the risk situation in which they found themselves, due to the constant threats of landslides, or due to the impact that this occupation had on the landscape of the listed site.

---

[9] The analysis of the housing projects carried out in the historic center of Salvador is based on the deeper one previously developed by Nivaldo Andrade Junior (2015, 2021).

[10] The Agreement was signed as a consequence of the International Design Seminar: Urban Renewal and Town Culture, promoted in 2001 at Faufba, with the participation of professors and students from Brazilian, Italian and Chilean universities and sponsored by the Government of the State of Bahia, having the areas between the streets of Pilar and Julião and Avenida Jequitaia as an intervention area. Just after the agreement was signed, in 2003, the Pilar Urban Requalification Laboratory (RUP) was created, within the scope of Faufba, to elaborate the plan and a series of architectural projects for the area. Both the seminar and the RUP were conceived and coordinated by Profa. Esterzilda Berenstein de Azevedo.

Fig. 10.14 Panoramic view of the Pilar 1 Housing Complex. Source: RUP/FAUFBA Collection

Fig. 10.15 Detail of the access staircase to the apartments that connects the two housing blocks—Pilar 1 Housing Complex. Source: RUP/FAUFBA Collection

The Pilar 1 Housing Complex (see Figs. 10.14 and 10.15) was finally opened in 2012, after a series of changes in the project, including the decision to demolish the existing concrete structure and build three *ex novo* apartment blocks. The Pilar 1 Housing Complex would be the only project prepared in the RUP/FAUFBA and

Conder agreement to be effectively built. The other projects, developed between 2005 and 2008 and corresponding to the restoration and recycling of ruined properties or to new buildings to be built on vacant lots on Julião street and Caminho Novo do Taboão, would create hundreds of housing units to shelter the resident population in the area. In addition to the projects not having been executed, many of the townhouses that should have been restored and requalified were ruined to such an extent that, in 2010, they had to be propped up, as happened with those located on Julião street, between the numbers 01 and 09. The adoption of shoring as a strategy by the preservation agencies has only delayed the collapse of these buildings, since the metal struts also degrade, and, in the absence of actions aimed at the effective recovery of the buildings, they end up transforming, after a few years, in an overload for already damaged structures.

In 2007, a new phase began in the management of the historic center of Salvador by the State Government, adopting a new policy, supposedly less focused on tourism and more on its inhabitants. To mark this new moment, the government of the State of Bahia hired the architectural office Brasil Arquitetura, settled in São Paulo by former associates of Lina Bo Bardi—among them Marcelo Ferraz, co-author of the Pilot Project of Ladeira da Misericórdia—to develop a requalification project for Vila Nova Esperança (see Figs. 10.16 and 10.17), a community located at the back of the old Faculty of Medicine of Brazil, a few meters from Terreiro de Jesus, one of the most important squares of the historic center of Salvador. Formed by 49 families, "the Vila Nova Esperança community is notably one of the last strongholds of resistance of the black population in the Historic Center," being, according to one of the residents, "the only community that remained in the entire revitalization process do Pelourinho" conducted by the State Government in the 1990s (Teixeira & Espírito Santo, 2009).

Despite having been developed with the intense participation of the community and having received the first place in the Prize of the Institute of Architects of

Fig. 10.16 Sketch of the urban insertion of the *Vila Nova Esperança* Project. Source: *Brasil Arquitetura* Collection

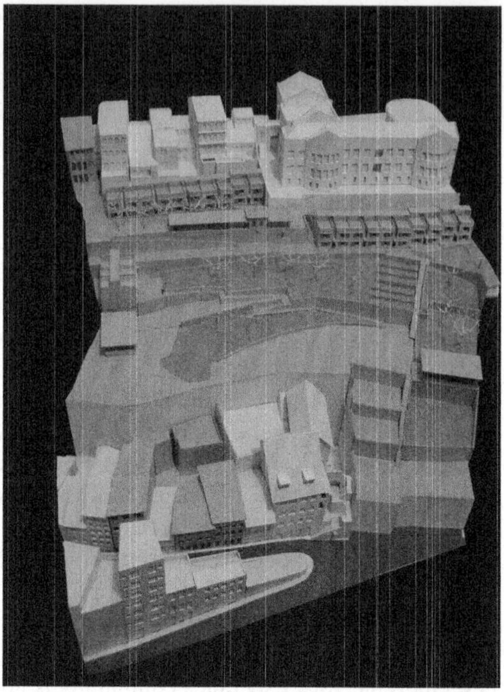

Fig. 10.17 Vila Nova Esperança Project physical model. Source: *Brasil Arquitetura* Collection

Brazil—São Paulo Department (IAB-SP) in 2008, in the category "Public Social Housing," the Brasil Arquitetura's proposal for Vila Nova Esperança had its construction started only in 2012 and the works were stopped a few months later. In 2018, a new bid was carried out to continue the execution of the works, that are expected to be finished in 2021.

10.6.6 The Villa Sanhauá Experience in João Pessoa

Located in the historic center of João Pessoa, a listed site by the Institute of Historic and Artistic Heritage of the State of Paraíba (IPHAEP) in 2004 and by the National Institute of Historic and Artistic Heritage (Iphan) in 2009, the Villa Sanhauá corresponds to the rehabilitation of a set of eight ruined detached houses carried out by the local municipal government. The project, designed by architects Raíssa Monteiro and Yuri Duarte, of the Municipal Housing Office (SEMHAB) of the Municipality of João Pessoa, under the coordination of Pascal Machado, then director of Housing Programmes, was drawn up between 2013 and 2014. The works were concluded in 2018 and included 17 housing units and, on the ground floor, commercial activities related to the creative economy, aiming to contribute to income generation for the house's maintenance.

Initially, the project would be executed through the *Minha Casa Minha Vida* Program (PMCMV), in the range of families with average income between US$ 800.00 and US$ 1600.00. It would correspond to PMCMV first experience related to the conversion of historical buildings into housing. However, PMCMV was conceived and formatted for new buildings and, therefore, did not contemplate the specificities of this type of project. The various bureaucratic and operational obstacles to the project feasibility put by the financing agent of the PCMVM, the national bank Caixa, led the Municipality to decide to fund the operation with its own resources (Machado et al., 2019, p. 142).

The Cultural Historical Assets Protection Council (CONPEC), IPHAEP's advisory agency, had classified these eight old buildings as "Partial Conservation," for preserving "part of their spatial, structural, volumetric, typological and decorative original characteristics." In this way, any intervention in the buildings should contemplate the "preservation of the original roofs and the adequacy of the changed ones to the traditional typologies," the "preservation and [...] rehabilitation of the original typological composition of the openings, doors and windows on the buildings facades" and the "repair or adaptation of the internal spatial distribution and the strictly necessary roof to improve the conditions of stability, health, habitability, ventilation and insolation" (apud Machado et al., 2019, p. 142). According to the project authors, this last determination corresponded to "an opportunity to justify interventions to enable projects like this" (Machado et al., 2019, p. 142).

Five of the eight old buildings preserved only the external walls, of the original structure, while another two also preserved their roofs and a third also had a concrete slab. In general, the remaining buildings walls were in an advanced state of degradation, presenting structural pathologies due to maintenance lack and large exposure to weather. In the facades, the *art déco* ornaments were reasonably preserved, which allowed their restoration (Machado et al., 2019) (see Fig. 10.18).

According to the project authors, the main design challenges were related to the roof preservation in its original format, usually with three faces, and to the narrow and deep old buildings adaptation in residential units with adequate ventilation and lighting conditions.

The solution found by the architects was to create an internal courtyard integrating the housing units and ensuring the natural lighting and ventilation needs (see Fig. 10.19). This continuous courtyard corresponds to the only area where the original roof, reconstructed in colonial type ceramic tiles, is interrupted, and separates each old townhouse in two housing units sets: those in the front, with openings in the main facade and in the courtyard, and those in the background, illuminated and ventilated by the courtyard and the small backyards kept on the back of the lot.

All housing units therefore have cross ventilation and have their accesses through the courtyard, which is traversed by a continuous metal walkway. This is a variation of the architectural solution adopted in the adaptation projects of some old townhouses in social housing, in the historic center of Salvador, elaborated a few years before by the RUP/FAUFBA team.

Fig. 10.18 *Vila Sanhauá's* facade after the intervention. Source: Author's Collection

Fig. 10.19 Detail of a *Vila Sanhauá's* apartment with a mezzanine. Source: Author's Collection

According to the authors of the Villa Sanhauá architectural design, the courtyard

> also creates a common access hall that integrates the apartments in a single neighborhood unit, a design strategy that rescues the sense of 'village' inside, keeping the exteriority of the houses. The courtyard and the metal walkway that crosses the townhouses, besides connecting the new neighbors, ensures the destination for gardens areas. (Machado et al., 2019, p. 143).

The window frames and balconies were also recovered according to their original design, while parts of the cement tiles that were in good condition were restored. The 17 housing units, with areas between 40 and 73 m^2, have two bedrooms, living room, kitchen, bathroom, and laundry room; one of the units was designed to house disabled people. The ground floor has six commercial spaces, with areas between 40 and 131 m^2, with direct access to the street, and a room for institutional use. Vertically, the floor levels followed pre-existing buildings, to avoid the intersection of the new floor slabs on the historic facade's windows. Because of the height of the ceiling in the upper floor apartments, it was possible to insert a mezzanine and increase the useful area of the unit (see Figs. 10.20 and 10.21).

Inaugurated in June 2018, Villa Sanhauá is under Municipality management and adopts the social leasing system, aiming to prevent the complex from becoming disputed by the real estate market and the area from going through a process of gentrification. Through this system, units are granted to beneficiaries chosen through a public call, which prioritized the selection of artists, artisans and other culture professionals who demonstrate already reside or work in the historic center of João Pessoa. In 2021, the monthly rent, partially subsidized, is only US$ 60.00, which, in an economic crisis period such as Brazil goes through, can be fundamental to ensure the permanence of residents in the historic center.

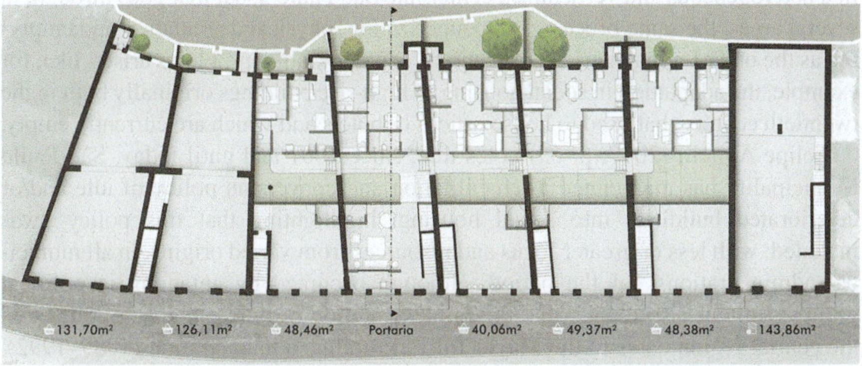

Fig. 10.20 Ground floor plan of *Vila Sanhauá*. Source: Architects Raíssa Monteiro and Yuri Duarte Collection

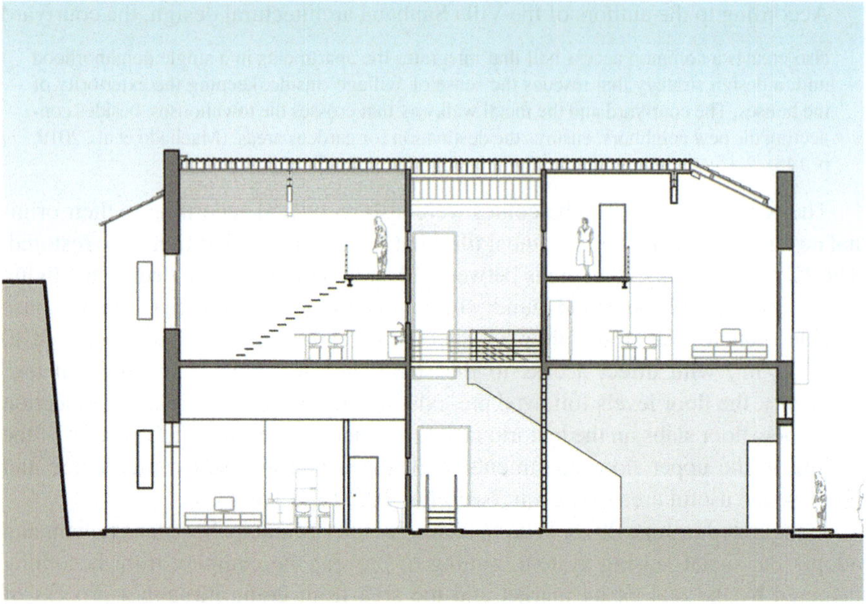

Fig. 10.21 Cross section showing the project solution for the roof, the mezzanines and courtyard—*Vila Sanhauá* Project. Source: Architects Raíssa Monteiro and Yuri Duarte Collection

10.6.7 The Challenge of Converting High-rise Office Buildings in São Paulo

In a heterogeneous and verticalized center like São Paulo's, but that goes through, in several areas, the same processes of pauperization, physical degradation and emptying as the other Latin American metropolis, another kind of problems arises, like, for example, the adaptation in social housing of high-rise buildings originally built in the twentieth century that used to house offices or hotels and which are currently empty.

Felipe Anitelli (2017, p. 65) notes that, since 2001 and until today, São Paulo Municipality has maintained a rehabilitation and conversion policy of idle and/or deteriorated buildings into social housing, highlighting that this policy "was included, with less or greater focus and resources from varied origins, in all municipal administrations" of the period, although occupied by antagonistic political groups. Anitelli also recalls that the implementation of public housing policies in the central region dates to the Mayor Luiza Erundina administration (1989–1992), as part of the Brazil's redemocratization process, after the military dictatorship. These experiences, however, were discontinued in the 1990s.

In the last 20 years, the financing social housing models in the central area of São Paulo have been the most diverse: from rent with possibility of purchase to the *Minha Casa Minha Vida* Program, in the modality Entities, with shared management with residents. Some properties had their units rented by the Social Leasing Program. According to Anitelli (2017, p. 66),

> [...] the few and discontinued rehabilitation experiences certainly made it impossible to establish methodological procedures that could be (re)applied and (re)evaluated in each new rehabilitation, thus preventing the identification of the main recurrences, problems, unforeseen, costs, waste, demands, etc. It must have also prevented the improvement of this activity and the professionals involved, both in the public and private sectors. [...] On the other hand, there are 1351 families who now live in rehabilitated buildings, a number that can more than double when (and if) all properties already expropriated or in expropriation process are rehabilitated. Over the years, the opportunities and amenities existing in the center should be increasingly present in the beneficiaries' daily lives. Finally, if this is associated with other public policies to combat social inequality, such as the *Bolsa Família* Program or facilitated access to public universities, these people can enter society under more balanced and stable conditions.

One of the most paradigmatic examples of this policy in São Paulo is the Riachuelo Building rehabilitation. Built between 1942 and 1945, with 17 floors, it originally housed offices and the social activities of the Association of Trade Employees of São Paulo. After some time abandoned, it was occupied by homeless families in the 1990s until, in the Mayor Marta Suplicy term (2000–2004), was included in the Living Downtown Program, aimed at promoting housing in the central area through a social leasing system. The rehabilitation of the building was promoted by the Metropolitan Housing Company of São Paulo (Cohab), with a project by the architectural office Paulo Bruna Arquitetos, and was completed in 2008 (see Fig. 10.22). According to the authors of the project,

Fig. 10.22 Riachuelo Building's facade after the intervention. Source: Paulo Bruna Architects Associated Collection

> The recovery sought to respect the original constructive logic [...]. The solid brick walls that separated the different offices were kept since they are part of the structure's bracing, so basically every office was turned into an apartment, in a total of 120 units, ranging from 301 square foot to 516 square foot of floor space. Many of them include private balconies and balcony doors, and those were kept because the façade has been listed as historical heritage. The apartments were eventually sold and their residents value and maintain the building in an exemplary way. (Bruna & Gouveia, 2017, p. 195).

10.7 Final Considerations

The Ecuadorian architect Fernando Carrión argues that in the city, it is up to the historic center to play the role of "public space *par excellence,*" thus being "the fundamental element of social integration and structuring of the city." He remembers that historic centers are, moreover, "civic places in which the invisible society becomes visible and in which otherness is generated" (Carrión, 2005, pp. 54–55).

The implementation of public policies that privilege tourists to local inhabitants in Latin Americans historic centers, held since the 1960s and 1970s with the support and patronage of international organizations such as UNESCO and OAS has led to the gentrification of these areas, expelling and driving away the most vulnerable, and converting the *place of otherness* into *non-places for immediate consumption*: the *ice cream city* that Lina Bo Bardi did not want the see the historic center of Salvador transformed into. Quoting Fernando Carrión (2003, p. 40) again,

> The historic centers are characterized by heterogeneity because they come from it—it was, in a big moment, the whole city and because without it they die. Social, economic and cultural heterogeneity is a condition of its existence; hence, if an homogenizing proposal is planned, it would end up losing centrality and reducing time, space and citizenship condition. A historic center intended only for certain activities, such as tourism, or converted in the habitat of poverty, will eventually become a periphery.

For many years, social housing in historic centers has been seen as a taboo theme, incompatible with tourism. After the pioneering (and abandoned) experiences in the 1980s, a serie of interventions, some of exceptional quality, started to be implemented in several Latin American historic centers in the 1990s and especially in the last two decades.

Even if these projects are almost always limited to a few dozen housing units and take many years to complete—and some are not even completed or occupied, as we have seen in this article—, they show the viability of social housing in historic centers and its compatibility with other uses and activities, including tourism. These experiences also place several specificities, in the stages of project and of construction as well as in its management aspects, which differentiate them from other social housing projects, carried out in contexts with less historical complexity or without such significant cultural values. But also differentiate them from other heritage intervention projects, considering that they must associate principles inherent to social housing projects—as constructive rationality, cost-effectiveness and durability of materials—with the conservation of architectural values of the buildings that are being intervened.

Regarding management, one question that emerges is the social location system, adopted in some of the examples analyzed, such as Villa Sanhauá, in João Pessoa, and some projects of the Living Downtown Program, in São Paulo. Experience shows that, in programmes where residents become the owners of the properties in which they live, they often may have difficult to pay all the costs involved, such as condominium taxes, water and energy bills, etc., so they can end up selling the property. Besides minimizing this impact, social leasing helps to prevent that the real estate market, in its voracity, acquire the rehabilitated properties, promoting a gentrification process that expels those whom the intervention was intended to meet. The social leasing system, however, needs to impose itself in opposition to the traditional properties system, with decades of prioritization and extremely mobilizing in countries characterized by constant economic instability—a context in which owning a roof, in a way, represents a safety guarantee in times of work shortage.

Still in the scope of management, one aspect to be highlighted is the residents' participation in the project and in the work execution, through the Uruguayan model of mutual aid cooperatives (even in individually listed buildings, as in the case of COVICIVI II, in Montevideo) whether through the residents training in services related to construction, participating directly in the building process of their own homes (as in the case of the La Muralla Complex, in Lima). In the latter case, the perspective of improving life quality with the new housing is enhanced: beyond the new home, these experiences bring the possibility of better job opportunities because of the professional qualification obtained.

About architectural projects, as noted by Raúl Vallés and Alina del Castillo,

> This action model poses new challenges to the project and management. The project logic is no longer generated by the repetition of a housing unit. The conditions of pre-existence force us to think from the whole and the projects respond more to its singularities than to preconceived types. (Vallés & Castillo, 2015, p. 72).

The analysis of social housing projects in old buildings located in Latin American historic centers allows to identify certain recurrences. For example, faced with the need to create smaller spaces with low ceilings in buildings that used to have large and high rooms, the creation of internal courtyards has been a recurring solution adopted to ensure ventilation and lighting to all rooms. The analysis of adaptation projects executed into narrow and relatively deep townhouses, originally characterized by dark rooms with no ventilation, such as those of Julião street and Caminho Novo do Taboão, in Salvador (2005–2008), and those of Villa Sanhauá, in João Pessoa (2013–2018), shows how the creation of internal courtyards is recurrent, in these cases crossed by metal walkways that guarantee access to the units to the bottom.

The experiences held in São Paulo in the last 20 years bring other design challenges, related to the conversion of obsolete high-rise commercial buildings into social housing. Projects should solve problems such as insufficient vertical circulation, reduced area of vertical surfaces facing open spaces and inadequate electrical and hydraulic installations. The urgency in implementing programs and developing projects to face this challenge is directly proportional to the number of idle office towers existing in the large Latin American cities' central areas: in Brazil, only the Federal Government has more than 10,000 vacant properties, which could, in many cases, be converted into housing (not just social housing), optimizing the use of installed infrastructure, avoiding urban sprawl and reducing displacements. And if we remember tragedies like that of the Wilton Paes de Almeida Building, in São Paulo's central area, which belonged to the Union and was occupied by 146 families after years of abandonment, and its collapse after a fire in May 2018, resulting in seven deaths, this urgency is even greater.

If, as commented by McGuirk, the gigantic modern housing complexes represented the Latin American architectural experimentation between the 1940s and the 1960s, and more recently this experimentation was identified with other approaches, as the incremental housing of Elemental, in Chile, or some interventions in the slums of Rio de Janeiro and Medellín, it is necessary to recognize that adaptation projects in historic centers carried out in recent decades in Latin America are also part of this investigative and experimental tradition.

The social inequality that characterizes Latin American countries, aggravated by the economic and social crisis, makes housing precarious for the most vulnerable strata of the population. A good part of the population of these countries resides in areas lacking basic urban infrastructure, with insufficient or non-existent sanitary sewage, inadequate paving and lighting, and scarce supply of transport and services; the dwellings do not reach the minimum parameters of habitability and healthiness and have small and/or shared rooms, with inadequate ventilation and lighting. The Covid-19 pandemic has brought to light the urgent need to expand social protection for the most vulnerables. The projects presented here, developed in the pre-pandemic period, acquire even more importance in the current context and demonstrate how good architecture, inserted in historic centers, can contribute to social justice, ensuring quality of life, creating decent housing and promoting the empowerment of communities commonly marginalized by public action.

References

Andrade Junior, N. V. (2015). "Habitação de interesse social em centros históricos: Experiências e desafios em Salvador (e um contraponto *limeño*)". In M. A. A. de Filgueiras Gomes, W. Ludeña Urquizo, (Orgs.). *Diálogos metropolitanos Lima/Salvador: processos históricos e desafios do urbanismo contemporâneo* (pp. 149–182). Salvador: Edufba.

Andrade Junior, N. V. (2021). "Current challenges and risks for preservation of the historic center of Salvador". In R. Christofoletti, & M. Olender (Orgs.). *World Heritage Patinas: actions, alerts and risks* (pp. 37–64). Cham, Suíça: Springer.

Anitelli, F. (2017). "100 Vezes Habitação Social: Edifícios reabilitados ou com potencial de reabilitação na região central de São Paulo". Cidades, Comunidades e Territórios, Lisbon, n° 35, dec. https://www.researchgate.net/publication/322280188_100_Vezes_Habitacao_Social_Edificios_reabilitados_ou_com_potencial_de_reabilitacao_na_regiao_central_de_Sao_Paulo. Accessed 10 Dec 2020.

Azevedo, R. M. (1994). "Será o novo Pelourinho um engano?" Revista do Patrimônio, Rio de Janeiro, n° 23, pp. 131–137.

Azevedo, P. O. (2001). "La lenta construcción de modelos de intervención en centros históricos americanos". In F. Carrión (Org.). Centros históricos de América Latina y el Caribe. Quito: Unesco/BID/Ministério de Cultural y Comunicación de Francia/Flacso – Sede Ecuador, pp. 297–315.

BAHIA (2010). Governo do Estado. Secretaria de Cultura, Fundação Pedro Calmon.

Baravelli, J. E. (2006). O cooperativismo uruguaio na habitação social de São Paulo: das cooperativas FUCVAM à Associação de Moradia Unidos de Vila Nova Cachoeirinha. Dissertação (Mestrado em Arquitetura e Urbanismo) – Faculdade de Arquitetura e Urbanismo, Universidade de São Paulo, São Paulo.

Bardi, L. B. (2008). Lina Bo Bardi. São Paulo: Instituto Lina Bo e P.M. Bardi.

Bruna, P. J. V., & Gouveia, S. M. M. (2017). "Requalificação do Edifício Riachuelo no Centro Histórico de São Paulo". Revista CPC, São Paulo, n. 22, pp. 193–216. https://www.revistas.usp.br/cpc/article/view/119239/127831. Accessed 15 June 2021.

Cañellas, E., Colella, V., Da Representaçao, N. (2008). "Patrimonio urbano y vivienda social en el centro de Buenos Aires. Nuevas perspectivas y conflictos". In C. Torti, & J. Piovani (Comps.): V Jornadas de Sociología de la UNLP: Cambios y continuidades sociales y políticas en Argentina y la región en las últimas décadas. Desafíos para el conocimiento social. La Plata: Universidad de La Plata, 2008. http://www.memoria.fahce.unlp.edu.ar/trab_eventos/ev.5954/ev.5954.pdf. Accessed 15 May 2021.

Carrión, F. (2003). "Ciudad y centros históricos: centros históricos y actores patrimoniales". In: M. Balco, et al. La ciudad inclusiva, cuadernos de la CEPAL, 88, Santiago, CEPAL/Naciones Unidas/Cooperación Italiana, pp. 129–152.

Carrión, F. (2005). "El centro histórico como objeto de deseo". In F. Carrión, & L. Hanley, Lisa (Eds.). Regeneración y revitalización urbana en las Américas: hacia un Estado estable. Quito: Flacso-WWICS-Usaid, pp. 35–57.

Cecilio, M. (2015). "El cooperativismo de viviendas. Proceso de gestación". In R. Vallés, & A. del Castillo (Eds.). Cooperativas de vivienda en Uruguay. Medio siglo de experiências. Montevideo: Facultad de Arquitectura de la Universidad de la República, pp. 23–36.

Cervellati, P. L., & Scannavini, R. (1976). *Bolonia: politica y metodología de la restauración de centros históricos*. Gustavo Gili.

Cervellati, P. L., Scannavini, R., & De Angelis, C. (1981). *La nouvelle culture urbaine: Bologne face à son patrimoine*. Éditions du Seuil.

Del Pino, M. I. (2010). Centro histórico de Quito. Una centralidad urbana hacia el turismo. Quito: Flacso Ecuador: Ediciones Abya-Yala.

Gordilho Souza, A. (2010). "Ocupação Urbana e Habitação". In Bahia. Governo do Estado. Secretaria de Cultura. Escritório de Referência do Centro Antigo. Unesco. Centro Antigo de Salvador: Plano de Reabilitação Participativo. Salvador: Secretaria de Cultura do Estado da Bahia, Fundação Pedro Calmon, pp. 72–101.

Guevara, T. A. (2011). "¿La ciudad para quién? Políticas habitacionales y renovación urbana en la Boca". Cuaderno Urbano. Espacio, cultura, sociedad, núm. 11, pp. 81–99, Universidad Nacional del Nordeste. https://www.redalyc.org/jatsRepo/3692/369236772004/html/index.html. Accessed 20 June 2021.

ICOMOS. (1967). International Council on Monuments and Sites. The Norms of Quito – 1967. Final Report of the Meeting on the Preservation and Utilization of Monuments and Sites of Artistic and Historical Value held in Quito, Ecuador, from November 29 to December 2, 1967. Paris: ICOMOS, 1967. https://www.icomos.org/en/resources/charters-and-texts/179-articles-en-francais/ressources/charters-and-standards/168-the-norms-of-quito. Accessed 19 May 2021.

Iglesias, E. (2008). "Organización social y gestión cooperativa". In Una historia con quince mil protagonistas. Las Cooperativas de vivienda por ayuda mutua uruguayas. Montevideo: Intendencia Municipal de Montevideo; Junta de Andalucía/Consejería de Obras Públicas y

Transporte / Dirección General de Arquitectura y Vivienda; Agencia Española de Cooperación, pp. 44–53.

Jiménez Campos, L. (2020). "Renovación urbana en el Centro Histórico de Lima: una apuesta por la vivienda social". In W. Ludenã Urquizo, & J. C. Huapaya Espinoza (Orgs.). Territorios, ciudades y arquitectura sur-sur: procesos históricos y desafíos, Diálogos Metropolitanos Lima | Salvador. Lima, Salvador: Fondo Editorial de la Pontificia Universidad Católica del Perú; Centro de Investigación de la Arquitectura y la Ciudad; Programa de Pós-Graduação em Arquitetura e Urbanismo da Universidade Federal da Bahia, pp. 239–255.

Jiménez Campos, L. (2021). "Reabilitação Urbana no Centro Histórico de Lima: uma aposta pela habitação social". In: Ciclo de palestras da disciplina Sítios Históricos Urbanos: análise, gestão e intervenção – aula 9 do Programa de Pós-Graduação em Arquitetura e Urbanismo da Universidade Federal da Bahia, 6 May 2021. https://www.youtube.com/watch?v=_5faGOBIBVc&t=5580s. Accessed 18 June 2021.

Land, P. (2015). The Experimental Housing Project [PREVI], Lima. Design and technology in a new neighborhood. Bogotá: Universidad de los Andes, Facultad de Arquitectura y Diseño, Departamento de Arquitectura, Editora Uniandes.

Lopes, J. M. A. (2018). "Nós, os arquitetos dos sem-teto". Revista Brasileira de Estudos Urbanos e Regionais, São Paulo, v. 20, n° 2, pp. 237–253, May–Aug.

Machado, P., Lopes, Y. D., & Monteiro, R. G. (2019). "Reabilitar para re-habitar: a trajetória dos oito casarões da Rua João Sussuna". In A. S. de Moura, P. F. de Oliveira Rossi (Orgs.). Morar, viver e (re)existir nas cidades. Cabedelo, PB: Editora IESP, pp. 138–146.

McGuirk, J. (2014). Radical cities: across Latin American in search of a new architecture. London: New York: Verso.

Montaner, J. M., & Adriá, M. (Textos) (2013). JSa. Encajes Urbanos | Urban Interlacing. Javier Sánchez 2004–2013. México, D.F.: Arquine: Consejo Nacional para la Cultura y las Artes.

Monteiro, J. C. C. S. (2015). "Habitação de interesse social em cenários de revalorização urbana: considerações a partir da experiência carioca". Cadernos Metrópole, São Paulo, v. 17, n° 34, pp 441–459, Nov. https://www.scielo.br/j/cm/a/pzRGHvm4bKnTLYZWtH9X9Pj/?format=pdf&lang=pt. Accessed 15 May 2021.

Municipalidad de Buenos Aires. (1991). Rehabilitación Manzana de San Francisco. Proyecto y Gestión. Buenos Aires: Municipalidad de la Ciudad de Buenos Aires; Agencia Española de Cooperación Internacional; Junta de Andalucía/Consejería de Obras Públicas y Transportes/ Dirección General de Arquitectura y Vivienda.

Parent, M. (2008). "Proteção e valorização do patrimônio cultural brasileiro no âmbito do desenvolvimento turístico e econômico". In C. F. B. Leal, (Org.). As missões da Unesco no Brasil: Michel Parent. Rio de Janeiro: IPHAN, pp. 35–191.

Quito. Distrito Metropolitano de Quito. (2000). El Fondo de Salvamento del Patrimonio Cultural, 1996–2000. Quito: Distrito Metropolitano de Quito – Fonsal.

Quito. Ilustre Municipio de Quito. (1992). El Fondo de Salvamento. Ilustre Municipio de Quito, 1988–1992. Quito: Ilustre Municipio de Quito – Fondo de Salvamento.

Suárez Pareyón, A. (2004). "El centro histórico de la Ciudad de México al inicio del siglo XXI". Revista Invi, Santiago, Chile, v. 19, n° 51, p. 75–95, Aug. https://revistainvi.uchile.cl/index.php/INVI/article/view/61927/66259. Accessed 15 May 2021.

Teixeira, N. A., Espírito Santo, M. T. G. (2009). "A ZEIS de Vila Nova Esperança: habitação de interesse social no Centro Histórico de Salvador (Pelourinho/BA)". Revista VeraCidade, Salvador, vol 8, pp. 21–35, Mar.

Thomasz, A. G. (2016). "Los nuevos distritos creativos de la Ciudad de Buenos Aires: la conversión del barrio de La Boca en el 'Distrito de las Artes'". EURE (Santiago), vol. 42, n° 126, Santiago, May. https://scielo.conicyt.cl/scielo.php?script=sci_arttext&pid=S0250-71612016000200007. Accessed 10 May 2021.

Vallés, R. (2008). "Reciclajes". In: Una historia con quince mil protagonistas. Las Cooperativas de vivienda por ayuda mutua uruguayas. Montevideo: Intendencia Municipal de Montevideo; Junta de Andalucía / Consejería de Obras Públicas y Transporte / Dirección General de Arquitectura y Vivienda; Agencia Española de Cooperación, pp. 152–165.

Vallés, R. (2015). Una mirada al sistema cooperativo de viviendas en Uruguay. In R. Vallés & A. de Castillo (Eds.), *Cooperativas de vivienda en Uruguay. Medio siglo de experiências* (pp. 15–22). Facultad de Arquitectura de la Universidad de la República.

Vallés, R., Árias, D., & Solanas, M. (2011) "Habitar el patrimonio. La cooperativa de vivienda por ayuda mutua como medio de gestión y desarrollo edilicio y urbano". Revista América Patrimonio, Santiago de Chile, pp. 29–42.

Vallés, R., & Castillo, A. (2015). Cooperativas de vivienda en Uruguay. Medio siglo de experiências. Montevideo: Facultad de Arquitectura de la Universidad de la República.

Chapter 11
Designing for Vulnerabilities. The Definition of Public Spaces as a Strategy for a Sustainable Renovation of Beijing Heritage Sites

Eugenio Mangi and Hiroyuki Shinohara

11.1 Old Beijing's Vulnerabilities

The conservation and renovation of heritage sites in China is a topic that has sparked several debates and reflections among scholars, policymakers, and practitioners since the first half of the twentieth century (Xie et al., 2020)[1]. The capital city of Beijing, due to its political, cultural, and symbolic roles in the country, and thanks to the extension and relevance of its inhabited heritage sites, has always been at the center of this century-long debate. To a certain extent, it can be said that this is the city where the political forces of the Maoist era first (Tao, 2016), and the pressures of the massive economic development after 1978 (Junhua 1997; Fang & Zhang, 2003) crushed the weak resistance of the heritage conservation and community cohesion principles, causing a series of fragilities that are still affecting both the built and the social dimensions. After the foundation of the People's Republic of China in 1949, Beijing's historical residential areas have undergone several physical and social transformations that can be roughly divided into three main periods (Yu, 2017).

[1] In this regard, the documentation and investigation work about the traditional Chinese architecture that was conducted by Liang Sicheng (1901–1972) and his wife Lin Huiyin (1904–1955) between 1920 and 1940 (Liang, 2014; Lin, 2011) is exemplary.

E. Mangi (✉)
Department of Architecture and Built Environment, University of Nottingham Ningbo China, Ningbo, China
e-mail: eugenio.mangi@nottingham.edu.cn

H. Shinohara
Chinese University of Hong Kong, Hong Kong, China

E. Giorgi et al. (eds.), *Design for Vulnerable Communities*, The Urban Book Series,
https://doi.org/10.1007/978-3-030-96866-3_11

11.1.1 Jingzu Policy

The first epoch, which spans from the 1950s to the 1960s, is characterized by the implementation of the so-called *Jingzu* policy. The expression *Jingzu*, which means proxy rent in Mandarin, was referred to the nationalization of the existing housing stock and the subsequent reallocation of the units operated by the local governments (Li, 2004). In Beijing, this reform meant that the privately owned *siheyuan*[2], which were normally occupied by one extended family, had to be nationalized, divided into multiple units, reallocated, and shared among different families.[3] The aim of the policy was to tackle the lack of residences destined for the migrant workers who were employed by the recently established state-owned factories and, consequently, to transform the Chinese capital from a "city of consumption into a city of production" (Wang, 2011, p. 85)[4].

11.1.2 From Quadrangles to *Dazayuan*

The second period that covers the Cultural Revolution in the 1960s and 1970s is characterized by the metamorphosis of *siheyuan* into the so-called *dazayuan*[5]. The traditional residences were progressively occupied by more peasants' families who filled in the courtyard layouts with temporary and low-quality structures to increase the available living spaces. This phenomenon that lasted until the recent years boosted an acute overcrowding of the Old Beijing[6] that, together with the lack of basic infrastructures, caused a further deterioration of the conditions of these

[2] *Siheyuan* is the name of the traditional Chinese courtyards, or quadrangles, that define the urban morphology of the Beijing's historical city center (Knapp, 2005). They are composed of a series of courtyards organized according to a north-south central axis and surrounded by separate buildings connected through corridors, porches, and gates. "The *Siheyuan* residence of Beijing embodies the traditional Chinese values of Confucianism, honouring the elders, spatializing hierarchical social and domestic orders, and safeguarding the chastity of women (Yu, 2017, p. 1340)." The access to the traditional courtyards is granted through the *hutong* alleys, which run east-west and that are interconnected by the *dajie,* a series of north-south large lanes destined to faster traffic. The traditional neighborhoods of Beijing are also referred as *hutong*.

[3] "Production first, livelihood second" was one of the official slogans at that time (Xu & Yang, 2009).

[4] "Only when production in the cities is restored and developed, when consumer-cities are transformed into producer-cities, can the people's political power be consolidated. Other urban activities, for example, Party organization, public administration, trade unions and other people's organizations, culture and education, and the suppression of counterrevolutionaries, new agencies, newspapers and broadcasting stations—all these activities revolve around and serve the central task of production and construction (Mao, 1966, p. 365)."

[5] In Mandarin, *dazayuan* literary means multifamily compounds. Over the years, the term also acquired a negative nuance and became synonym of messy and crowded compound (Zhang, 2015)

[6] Old Beijing, or *Lao Beijing* in Mandarin, is another way to refer to the historical areas in the capital (Yu, 2017).

quarters. In this period, the nationalization process of private houses suffered a sudden halt due to the turbulent events of the Cultural Revolution leaving several landlords in a property limbo. This situation was addressed in the 1980s, when the *si-fang-zhu*, or private homeowners, were allowed by the government to retain their private housing ownership. "It was estimated that about a third of traditional houses in inner city Beijing were owned by *si-fang-zhu*" by that time (Fang, 1999 cited in Fang & Zhang, 2003, p. 161).

11.1.3 The *Hutong* in the Post-Reform Era

The last period, which spans 40 years from the opening of China in 1978 to the contemporary era, has witnessed two different attitudes in dealing with the built heritage and related local communities in Beijing and, more broadly, in the whole country. In 1990, the People's Government of Beijing Municipality promoted the Old and Dilapidated Housing Redevelopment Program (ODHRP) to tackle the quasi-slum conditions of many housing compounds, including the traditional courtyards, and the chronical shortage of residential units. The introduction of the ODHRP was made by two major economic reforms aimed, among other targets, at boosting the national real estate sector (Wu, 2015). The first reform is the Land Administration Law of the People's Republic of China that separated the land use rights from its ownership, that remained of the state in the cities and collective in the rural areas at the end of the 1980s (National People's Congress, 1988). The second one is the progressive but fluctuating devolution of administrative and fiscal powers to the local authorities operated by the central government to grant them more financial independency (Cai & Treisman, 2006).

The targets of the OHDRP were to tackle those environmental and social issues that were affecting Beijing inner city neighborhoods for several decades through private-public partnerships and the support of enormous foreign investments.[7] In reality, they "quickly became a large-scale speculative form of development involving massive [institutional] demolition and ruthless displacement, resulting in enormous social and cultural costs" (Fang & Zhang, 2003, p. 149)[8]. The implementation of the commodity housing stock and the update of the city's infrastructures can be considered among the main causes of the disappearance of roughly 7000 *hutong* alleys and related local communities in the last 25 years (Bideau & Yan, 2018)[9].

[7] "Between 1990 and 1997, 68.1% of total investments in ODRHP in Beijing's inner-city districts came from foreign investments" (Shin, 2009a, p. 2821).

[8] It is estimated that between 1991 and 1999 the total number of residents whose dwellings were demolished during the implementation of the ODHRP were around 17,000 (Jie, 1997; Shin, 2009a; The State Council of the People's Republic of China, 2001).

[9] Beijing hosted the Summer Olympic Games in 2008. If, on the one hand, this event was the chance to affirm the Chinese capital's global vocation, on the other it greatly affected both the outskirts and the inner neighborhoods of the city. Large-scale demolitions and mass relocations

At the beginning of 2000s, the central government started to enhance the Cultural and Creative Industries (CCI) as an important part of the new market economy, and it released the Outline of the National Cultural Development Plan During the 11th Five-Year Plan Period in 2006 to clarify and map the different CCI macro-categories (Liang & Wang, 2020; The State Council of the People's Republic of China, 2006). Moreover, the Chinese urban development strategies shifted from a fast-paced urbanization process to the so-called New Urbanization Plan (NUP) during the second decade of 2000. The NUP aims at optimizing the use of the land and at limiting the growth of the cities "beyond the means of their natural resources and environment" (Xinhua, 2016), while providing affordable housing to low-income residents through the regeneration and renovation of the dilapidated downtown urban areas (Central Committee of the Chinese Communist Party, 2020; Chu, 2020). Besides these aspects, the NUP model highlights a renewed central role of the heritage preservation, adopting the "principles of giving priority to protection, rescue first, reasonable use, and strengthening the management" (EU-China Policy Dialogues Support Facility II, 2015, p. 25).

11.1.4 Approach to the Beijing's *Hutong* Vulnerabilities

In this context, and with the success achieved by creative clusters such as the 798 Art Zone[10] at the beginning of 2000, the concept of Beijing as a global cultural city started to gain traction. In fact, together with the construction of dedicated CCI parks, the local authorities initiated a process to encourage and install creative factories in the *hutong* alleys attracting several small and medium enterprises related to culture production and promotion. In this regard, we can read the government's intention of taking advantage of the existing spaces and creative transformation of old buildings to construct new urban cultural spaces. (Office of the People's Government of Beijing Municipality, 2017, p. 41). This trend, which is still unfolding, is leading to a progressive privatization and commodification of the historical neighborhoods that are conceived as commercial resources to be exploited rather than assets to be protected and enhanced through a delicate balance between economic development, heritage preservation, and community enhancement (Su, 2015).

This is causing an inevitable raise of the living costs in the area and it is driving a permanent modification of the morphological characters of the urban fabric, irreparably altered to obtain the maximum economic profit in the shortest possible time (Martínez, 2016; Wu et al., 2015). Moreover, the proliferation of private and

were systematically implemented to build new infrastructures and to update the old ones, while a generalized urban renewal was realized at the expense of affordable housing and public space (Shin, 2009b; Shin & Li, 2013).

[10] McCarthy and Wang (2016) provide a comprehensive overview of the development of Beijing's 798 Art Zone that started as an artists' enclave and quickly became a cultural hub supported and monitored by the government institutions.

privately managed cultural and touristic amenities is inevitably shrinking the space for the installation of affordable public facilities and places for the betterment of the communities' everyday life, undermining the social cohesion of the area (Miao, 2011).

If we add that other urgent issues are still unsolved, such as the chronic shortage of basic infrastructures (Yan et al., 2021) and the overcrowded conditions of the remaining traditional courtyards (Conservation Plan for Twenty-Five Beijing Old City Historical and Cultural Conservation Areas, 2016), it is evident how the Old Beijing's neighborhoods are characterized by a series of vulnerabilities that are threatening both the social and the built environment, causing the inevitable decay of the communities and the erosion of the heritage sites.

With these considerations, and the awareness of the many implications that this study involves, the chapter intends to approach the abovementioned fragilities through the lens of the design inquiry supported by the heritage value assessment, and to elaborate conceptual design proposals that can—partially—clarify these problems and address the related vulnerabilities. In other words:

> 'Designerly' research is understood not as problem-solving, but as questioning — reformulating problems and forming insights. Conceptual design is equated with provisional synthesis of several factors and at multiple scales; it is also able to overcome antitheses that are insolvable in theory. (Loeckx & Shannon, 2004, p. 165).

This chapter draws on the authors' first-hand experience on two heritage renovation projects developed between 2016 and 2018 in Beijing to provide insights about potential and sustainable directions to preserve the heritage integrity, while improving the residents' living conditions and enhancing the community cohesion through the introduction of public spaces and collective facilities in Beijing's historical neighborhoods. In fact, as stated in the *United Nations Settlement Program*, "good public spaces enhance community cohesion and promote health, happiness, and well-being for all citizens as well as fostering investment, economic development, and environmental sustainability" (United Nations Human Settlements Programme, 2021).

11.2 Theoretical Framework

In this section we will introduce the theoretical foundations that underlie our research and that serve as methodological approach of the study. The renovation of Beijing *hutong* neighborhoods presents a high degree of complexity due to several reasons that go well beyond the scope of this text. At the same time, we deem important to highlight a few of them to further clarify the context and the aims of this study.

Firstly, a large constellation of stakeholders is generally involved in the decision-making and realization processes of the renovation projects, and they often pursue different and even contrasting interests. For example, the district authorities are generally more keen on improving the physical appearance and the economic development of the neighborhoods to pursue their own political agenda, while the residents chase an enhancing of their living conditions or, when the relocations are inevitable, they look for an appropriate economic settlement to leave the area (Zhang et al., 2020).[11]

Secondly, the heritage areas are characterized by several structural deficiencies, such as the crumbling and overcrowded conditions of the residential units, the lack of adequate sanitation services, and the absence of proper public spaces and services for the community. These fragilities are deeply rooted in the urban evolution of the neighborhoods, and they are aggravating their social and physical environments (Lanchun & Yang, 2016).

11.2.1 The Renovation of Beijing's *Hutong* as a Wicked Problem

If we add the commodification of the neighborhoods and the generally deprived conditions of the inner-city communities[12] to the abovementioned issues, it is clear how a holistic and sustainable regeneration of Beijing historical areas becomes what Rittel and Webber define as wicked problem (Rittel & Webber, 1973). To further clarify the object of the research, we borrow Alford and Head's framework that classifies various degrees of wickedness (Alford & Head, 2017). The case of Old Beijing's renovation most probably falls into what they categorize as "politically turbulent problem." It encompasses wicked problems characterized by a certain clarity in the formulation of the issues but lacking a clear solution and the multiple parties involved in the problem pursue conflicting interests and values (Alford & Head, 2017, p. 402).

11.2.2 The Research by Design Framework

According to Roggema (2017), the wicked problems need to be approached through a counterintuitive research method that allows the elaboration of possible scenarios "creating the freedom to move with the proposals in uncharted territory, and producing new insights and knowledge interesting and useful for a wide audience" (p. 15). The research by design framework becomes a very suitable tool to deal with wicked

[11] These topics were raised during the unstructured interviews that the authors conducted with the different stakeholders involved in the projects in June 2016 and August 2017. The data collected during these interviews and the value typology that emerges from them are illustrated in the following paragraphs.

[12] The communities are generally composed of a mix of old and new rural migrants (Wu, 2012, 2016)

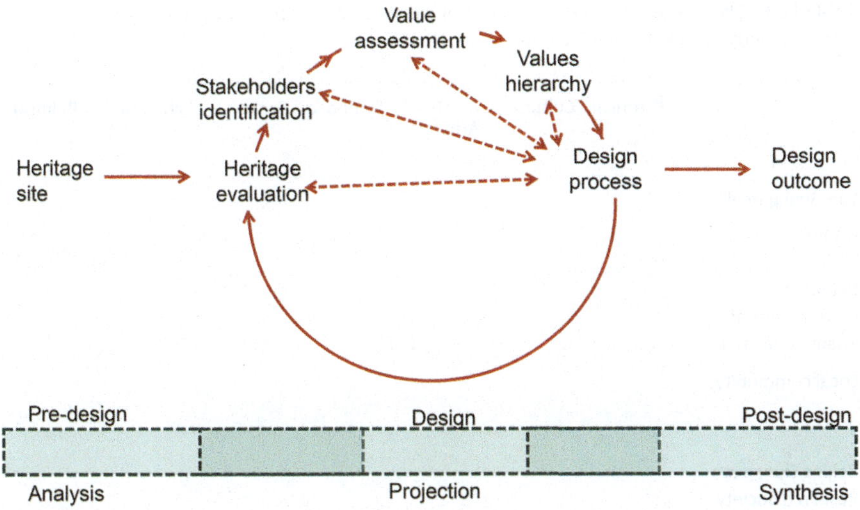

Fig. 11.1 A diagram representing the integration between the research by design framework and the system of values (Source: the authors)

problems because it provides "a philosophical and normative basis for the design process" (Roggema, 2017, p. 15) and it allows to perform creative jumps to clarify the initial dilemma and to explore potential solutions. Design research can be methodologically devised to conceive a gap between the existing situation and the new circumstances (Rust, 2004). Following the research by design framework, the investigation is structured according to the 3 phases identified by Roggema (2017)—pre-design or analysis, design or projection, and post-design or communication, which will be further explained in detail in the following paragraph (Fig. 11.1).

11.2.3 The Provisional Typology of Heritage Values

Considering that the subjects of the study are heritage sites and the necessity of a value assessment, we employ the Provisional Typology of Heritage Values elaborated by Mason (2002) for the data collection of the pre-design phase and the subsequent decision-making process. The choice of Mason's system in our study is supported by four reasons. Firstly, it is a typology, a way to classify, and typologies "constitute a first-order research tool, ordering and organizing knowledge so that research builds on itself—it keeps practitioners from having to continually reinvent the wheel" (Mason, 2002, p. 9). Secondly, in his system of values, Mason encompasses the most relevant typologies that were defined by scholars and institutions throughout the twentieth century (Mason, 2002, p. 9). Thirdly, he makes a clear distinction between sociocultural and economic values to highlight the importance of the first ones (Mason, 2002, p. 10). Finally, being provisional, it can be easily

Table 11.1 The stakeholders and the system of values emerging from the pre-design phase for the first case study, the Dongguan New Assembly House

	Historical	Cultural	Aesthetic/ Artistic	Symbolic	Social	Economic[a]	Political
Yan Guang Real Estate						■	■
People's Government of Xuanwu district	■	■	■		■	■	■
Local community	■	■	■	■	■		
Beijing Dongguan Research Society	■	■	■	■	■		

[a]On the X axis are presented the assessed values; on the Y axis the stakeholders identified during the research (Source: the authors). The economic value is considered as one single category to simplify the data collection and to avoid potential misinterpretations

adapted and implemented according to different contextual conditions. The values we decided to assess are the historical, cultural, aesthetic/artistic, social, and economic ones. Moreover, while Mason does not include the political value, we decide to assess it because this is an unavoidable dimension in China (Table 11.1).

11.3 The Case Studies

In this section we introduce the two case studies employed for the application of the theoretical framework and which present the typical Old Beijing vulnerabilities. These are the Dongguan New Assembly House in Xuanxi North, an area of Xuanwu district, and the block composed of the courtyards #28 #26 #33 #22 in Dashilar, which is located in Xicheng district (Fig. 11.2).

11.3.1 Selection of the Case Studies

Four main reasons are behind the choice of these sites and that, in some way, make them comparable. Firstly, the land use rights of both areas belong to two State Owned Enterprises (SOEs): the Yan Guang Real Estate, which manages the entire Xuanxi North subdistrict, including the Dongguan New Assembly House (Sunshine

Fig. 11.2 Location of the case studies in downtown Beijing. 1 Dongguan New Assembly House in Xuanxi North—Xuanwu District; 2 #28 #26 #24 #22 courtyards in Dashilar—Xicheng District (Source: authors' edition based on Google Earth map)

Xinyie Real Estate, n.d.), and the Beijing Guangan Holding Co., Ltd., which is in charge of the Dashilar renovation projects, their promotion and management (Beijing Guang Holdings Limited, 2020). Secondly, the extension of both areas is around 1000 m^2 and, according to the collected data, their construction dates to the beginning of the twentieth century. Thirdly, the occupancy conditions of the two sites are different. In fact, while most of the residents of the Assembly House had already been relocated and the remaining ones were in process of moving out, the block of courtyards was still inhabited by several families at the time of the research.[13] This enables us to apply the framework to two different scenarios and draw a comparison between the proposals to understand how the contextual conditions and the systems of values are differently implemented to address the sites' vulnerabilities.

Finally, both sites were the objects of two heritage renovation projects that were peer-reviewed through public events. In detail, the Assembly House was at the center of a design competition (Akiact, 2016), while the courtyard block proposal was commissioned for an exhibition in conjunction with the 2017 Beijing Design Week

[13] Altogether, around 18 families were still living in the #28 #26 #33 #22 courtyards, while 10 had already been relocated at the time of the project (see Fig. 11.9).

(Dashilarproject, 2017). These initiatives were both organized and promoted by the SOEs and provided us with the opportunity to communicate and present the results of the design research to the stakeholders and to a wider public.

11.3.2 The Dongguan New Assembly House in Xuanxi North and the #28 #26 #33 #22 Courtyards in Dashilar

The Dongguan New Assembly House is in Xuanxi North subdistrict in Beijing, it is inscribed as Historic Appearance Coordination Zone, and it is a Registered Immobile Cultural Heritage (Beijing Municipal Commission of Planning and Natural Resources, 2020). It abuts with Guang'an Hutong on the east side, and with Shangxie Jie on the North side (Figs. 11.3, 11.4, and 11.5). According to an inscription reported on the 1919 stone stele *Record on the Dongguan New Assembly House* (1919)[14], and to the information provided by Lun Zhiqing, president of the Beijing Dongguan Research Society, the area was purchased in 1900 by a Beijing education official who was born in Dongguan, Guangdong Province, from a Japanese language school. The construction of the Assembly House, which was designed as an additional space[15] for the accommodation of Dongguan-born officials and students in Beijing, started in 1911 but it was interrupted by the Xinhai Revolution. In 1914–1915 its urban layout was modified according to the advice of a *Fengshui* master, and 2 new courtyards were added, together with other architectural elements (Record on the Dongguan New Assembly House, 1919). In 1960, in conjunction with the implementation of the *Jingzu* policy, several linear buildings were added to accommodate the migrant workers on the northern, on the eastern and on the southern borders of the site and the Assembly House was transformed into a residential compound. In the following decades, all the courtyards were progressively filled in with temporary shelters by the incoming migrants, causing the alteration of the original layout. The last modification happened in 2008, when the Guang'an Hutong was realized and one third of the historical buildings was demolished to make space to the new road. The second case study is the urban block composed of the #28 #26 #33 #22 traditional quadrangles which are located in Yaowu Hutong in Dashilar subdistrict (Fig. 11.6). In contrast with the Dongguan New Assembly House, which was a relevant gathering point for businessmen and scholars, these were private residences characterized by a single-courtyard layout (Fig. 11.7).

[14] The object was moved to Beijing Stone Carving Museum in the 1970s while a rubbing of its inscription is conserved in the National Library of China in Beijing, and it is reported in the references.

[15] Another Dongguan Assembly House was already present in the area at that time.

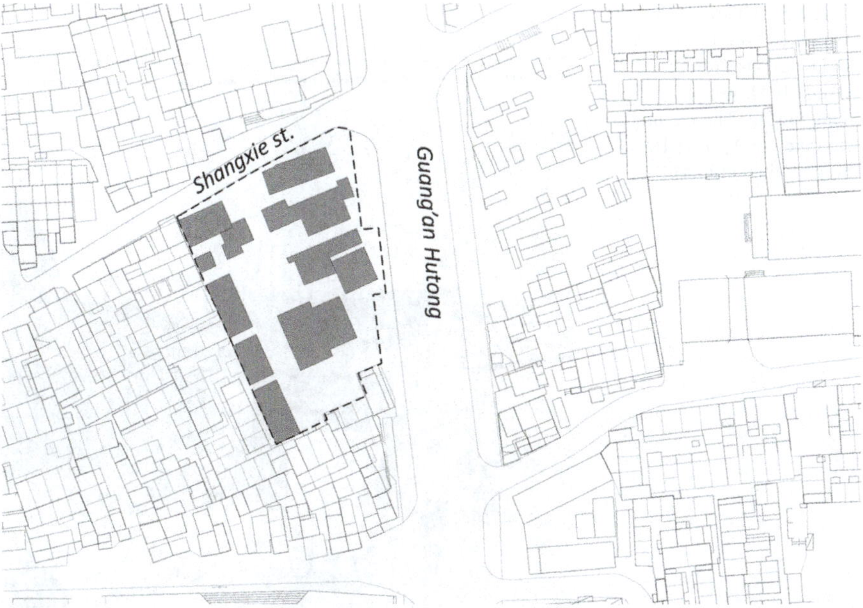

Fig. 11.3 Plan of the Dongguan New Assembly House as it is recorded by the cadaster map (Source: the authors)

Fig. 11.4 A view of the Dongguan New Assembly House from the Shangxie Jie and Guang'an Hutong crossroad. The picture was taken in April 2016 (Source: the authors)

Fig. 11.5 A view of the historical entrance gate of the Dongguan New Assembly House on Shangxie Jie. The picture was taken in April 2016 (Source: the authors)

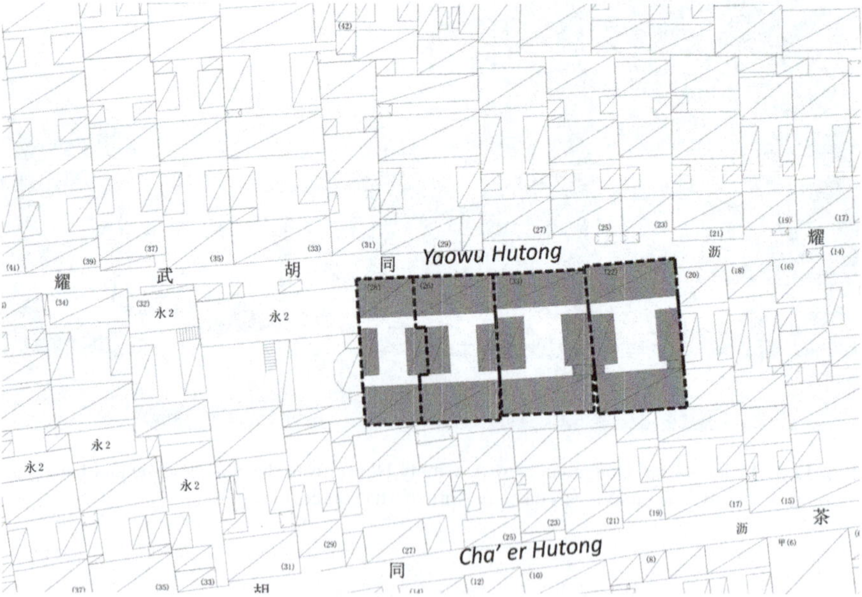

Fig. 11.6 Plan of the #28 #26 #24 #22 courtyards block in Dashilar. The floorplan is based on the documents that were provided by the SOE (Source: the authors)

Fig. 11.7 A view of the #28 courtyard interior space. The picture was taken in August 2017 (Source: the authors)

11.4 Implementation of the Theoretical Framework

In this section, we illustrate the application of the broader research by design framework to the case studies and how it is integrated with the system of heritage values. The development of the research is structured according to the three moments of analysis, projection, and communication as proposed in Roggema's framework (2017). At the same time, it is important to highlight that the research by design method is not a linear and mechanical process as it can appear in the form of a written essay, but there is a continuous communication "between the inquiring and the propositional [phases]" (Thomsen & Tamke, 2009, p. 1). In other words, analysis and projection feed into each other in a cyclical and iterative way to define a holistic proposal that will be disclosed in the communication phase to a wider public (Fig. 11.1).

11.4.1 The Pre-Design Phase

The pre-design or analysis phase is the moment when the questions are formulated, the data are collected, and it is possible to envision potential creative jumps towards the design directions. In our case, the data collection is informed by the *Provisional*

Typology of Heritage Values to establish a hierarchy that would guide the design phase decisions. Considering that this system of values "is offered as a point of departure and discussion" (Mason, 2002, p. 14), we decide to adapt it according to the specific social, political, economic, and environmental context we are investigating. The definition of values is mainly informed by a qualitative ethnographic approach through an investigation on both sites and related communities that was carried on between April 2016 and October 2018. This choice is reinforced by the fact that "ethnographic approaches are broader and include the historical, as well as the social and political, context of the site as a means of understanding contemporary sociocultural patterns and cultural groups" (Low, 2002, p. 32).

We started with an archival research about the sites through documents, historical and cadastral maps coupled with Google Earth Historical Imagery, to understand the urban evolution of the sites and that of the surrounding context.

The fieldwork that included an extensive survey and mapping of all the buildings was carried on especially for the Assembly House, while the courtyards in Dashilar could not be extensively studied due to residents' reluctance to allow any visitor to access the site. Nonetheless, the SOE provided accurate drawings and a census of the available and occupied units in the quadrangles. The fieldwork also consisted of a mapping of the existing public facilities and programs surrounding the areas to understand what functional solution could be integrated in each specific context.

In parallel, we interviewed the main stakeholders who were directly involved in the renovation projects. For the Assembly House, the group included the People's Government of Xuanwu district and the SOE, some residents of Xuanxi North, and the president of the Beijing Dongguan Research Society that was the initial owner of the site throughout the first half of the last century. For the #28 #26 #33 #22 quadrangles in Yaowu Hutong we identified the Beijing Guangan Holding Co., Ltd., the People's Government of Xicheng district, the residents living around the area and the families who were still in the courtyards. The interviews with the institutional subjects were in the form of semi-structured interviews and took place during Q&A meetings that were organized by the SOEs with the participation of a few government officials.

On the other side, unstructured interviews were conducted with members of the community, with the president of the Beijing Dongguan Research Society, and with some residents of the courtyards during the fieldworks.[16]

The typology of values that emerged from the overlapping of archival inquiry, urban mapping, site investigation, and stakeholders' interviews encompassed 7 core values—historical, cultural, aesthetic, political, symbolic, social, and economic—and each of them assume different weights in the two sites. For the Assembly House (Table 11.1), the historical, cultural, and symbolic values were inherent to its history and to the relevance it had in the past as gathering point for the Dongguan-born

[16] The interviews with Xuanxi North residents and with Mr. Lun Zhiqin, the president of the Beijing Dongguan Research Society, took place on 15 and 24 April 2016, while the meeting with the SOE was on 5 June 2016. In Dashilar, the interviews with the residents and with the took place in August and September 2017.

businessmen and scholars in the capital city. These were predominantly championed by the members of the community[17] and by the Beijing Dongguan Research Society. The artistic/aesthetic value lies in the fact that, according to our discoveries, the compound was probably the only in Beijing with buildings in Dongguan style from that period. The local authorities and the SOE stressed the importance of the political and economic dimensions of the renovation project, envisioning a rejuvenated Xuanxi North subdistrict where commercial and touristic activities could guarantee the area's development and social stability. On the other side, the interviews with the residents touched several aspects that were embedded in the social value of the heritage site. Being the community mostly composed of elderly people who could not access any close by aggregation space and of young migrant families who work the entire day and could not look after their children after the school time, they lamented a severe lack of accessible and safe public facilities and outdoor spaces for neighborhood activities.

The case of the #28 #26 #24 #22 courtyards outlined a different scenario. In fact, the political, economic, and social values resulted more relevant than the historical, cultural, artistic/aesthetic, and symbolic ones. This is due to two main factors. On the one hand, the four courtyards had apparently limited artistic and historical relevance; on the other hand, the presence of families in the quadrangles mainly raised concerns about the standards of their living spaces. This aspect was also stressed by some individuals who expressed the need of enjoying decent—in terms of quality and size—spaces and basic services. The SOE and the local government officials stressed the preeminence of the political and economic values in view of the promotion of Dashilar as creative and cultural cluster to contribute to the social cohesion of the neighborhood and to improve its physical appearance and perception.

11.4.2 The Design Phase

In the design phase "the process of moving between the exterior and the interior of making, creates a conversation between the dissective nature of analysis and critical assessment and the creative nature of proposition and result" (Thomsen & Tamke, 2009, p. 3). In this section, we illustrate how the vulnerabilities of heritage erosion and the community weakening are addressed through the design process and how the values are embedded in the conceptual proposals. We focus on two aspects: the determination of the functional program and the redefinition of the sites' urban layouts.

In terms of program, the main element that characterizes the conceptual proposals is the provision of generous outdoor public spaces underlying the intention to open the courtyards to the communities. Here, the social value of the heritage

[17] It is worth to note that the residents demonstrated a comprehensive understanding of the history and of the landmarks sites of Xuanxi north neighborhood in the unstructured interviews.

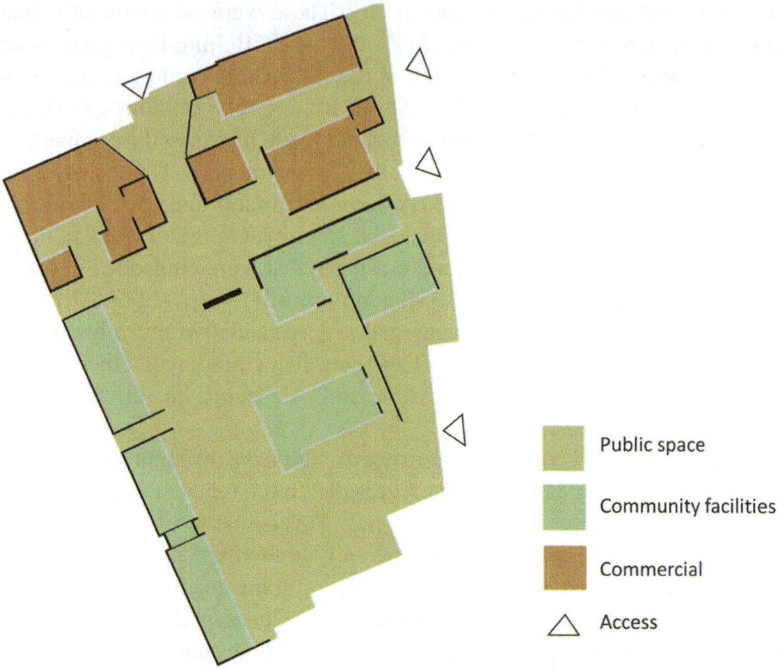

Fig. 11.8 A diagram that represents the functional program of the Dongguan New Assembly House. The outdoor public space is the prevalent element of the layout (Source: the authors)

acquires a central role to "enable and facilitate social connections, networks, and other relations in a broad sense" (Mason, 2002, p. 12). This solution might seem rhetoric and utopian in a context such as the Beijing *hutong* area, where the living conditions are still precarious, the value of the land is exorbitant and other infrastructural matters need urgent solutions. Instead, the necessity of safe and free of charge public areas has become a very relevant issue in the newly developed contemporary Chinese cities, especially for the low- and middle-income residents. In this regard, Miao (2011) points out that "the Chinese urban renewals have fulfilled the government and corporations' needs for economic expansion. However, the same cannot be said of public spaces used daily by ordinary residents" (p. 180). This need of public space becomes clear when we analyze the popularity achieved by renovation projects such as the Micro Yuan'er in Dashilar by Standard Architecture. Here a traditional quadrangle has been entirely preserved and transformed into a public facility that is managed by members of the community to promote social activities (Bayndrian, 2018). The idea to open the outdoor areas of the courtyards to the public also finds its theoretical support in the 13 principles that inform the Chinese traditional architecture and urban design illustrated by Miao (1990). In the sixth principle, the author demonstrates that public nodal spaces for gathering and collective activities did exist in Chinese traditional cities. Most of them were semi-private quadrangles that were associated with guild houses or

Fig. 11.9 A comparison between the existing and the proposed functional layouts of the #28 #26 #24 #22 courtyards. The diagrams display the available units (AV) after the residents' relocation and the families who were still living in the quadrangles (the number on the left of the hyphen is the courtyard; the number on the right is the family). Also in this case, the public space is the main outdoor functional element (Source: the authors)

temples (Miao, 1990, p. 18)[18]. These spaces were connected by the street that was the only—linear—public space in the Chinese traditional urban design. Starting from these references, the public space of the two sites is treated in slightly different ways.

In the Assembly House, the open area can be entirely accessed through four points, the historical main gate on Shangxie Jie and three new entrances on Guang'an Hutong (Fig. 11.8). In the #28 #26 #24 #22 courtyards, the quality of the public space is conditioned by the assumption that relocations should be avoided and that the presence of families still living in the buildings should be maintained. For these reasons, we define different degrees of publicness to respond to the residents' concurrent needs of open-air areas and privacy. A small gathering space and a sitting area close to the main entrance of each quadrangle faces Yaowu Hutong while the interior of the courtyards can be accessed only by accompanied visitors. The social value is also implemented through the introduction of community indoor facilities in the functional layouts of both sites.

[18] The sketch by Miao illustrates very clearly how the semi-private courtyards are connected to the linear public space of the lanes (1990, Fig. 18).

Fig. 11.10 The physical model of the Dashilar proposal that illustrates the relation between Yaowu Hutong and the interior public spaces of the #28 #26 #24 #22 courtyards (Source: the authors)

In the Dongguan New Assembly House, a community library and a children day-care space are installed in the three side buildings of the main courtyard, while the other historical structures are destined to exhibition and gathering areas. Small economic activities are installed along the north border of the site, which is the most exposed to pedestrian and vehicular traffic, facing Shangxie Jie and Guang'an Hutong. In the courtyards' block, small indoor gathering spaces are located at the entrance gate to protect the families' privacy. At the same time, the residences are rearranged according to two principles: minimize the residents' relocation, even among the 4 courtyards, and provide basic units composed of kitchen, living, bathroom, and bedroom while keeping the buildings' original structural layout (Figs. 11.9 and 11.10).

The second aspect of the renovation projects we highlight is how the historic, symbolic, artistic/aesthetic values have informed the urban layout solutions for the built heritage. The assumption we make for both case studies is to preserve the historical layouts and buildings integrity where possible. In the Assembly House, the reconstruction of the site evolution has enabled us to identify the original structures dating back to the beginning of the twentieth century and those that were erected later, in the Socialist and in the post-1978 periods. This allowed us to understand which structures should be kept and restored, both at the urban and architecture levels, and can be demolished and replaced according to outlines defined by the property cadaster maps.

11.4.3 The Post-design Phase

In the post-design phase the new knowledge produced during the projection is manifested to a broader audience to collect feedbacks from the stakeholders who have interests in the renovation projects. For these case studies, the post-design phase consists of a public presentation of the conceptual proposals and findings of the research by design to the abovementioned stakeholders aimed at receiving feedbacks and comments. In the case of the Dongguan New Assembly House, the project underwent 3 different steps of assessment: an open vote on an online platform, an assessment conducted by Beijing heritage scholars and design experts, and the evaluation by the SOE and representatives of the local government at the end. The proposal was appreciated by the entire group of stakeholders because, while actively protecting the built heritage, it enhanced the social role of the area while including the economic values promoted by the institutions.

For the 4 courtyards, the renovation project was exhibited during the 2017 Beijing Design Week in October 2017, but neither the institutional stakeholders provided any feedbacks, nor there was chance to have an open presentation or discussion with the community and the families of the courtyards. Therefore, limited insights and research results could be elaborated in this case.

11.5 Conclusions

The research by design method is a powerful tool to address some of the Old Beijing's vulnerabilities, namely the progressive community weakening and the heritage erosion, by means of the definition of a typology of values. The two case studies identify the tangible need of public spaces for daily community activities and the urgency of heritage conservation and enhancement in the *hutong* neighborhoods, in addition to the economic revitalization that embeds strong political values promoted by the institutional stakeholders.

The implementation of public space as a strategy to address Old Beijing's vulnerabilities implies broader considerations concerning the current renovation practices in the area. According to the study here illustrated, it emerges how the local institutions still conceive the *hutong* and their communities as two separate entities. The Standing Committee of the Beijing Municipal People's Congress states in 2021

> While strengthening the protection of famous historical and cultural cities, we should pay attention to improving the living environment. The District People's government guides and encourages the owners and users of houses in historical and cultural blocks, traditional cottage areas and characteristic sites to voluntarily improve their living conditions through the implementation of rent withdrawal and house replacement (Standing Committee of the Beijing Municipal People's Congress, 2021).

Despite the central government issued four relevant policies to enhance public participation in the planning and redevelopment processes in China, namely the

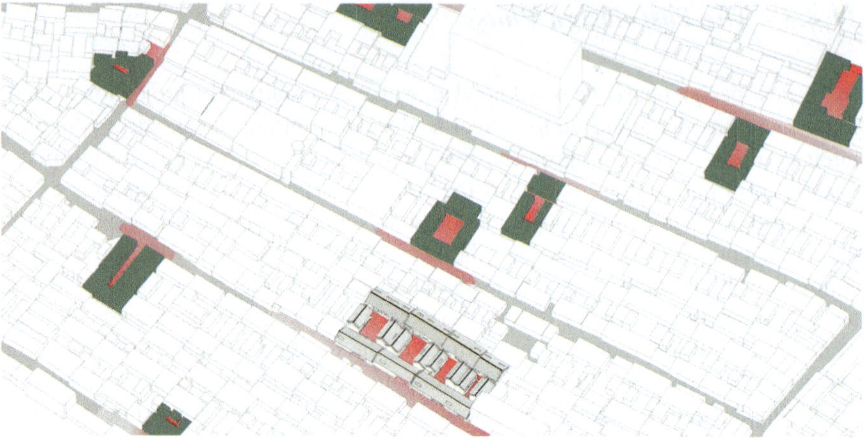

Fig. 11.11 An exploratory scenario in which the interior areas of the quadrangles are transformed into public spaces. The diagram is based on the #28 #26 #24 #22 courtyards proposal (Source: the authors)

Property Rights Law in 2007, the Urban & Rural Planning Law and the Regulation on the Preservation of Famous Historic-Cultural Cities, Towns and Villages in 2008 and the Regulation on Expropriation & Compensation of Housing on State-owned Land in 2011 (Zhang et al., 2020), the compensation—clearance—relocation practice seems to be still the preferred option at the Beijing local level. To this point we should remind what UNESCO states about the necessity to actively involve "the widest possible participation of communities, groups and, where appropriate, individuals" (2003) to create, maintain and preserve heritage sites. The communities' involvement can be enhanced through a mixed approach where the values and interests of the institutional stakeholders are integrated with the needs of the community and this is a more effective way to identify, formulate and tackle the vulnerabilities of a specific urban area, as demonstrated by other experiences in similar Chinese and overseas contexts (Kato, 2006; Ni & Mingjie, 2017; Zhang et al., 2020). The introduction of the public dimension into the urban fabric of the *hutong* areas concretely and comprehensively addresses the needs of the stakeholders, as demonstrated by the general appraise of the Dongguan New Assembly House renovation project during the post-design phase. In this case, the public space becomes the common ground where economic activities and social facilities can coexist in a logic of heritage protection and promotion. And this indicates that "numerous small courtyards within urban blocks may provide a more practical and culturally sensitive answer to the need for nodal public spaces" (Miao, 2001, p. 278) and it can be an effective strategy to regenerate the contemporary Old Beijing (Fig. 11.11).

Although the research by design framework benefits the addressing of the sites' vulnerabilities, the comparison between the two proposals also highlights its shortcomings. In the Dongguan New Assembly House, the phases of the research by design have been fully implemented, from the pre-design to the communication, and they have provided meaningful insights about the fragilities of the neighborhood. On the contrary, the proposition for the #28 #26 #24 #22 courtyards is half-accomplished because of the lack of reliable and accessible data and the reluctance of the stakeholders to participate in the process. Eventually, this project became more like a conventional design practice proposition, where the solution mainly responds to the brief provided by the institutions and it is based on the designers' subjective experience and judgment about how the scheme should be (van de Weijer et al., 2014). To conclude, it emerges how the full engagement of the stakeholders is a necessary condition to identify the problems and consequently, to develop comprehensive proposals that clarify and address the vulnerabilities through the research by design method.

Acknowledgments The authors would like to thank the editors of the book for their help in the publication process. Eugenio Mangi would like to thank Cui Jinze (崔金泽) for his unique contribution during the investigation and project development of the Dongguan New Assembly House and Lun Zhiqing (伦志清), president of the Beijing Dongguan Research Society in 2016 and 2017, for his precious help.

References

Akiact. (2016, 04 12). *Open Call: Xuanxi North Courtyard Reconstruction - An International Conceptual Design Competition*. Retrieved 07 21, 2021, from Archdaily: https://www.archdaily.com/785449/open-call-xuanxi-north-courtyard-reconstruction-an-international-conceptual-design-competition

Alford, J., & Head, B. W. (2017). Wicked and less wicked problems: a typology and a contingency framework. *Policy and Society, 36*(3), 397–413.

Bayndrian, J. (2018, 04 05). *Big Messy Courtyard: Micro Yuan'er*. Retrieved 07 28, 2021, from Assemble Papers: https://assemblepapers.com.au/2018/04/05/big-messy-courtyard-micro-yuaner/

Beijing Guang Holdings Limited. (2020). *Company Summary*. Retrieved 07 19, 2021, from Guangan Holding: http://www.gaholding.com.cn/AboutUs.aspx?nid=2

Beijing Municipal Commission of Planning and Natural Resources. (2020, 09 01). *The functional core area of the capital | Strengthen the overall protection of the old city and build a model area for the promotion of Chinese civilization*. Retrieved 08 08, 2021, from http://ghzrzyw.beijing.gov.cn/zhengwuxinxi/zcfg/zcjd/202009/t20200901_1994664.html

Bideau, F. G., & Yan, H. (2018). Historic Urban Landscape in Beijing the Gulou project and its contested memories. In M. Christina & M. Svensson (Eds.), *Chinese cultural heritage in the making. experiences, negotiations and contestations* (pp. 93–117). Amsterdam University Press. Retrieved from http://www.jstor.org/stable/j.ctt2204rz8.7

Bozhen, Z. (1919). Record on the Dongguan New Assembly House [stone rubbing – ink on paper] Beijing: National Library of China. Retrieved 04 05 2022 from http://read.nlc.cn/allSearch/searchDetail?searchType=all&showType=1&indexName=data_418&fid=10646.

Cai, H., & Treisman, D. (2006). Did Government decentralization cause China's economic miracle? *World Politics, 58*(4), 505–535. https://www.jstor.org/stable/40060148

Central Committee of the Chinese Communist Party. (2020, 12 01). *Proposal of the Central Committee of the Chinese Communist Party on Drawing Up the 14th Five-Year Plan for National Economic and Social Development and Long-Range Objectives for* 2030. Retrieved

07 22, 2021, from Center for Security and Emerging Technology: https://cset.georgetown.edu/wp-content/uploads/t0237_5th_Plenum_Proposal_EN-1.pdf
Chu, Y.-W. (2020). China's new urbanization plan: Progress and structural constraints. *Cities, 103*, 102736. https://doi.org/10.1016/j.cities.2020.102736
Conservation Plan for Twenty-Five Beijing Old City Historical and Cultural Conservation Areas. (2016). Chinese. *Law & Government, 48*(3), 245–253. https://doi.org/10.1080/00094609.2016.1067558
Dashilarproject. (2017, 10 01). *2017 Dashilan design community!* Retrieved 07 21, 2021, from Dashilar: https://mp.weixin.qq.com/s/TjCx8qM3Iv9JP_OH8fPm-g
EU-China Policy Dialogues Support Facility II. (2015, September). *Mapping the EU-China Cultural and Creative Landscape. A joint mapping study prepared for the Ministry of Culture (MoC) of the People's Republic of China and DG Education and Culture (EAC) of the European Commission.* Retrieved 07 14, 2021, from European Commission: https://ec.europa.eu/culture/sites/default/files/2020-07/eu-china-cultural-mapping-2015.pdf
Fang, K. (1999). Housing relocation and housing property in Beijing. *China Lawyers, 5*, 33–36.
Fang, K., & Zhang, Y. (2003, August). Plan and market mismatch: Urban redevelopment in Beijing during a period of transition. *Asia Pacific Viewpoint, 44*(2), 149–162.
Jie, Z. (1997). Informal construction in Beijing's old neighborhoods. *Cities, 14*(2), 85–94. https://doi.org/10.1016/S0264-2751(96)00046-7
Junhua, L. (1997). Beijing's old and dilapidated housing renewal. *Cities, 14*(2), 59–69. Retrieved from https://www.sciencedirect.com/science/article/pii/S0264275196000431
Kato, K. (2006). Community, connection and conservation: intangible cultural values in natural heritage—the case of Shirakami-Sanchi world heritage area. *International Journal of Heritage Studies, 12*(5), 458–473. https://doi.org/10.1080/13527250600821670
Knapp, R. G. (2005). *Chinese houses: the architectural heritage of a nation*. Tuttle Pubishing.
Lanchun, B., & Yang, S. (2016). Regeneration of historic area with social orientation: investigation and analysis of three historic areas in Beijing. *International Review for Spatial Planning and Sustainable Development, 4*(1), 91–105. https://doi.org/10.14246/irspsd.4.1_91
Li, G. (2004). *On the socialist transformation of private rental housing in China's cities after the founding of the People's Republic of China*. Retrieved 03 04, 2021, from on the ownership of historical real estate: http://www.chinarealestatelaw.com/data/365-18.asp
Liang, S. (2014). *Chinese architecture: Art and artifacts*. Cengage Learning Asia.
Liang, S., & Wang, Q. (2020). Cultural and creative industries and urban (re)development in china. *Journal of Planning Literature, 35*(1), 54–70. https://doi.org/10.1177/0885412219898290
Lin, W.-C. (2011). Preserving China: Liang Sicheng's Survey Photos from the 1930s and 1940s. *Visual Resources, 27*(2), 129–145. https://doi.org/10.1080/01973762.2011.568167
Loeckx, A., & Shannon, K. (2004). Qualifying urban space. In A. Loeckx, K. Shannon, R. Tuts, & H. Verschure (Eds.), *Urban Trialogues. Visions_Projects_Co-Productions. Localising Agenda 21* (pp. 156–166). UN Habitat - K.U. Leuven Post Graduate Centre Human Settlements.
Low, M. S. (2002). Anthropological-ethnographic methods for the assessment of cultural values in heritage conservation. In M. De la Torre (Ed.), *Assessing the Values of Cultural Heritage* (pp. 31–51). The Getty Conservation Institute. Available at https://www.getty.edu/conservation/publications_resources/pdf_publications/pdf/assessing.pdf
Mao, T.-T. (1966). Report to the Second Plenary Session of the Seventh Central Committe of the Comunist Party of China. In *Selected works of Mao Tse-Tung*. Beijing: Peking Foreing Language Press.
Martínez, P. G. (2016). Authenticity as a challenge in the transformation of Beijing's urban heritage: The commercial gentrification of the Guozijian historic area. *Cities*, (59), 48–56. https://doi.org/10.1016/j.cities.2016.05.026
Mason, R. (2002). Assessing values in conservation planning: methodological issues and choices. In M. De La Torre (Ed.), *Assessing the values of cultural heritage* (pp. 5–30). The Getty Conservation Institute. Available at https://www.getty.edu/conservation/publications_resources/pdf_publications/pdf/assessing.pdf

McCarthy, J., & Wang, Y. (2016). Culture, creativity and commerce: trajectories and tensions in the case of Beijing's 798 Art Zone. *International Planning Studies, 21*(1), 1–15. https://doi.org/10.1080/13563475.2015.1114446

Miao, P. (1990). Essence of tradition – The 13 characteristics of Chinese Traditional Architecture. *Journal of Building and Planning National Taiwan University, 5*(1), 57–72.

Miao, P. (2001). Design with high-density: A Chinese perspective. In P. Miao (Ed.), *Public places in Asia pacific cities current issues and strategies* (pp. 273–294). Kluwer Academic Publisher.

Miao, P. (2011). Brave New City: Three problems in Chinese urban public space since the 1980s. *Journal of Urban Design, 16*(2), 179–207. https://doi.org/10.1080/13574809.2011.548980

National People's Congress. (1988). *Land Administration Law of the People's Republic of China.* Retrieved 05 30, 2021, from National People's Congress: http://www.npc.gov.cn/zgrdw/englishnpc/Law/2007-12/12/content_1383939.htm

Ni, M., & Mingjie, Z. (Eds.). (2017). *Open your space: design intervention for urban resilience.* Tongji University Press.

Office of the People's Government of Beijing Municipality. (2017). *Beijing cultural and creative industries white paper 2017. State-owned cultural assets supervision and administration.* University of China.

Rittel, H. W., & Webber, M. M. (1973). Dilemmas in a general theory of planning. *Policy Sciences, 4*(2), 155–169.

Roggema, R. (2017). Research by Design: Proposition for a methodological approach. *Urban Science, 1*(2), 1–19. https://doi.org/10.3390/urbansci1010002

Rust, C. (2004). Design enquiry: Tacit knowledge and invention in science. *Design Issues, 20*(4), 76–85. https://www.jstor.org/stable/1512004

Shin, H. B. (2009a). Residential redevelopment and the entrepreneurial local state: The implications of Beijing's shifting emphasis on urban redevelopment policies. *Urban Studies, 46*(13), 2815–2839.

Shin, H. B. (2009b). Life in the shadow of mega-events: Beijing Summer Olympiad and its impact on housing. *Journal of Asian Public Policy,* 2(2), 122–141. https://doi.org/10.1080/17516230903027872

Shin, H. B., & Li, B. (2013). Whose games? The costs of being "Olympic citizens" in Beijing. *Environment & Urbanization, 25*(2), 559–576. https://doi.org/10.1177/0956247813501139

Standing Committee of the Beijing Municipal People's Congress. (2021, 01 28). Retrieved 08 10, 2021, from Explanation on the "Regulations on the Protection of Beijing's Historic and Cultural Cities (Draft)": http://www.bjrd.gov.cn/zyfb/bg/202101/t20210128_2239244.html

Su, X. (2015). Urban entrepreneurialism and the commodification of heritage in China. *Urban Studies, 52*(15), 2874–2889.

Sunshine Xinyie Real Estate. (n.d.). *About Us.* Retrieved 07 19, 2021, from Yang Guang Real Estate: http://www.yangguangxinye.com/aboutUs/

Tao, Z. (2016). Building big, with no regret. From Beijing's "Ten Great Buildings" in the 1950s to China's Megaprojects Today. In L. Jie & E. Zhang (Eds.), *Red legacies in China. Cultural afterlives of the communist revolution* (pp. 56–87). Harvard University Asia Center.

The State Council of the People's Republic of China. (2001, June 04). *Congressional Executive Commission on China.* Retrieved 05 30, 2021, from Urban Housing Demolition and Relocation Management Regulations (CECC Full Translation): https://www.cecc.gov/resources/legal-provisions/urban-housing-demolition-and-relocation-management-regulations-cecc-full#body-chinese

The State Council of the People's Republic of China. (2006). *Outline of the National Cultural Development Plan during the 11th Five-Year Plan Period in 2006.* Retrieved 07 13, 2021, from http://www.gov.cn/xinwen/2016-03/17/content_5054992.htm

Thomsen, M. R., & Tamke, M. (2009). Narratives of making: thinking practice led research in architecture. *Communicating (by) Design 2009,* (pp. 1–8 Available at https://adk.elsevierpure.com/en/publications/narratives-of-making-thinking-practice-led-research-in-architectu).

UNESCO. (2003). *Intangible Cultural Heritage*. Retrieved 08 10, 2021, from Text of the Convention for the Safeguarding of the Intangible Cultural Heritage: https://ich.unesco.org/en/convention#art15
United Nations Human Settlements Programme. (2021). *Public Spaces for All*. Retrieved 06 06, 2021, from UN Habitat for a Better Urban Future: https://unhabitat.org/public-spaces-for-all
van de Weijer, M., van Cleempoel, K., & Heynen, H. (2014). Positioning research and design in academia and practice: A contribution to a continuing debate. *Design Issues, 30*(2), 17–29. http://www.jstor.com/stable/24266946
Wang, J. (2011). *Beijing record: A physical and political history of planning modern Beijing*. World Scientific.
Wu, F. (2012). Neighborhood attachment, social participation, and willingness to stay in China's low-income communities. *Urban Affairs Review*, 547–570. https://doi.org/10.1177/1078087411436104
Wu, F. (2015). *Planning for growth: urban and regional planning in China*. Routledge.
Wu, F. (2016). State dominance in urban redevelopment: Beyond gentrification in urban china. *Urban Affairs Review*, 631–658. https://doi.org/10.1177/1078087415612930
Wu, B., Liu, L., & Morrison, A. M. (2015). The evolution and space patterns of hutongtels in Beijing historic districts. *Journal of Heritage Tourism, 10*(2), 129–150. https://doi.org/10.1080/1743873X.2014.985225
Xie, S., Gu, K., & Zhang, X. (2020). Urban conservation in China in an international context: Retrospect and prospects. *Habitat International, 95*, 102098.
Xinhua. (2016, 02 21). *China outlines roadmap to build better cities*. Retrieved 03 15, 2021, from China.org.cn: http://www.china.org.cn/business/2016-02/21/content_37840593.htm
Xu, M., & Yang, Z. (2009). Design history of China's gated cities and neighbourhoods: Prototype and evolution. *Urban Design International, 14*, 99–117. https://doi.org/10.1057/udi.2009.12
Yan, R., Cheng, S., Chen, J., Li, X., Sharma, S., Uddin, S. M., Mang, H.-P., Chen, C., Li, Z., Li, T., & Wang, X. (2021). Operating status of public toilets in the Hutong neighborhoods of Beijing: An empirical study. *Journal of Environmental Management, 87*, 112252. https://doi.org/10.1016/j.jenvman.2021.112252
Yu, S. (2017). Courtyard in conflict: The transformation of Beijing's Siheyuan during revolution and gentrification. *The Journal of Architecture, 22*(8), 1337–1365.
Zhang, D. (2015). Courtyard houses of Beijing: Lessons from the renewal. *Traditional Dwellings and Settlements Review, 27*(1), 69–82. https://www.jstor.org/stable/24720001
Zhang, L., Lin, Y., Hooimeijer, P., & Geertman, S. (2020). Heterogeneity of public participation in urban redevelopment in Chinese cities: Beijing versus Guangzhou. *Urban Studies, 57*(9), 1903–1919. https://doi.org/10.1177/0042098019862192

Chapter 12
Urban Form and Social Vulnerability in Shanghai: A Comparative Study of Hongkou District Before and After the 1990s Urban Renewal

Carmelo Ignaccolo and Yuqi Zhai

12.1 Introduction

The availability of shops and amenities within walking reach from homes or workplaces has become an essential aspect in achieving vibrant and socially inclusive neighborhoods with high environmental, social, and economic performances (Sevtsuk, 2014). Western cities—especially in Europe—have recently started exploring urban development models in which all city residents can meet most of their needs within a short walk or bicycle ride from their homes. However, such development models are often hampered by the physical legacy of mid-twentieth-century urban transformations such as mono-functional blocks, urban renewal, and suburbanization.

At the end of the twentieth century, the global spread of large-scale shopping centers started transforming traditional retail spaces beyond suburbs, by reaching downtown metropolitan areas. As a result of this spatial shift, shopping malls in city centers started performing as city-wide commercial destinations rather than serving locals living in their immediate adjacency. In other words, these radical transformations of urban morphology altered traditional low-rise mixed-use urban blocks, changing the neighborhood level supply-demand balance. However, while some of these enormous commercial buildings continue to flourish (thanks to high-performing public transit networks or urban highways), others seem to experience an irreversible decline.

In light of the above, this paper focuses on the effects of morphological alteration of commercial blocks in the communities' livelihood of Shanghai's Hongkou

C. Ignaccolo (✉)
Department of Urban Studies and Planning, MIT, Cambridge, MA, USA
e-mail: cameloi@mit.edu

Y. Zhai
College of Architecture and Urban Planning, Tongji University, Shanghai, China

E. Giorgi et al. (eds.), *Design for Vulnerable Communities*, The Urban Book Series,
https://doi.org/10.1007/978-3-030-96866-3_12

district. Driven by the rapid urbanization experienced by the city of Shanghai in the 1990s, thousands of traditional low-rise buildings were demolished to make space for large-scale commercial facilities and office towers. Hongkou stands as an emblematic example of such transformation. However, the commercial development enabled by government-led commercial renewal plans did not bring long-term revitalization. As a result of contemporary challenges regarding the regeneration of large-scale urban shopping malls in Shanghai, this research contributes to the theoretical discussions on how distribution patterns of commercial units affect the development of commercial districts. In doing so, this study argues that modernist urban redevelopment models have impoverished the existing network of small local businesses by either homogenizing the commercial offer of a neighborhood or, on an urban morphology standpoint, promoting block-wide shopping malls. These changes in the built environment were mirrored by a paradigmatic shift in the socioeconomic fabric of Hongkou where low-income communities had to succumb to the development pressure along North Sichuan Road, motivated by rising market values (Fig. 12.1).

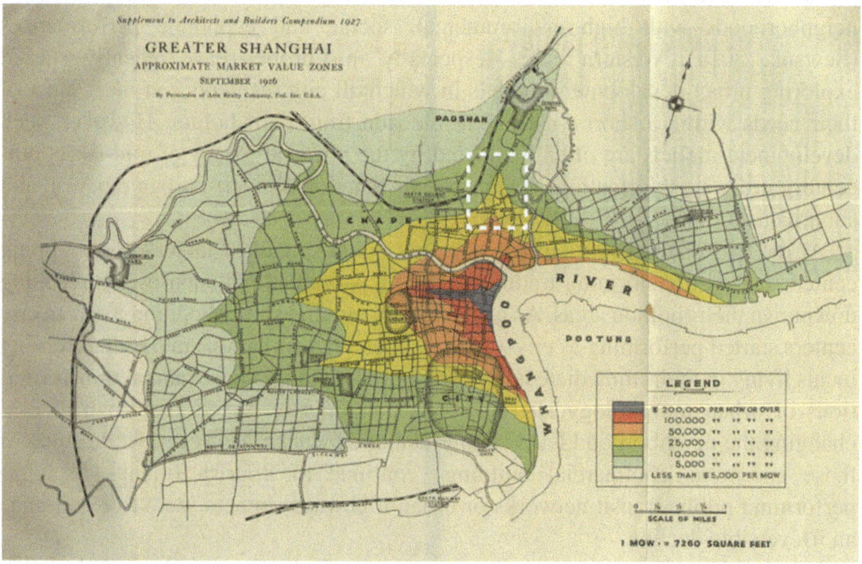

Fig. 12.1 Greater Shanghai—Approximate land value zones (Asia Realty Company, Fed. Inc. USA)—September 1926—Copyright@Virtual Cities Project (Institut d'Asie Orientale). The authors have drawn a white box to identify the central part of Hongkou district

12.2 Literature Review on Distribution Patterns of Commercial Activities

Scholars have widely studied the spatial distribution patterns of commercial activities and how shopping behaviors depend on space. In 1935, Christaller postulated the central place theory and established two fundamental concepts in terms of access range and threshold demand: "the range is defined as the maximum distance consumers are willing to travel for goods; and the threshold of a good implies the minimum amount of demand that must exist in an area for a store to be economically viable (Craig et al., 1984)." Stahl, for example, indicates that retailers tend to position their stores "as closely as possible to the consumers demanding their commodity bundle" (Stahl, 1987). In a perfectly uniform market under the central place theory, the sellers will be equidistant from neighboring stores and serve an equally sized hexagonal market area (Fig. 12.2) (Craig et al., 1984). Thus, centrality theory promotes a microeconomic approach to the analysis of retail locations. Early application of the centrality theory includes the comprehensive research of Isard (1956), Stahl (1987), Vandell and Carter (1993), and Eppli and Benjamin (1994). These scholars investigated the spatial factors determining the location of retail stores under a simplified theoretical condition rather than complex urban conditions. Also, Huff (1963), O'Kelly (1981), Achabal et al. (1982), and Craig et al. (1984) explored different mathematical models to predict potential store locations. However, studying retail environments often requires a more in-depth socio-economic, cognitive and cultural understanding than geometric rules that lack human behavioral considerations.

The centrality theory, in general, stimulated many empirical experiments on towns and market centers (Craig et al., 1984; Dacey, 1964). For example, Hillier has addressed morphological studies in city environments based on centrality theory through space syntax research (Hillier, 1996; Hillier and Hanson, 1984). His work

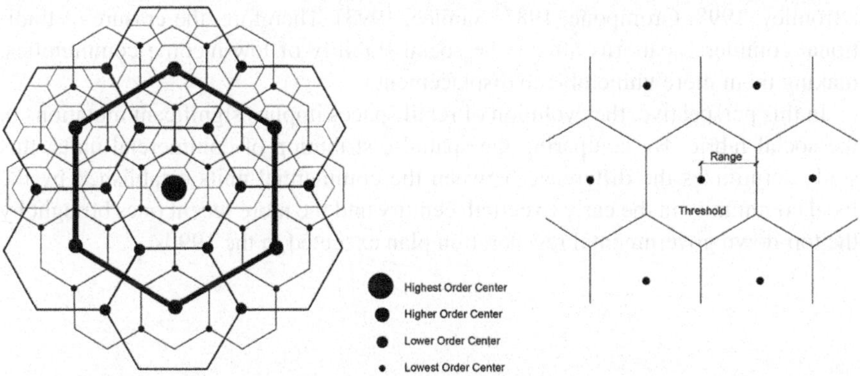

Fig. 12.2 Left: Overlapping variable-size market areas for different types of goods. Right: Identical market areas for identical goods according to the Central Place Theory. Image redrawn based on Walter Christaller's central place theory. Author YUQI ZHAI

focuses on the notions of visibility and integration within the physical structure of the built environment. Waldfogel, Meltzer, and Schuetz have conducted other applications of central place theory. More specifically, they started testing how location directly affects patronage and revenues of retail establishments and, therefore, how it constitutes an essential part of a retailer's production function (Meltzer & Schuetz, 2012; Waldfogel, 2008). Recently, several studies have focused on further examinations of the centrality theory through a detailed analysis of spatial distribution patterns of commercial activities in dense urban contexts. For instance, in 2017, Piovani demonstrated how in London, the locations of well-known retail clusters significantly correlate with the location of road intersections, which in turn attract retail activity (Piovani et al., 2017). Although the application of central place theory on commercial patterns has been studied for more than 80 years, measuring how morphological and typological transformations of commercial units affects the development of specific commercial districts is still unclear.

Centrality can guarantee easy accessibility from both immediate surroundings and more distant locations (Porta et al., 2009). Thus, a central place tends to attract more customers and has a more significant potential to evolve into a social catalyst (Sevtsuk, 2014). In other words, centrality provides an essential base for discussing commercial growth patterns. Central locations in urban areas can sustain higher densities of retail and services, and they are also crucial factors for supporting the formation and vitality of urban "nodes" (Newman & Kenworthy, 1999).

Both traditional and modern commerce nodes are vital elements to form a vibrant neighborhood. However, different retail patterns attract different groups of consumers. For example, while conventional retail facilities (such as grocery shops, cafes, convenience stores) attract customers from all socioeconomic groups, commerce center (such as shopping malls) primarily relies on a high-status clientele (Thomas and Bromley, 2000). Specifically, because of the paucity and unpredictability of financial resources, as well as the lack of refrigerators or privately owned transportation means which mitigates against bulk purchases, low-income groups purchase small quantities frequently, and they rely more on small neighborhood shops (Bromley, 1998; Grompone, 1985; Samiee, 1993). Therefore, the erasure of traditional commerce patterns affects the social stability of low-income communities, making them more vulnerable to displacement.

In this perspective, the evolution of retail spaces implies significant mutations in the social fabric. By comparing the spatial distribution of commercial units, this study scrutinizes the difference between the commercial units established by the local community in the early twentieth century and the more recent ones brought by the top-down governmental regeneration plan executed in the 1990s.

12.3 Historical Context of Commercial Activities in Hongkou District

12.3.1 The Development of Commercial Activities in Hongkou District

In the 1920s, North Sichuan Road in Hongkou district hosted the vast majority of family-run commercial businesses in Shanghai. Given the commercial history of this neighborhood, investigating the spatial transition of commercial units in North Sichuan Road is crucial to understand how large-scale commercial retail has changed the spatial structure of this formerly traditional area (Fig. 12.3).

In the eighteenth century, the Hongkou district used to be a suburb of Shanghai's old town. Later, in the 1900s, along with constructing the bridge over Suzhou Creek and the Woosung Railway, Sichuan Road became the most significant connection between the northern and southern banks of Suzhou Creek. During the 1920s, the north end of Sichuan Road became the third-largest street in Shanghai (after Nanjing Road and Huaihai Road). Japanese expatriates moved into the area and opened factories near the Suzhou river around the south end of Sichuan Road. Thanks to the convenient transportation accessibility provided by Sichuan Road, an increasing number of shops and inhabitants gathered in its north end. In 1940, the number of Japanese stores around North Sichuan Road increased by 823%: from the 65 stores registered in 1938 to 600 stores, accounting for nearly 90% of the total number of shops in the district (Penhui, 2008). Because of this rapid commercial development

Fig. 12.3 Hongkou district in Shanghai. Author CARMELO IGNACCOLO

driven mainly by Japanese developers, North Sichuan Road was then named "Little Tokyo," and it was also recognized as a commercial district by the local planning authorities.

Following the establishment of the People's Republic of China (1949), retail stores, theaters, teahouses, new grocery shops, and fresh food markets were plugged into the traditional urban morphology to meet residents' daily needs. Initially, most of these new shops were built along Sichuan Road. However, in the following years, they started spreading within those minor streets intersecting Sichuan Road, acting as pivotal nodes for community interaction along residential streets.

In the 1980s, just after China's economic reform, Hongkou district attracted various retail businesses in the commercial district of North Sichuan Road. The existing neighborhood shops remained in the original locations, while some medium-sized malls sprouted up in street intersections. These changes were part of a district-wide commercial plan that supported several large-scale commercial developments, including seventh Department Store, International Commercial Shopping Mall, Yuandong Household Electrical Appliances, Kaifu Commercial Building, Hongkou Mall, Duolun Commercial Building, Fashion Commercial, Fudu Mall, Hongye Commercial, the eighth Huaqiao Store, Boyang Store, Dongbao Mall and dozens of merchants stores, such as Hong Kong Hair Salon, Dechang Suit Store, Vienna Leather Shoes, Guangmao Roast Duck and Delicious Tobacco. In contrast to western countries, automobiles were not very popular in Shanghai during the 1980s; therefore, large-scale commercial hubs on both sides of North Sichuan Road generated a pedestrian-friendly shopping area that responded to the local community's needs.

In the 1990s, however, new large-scale commercial typologies replaced entire traditional neighborhood blocks. Residents could not afford to relocate within the same neighborhood where newly constructed condos with two-bedroom apartments were out of their price range. For the vast majority of Hongkou residents, there was no other option but relocating to the city's outskirts, leaving behind their original social networks and small-sized housing units (Wu, 2004). As a result, many residents were forced to move out of their neighborhoods and relocate to the city's outskirts. At the same time, thousands of migrants from Chinese extra-metropolitan areas became the new residents of Hongkou district. Most of them lived in the inexpensive shared accommodation units of traditional *lilong* houses. However, despite the spatial proximity to low-income communities, the newly built commercial centers chose a business model oriented towards luxury brands to gain city-wide commercial popularity. This business format disregarded the daily needs of Hongkou's new residents who could not afford high-end products. Thus, these shopping malls relied on high-income consumers across the city. In other words, urban transformations in Hongkou district triggered major socioeconomic shifts: local residents had to relocate to the outskirts of Shanghai, and low-income migrant workers moved to some poorly maintained residential buildings. Newcomers, while initially benefiting from the availability of low-skilled jobs in the newly developed malls (such as cleaning and cooking services), have remained with no access to walking-distance street commerce they could potentially afford.

Among the follow-on effects of these neighborhood changes, large-scale shopping malls brought a rapid increase in the real estate land value (11 billion RMB in 1994, 12.72 billion RMB in 1995, and 13.65 billion RMB in 1996). This steep increase made Hongkou district the second most profitable commercial district in the Shanghai downtown area after the Nanjing Street district. North Sichuan Road, for example, has an average daily flow of approximately 900,000 people, and together with Nanjing Road, Huaihai Road, Jinling Road, and Yuyuan Shopping Mall, it forms a Shanghai-level commercial center, commonly known as "The Four Streets" (Desvaux et al., 2002). Even in the early 2000s, along with the continuous construction of large-scale shopping malls, the development of North Sichuan Road Shopping Center has shown positive financial results, especially in terms of the rapid growth of retail sales.

In 2006, however, the North Sichuan Road commercial district witnessed a substantial decline in its business performance. Scholars have argued that there were mainly two reasons for this sudden decline: first, two mega-commercial centers (Daning International in Jingan district and Wujiaochang in Yangpu district) became the major shopping destinations for high-income customers for the entire city of Shanghai; second, more medium-sized community shopping malls opened in the periphery of the city. Consequently, the increasing residential community in the outskirt of Shanghai stopped going to North Sichuan Road for their daily shopping. Under these business circumstances, the 1990s commercial facilities in Hongkou did not maintain their customer attractiveness. Without entertainment facilities and local residents' demand, the now so-called "old-fashioned 1990s retail malls" in Hongkou district fell into decline. Nowadays, compared to other large-scale commercial centers, the North Sichuan Road shopping options are not as well regarded as those located in other commercial districts of Shanghai.

12.3.2 A Comparison of Commercial Patterns in Hongkou District before and after the '90s

The traditional commercial units of North Sichuan Road were mostly small-scale shops arranged according to a linear layout. Maps of business licenses in 1937 reveal the location of small offices, restaurants, and retail stores. Furthermore, it is utterly apparent that most of them were located along the streets, and some of them were even in the inner part of residential blocks. Most of these small shops were managed by individuals and enterprises from Hongkou district. Thanks to a tight connection between the sellers and the clients, stores kept having a profitable business throughout the mid-twentieth century.

The 1980s improvement strategy of commercial units maintained the location of the original store network and supported the construction of medium-sized malls as a strategic upgrade of the traditional commercial network (Table 12.1). In general, the sellable goods and the distribution of commercial facilities along the streets

Table 12.1 Comparison of commercial units between 1940 and 1980

Attributes	1940	1980
Business form	Retail facing the street	Shopping mall
Business owner	Private retail and enterprises	Large real estate development company
Spatial layout	Linear along the street	Multistory large-scale building
Operation model	Individual or enterprises operation	Subletting mode
Business content	Retail, service	Retail, service, entertainment
Property management rights	Individual or enterprises	Managed by the developer
Funding sources	Individual or state-owned properties	Quoted company (finance)

Source: Table redrawn based on *Table 1.1* in Penhui (2008). *The Effect of Shopping Center on Society Network of North Sichuan Road*. Tongji University; College of Architecture and Urban planning, master thesis

were generally appreciated by the local community: North Sichuan Road was unquestionably the most popular commercial street in Shanghai around the 1940s and during the 1980s.

However, by the early 1990s, due to the inability of commercial facilities to meet the needs of recently urbanized consumer groups and the competition of modern commercial districts (such as the neighboring Xujiahui), the commercial district of North Sichuan Road gradually started losing its economic vibrancy. The commercial facilities from the early twenty-first century were usually designed as city-block-wide shopping malls and operated under a unified management framework guided by foreign corporations or companies outside the Shanghai region. Because of the corporate-driven development, early twenty-first-century shopping malls targeted a very wealthy clientele in search of famous fashion brands, cinemas, and gourmet restaurants. In general, commercial facilities from the 1990s did not trigger a long-term improvement of the commercial attractiveness of the neighborhood. Instead, they generated a mid-scale commercial typology that is now in search of a new identity.

12.4 Methods

12.4.1 Case Studies along Sichuan Road in Hongkou District

This research uses two blocks in Hongkou district as selected case studies where to compare (1) the urban morphology and (2) the distribution of commercial facilities before and after the 1990s urban renewal (Fig. 12.4). More specifically, the selected blocks are located in the southern part of Hongkou district: the first block (A) is enclosed by Sichuan Road and Duolun Road; the second block (B) is bounded by

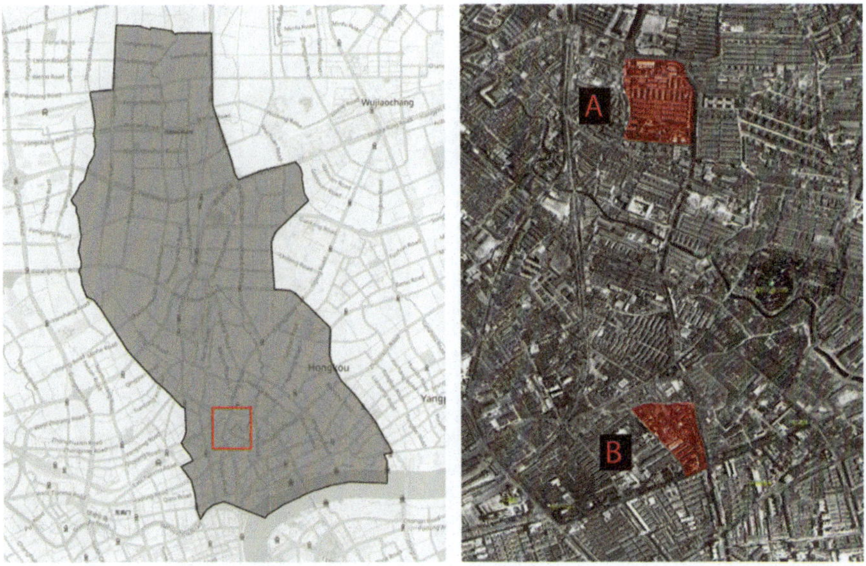

Fig. 12.4 Left: Hongkou district. Right: selected blocks for case study analysis in Shanghai. Author CARMELO IGNACCOLO

Qiujiang Road, Zhongzhou Road, and North Sichuan Road. Regardless of their mutual proximity (1.13 km) and their similar dimensions (0.97 sq^2 for block A and 0.88 sq^2 for block B), the two blocks experienced opposite urban transformations during the urban renewal.

Block A has almost entirely kept its original urban morphology made of low-rise *lilong*: rows of two-story housing units are still located in the core of the blocks (E-W orientation). Their neighboring housing units have the same typology but different orientation (N-S). In general, the block is divided into three different areas: the northern part facing Sichuan Road comprises 5–7 floors residential buildings with commercial units on the ground floor. The central region has kept its original *lilong* urban configuration. The southern part, instead, has undergone some changes in its building typology: the tabula rasa of small-scale housing units and vacant lots offered precious development space for large commercial buildings both facing Sichuan Road and the southern part of Duolun Road. The large east-west plot on the southern side of the block used to host offices and printing machines for a newspaper company. While in 1937 this plot was mostly vacant except for the newspaper company building, it experienced an intensive infill development process during the 1990s. Three six-story condos occupy the eastern and western edges of the plot, hiding the southern part of residential low-rise houses (Fig. 12.5).

Differently from Block A, Block B experienced a radical transformation of its building types and urban form. In the late 1990s, a curvilinear shape tower with commercial facilities on the ground floor was built at the intersection of Sichuan Road and Qiujiang Road (Fig. 12.6). By 2018, the rest of the block, which used to

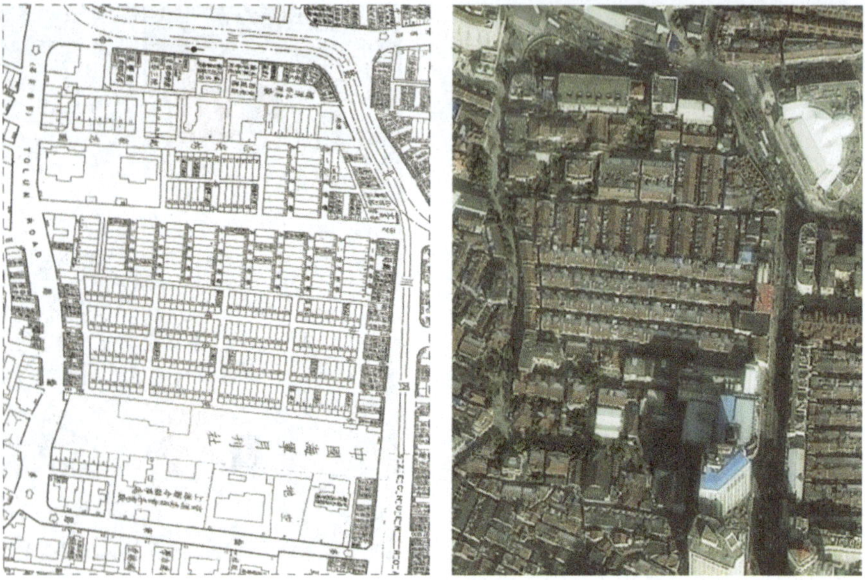

Fig. 12.5 Left: 1937 Shanghai map of Block A. Right: 2018 satellite imagery of Block A. Author CARMELO IGNACCOLO

Fig. 12.6 Left: 1937 Shanghai map of Block B. Right: 2018 satellite imagery of Block B. Author CARMELO IGNACCOLO

encompass two-story small mixed-use typologies, was erased in favor of the 35-floor One Prime Tower built in 2010 and the shopping mall located at the footsteps of the tower. The northern part of the block hosts a gated residential compound of two 20-floor towers. In addition to an overall change in the building typology, the block has experienced a severe street design transformation. Since 2009, the newly built Hengshui Road has divided the block into two distinct areas: the northern part with a much more defined residential character and the southern part with a predominant mixed-use character. Because of the different development outcomes, these blocks are ideal case studies to unpack the evolution of commercial units' types and their distribution patterns in relation to changes in urban form.

12.4.2 Data Overview

The project employs two novel datasets: the 1937 high-resolution map of business licenses in Hongkou district (Shanghai) and the Gaode Points of Interest (POIs) data extracted through the Gaode API in September 2018. The authors have georeferenced the 1937 map using the UTM 51N projection system in Geographic Information System software (ESRI ArcMap). By detecting and translating Chinese characters contained in some of the 1937-map plots, it was possible to classify commercial units located in Block A and B into eight types (café and teahouse, public service, office, restaurant, clothing store, convenience store, rickshaw and taxi, and craftsmen workshop). Commercial units and their attributes were digitized and converted into georeferenced shapefiles of incident points (Fig. 12.7).

The second dataset contains the 2018 Gaode georeferenced POIs for the Hongkou district. This dataset includes more than 58,000 observations for the Hongkou area. Given the research focus on two specific blocks in Hongkou, the two selected samples contain 372 items in Block A and 532 items in Block B. Gaode data items have several attributes for each commercial unit, such as address, unit type, and customer ratings. In line with the aim of this research, the authors have omitted Gaode points containing attributes about ATMs, entrances to residential units, and parking lots. As a result of this data filtering, Block A and B, respectively, count 268 and 397 commercial units. Taking into consideration the comparative scope of this research on commercial types and their locations, the authors have grouped the 1937 and 2018 data into fourteen categories: café and teahouse, public service, office, restaurant, clothing store, convenience store, book-café and souvenir shops, hotel and entertainment, museum, antique store, jewelry and beauty, tech. and electric appliances, photo studio, and education.

In 1937 Block A encompassed mainly convenience stores (28.7%), restaurants (20.2%), and office spaces (23.4%) (Table 12.2). The block edge along North Sichuan Road used to host more commercial units than the western edge along Duolun Road. This antipodal distribution might be related to lower rates of pedestrian traffic along Duolun Road, which still acts as a minor and narrower infrastructural link compared to still heavily trafficked North Sichuan Road.

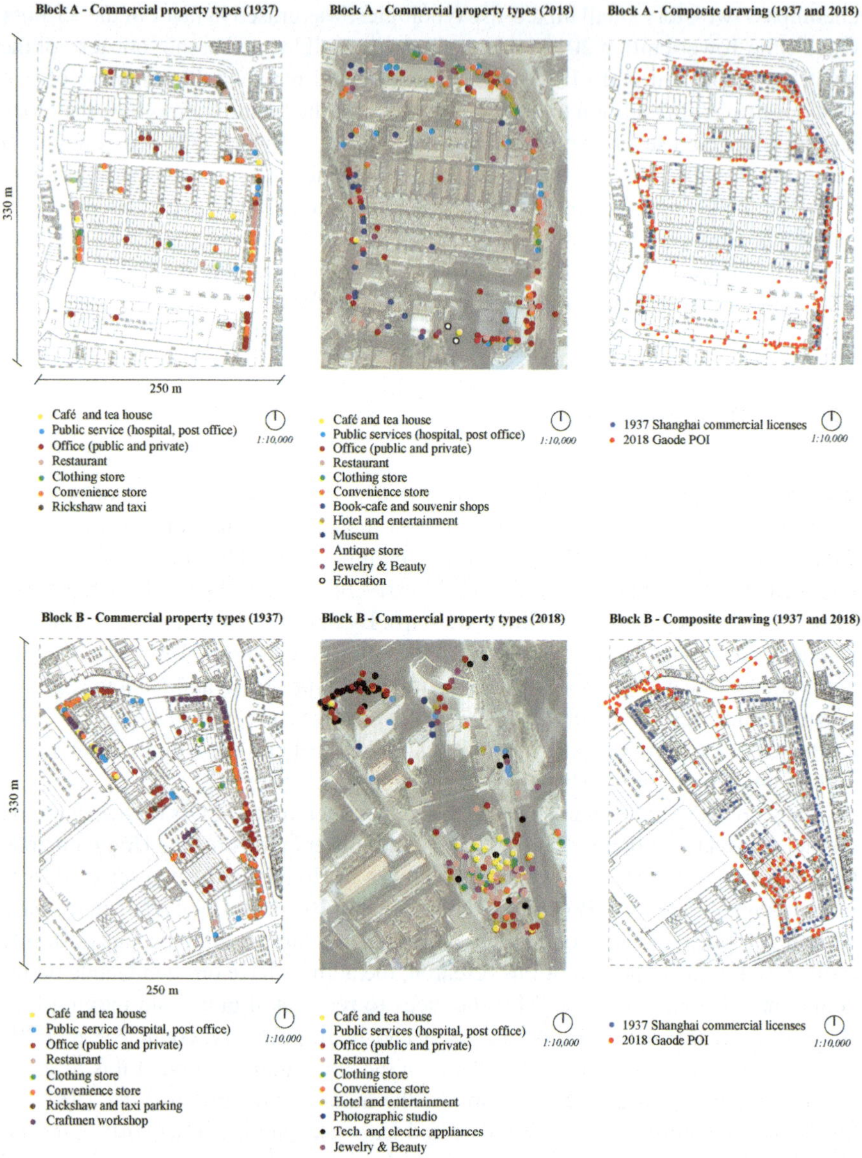

Fig. 12.7 Top-left: commercial property types in Block A (1937); top-center: commercial property types in Block A (2018); top-right: composite drawing with commercial properties (1937, blue; 2018, red); bottom-left: commercial property types in Block B (1937); bottom-center: commercial property types in Block B (2018); bottom-right: composite drawing with commercial properties (1937, blue; 2018, red). Author CARMELO IGNACCOLO

Block B mainly had an industrial character in 1937. Commercial property types at the intersection of North Sichuan Road with Qiujiang Road included carpenters, metal craftsmen, and plumbers (21.26%) (Table 12.3). Similar to Block A, Block B

Table 12.2 Types of commercial units in Block A (1937 and 2018)

Types	1937 (#)	1980 (%)	2018 (#)	2018 (%)
Café and teahouse	9	9.6	10	0.8
Public service	5	5.3	31	11.9
Office	22	23.40	66	25.4
Restaurant	19	20.20	43	16.5
Clothing store	6	6.40	10	3.8
Convenience store	27	28.70	42	16.2
Rickshaw and taxi	6	6.40	/	/
Antique store	/	/	10	3.8
Book-cafe and souvenir	/	/	19	7.3
Education	/	/	6	2.3
Hotel and entertainment	/	/	8	3.1
Jewelry and Beauty	/	/	18	6.9
Museum	/	/	5	1.9
TOTAL	*94*	–	*268*	–

Authors CARMELO IGNACCOLO and YUQI ZHAI.

Table 12.3 Types of commercial units in Block B (1937 and 2018)

Types	1937 (#)	1980 (%)	2018 (#)	2018 (%)
Café and teahouse	6	4.72	32	8.06
Public service	10	7.87	19	4.79
Office	39	30.71	151	38.04
Restaurant	4	3.15	53	13.35
Clothing store	3	2.36	16	4.03
Convenience store	36	28.35	26	6.55
Rickshaw and taxi	2	1.57	/	/
Craftsmen workshop	27	21.26	/	/
Tech. and electric appliances	/	/	52	13.1
Education	/	/	7	1.76
Hotel and entertainment	/	/	8	2.20
Jewelry and Beauty	/	/	27	6.8
Photo studio	/	/	6	1.51
TOTAL	*127*	–	*397*	–

Authors CARMELO IGNACCOLO and YUQI ZHAI.

had many office spaces (30.71%) and convenience stores (28.35%), but it lacked the network of cafés and teahouses dotting Block A. From a spatial distribution standpoint, office spaces do not seem to rely on being located along heavily trafficked street fronts. Craftsmen workshops and convenience stores, instead, are highly dependent on street front exposure. Figure 12.7 (bottom-left), in fact, shows two clusters of office spaces in the inner part of the block.

12.5 Data Analysis and Results

12.5.1 Data Analysis

This study employs a two-step data analysis process. First, it examines the spatial distribution of commercial units throughout time; second, it investigates whether the typological composition of commercial units has changed in both blocks between 1937 and 2018.

The spatial distribution research borrows the methods from Mitchell (2005) and Anselin (1995), where (1) Point Density analysis is used to determine the density of incident point data and (2) Getis-Ord Gi* index is used to render "hot spots" and "cold spots" of spatial features.

The Point Density tool calculates the density of point features around each output raster cell. Conceptually, a neighborhood is defined around each raster cell center, and the number of points that fall within the neighborhood is totaled and divided by the area of the neighborhood. Borrowing the methodology from Silverman (1986), the neighborhood radius is the shortest of the width or height of the output extent divided by 30. The radius utilized in this study is 8.78 m.

The hot spot analysis was carried out using a fishnet shapefile generated according to the extension of the selected blocks as bounding polygons containing the incident point data of commercial units. The aggregation process resulted in 355 squares (10m edge) weighted by the number of commercial units contained within each square's perimeter. The dimension of the square units was calculated through the ANN method (Average Neighbor Distance): for all of the unique location points excluding locational outliers, the ANN is computed by summing the distance to each feature's nearest neighbor and dividing by the number of features (n).

In regard to shifts in street commerce types, this study calculates percentages of commercial spaces as fractions of the whole sample of commercial units in each block. It then quantifies the percentage change across 80 years by comparing the 1937 typological composition of commercial units to the one of 2018. In doing so, it engages with path-dependence theory on the persistence and change of street commerce clusters. More specifically, it demonstrates that the legacy of pre-development commercial licenses (1937) persists in the spatial and typological arrangements of some commercial units in contemporary Shanghai.

12.5.2 Results

The results of the data analysis show how changes in urban morphology have triggered re-arrangements in the distribution of commercial units from 1937 to 2018. In Block A, where the configuration of the urban block has not changed throughout the last 80 years, the distribution of commercial units is still mainly centered along North Sichuan Road. However, the Point Density analysis reveals how the density of commercial units has increased in two street intersections: Sichuan Road and Shanyin Road (northwest corner) and Sichuan Road and Duolun Road (southwest corner). These results are in line with the existing literature on commercial units' distribution which expects store owners to locate their businesses at points of maximal demand, such as street intersections (Stahl, 1987). More specifically, on the northeast corner of Block A, the average density of commercial units has almost doubled (154% increase, from 240 shops/ha in 1937 to 618 shops/ha in 2018). In general, no changes in the urban form of Block A have resulted in a higher density of commercial units throughout time with specific peaks in the adjacencies of street intersections (Fig. 12.8).

The comparison between commercial property types in 1937 and 2018 sheds light on the persistent cultural and historic character of Block A. In fact, while in 1937 there had been approximately nine café and teahouses, in 2018, the neighborhood counted ten café and teahouses and 19 book-café and souvenir stores. The original functions of twentieth-century teahouses have been gradually replaced and enhanced by many book-cafés dotting the neighborhood today. However, in terms of shops serving daily needs, this study reveals a steep decrease in convenience stores (−77%). This result seems in line with the literature on amenity-districts where amenities for visitors are often prioritized over residents' needs. The presence of amenities fostering social interactions combined with antique stores and museum spaces has made Duolun Road a prominent cultural district of Shanghai. This transformation started in 1998, when the local government launched a regeneration in order to counter degrading infrastructures and poorly maintained *lilong*—usually occupied by low-income households. The project focused on conserving and restoring historic buildings (such as the Xi Shi Bell-Tower) by turning them into museums, galleries, cafés, or craft shops.

Unlike Block A, Block B has experienced a radical change in its morphology throughout the last 80 years. Because of these extreme changes and, therefore, the impossibility of comparing commercial units on the same street network, this study

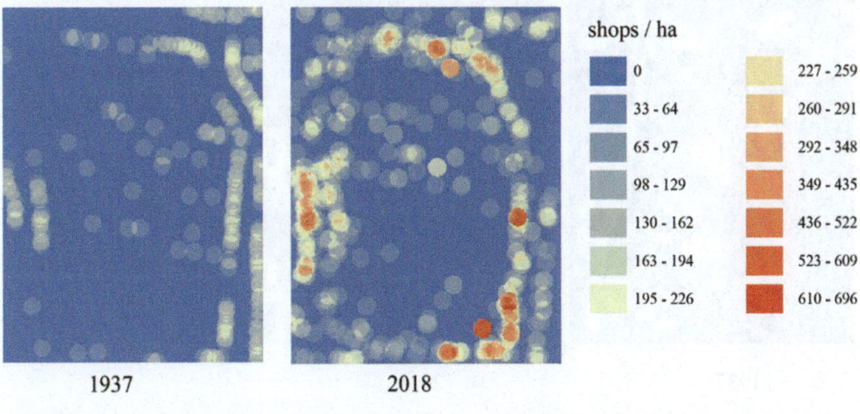

Fig. 12.8 Comparative point density analysis of commercial units between 1937 and 2018 (shops/hectares). Author CARMELO IGNACCOLO

utilizes an optimized hot spot analysis which creates a map of statistically significant hot and cold spots using the Getis-Ord Gi* statistics. The local Getis-Ord Gi* evaluates the spatial dependency effect of the number of commercial units within 10m X 10m polygonal features. It uses a calculation that considers each of the features' values and those of its neighboring ones. As a result of this operation, the hot spot analysis gives a z-score for each feature in the dataset, which can be significantly positive or negative with a minimum confidence level of 90%.

The results show a statistically significant hot spot of commercial units in the southeast corner of the block where the "Shengbang International Plaza" is located (Sichuan Road and Wujin Road). The Gi* index in this area of the block reflects a 90% confidence interval with values comprised between 3.51 and 2.81 (Fig. 12.9). This peak might be a result of changes in both building typology and in block configuration. In fact, this block has two shopping malls: "One Prime" and "Rose Plaza." Rose Plaza is a gated residential community with commercial facilities, and One Prime is a large shopping mall.

From a typological standpoint, the longitudinal comparison of commercial units in 1937 and 2018 reveals interesting results. In 1937, Block B was generally considered an industrial neighborhood rich with craftsmen and industrial makers. This research indicates that part of this industrial legacy has persisted into the contemporary block, despite aggressive urban transformations. In fact, the northwest part of the block facing Qiujina Road and Zhongzhou Road hosts a large mall containing dozens of small electronic stores. Thus, while craftsmanship has transitioned from steel/wood to electric appliances, the neighborhood has maintained—at least in the northwest corner—a maker culture. However, further investigation is required to provide evidence on a plausible ownership nexus between pre-development craftsmen and contemporary electricians and tech experts.

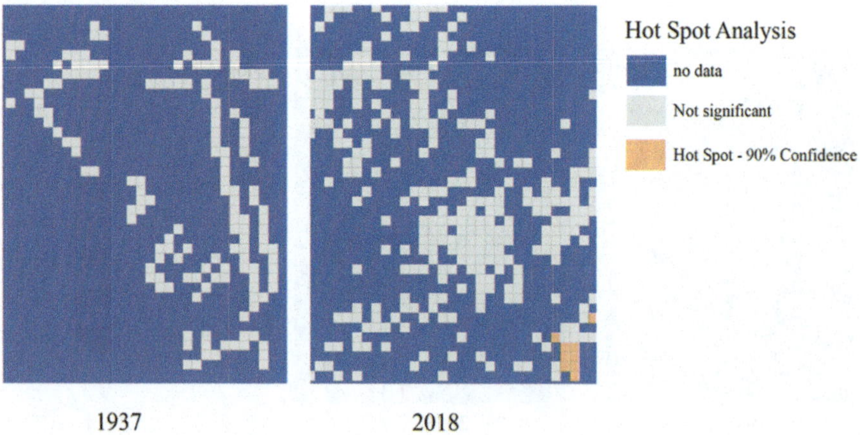

Fig. 12.9 Comparative hot spot analysis of commercial units between 1937 and 2018 in Block B. Author CARMELO IGNACCOLO

12.6 Discussion and Conclusions

Rapid urbanization in Shanghai has contributed to a drastic reconfiguration of the city morphology and to typological changes in residential and commercial units across the city (Whitehand & Gu, 2007). Low-rise traditional buildings were torn down to make space for 20-story condominiums—often developed within gated communities—and small family-run commercial activities were rapidly replaced by shopping malls strategically located at the intersections of heavily congested streets.

Even if the shopping malls constructed in the early-mid 1990s during the government-led urban renewal had a considerably profitable business for around 10 years, they are now experiencing a period of general economic decline. These negative trends are often caused by (1) a mismatch between shops and customers living in the same neighborhood and (2) high competition between the 1990s mall generation and the more glamorous and sophisticated twenty-first-century shopping malls. Within the scope of contributing to the literature debate on commercial streets and to the policy discussion on the future of the Hongkou district, this research has conducted a comparative study of the distribution and types of retail units from 1937 to 2018. This comparison provides empirical evidence of changes in locations and types of retail units; and it contributes to the theoretical debate on the relationship between people's livelihoods and street commerce.

The result of this comparative study is fourfold: first, it demonstrates how certain blocks managed to keep a vibrant streetscape with small shops spatially arranged on a much denser layout than in 1937 (Block A); second, it suggests that large-scale shopping malls have impoverished the formerly vibrant streetscape of Hongkou district by allocating shops into multistory malls; third, this study demonstrates that the legacy of 1937 neighborhood-retail identity tends to persist in the typological arrangements of contemporary Shanghai, despite major urban transformations; fourth, it proves that the rising number of amenities such as museums, book-café, and souvenir stores has come at the expense of local residents living in areas with an ever decreasing amount of convenience stores serving their daily needs (−77%).

Although the centrality theory has proven itself as a valuable tool for studying urban areas, little work has been done to apply its concepts to historical research on specific traditional districts in China. Moreover, this study employs a big-data research approach on both historical data (1937) and contemporary data on retail units in the city of Shanghai (2018). Thanks to a spatial comparison of commercial patterns recorded in 1937 and 2018 in the North Sichuan Road area, this study sheds light upon the morphological implications connected to the evolution of commercial patterns in downtown Shanghai. Further research could illuminate the current performances of shopping malls and small retail units through Gaode maps reviews and rating data. On a methodological note, a longitudinal model could provide a more granular understanding of businesses' performances throughout the selected 80-year time range.

Learning from the evolution of street commerce patterns is essential to address contemporary challenges related to the profitability of commercial facilities and the vulnerability of Hongkou's residents. Furthermore, from a policy standpoint, this study could contribute to a policy effort on re-branding the district of Hongkou, especially along Sichuan Road, and make it again attractive to investors and locals. Finally, by avoiding neighborhood-wide relocations of low-income groups and providing vulnerable communities with in situ alternative and affordable housing solutions, this study envisions the future of Hongkou as a socially inclusive and diverse commercial hub for both neighborhood residents and the whole city of Shanghai.

References

Achabal, D. D., Wilpen, L. G., & Mahajan, V. (1982). MULTILOC: a multiple store location decision model. *Journal of Retailing, 58*(2), 5–25.

Anselin, L. (1995). Local indicators of spatial association-LISA. *Geographical Analysis, 27*(2), 93–115. https://doi.org/10.1111/j.1538-4632.1995.tb00338.x

Bromley, R. (1998). Market-place trading and the transformation of retail space in the expanding Latin American City. *Urban Studies, 35*, 1311–1333. https://doi.org/10.1080/0042098984376

Craig, C. S., Ghosh, A., & McLafferty, S. (1984). Models of the retail location process: A review. *Journal of Retailing, 60*(1), 5–36.

Dacey, M. F. (1964). Modified poisson probability law for point pattern: More regular than random. *Annals of the Association of American Geographers, 54*(4), 559–565. https://doi.org/10.1111/j.1467-8306.1964.tb01784.x

Desvaux, G., Li, G., & Fenhirin, J. (2002). Shanghai shopping. (China Research). *The McKinsey Quarterly*, 17+. Gale Academic OneFile.

Eppli, M., & Benjamin, J. (1994). The evolution of shopping center research: A review and analysis. *Journal of Real Estate Research, 9*(1), 5–32. https://doi.org/10.1080/10835547.1994.12090737

Grompone, R. (1985). *Talleristas y Vendedores Ambulantes en Lima*. DESCO, Centro de Estudios y Promocion del Desarrollo.

Hillier, B. (1996). *Space is the machine: A configurational theory of architecture.*

Hillier, B., & Hanson, J. (1984). *The social logic of space*. Cambridge University Press.

Huff, D. L. (1963). A probabilistic analysis of shopping center trade areas. *Land Economics, 39*(1), 81. https://doi.org/10.2307/3144521

Isard, W. (1956). *Location and space-economy: A general theory relating to industrial location, market areas, land use, trade, and urban structure*. M.I.T. Press.

Meltzer, R., & Schuetz, J. (2012). Bodegas or bagel shops? Neighborhood differences in retail and household services. *Economic Development Quarterly, 26*(1), 73–94. https://doi.org/10.1177/0891242411430328

Mitchell, A. (2005). *Spatial measurements & statistics*. ESRI Press.

Newman, P., & Kenworthy, J. R. (1999). *Sustainability and cities: Overcoming automobile dependence*. Island Press.

O'Kelly, M. E. (1981). A model of the demand for retail facilities, incorporating multistop, multipurpose trips. *Geographical Analysis, 13*(2), 134–148. https://doi.org/10.1111/j.1538-4632.1981.tb00721.x

Penhui. (2008). *The Effect of Shopping Center on Society Network of North Sichuan Road.* [Master Degree Thesis]. Tongji University, College of Architecture and Urban planning.

Piovani, D., Molinero, C., & Wilson, A. (2017). Urban retail location: Insights from percolation theory and spatial interaction modeling. *PLoS One, 12*(10), e0185787. https://doi.org/10.1371/journal.pone.0185787

Porta, S., Strano, E., Iacoviello, V., Messora, R., Latora, V., Cardillo, A., Wang, F., & Scellato, S. (2009). Street Centrality and Densities of Retail and Services in Bologna, Italy. *Environment and Planning B: Planning and Design, 36*(3), 450–465. https://doi.org/10.1068/b34098

Samiee, S. (1993). Retailing and channel considerations in developing countries: A review and research propositions. *Journal of Business Research, 27*(2), 103–129. https://doi.org/10.1016/0148-2963(93)90018-K

Sevtsuk, A. (2014). Location and agglomeration: The distribution of retail and food businesses in dense urban environments. *Journal of Planning Education and Research, 34*(4), 374–393. https://doi.org/10.1177/0739456X14550401

Silverman, B. W. (1986). *Density estimation for statistics and data analysis*. Chapman and Hall.

Stahl, K. (1987). Chapter 19 Therories of urban business location. In *Handbook of regional and urban economics* (Vol. 2, pp. 759–820). Elsevier. https://doi.org/10.1016/S1574-0080(87)80005-6

Thomas, C. J., & Bromley, R. D. F. (2000). City-centre revitalisation: Problems of fragmentation and fear in the evening and night-time city. *Urban Studies, 37*(8).

Vandell, K., & Carter, C. (1993). Retail store location and market analysis: A review of the research. *Journal of Real Estate Literature, 2*(2), 13–45. https://doi.org/10.1080/10835547.1994.12090037

Waldfogel, J. (2008). The median voter and the median consumer: Local private goods and population composition. *Journal of Urban Economics, 63*(2), 567–582. https://doi.org/10.1016/j.jue.2007.04.002

Whitehand, J., & Gu, K. (2007). Urban conservation in China: Historical development, current practice and morphological approach. *Town Planning Review, 78*(5), 643–670. https://doi.org/10.3828/tpr.78.5.6

Wu, F. (2004). Residential relocation under market-oriented redevelopment: The process and outcomes in urban China. *Geoforum, 35*(4), 453–470. https://doi.org/10.1016/j.geoforum.2003.10.001

[illegible]

Chapter 13
Climate Urbanism in the Post-pandemic World: Mapping Vulnerabilities and Exploring Community Activism in East London

Giulio Verdini and Corinna Dean

13.1 Introduction

Major urban areas have been hit by the COVID pandemic in an unprecedented way, revealing old and emerging socio-economic and environmental vulnerabilities. London is no exception, especially due to long-lasting problems of housing affordability, deprivation, and the persistency of environmental management challenges typical of post-industrial global cities. On the other hand, the pandemic has also stimulated new bottom-up demands of quality public and green space, and urban facilities that will likely to last beyond this period.

The spreading of the global pandemic of COVID since early 2020 exposed cities and nation states to unprecedented challenges, determining a rapid reshaping of the global agenda of sustainability. While the UN Sustainable Development Goals and the new 'Urban Agenda' have already placed emphasis on tackling together social disparities and environmental challenges, COVID has demonstrated the need of more holistic approaches to take into account a more complex horizon of resiliency. This is the key message of the UN-HABITAT Report 'Cities and Pandemics: Towards a More Just, Green and Healthy Future' (2021). According to the report, the current global public health crisis has determined raising inequalities; has revealed structural problems of neighbourhoods, cities and regions in terms of their form, functions and effective governance; and requires now to envision a green recovery which cannot underestimate climate change, the next great new challenge the world will need to face in the years to come.

It is too early to understand whether efforts towards a more resilient urban future will generate a genuine green recovery, as advocated from various sectors of the civil society and governments, but there are already controversial aspects that

G. Verdini (✉) · C. Dean
University of Westminster, London, UK
e-mail: g.verdini@westminster.ac.uk; c.dean@westminster.ac.uk

E. Giorgi et al. (eds.), *Design for Vulnerable Communities*, The Urban Book Series,
https://doi.org/10.1007/978-3-030-96866-3_13

populate the debate towards the implementation of effective and inclusive low-carbon cities (The Guardian, 2020). Nevertheless, from the point of view of academic scholars which train architects and planners to design future low-carbon interventions in cities, it is essential to strengthen their analytical ability to understand such complexity, looking at how urban spaces materialise these contradictions and how it is possible to design places, envisioning context-specific and fine-grained opportunities for an effective transition towards sustainability.

This chapter, therefore, explores the development and application of a new integrated framework to understand and map city vulnerabilities, tested during an urban and landscape design studio (called 'Climate Urbanism Studio') targeting the study of the Lower Lea Valley in East London. This tool attempts to address together economic shocks, pandemics and climate challenges, revealing the narrow margin in the real world to effectively take on board such challenges in a meaningful way. The underlying assumption of this exercise is that measures to counter the exacerbated vulnerabilities in cities might not produce any relevant and transformative change addressing issues of climate justice, as advocated by the new set of theories of 'Climate Urbanism' (Castán Broto et al., 2020), but rather another version of the business as usual. The proposed framework supports this argument with evidence from the fieldwork implemented. On the other hand, there are promising bottom-up experiences showing the resilience of local community and their demand for more sustainable and inclusive neighbourhoods. In this respect the studio is taken as an opportunity to reflect on practices of sustainability and their potential scalability in a complex area like London. The ultimate goal is to reinforce the effectiveness of a new incipient climate pedagogy, which often risks to be narrowly confined to addressing pure sectoral environmental problems.

The Climate Urbanism Studio ran from January to April 2021. It is a semester-long studio for year 2 students of the undergraduate course BA Designing Cities: Planning and Architecture at the University of Westminster.[1] The brief states explicitly that '*The studio investigates how to build up climate resilient post-pandemic cities and neighbourhoods, capitalizing on existing learning from the COVID-19 outbreak*'[2]. The case study selected is the Lower Lea Valley in East London, stretching north to south from the Elisabeth Park in Stratford, to the Leamouth and the river Thames. The river coincides with the administrative border between the neighbourhoods of Tower Hamlets and Newham and, for this reason, it has remained relatively untouched by large property-led developments of the London Docks. It is an important ecological resource for this part of the city, with the potential to reinforce its role of social and environmental infrastructure for the surrounding residential areas. The area hosts a very diverse population, including some of the most deprived areas of the UK, for example, Aberfeldy Village in Poplar; new gentrified developments, such as Limmo Peninsula; and relatively underdeveloped areas hosting

[1] The BA Designing Cities course page: https://www.westminster.ac.uk/architecture-interiors-and-urban-design-courses/2021-22/september/full-time/designing-cities-planning-and-architecture--ba-honours and its related blog: http://blog.westminster.ac.uk/designingcities/

[2] Quoted from the Studio brief distributed to students at the beginning of the semester in early 2021.

creative industries and art communities, such as Trinity Buoy Wharf, both in the Lea Mouth of Tower Hamlet (Fig. 13.1).

In the next section a series of theories and concepts from the emerging field of climate urbanism will be introduced, relating environmental and climate challenges to wider issues of social inequality and economic deprivation. This will be followed by another section that explores the feature of the mentioned case study of East London, looking at its social and environmental history until recent, and not fully implemented, attempts to develop the Lea River Park since its conception in the late 2000s, as part of the Olympic Games works[3] (Nicholls, 2014).

The second part will instead introduce the overall methodology adopted to map the area, including the framework of sustainability to assess resiliency against all hazards and design climate actions. This will be followed by a series of real case studies from the Lower River Lea, utilised to inspire the studio explorations, and a discussion session that highlights the critical aspects to take into account when moving from theory to practice.

Fig. 13.1 East London with the area of the Lower Lea Valley object of study. Source: Authors, pictures: Giulio Verdini, based on UK digimap base (https://digimap.edina.ac.uk). © Crown copyright and database rights [2021] Ordnance Survey (100025252), reproduced with permission

[3] For more information about the River Lea Park: https://www.queenelizabetholympicpark.co.uk/the-park/venues/parklands-and-playgrounds/waterways-and-rivers/lea-river-park

13.2 Theory

13.2.1 Emerging Vulnerabilities in Times of Pandemic and How they Could (Or Should) Inform a Critical Climate Urbanism

In introducing 'Unsustainable Inequalities' Chancel argues that 'reducing inequality is inseparable from the attempt to fundamentally alter our relationship to the environment' (2020, p.1). While warning of potential social and institutional resistance to change, he argues that disadvantaged groups might benefit the most from environmental protection and the fight to climate change, in the long run. However, the battle to achieve that should be played both in the national, or supra level, policy arena, and at the local level. This would avoid selfish local behaviours that, despite promising attempts to initiate transition towards sustainability, would replicate locally broader inequalities. It is the case, for instance, of the increasingly debated topic of green gentrification that, in the absence of equity-oriented public policies, would accelerate the displacement of low-income communities in the face of urban greening improvements (Gould & Lewis, 2016).

The aspect of the quantitative improvement of the provision of urban green, which is important in the emerging field of climate urbanism, is even more problematic in light of the recent COVID pandemic. Urban green and, in general, quality public space has been advocated as an essential component to heal the wounds of this health crisis, particularly from the point of view of communities' well-being as a series of recent publications from very different contexts are showing (Berdejo-Espinola et al., 2021; Mayen Huerta & Utomo, 2021). In this respect, it is legitimate to believe that the risk for any climate-resilient post-recovery strategies would be to exacerbate existing social inequalities, if the equity dimension is not taken seriously into account in the first place.

Overall, climate urbanism is associated with the new emerging trend of cities to promote themselves as 'viable and appropriate sites of climate mitigation and adaptations' and to protect them 'from the hazards associated with climate change' (Long & Rice, 2019, 993). However, such discourse has essentially gained ground as a way to protect urban economies, but has failed so far to provide enough elements to appreciate the social justice impact of climate change. It is in this spirit that a recent book on climate urbanism has been published, in the attempt to put forward a more critical research agenda (Castán Broto, et al. 2020).

Climate urbanism should therefore favour the implementation of both urban mitigation and adaptation strategies to climate change, while having the ambition to be truly transformative, ethical and democratic (Long et al., 2020). For this purpose a series of potential pathways of climate urbanism are highlighted: reactive, entrepreneurial and transformative. Reactive climate urbanism refers to actions to reduce ex post noticeable impacts of climate change; entrepreneurial refers instead to actions that assume the fight to climate change as a way to boost city competitiveness; and, finally, transformative refers to actions that are capable to alter the status quo by mobilising multiple stakeholders in a more inclusive way, addressing the unequal distributions of climate costs, especially among vulnerable communities (Castán Broto, et al., 2020). While intuitively cities will need to embrace a meaningful combination of each pathway, it is from its power to generate collective projects around climate change, engaging local communities and the diversity of civil society, that climate urbanism can unveil its transformative and inclusive potential.

In this respect, ensuring wider and meaningful engagement of a different variety of stakeholders in climate urbanism projects is surely a *conditio sine qua non* for building up more inclusive approaches, tackling more effectively emerging vulnerabilities. It would not, however, necessarily ensure equity, as initially mentioned, especially if such practices are disjointed by genuine inclusive public policies and practices, with the risk of participation being a vehicle to give only voice to the powerful ones or to misrepresent the marginal ones (Beebeejaun, 2006). In other words, the risk of the local trap that assumes the local scale as inherently democratic should be avoided, and a careful assessment of the local conditions should be made (Purcell, 2006).

The COVID pandemic has exacerbated even further inequalities in cities. The pandemic has in fact only relatively impacted white collars jobs, that could be replaced by safer home working, but it has disproportionally impacted low-income jobs crucial to carry on some essential works. In other words, 'the impact of the virus has diverged according to geography and social class, with the least privileged people and places normally seeing the worst effects' (Florida et al., 2021, p.4). The potential legacy of the new structure of work together with the new needs of households, assuming a part will remain as it is now, may further increase the quest for suburban and greener locations. However, this digital-led centripetal force might have an impact only at the microgeography scale of cities, but not necessarily questioning agglomeration forces as a whole (Florida et al., 2021). Therefore, while knowledge workers and creative will still gravitate around cities, delivering their work in hybrid forms, the rest, on the contrary, will be even more precarious or made redundant by the accelerated digitisation imposed by new economy. This is something that very likely is going to stay in place for long, and it has already transformed cities, as the case study of East London will show later on.

Climate urbanism will be increasingly implemented in every city, given the emphasis on city climate actions that is made in international and national agendas especially in light of the UN Climate Change Conference (Cop26) in Glasgow.[4] However, the way in which it will be interpreted and adapted at the local level will make a huge difference, from the point of view of prioritising actions, generating more or less holistic frameworks, and eventually tackling effectively inequality.

The capability of reducing inequalities will be, in the long run, the only way to make any climate policy really politically and socially viable, therefore missing the opportunity of being socially inclusive, would prevent or at least affect its overall efficacy (Chancel, 2020). However, while there is a great deal of advocacy for

[4] For more information about COP26: https://ukcop26.org

setting up effective climate resilient post-pandemic strategies (UN-HABITAT, 2021) this might be in risk to remain an optimistic rhetoric. The idea so largely acclaimed that green recovery should be a way to emerge from the crisis better equipped, could hide the risk of not being able to tackle effectively inequality in the long run. Additionally, the current pandemic has not only exacerbated social inequality in a generic way, but it has raised even further people's exclusion by class and ethnic connotation (Florida et al., 2021). This seems to be the biggest challenge.

The framework proposed in the second part of this chapter is an assessment tool to help design and hopefully build climate-resilient post-pandemic strategies, potentially applicable both in pedagogy and practice. It attempts to analyse the various dimensions of climate urbanism (Castán Broto, et al. 2020) together with emerging vulnerabilities. However, the result of this exercise cannot be discussed in abstract terms. This is imperative in any attempt to put forward new and genuinely locally informed climate pedagogies and practices, where place-based understanding of problems is crucial to determine concrete ways forward (Verdini et al., 2019). It is for this reason that a case study approach is utilised here. In the next chapter the context of East London will be introduced and methodologically the framework will be applied to such context.

13.2.2 The Case of East London: A Socioeconomic and Environmental History of Vulnerabilities

This section will introduce the relevance of East London to a debate on climate urbanism with a focus on the Lee Valley, which is undergoing major acceleration of urban development resulting in contested urban trajectories. A consideration of this can be positioned in relation to Doreen Massey's argument which calls for a reinvigoration of the spatiality of our cities, as discussed in *For Space* (2005). With a focus on London she argues that political and social issues are battled out within, 'a confrontation between imaginations of the city' (Massey, 2005, p. 157). So despite London being a successful city in terms of finance as exemplified in the physical manifestation of the City and its economy, it is because of this, not despite, that we still have great areas of poverty and exclusion. If we are to respond to Massey's proposition to reimagine the city, in response to the River Lee, there is a cacophony of social identities and rich typologies expressed within its diversity of spatial conditions, informal occupancies (a lot recently cleared to make way for the Olympics), occupying the spatial margins of the city, and houseboat dwellers, all of which could drive an enlightened way of reimagining how space is shaped. The architect Mann writes in 'Bastard Countryside' (2003) about mixed landscapes, describing the history of the industrial hinterland, celebrating the non-pedigree typography of the site. Our approach to the site therefore introduces to current thinking about how to engage with the non-human. Clark quotes, '*What seems to underpin the new cosmopolitan environmentalism … is the premise that left to itself, nature is docile;*

it maintains its given forms and positions. Culture on the other hand, is seen to be inherently dynamic' (Clark, 2002, p. 107).

In our current climate this is no longer the case, and as we highlight in the case studies, nature and its assemblages contribute to a balanced system of ecology between water, land and interstitial spaces, a natural habitat of reedbeds is used to filter out nitrates, and other nature-based solutions are being reintroduced. By considering non-human agency of active agents it is possible to engage with new forms of knowledge making. Bennet, the political theorist, argues for a more dynamic involvement with the environment and the need to recognise the active participation of non-human forces in events (Bennett, 2010).

Rivers have always played an important part in the development of urban conglomerations encompassing a variety of functions and uses, related to agriculture, industrial locations, and forms of transportations, and witnessing, in its evolution, a balance between natural ecologies, history and culture. The history of the River Lea is primarily one of industrialisation. It performed the role as London's back goods yard, providing connections for developing industry, supporting agricultural areas for growing foodstuffs as well mills, and most recently light industry. This has left a negative ecological legacy of ground and water pollution, and altered and engineered river ways, which has had a detrimental effect on the natural flood plains. The River Lea is the most engineered river in England (Environmental Agency, 1947). In 1424 an Act was issued to improve the navigability of the river. The Act constituted a commission drawn from local landowners who were responsible for implementing improvement works funded by tolls. In 1571 a second Act was passed with the intention of increasing the supply of grain into the City, by improving navigation on the Lower Lea. These works included the construction of water cuts and tow paths. In the seventeenth century disputes recorded between the mill operators and Commissioner of Sewers, accused them of penning back the water and causing flooding and the silting up of the river channels (London Borough of Newham, 2006). The mouth of the River Lee is marked by the remnants of the East India Dock Basin, which stood at the head of the colonial shipping trade, to east, the exoticism of the trade stuffs marred by the slave exploitation and associated insidious practices. This dock area remains an abandoned relic of the past. The entire East London, especially along the prime locations of the Thames, has been for long object of massive urban regeneration processes, reverting, in some cases, the entire image of this area, from a former and deprived post-industrial wasteland, to a global hub of finance and services, especially in the area of Canary Wharf, on the West side of the River Lea, with concerted attempts to bring property development to the East, in the area of Royal Docks (FT, 2021a).

Most recently there has been a focus on the area of Stratford, as a site for regeneration with the staging of the London 2012 Olympics and the legacy of the event (Nicholls, 2014). The creation of this new urban quarter with swathes of landscaping was intended to usher in new job creation, and increased mixed tenure housing to readdress the high levels of overpopulated spaces. The reality is that the opportunities has been limited, economic deprivation is still high and public health discrepancies persist. In terms of the physical landscape, the ecological planting around the

Olympic Park is compromised by the existing inadequacies of the Thames Water sewerage system, evidenced after heavy rainfall with visible raw sewerage and debris being pumped into the wildlife ponds (Thames Waterkeeper). The combined sewerage overflow system permits raw sewerage to be pumped directly into rivers, with nitrates and heavy rainwater runoff from adjacent road surfaces.

During the pandemic the areas immediately east of the River Lea, with their persisting high level of social deprivation, were reported in the Financial Times describing the three boroughs—Barking and Dagenham, Redbridge and Newham, as the 'Covid Triangle'. According to the article: '*The restricted residential tenures expressed in the highest number of multiple occupancy per capita in London and zero hours contracts have been exasperated by Covid, and this is under the shadow of the Olympic Legacy which was up held as an intervention which would have a long term ameliorative effect across classes to improve health, employment and housing*' (FT, 2021b). The article drew focus to the discrepancy between white collar workers able to work from home though digital platforms in contrast to the zero hour employees working in hospitality, catering, cleaning, or as taxi drivers and other insecure employment. Overall the River Lea crosses some of the most deprived areas of London as Fig. 13.2 shows. Therefore to understand the context of our approach to climate urbanism it is important to realise the existing levels of deprivation, before testing tools and creating layers of understanding and responses.

To reveal the interrelations between the urban actors involved in shaping future identities and visions for the sites, a number of different sources were used: empirical data collection, grey literature and local reports, and engagement with key local actors. A major source of information is the publication Cinderella River Report to respond to the Lee Valley Hydrocitizenship project, a combined academic and community participation research project to challenge societal assumptions that the river is merely a utility (Read, 2017). Key actors consulted included the non-statutory bodies operating in the area, such as the Canal & Rivers Trust, which has been mandated by government to maintain England's river and canals as well as engage the public with the heritage and ecology of waterside sites.

One of the key proposed interventions into the area was to extend the existing Lee Valley Regional Park southwards from the Olympic Park at Stratford to the River Thames at East India Dock Basin. This was in response to a historical proposal by Patrick Abercrombie in the 1944 London Plan, to connect the Green Belt to the Thames. The project client, the London Thames Gateway Development Corporation stated their commitment to create a new urban park of regional importance, named Lea River Park. This was stated in a document delivering the principles of the design. Due to successive changes in local government, the phased development suffered numerous funding delays. The patch work state of the land around the Lea existed as dead-ends, fractured routes and off-limits industrial edges.

Tom Holbrook from 5th Studio, main consultant of the Lea Valley Park project, describes the practice's approach to project, then titled 'the Fatwalk', as installing '*vanguard top-down infrastructure moves that will enable further, bottom-up projects*' (Wainwright, 2010, p.13). Here he refers to the introduction of a foot bridge crossing over the A13 a major east/west artery and two other crossings. The Lower

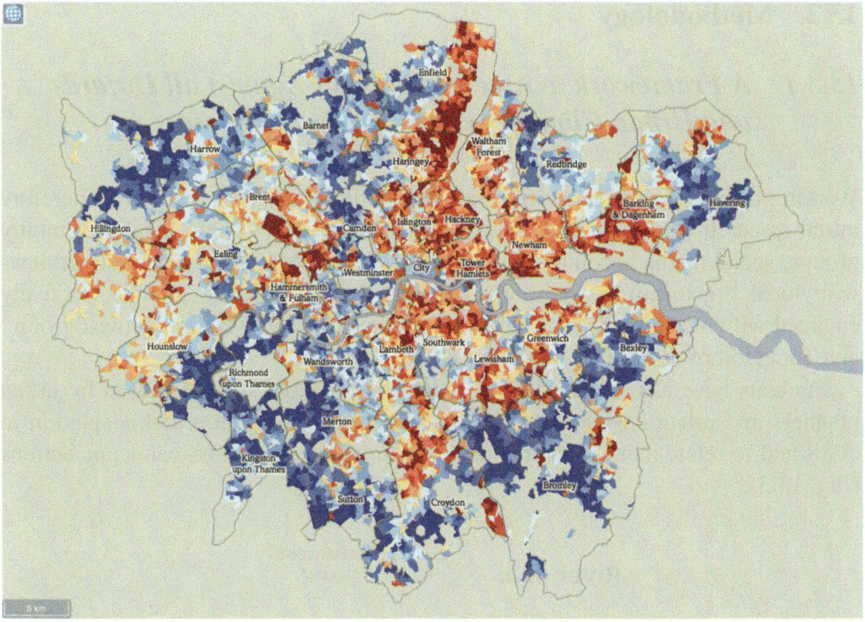

Fig. 13.2 Map showing the English index of multiple deprivation (rebased for London), 2019. In dark red the most deprived decile and in dark blue the least deprived. The index takes into consideration income, employment, crime, living environment, education, health and barriers to housing and services. Source: Trust For London, from UK Open Government Licence (Public Domain). Available at: https://www.trustforlondon.org.uk/data/index-multiple-deprivation-2019-rebased-london/

Lea runs from Stratford High Street, which neighbours the Olympic District and the head of the river which runs into the Thames.

On winning the competition for the Design Framework in 2006, 5th Studio's preliminary work created a description of the remarkable existing landscape, in order to value and distinguish it from the Olympic Park design which took a tabula rasa site after a series of clearing the site of allotments and east London activities of small-scale business. The design direction was to work with what was there and what is latent in the peripheral and marginal zones. Carefully mapping the existing characteristics, the object was to create links between and to the river's bank. Connections running from east to west are poorly serviced leaving residents who live nearby with blockages both physical and imagined to reach the bank. There is little access to open space '*statistically speaking, each inhabitant of the adjacent residential areas has only a third of the open space recommend for London*' says Holbrook (Meyer and Schlaich, 2012).

13.3 Methodology

13.3.1 A Framework to Assess Resiliency Against all Hazards and Urban Climate Actions in Design Studios

Working during the pandemic, the site was initially analysed remotely. Therefore initial research was desk based which involved researching into local community groups, social media accounts, and official websites. Moreover, remote interviews with local organisations were organised as well as meeting members actively engaged with the site such as Celia Coram from Save Lea Marshes activist group,[5] and Tom Holbrook from 5th Studio.[6]

Students have also been asked to reflect on potential strategies used by urban planners and urban designers to facilitate integrated mitigation and adaptation in cities and to relate them to the building of climate resilient post-pandemic actions (Fig. 13.3).

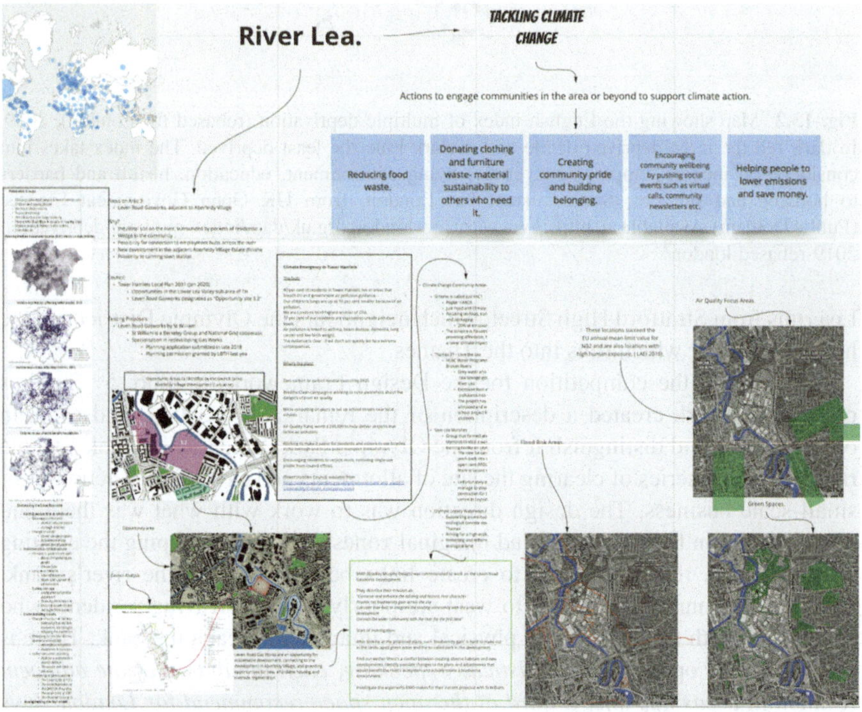

Fig. 13.3 Strategies to tackle climate change. Analysis from the Studio Miro Board. Source: Students of BADC Year 2

[5] For more information about Save the Lea Marches: https://www.saveleamarshes.org.uk

[6] For more information about the architectural and urbanism practice fifth Studio https://www.5thstudio.co.uk

Strategies taken in consideration, according to the *Second Assessment Report of the Urban Climate Change Research Network* (Raven et al., 2018), are the following:

1. Reducing waste heat and greenhouse gas emissions through energy efficiency, transit access, and walkability.
2. Modifying form and layout of buildings and urban districts.
3. Use of heat-resistant construction materials and reflective surface coatings.
4. Increasing vegetative cover.
5. SUDs (sustainable urban drainage systems)

In our studio students have primarily reflected on point 1 (transit access/walkability) and 4 (vegetative cover), touching also point 2 and 5, especially in their group analysis.

Consequently, an analytical framework was proposed and suggested to guide the collection of data, cases, practices, and precedents (Fig. 13.4). The framework enabled students to design climate resilient post-pandemic interventions, by understanding challenges as highlighted by recent international organisations' guidelines and mapping social, environmental and economic vulnerability (UN-HABITAT, 2021). It proposes to envision opportunities by considering and critically selecting one or more actor in their investigation (local communities, local government, private sector), active in their study area of the Lower Lea River. This is aligned with the three suggested pathways of climate actions (Castán Broto et al. 2020).

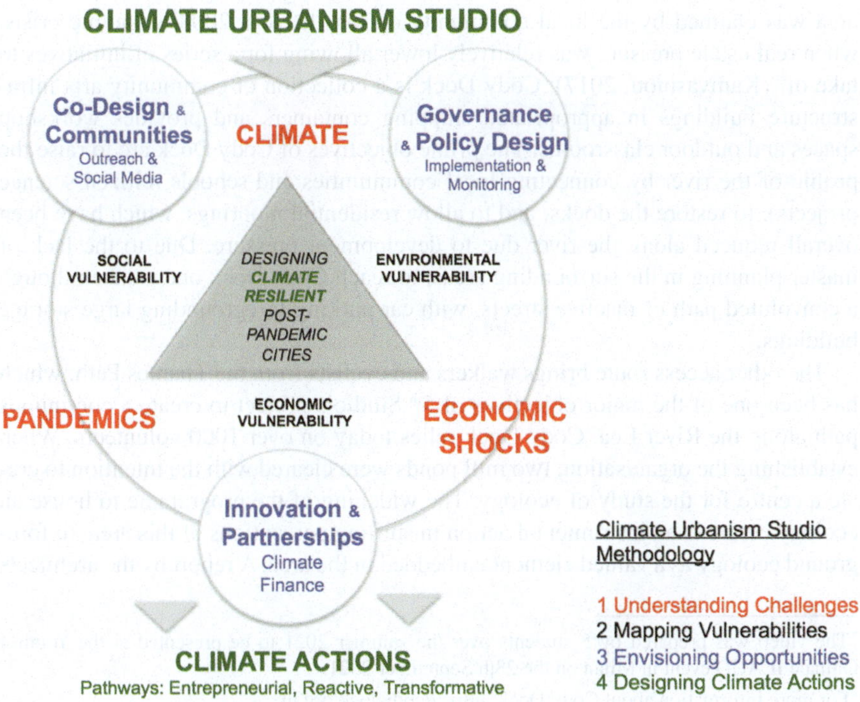

Fig. 13.4 The climate urbanism studio framework. Source: Authors

13.4 Case Study and Findings

13.4.1 The Lower Lea Valley in East London: Studio Explorations

The following section selects examples of the chosen students' real case studies, used to suggest tentative design strategies, and addressing key general and more specific questions debated in the studio. In particular: What are the social and cultural barriers in terms of accessing and appropriating green and blue public spaces across the River Lea? Are existing forms of community engagement truly inclusive, taking into account a wider range of vulnerable people? And more specifically, are there synergies between new bottom-up demand of quality space induced by the pandemic and incipient forms of transformative and just climate urbanism? Are there instead any associated risks in terms of inclusion?

To address the above a series of interviews were set up with key members of three local communities during the studio period and beyond in preparation of a video about the Lea Valley[7]: Cody Dock, the Levan Road Gasworks, and Trinity Buoy Wharf (Fig. 13.5).

Cody Dock is a community-based organisation, named after the original dock, built in the eighteenth century.[8] The site is leased to a community organisation, founded and directed by Simon Myers, and sits in the Lower Lea Valley amongst a number of serviced buildings of London, such as large storage warehouses. This area was claimed by the local community during the post 2008 economic crisis, when real estate pressure was relatively lower allowing for a series of initiatives to take off (Kamvasinou, 2017). Cody Dock is a collection of community arts infrastructure buildings in appropriated shipping containers and provides workshop spaces and outdoor classrooms. The prime objectives of Cody Dock are to raise the profile of the river by connecting local communities and schools (citizen science projects), to restore the docks, and to allow residential moorings, which have been overall reduced along the river due to development pressure. Due to the lack of master planning in the surrounding areas, to reach Cody Dock one is taken though a convoluted path of inactive streets, with car parking foregrounding large storage buildings.

The other access route brings walkers and cyclists from the Thames Path, which has been one of the major objectives of 5th Studio's project to create a continuous path along the River Lea. Cody Dock relies today on over 1000 volunteers. When establishing the organisation, two mill ponds were cleared with the intention to create a centre for the study of ecology. The widening of the programme to house an ecology centre is an instrumental action in shifting perceptions of this area, to foreground ecology as a valued element embedded in the area. A report by the architects

[7] The video was prepared by 5 students over the summer 2021 to be presented at the Youth 4 Climate ILAUD event in Milan on the 28th September 2021.

[8] For more information about Cody Dock: https://codydock.org.uk

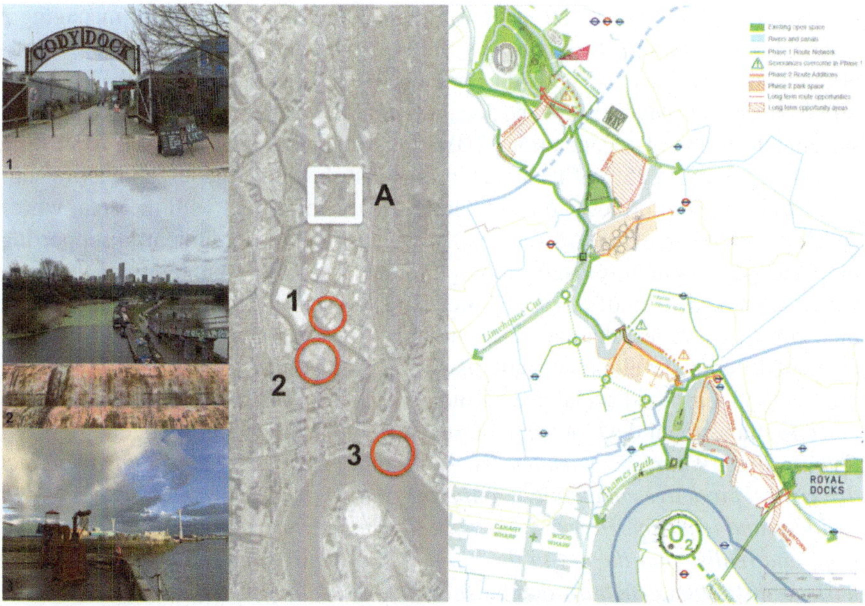

Fig. 13.5 The three areas selected on the left side Google map are: (1) Cody Dock; (2) the Levan Road Gasworks; and (3) Trinity Buoy Wharf. Number A is the Channelsea Island that was also discussed in the studio. On the right side there is the River Lea Framework 2007, as designed by 5th Studio. (Full map available at: https://www.5thstudio.co.uk/projects/lea-river-park-the-leaway/). Source: Authors composition, pictures: Giulio Verdini

Witherford Watson Mann of the Upper Lea Landscape Strategy (2008) commissioned by a wealth of stakeholders from London Development Agency to Thames Water, states that the diversity of landscape identities and mixed characters is celebrated alongside the acknowledgement of the piecemeal nature of disconnected developments and ad hoc industrial sites, '*this simply leads to a form of mutual degradation, where one neglected or defensive edge becomes the cause of another*' (WWMA and JCLA, 2008, p. 18). According to the interviews realised, Cody Dock has been object of rediscovery from the local inhabitants, especially during the partial lockdown. Due to limited possibility to travel and commute to central London locations, the River Lea has been an important resource for local people, who in some cases have become aware of Cody Dock and its initiatives for the first time. However, while the community initiatives promoted at Cody Dock have been set up in an inclusive and open way, the area, although located in Newham, is secluded from the rest of borough, being surrounded from industrial areas, and resulting to be more accessible towards Tower Hamlets. While this is not a clear indication of the social diversity of users, given the lack of data, it is intuitive the industrial area creates a barrier towards the more multicultural and deprived core of Newham.

The second case study is the arts organisation Trinity Buoy Warf (TBW), which is located adjacent to the East India Dock Basin, in an environmentally fragile area of London.[9] The director, Simon Cowan, talks about the challenge of managing a former brownfield site surrounded by industry wastelands against the recent proposal of construction of 55,000 housing units. He discusses the role of the organisation as a pivotal player in placing local community at the core of on-going more sustainable forms of regeneration. TBW has a unique model where it is let on a 124 years lease and 25% of its annual rent is distributed by the leaseholder to support local arts. The project upholds a long-term involvement with the area, where development is carried out incrementally establishing a commitment in supporting the local arts community over a long period of time. The complexity of the area, which is markedly divided between the characteristic of the Upper Lea Valley and the Lower Lea Valley, demonstrates the local divisions along the River Lea, as a result of different development patterns. Similar to Cody Dock, although profoundly different in their broad social and cultural aim, this area represents a form of resistance to mainstream development, as stated in the mentioned 'Cinderella Report': '*Trinity Buoy Wharf is unashamedly, if alternatively, marketed. This is a project driven by a proactive community group, who, although they have no money, are high on ideas and idealism. However, they operate in the context of inexorable development and realise that the key to the promotion of their plans lies in their ability to harness institutional support and partnership, which they manage with tremendous ingenuity and determination*' (Read, 2017, p. 17). During the pandemic, however, this area has suffered a real desertification, given the impact of COVID on art activities and cultural industries, as in many other similar contexts (UNESCO, 2021). Therefore, this is today questioning the future development of the site as an inclusive cluster of arts and culture, in a relatively marginal and more affordable area of London. The risk for TBW is to gradually fall into unavoidable residential-led pressure, like the neighbouring scheme of Limmo Peninsula, given the environmental attraction of the Lea Mouth.[10]

The third case here considered is the regeneration area of Leven Road Gasworks, a former industrial infrastructure of the decommissioned Gas Works in Tower Hamlet, on the West side of the River Lea, which had been earmarked for residential development by the London Based developer St William.[11] The rhetoric broadcasted by the developer is to act as an opportunity for sustainable development, connecting to a deprived area of Aberfeldy Village and providing opportunities for new affordable housing and riverside regeneration. This is set against the council's objective Climate Emergency in Tower Hamlets in which 40% of residents in Tower Hamlets live in areas which breach EU and government air pollution guidance, linked to

[9] For more information about Trinity Buoy Wharf: https://www.trinitybuoywharf.com

[10] For more information about the residential-led scheme of Limmo Peninsula: https://limmopeninsula.co.uk

[11] For more information about Leven Road Gaswork regeneration project: http://www.levenroadgasworks.co.uk

asthma, heart disease, and dementia. Optimistically, the council has set a target to be Zero carbon neutral by 2025.[12]

The developer states their aim to connect the wider community with the river for the first time and to conserve and enhance the existing and historic river characteristic, although these statements are difficult to quantify. A closer look to the plan highlights the segregated vision for the site from the neighbouring buildings, and minimal landscape intervention along the river for the pure sake of beautification of the site. The master plan appeared not to engage with the wider master planning issues and the Design Code, as this area was originally designated as a park, due to the lack of green space provision in the Lower Lea Valley. Due to constant changes in funding bodies and lack of overall strategic vision, the Design Code is not consulted. The largest obstacle to the work carried out by 5th Studio is in Holbrook's words the lack of '*constituency*' to carry the project through so that it has long-term supporters and is kept visible. This seems to be a constant issue in terms of operating on the larger scale level, as sites cross council boundaries, and with it stakeholder boundaries and interests. The intricacies and cross-referenced stakeholders' values and objectives reveal the complexity in shaping the urban landscape, and the risk that, in the absence of coordinated development, it is difficult to pursue the common good for the area.

An additional key observation important to add is that the pace of on-going urban transformation, potentially accelerated by COVID, might impact on the existing lack diversity of spaces' use. This arose from the 3-year study of communities around the River Lee, discussed in the Cinderella River Report, that focused on the problems of diversity and engagement with public and semi-public green spaces within the River Lea, stating that '*the current programme of management is paternalistic: volunteers accept and embrace a particular culturally indexed view of landscape, access, what is parkland, biodiversity etc. This may not be shared across all ethnic groups, classes and age ranges*' (Read, 2017, p. 158). In order to engage more thoroughly with levels of access to green/blue spaces it was not fully possible to respond to this issue, due to limited data available. Moreover, data collection from the council's website of QGIS mapping tended to be too generic and denoted ethnicity relating only to residential areas. In order to engage fully with this topic, qualitative interviews and observations would need to take place. This is an aspect that will need to take into account to monitoring trends of inclusions (or decreased diversity in accessing public and green space) across East London.

The final consideration, touched during the studio, was about climate urbanism and whether, and under which conditions, this can really help reimagine urbanism as a site of meaningful negotiation between the environment, public space and lived experience. In responding to the above question, it is worth noting that the River Lea remains a green/blue infrastructure with a certain level of accessibility, and a series of very diverse community-led initiatives that have preserved its environmental

[12] See: https://www.towerhamlets.gov.uk/lgnl/environment_and_waste/Sustainability/Climate_emergency.aspx

value. However, piecemeal developments are threatening the residual flexibility of the space, imposing a more rigid and privatised use. There are, nevertheless, areas still relatively untouched, such as the case of the ecological intervention of Channelsea Island in the River Lee.[13] This intervention, promoted by the charity Canal & River Trust (CRT),[14] solely ensures remote access, therefore allowing the natural conditions of the site, an island within the canal system, linked only to the riverside by a disused railway bridge, to flourish. The ecology of the river has been abused and still remains heavily polluted. Through looking at nature-based solutions, values are placed on the ability of the existing reedbeds alongside extensive work carried out by the ecology department of CRT to replant the verges along the River which have been artificially constructed causing the loss of natural habitats and root systems crucial to stabilising the River banks. It is an intervention that aims to reinstate the so-called third landscape, the space left over by man for nature to evolve (Clément, 2004). Spaces like this are rare in the area, and their conservation beyond speculative appetites could help introduce different and more inclusive narratives to the River Lea.

13.5 Discussion and Conclusion

East London, and particularly the area along the River Lea, is an area of persisting inequality. The impact of the pandemic has strongly hit this part of the city and it will be very likely to increase such trends in the near future.

From an environmental history point of view, the River Lea is an extremely fragmented and highly engineered natural space, showing an unsustainable pressure on the area of years of piecemeal developments, pollution, illegal and unregulated waste discharge into the river. Those developments have also favoured processes of gentrification in the area. It is only in recent years that advocacy has been made to revert this image, especially with the proposal of realising the Lea River Park, on the side of the large-scale investments for the Olympic Games in 2012. Currently, the area is suspended between the difficulties of implementing the park, further eroded by new developments, and new emerging bottom-up forces that are claiming for a new life of the river nurturing its ecology, alongside a greater provision of the green space and paths. This has been amplified during the pandemic when most Londoners have rediscovered the River as a space for their well-being, especially during the lockdown. It is promising as advocacy for improving access to green space has been probably one of the most recurring consequences of the pandemic.

The pandemic therefore seems to have accelerated a gradual process of community re-appropriation of the area, which has set up an incipient bottom-up environmental agenda of care of the river. It is indeed a positive aspect that however needs to be carefully contextualised. From the point of view of climate urbanism, which

[13] For more information about Channelsea Island: https://canoelondon.com/exploring-londons-channelsea-river/

[14] For more information about the Canal and River Trust: https://canalrivertrust.org.uk/about-us

seeks to decarbonise urban transformation and regeneration processes in cities, it is crucial to understand who can benefit from these actions and their overall impact. The climate urbanism studio presented in this chapter helped map and correlate vulnerabilities and insurgent practices of environmental activism. The three cases presented show interesting experiences. Cody Dock is an area where bottom-up grass roots organisations are setting up social enterprises, advocating for more sustainable and inclusive practices. They also witnessed a huge increase of use of the river walk and their facilities as a result of the pandemic and the rediscovery of local communities. Trinity Buoy Wharf is an art and cultural cluster that needs to redefine its identity in the post-COVID period. It is located in an environmentally fragile area, in risk of being object of residential speculation. The area of Leven Road Gasworks is a heritage area 'waiting to be transformed', which could easily fall into the development trap of other regeneration projects in London. Lastly, the additional case of Channelsea Island is an area without human access, that could be utilised as wildlife area and nature reserve.

Overall these existing interventions challenge the pro-development housing model so far perpetuated demonstrating a variety of enforced environmental improvements that have the ambition to conserve the residual natural environment. However, one might legitimately ask to what extent this can be conducive of more inclusive urban transformation in the post-pandemic period, reinforcing instead existing social divide in the area, and consequently whether climate urbanism can further generate urban (green) gentrification. The last question posed in the studio on whether, despite all, climate urbanism can help to reimagine urbanism as a site of negotiation between the environment, public space and lived experience is a problematic one. It is evident that the River Lea is a space of increasing activism, but it is still too early to assess whether it can be truly transformative. The risk, in fact, is that the new green demand in the post-pandemic period will come mostly from those who can afford it, accelerating the development pressure, and resulting in small scale green aesthetic adjustments, rather than in substantial improvement of green space provision that a serious climate agenda would require and deprived areas would need.

References

Beebeejaun, Y. (2006). The participation trap: The limitations of participation for ethnic and racial groups. *International Planning Studies, 11*(1), 3–18.

Bennett, J. (2010). *Vibrant matter: A political ecology of things*. Duke University Press.

Berdejo-Espinola, V., Suarez-Castro, A., Amano, T., Fielding, K., Rui Ying, R., & Fuller, R. (2021), People and Nature, *3*, 597–609.

Castán Broto, V., Robin, E., & While, A. (2020). *Climate urbanism. Towards a critical research agenda*. Palgrave Macmillan.

Chancel, L. (2020). Unsustainable inequalities. In *Social justice and the environment*. The Belknap Press of Harvard University.

Clark, N. (2002). The demon-seed: Bioinvasion as the unsettling of environmental cosmopolitanism. *Theory, Culture and Society, 19*(1–2), 101–125.

Clément, G. (2004), Manifeste du Tiers Paysage, Artlibre. Available at: https://www.caue-nord.com/SPASSDATA/attachments/2006_10/06/5f7f3d9a6fac7-d43183.pdf

Environment Agency. (1947). *River lee floods 50 years on.* Environmental Agency.

Financial Times. (2021a), 40 Years on: My Docklands, 13 June 2021.

Financial Times. (2021b), Inside the 'Covid Triangle': a catastrophe years in the making, 4 March 2021. Available at: https://www.ft.com/content/0e63541a-8b6d-4bec-8b59-b391bf44a492

Florida, R., Rodríguez-Pose, A., & Storper, M. (2021). Cities in a post-Covid world, Urban Studies, Ahead of Print.

Gould, K., & Lewis, T. (2016). *Green gentrification. Urban sustainability and the struggle for environmental justice*. Routledge.

Huerta, M. C., & Utomo, A. (2021). Evaluating the association between urban green spaces and subjective well-being in Mexico city during the COVID-19 pandemic, Health & Place, 70, ahead of press.

Kamvasinou, K. (2017). Temporary intervention and long-term legacy: lessons from London case studies. *Journal of Urban Design, 22*(2), 187–207.

London Borough of Newham. (2006). *Three mills conservation area. Character appraisal and management proposals*. London Borough of Newham.

Long, J., & Rice, J. (2019). From sustainable urbanism to climate urbanism. *Urban Studies, 56*(5), 992–1008.

Long, J., Rice, J., & Levenda, A. (2020), Climate urbanism and the implications for climate apartheid, in Castán Broto et al., ibidem, 31–49.

Mann, W. (2003). Bastard Countryside, Scroope 15, Cambridge Architecture Journal, Cambridge University School of Architecture.

Massey, D. B. (2005). *For space*. Sage Publications.

Meyer, F., & Schlaich, C. (2012), Saving the landscape. London 2012, Bauwelt, 24. Available at: https://www.bauwelt.de/themen/bauten/Saving-the-landscape-2154874.html

Nicholls, A. (2014). London 2012 legacy: Olympic Park waterways, Civil Engineering Special Issue, 167, Issue CE6, Nov. 2014.

Purcell, M. (2006). Urban Democracy and the Local Trap, *43*(11), 1921–1941.

Raven, J., Stone, B., Mills, G., Towers, J., Katzschner, L., Leone, M., Gaborit, P., Georgescu, M., & Hariri, M. (2018). Urban planning and design. In C. Rosenzweig, W. Solecki, P. Romero-Lankao, S. Mehrotra, S. Dhakal, & S. Ali Ibrahim (Eds.), *Climate change and cities: Second assessment report of the urban climate change research network* (pp. 139–172). Cambridge University Press.

Read, S. (2017). Cinderella River, Commissioned by Hydrocitizenship, London: CDS. Available at: https://www.leevalley.org/uploads/2/6/4/2/26426437/cinderella_river_lr_simon_read.pdf

The Guardian. (2020). What could a good green recovery plan actually look like?, 7th November. Available at: https://www.theguardian.com/environment/2020/nov/17/what-could-a-good-green-recovery-plan-actually-look-like Accessed 22 June 2021.

UNESCO. (2021). Cultural and Creative Industries in the Face of COVID-19: An Economic Impact Outlook, Paris: UNESCO. Available at: https://en.unesco.org/creativity/publications/cultural-creative-industries-face-covid-19

UN-HABITAT. (2021). *Cities and Pandemics: Towards a more just, green and healthy future.* UN-HABITAT.

Verdini, G., Alkhani, R., Beech, N., Cascone, P., Lamb, T., & Woltjer (2019), Globally-Informed City Climate Pedagogy & Practice: a position paper. Available at: http://blog.westminster.ac.uk/ccpp/ccpp-a-position-paper/

Wainwright, O. (2010). First look: Connectivity underpins 5th Studio's Lea Valley vision, Building Design, 26th November, p. 13.

WWMA & JCLA (Witherford Watson Mann Architects and John Cock Landscape Architects). (2008). From Edge to Common Ground – Upper Lee Valley Landscape Strategy, WWM Report.

Chapter 14
Facing Vulnerability: Sustainable Healthcare Design in the Global South

Marco Morandotti

14.1 Introduction and Scenario

Vulnerability gained a central role in understanding the condition of a system (both social and technological) or its predisposition to be damaged by different types of hazards, although "situations of vulnerability are not fixed and will change over time with changing circumstances" (UNHCR, 2016). Social vulnerability represents the inherent state of a community and comprises social, economic, political, and institutional factors (Lee, 2014).

The definition of contexts and situations of vulnerability is therefore itself dependent on many economic, social, and context-related variables. One of the aspects that is particularly relevant in several models in the assessment of the level of vulnerability of a community is the type of healthcare available. The Human Development Index (HDI), promoted by the UN Development Program in 1990, used life expectancy, education, and income as the main tools to rank countries in terms of performance. The HDI has been shown to be workable in highlighting the social progress achieved or expected in each country, particularly in relation to education and healthcare (Angeon & Bates, 2015). Furthermore, a negative relationship may be recognized between infant mortality rates and social vulnerability (Cutter et al., 2003) or in relation to limited access to healthcare (United Nations, 2015a).

A strong connection is also determined between vulnerability and sustainable development, as far as the Agenda for Sustainable Development and the related 17 Sustainable Development Goals are focused "particularly on the needs of the poorest and most vulnerable" (United Nations, 2015b).

M. Morandotti (✉)
Department of Civil Engineering and Architecture (DICAr), University of Pavia, Pavia, Italy

Committee for International Cooperation and Development (CICOPS), University of Pavia, Pavia, Italy
e-mail: marco.morandotti@unipv.it

E. Giorgi et al. (eds.), *Design for Vulnerable Communities*, The Urban Book Series,
https://doi.org/10.1007/978-3-030-96866-3_14

This is a priority that is reaffirmed in a cross-cutting way throughout the Agenda, demonstrating not only a programmatic focus on the needs of the most vulnerable individuals and communities, but also the multiplicity of approaches that can be implemented in this perspective.

In such a context, it is possible to assume that healthcare design strategies may have a positive impact on local communities from the perspective of reducing social vulnerability.

14.2 Methodology

For more than 20 years, the University of Pavia, through the CICOPS (Committee for International Cooperation and Development), has been promoting and supporting international collaboration with foreign universities. The Committee aims to promote cooperation activities between the University of Pavia and foreign universities. Final aim is to expand cultural, scientific and technical cooperation in order to extend the study of the social and economic problems of these countries and to contribute to the advancement of collective development. Experience in the field has shown that the development of joint research projects between scholars from different areas of the world can concretely contribute to the achievement of results of common interest, with a view to scientific and cultural collaboration and exchange.

The method applied in this specific study follows a systemic approach developed at the Laboratory of Science and Technology for Construction and Design (STEP) of the Department of Civil Engineering and Architecture (DICAr) at the University of Pavia since 2006, within didactic and research experiences. The approach relies on three main pillars: knowledge, feasibility, and sustainability (Morandotti & Besana, 2012) (see Fig. 14.1).

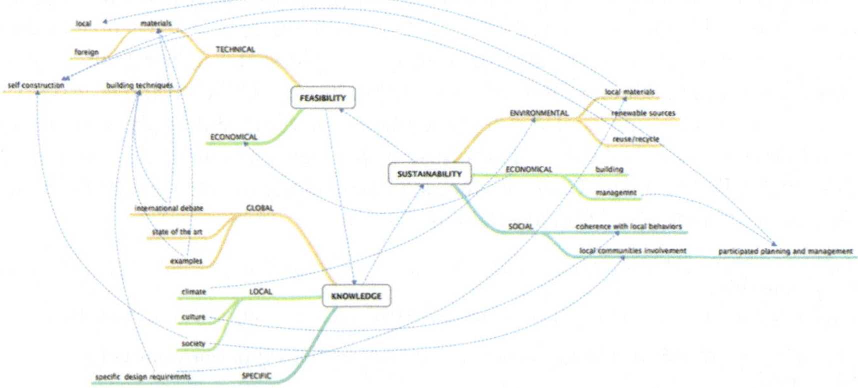

Fig. 14.1 Methodology diagram

The knowledge necessary in these cases relates to two different scales: global and local. The first obviously refers, as accurately as possible, to the knowledge from the design experiments carried out and the specific debate on the subject, as well as its evolution over the years. It constitutes an essential background to solutions, proposals, and attempts, whose mastery guarantees the possibility of comparing strategies, solutions, and contexts in order to stimulate specific solutions.

Inextricably linked to this second categorization are all aspects related to the context of specific intervention, which relate to a wide interaction of cognitive fields and necessary information. In fact, this "local" knowledge covers the socio-anthropological data of the context, such as the characteristics of traditional culture and the main economic reference data, in addition to the physical aspects related to the site.

The latter primarily concerns the climatic characterization of the context, including temperature, relative humidity and average diurnal rainfall (monthly, seasonally, and annually), the possible presence of prevailing winds and their average and maximum speeds, the local sun curve, the geomorphological characteristics of the site and the geological behavior of the soil. Aspects related to the technological scenarios available complete the analysis of the context, taking into account both the intrinsic characteristics of local materials and construction techniques, as well as specific building materials.

Climate-related topics should also be considered in the perspective of building's behavior related to its morphology and the adopted construction techniques.

Having acquired an adequate knowledge of the local context, an integrated system of constraints and suggestions can be obtained, which will be rooted in the specific response that the project provided. These responses should not only be consistent with the framework of specific contextual and economic constraints, but also with the set of specific needs related to the function to be set up.

Technical feasibility, or operational translatability of an architectural idea in a building organism, is crucial for any architecture in a wide range of intervention contexts. Design in the global south, however, raises specific issues that are particularly challenging in this regard. It is in fact a question of verifying the actual construction of the project, starting from the skills of local workers, which normally requires a significant realignment of the package of eligible techniques. Moreover, this is to be put in relation to the attempt to use, at the widest grade, materials available in the local context in order to avoid, or at least reduce, the behavior of materials and techniques coming from abroad and, as such, difficult to manage once the work is completed. The approach developed by the laboratory is precisely to limit the phenomenon of technological dependence regarding materials and construction techniques from local markets. The issue of durability and maintainability of the proposed design solutions is crucial from the perspective of not only the technical but also the economic sustainability of the project, hence the preferred choice of durable and yet easy to maintain construction technologies with limited costs and using local expertise.

The economic feasibility of the work is in fact an essential condition in determining the design options, as well as the technical-constructive aspects. This has not only to do with the requirement to respect the initial budget forecasts, but also to introduce the economic evaluation of technological choices as a real project discriminator. One line of active engagement is, for example, the search for constructive solutions at a lower cost than those traditionally used on the individual reference markets, allowing the possibility to reuse the amounts saved on the initial budget in technological upgrades to the system, such as by forecasting auto power plants (thermal and/or electric) through renewable supplies, or by forecasting meteoric water recovery systems with filtration and purification.

The control of the economic feasibility of the project is, on the other hand, one of the three possible and complementary constituents of the sustainability of intervention; the other two are the social and environmental dimensions.

At the same time, economic sustainability should consider both the effective eligibility of the building and the control of management costs. This second relevant topic refers both to the reduction of costs due to the maintenance of the asset, which should be minimized as far as possible, and any lower costs related to the production of energy (heat and electricity) gained from renewable sources.

Social sustainability means the ability to trigger, through the implementation of the project, some degree of collective participation in the phases of design and implementation, both in terms of participatory design and incentive to self-construction. When happening, the positive results are of two different orders. On the one hand there is a possible redistribution of resources at a local level, which in the case of a project managed by companies from other regions or countries would not occur. Moreover, and perhaps even more importantly, this determines the feeling the building really belongs to local community. This perception can spontaneously trigger more attention being given to the proper maintenance and management of the building. At the same time, in cases where a building can be self-built, it also determines a natural learning of techniques and methods of realization that can take root in the community, triggering emulative processes that multiply the positive effects of the original realization.

Finally, environmental sustainability means verify climatic adequacy of the building so that this can ensure the best levels of comfort expected, given the environmental conditions of the surrounding area, minimizing the necessity of air conditioning in order to achieve adequate indoor comfort results. The close relationship between the climatic investigations conducted in the analytical phase, the technological-constructive choices defined in the perspective of the technical feasibility of the intervention, and the degree of efficiency in the system's behavior determines its overall environmental sustainability.

In addition to indoor comfort at least two other elements may be considered within the topic of sustainable design. On the one hand, there is the self-sufficiency of the energy of the organism (complete or partial) through the generation of energy from non-renewable sources, and the recovery of rainwater. On the other hand, the use of recycled materials in the construction, which can have a positive impact on the allocation of the total resources available for the project, partly reducing construction costs.

Within the framework of this specific study, a critical analysis of the possible implementation of UN SDGs played a relevant role in terms of the general assessment of the effectiveness of the approach of local developing strategies.

In general terms, the study takes as its main benchmark the objectives and indicators of SDG11: "Make cities and human settlements inclusive, safe, resilient and sustainable." As the issues of global sustainability seem impossible to address without focusing on sustainability on an urban scale, SDG11 deals with both the city and the community scale. This is relevant not only from the perspective of achieving long-term developmental objectives, but also of directly impacting the quality of life among the rapidly increasing population of those living in urban contexts.

As known, the targets of SDG11 are ten and more precisely:

1. Ensure access for all to adequate, safe and affordable housing and basic services, and upgrade slums.
2. Provide access to safe, affordable, accessible and sustainable transport systems for all.
3. Enhance inclusive and sustainable urbanization and capacity for participatory, integrated and sustainable human settlement planning and management.
4. Strengthen efforts to protect and safeguard the world's cultural and natural heritage.
5. Significantly reduce the number of deaths and the number of people affected and substantially decrease the direct economic losses relative to global gross domestic product caused by disasters, including water-related disasters.
6. Reduce the adverse per capita environmental impact of cities, including by paying special attention to air quality and municipal and other waste management.
7. Provide universal access to safe, inclusive and accessible, green and public spaces.

    (a) Support positive economic, social and environmental links between urban, peri-urban and rural areas by strengthening national and regional development planning.
    (b) Increase the number of cities and human settlements adopting and implementing integrated policies and plans towards inclusion, resource efficiency, mitigation and adaptation to climate change, resilience to disasters, and develop and implement holistic disaster risk management at all levels.
    (c) Support least developed countries, including through financial and technical assistance, in building sustainable and resilient buildings utilizing local materials.

Many of these targets are thought to be applied at regional or national level, but they do in any case indicate a set of parameters that can support the definition of local action strategies, at least in terms of more or less consistency with global scenarios.

A particular case, however, is represented by the target 7(c) referring to the study of materials and techniques aimed at sustainability at building scale, within the global south scenario. The indicator defined to evaluate the target compliance is defined as the proportion of financial support to the least developed countries that is allocated to the construction and retrofitting of sustainable, resilient, and resource-efficient buildings utilizing local materials.

With regard to sustainable strategies to be adopted in such a context, the proposed methodology could suggest the use of the following:

1. Reuse/recycling of building materials.
2. Use of technological/constructive solutions aimed at achieving indoor environmental comfort conditions without the use (or with the minimum use) of climactic solutions such as roofing and ventilated façades; use of shielding and shading elements.
3. Use of solutions aimed at the recovery of rainwater for service functions/support to the activities planned in the building.
4. Integration of technological elements aimed at the production of electricity functional to the activities carried out in the building, or otherwise available to the local community.

14.3 Case Studies

Three different case studies, developed at different times within the research activity of the laboratory, are summarized below. Each example differs from each of the others regarding dimension, budget, and location, although they all follow the method previously described, as a common design approach.

The first example refers to the design of a medical dispensary in Kenya, about 70 km west of Malindi, near Tzavo Park.

The project started due to a proposal from the Policlinico San Matteo of Pavia, which was supported by the Diocese of Pavia answering a specific request by the local Diocese. The University of Pavia, with the Laboratory STEP project team (Marco Morandotti, Daniela Besana, and Francesco Maccarone), provided the technical support for the design and construction of the dispensary (Morandotti & Besana, 2014).

The project aims to fulfill the needs of the Chakama village community, consisting of about 2500 people. The isolation of this village from other settlements made the construction of a small residential care home, also to be used as a first aid and healthcare medical unit, a priority.

The main aim of the intervention was, therefore, to provide a small rural dispensary for local medical assistance and to direct the more serious cases to the larger neighboring hospitals, such as the Malindi General Hospital in Malindi.

The village of Chakama is located in a rural area about 70 km west from Malindi and has no transportation infrastructure, with the obvious difficulty of moving during the rainy season when unpaved paths become impassable. Due to the geographical and climatic conditions mentioned above, the nearest health service is often impossible to reach, especially by sick people, because of the almost total absence of public or private transport and adequate infrastructure.

Health and welfare are certainly major gaps in the Kenyan socio-cultural context, partly due to the fact that in recent years urbanization has been increasing in Kenya at a rapid pace. According to last data available, 46.5% of urban population live in slums (Begashaw et al., 2020).

This aspect is even more dangerous when the climatic situation of the Kenyan context is taken into account, especially that of neighboring Malindi. There is a general decline in rainfall during the main rainfall season of March-May. Drought is becoming more frequent and prolonged during the so-called "long rainy season." On the other hand, there is the generally positive trend of more rain from September to February. This suggests that the "short rain" (October-December) season is extending into what is normally a hot and dry period in January and February. This aspect has caused another phenomenon: the problem of desertification. Due to climate change and other human factors, desertification, the extent of arid and semi-arid land, is increasing, with severe consequences as social vulnerability due to agricultural crisis.

The water resources are unevenly distributed in both time and space. Climate change will worsen this already precarious situation as it affects the main hydrological components. Serious droughts have occurred in the last years. Major rivers show severely reduced volumes during droughts and many seasonal ones completely dry up. Malaria, cholera, Ebola, Lyme disease, plague, tuberculosis, sleeping sickness, yellow fever, and rift valley fever are some of the diseases that are expected to spread as temperatures rise and precipitation patterns change.

The starting point for the project was therefore a structure that primarily responds to the functional needs of the community, but at the same time could be easily recognized by other local communities and, from the perspective of a future network of healthcare units in the area, replicated. Therefore, the sustainable approach was attempted and emphasized, considering its environmental, economic and social dimensions, and actual technical and economic feasibility.

A vertical wall that becomes functionally the plants and technological plug of the whole structure characterizes the morphological layout of the dispensary. It allows for the division of the space into two areas: one, more public, just covered by a roof, and the other, more private, reserved for health services (see Fig. 14.2). This functional distribution is simple and clear and easily recognizable.

The open public space is paved and covered by a roof with a sunscreen function. Functionally, this space allows the family members of patients to rest in a separate space from the health area and which also acts as a waiting area for any patients who require treatment. Through a single entrance in the wall, patients can enter into the strictly medical area.

The dispensary's medical wing consists of modular spaces dedicated to care and is compatible with different building technologies according to the final budget.

The first space is a clinic for early medical evaluation and triage. Adjacent to it is located a medication room with first responders. The triage room also leads to a doctor's office and a room for the conservation and storage of medicines. The

Fig. 14.2 views of the dispensary of Chakama. Note the central red volume, which houses the power plants

dispensary is equipped with a room with toilets for both doctors and patients. Typologically, the areas containing the medical functions are standardized by a single over-wooden cover equipped with extruded eaves to limit overheating of the surfaces from solar radiation.

The dividing element between the two spaces is the plant technological space that allows the building to be almost self-sufficient in use and management. Some technical spaces, such as a tank for collecting rainwater, a filter for water purification, the generator and the battery for the domestic hot water, and the alternator and batteries for photovoltaic systems, are placed there.

While designing the dispensary, the aim was to achieve a high level of sustainability, concerning construction technologies and materials used.

The ability to use a simple technology without penalizing quality allows the project to be built with local resources and unskilled manpower (see Fig. 14.3).

Fig. 14.3 Chakama dispensary building site

This principle is therefore based on the concept of self-construction in which users are involved in the project. The theme of self-construction presents a number of advantages: it promotes education within the local community and contributes to the work group identity in which the sharing of effort and the enjoyment of the results obtained is a strong social cohesion. Furthermore, by contributing their efforts to achieve something, helps the population to accept the final result and therefore to actually use the building. Last but not least, the community acquires techniques and expertise from the work and can use them in the event of further construction needs.

Analogously, this also occurs for building materials. Working with materials specific to the location means getting both easy access and minimum costs as well as knowledge through using local labor. In fact, using sophisticated technologies or imported materials would be almost senseless because they'd increase the costs of construction and not help make the population independent during the management and use of the building. Materials which are not known by the workers inevitably cause difficulty for the maintenance of the building and make the community dependent on skilled workers.

The project was conceived as a prototype of spatial and functional quality, both as it is made with local materials and traditional building technologies, but also using recycled materials, commonly discarded materials readily available without cost. The project was therefore designed with modular proportions for the rooms to avoid wasting materials and increasing costs, but at the same time to be flexible in the choice of the room functions and the types of construction technology used.

Before choosing materials and technology, research on materials commonly used and found in the territory of Malindi was carried out, as previously discussed.

In general, a concrete framework is the most common structure typology, while the vertical walls are created with bricks made of coral blocks. Roofs are typically built with the so-called "makuti" technique. This traditional technique gives a lot of problems linked with safety and maintenance: it must be renovated every five years; it's highly inflammable, and in the case of fire, the temperature can exceed 800°.

Regarding materials, in Kenya, concrete is the most common material used for construction. It's currently produced in three major factories for both the local and export market. Domestic prices have steadily increased over time. The high cost of cement coupled with occasional shortages, high transport costs, and its unavailability in some remote parts of the country has adversely affected the cost of many cement-based buildings materials, like concrete blocks and mass and reinforced concrete slabs. It is also possible to find concrete blocks either factory or manually produced, but they depend on cement and are therefore expensive. They are, however, the most commonly used walling material for residential properties within Nairobi.

Natural stone (in the coastal area the so-called "coral block") is a cheaper walling material as compared to concrete blocks. It is also commonly used, especially in the surroundings of main towns. Finally, the most common roofing materials are tiles and galvanized corrugated iron sheets. Tiles are either clay or concrete.

According to the results of this technological research, design choices tried to stress the use of recycled materials, such as the traditional pallets used for commercial packaging as a building material. They have good properties regarding how they behave in use and also some geometric characteristics of the joists and their distances, which are very similar to frame structures such as balloon and platform frames.

Thus, the use of pallets would respond positively to a number of requirements in the dispensary project, such as the concept of self-construction (Foti, 1991). Working with light elements and small dimensions means using a simple technology easily learned by the local community and involving easy construction (May, 2010). Contextually, the realization of a structure dry assembled, using only mechanical joints and riveted or bolted connections, is very important in a context so poor of water. It also responds positively to the requirements of components' maneuverability, thereby allowing its handling by a single person and, therefore, simplifying the construction. The requirement of construction durability is satisfied by the material itself because it is a prefabricated element that has already been subjected to treatments that guarantee an acceptable life cycle and controls to be placed on the market. The wooden material is healthy because it does not emit harmful substances during the operating phase and does not constitute a danger to the health of users.

In conclusion, the performance characteristics that have accompanied the design choices, still in the initial feasibility study, were mainly referred to the following: modularity of the structure and its components; future expandability; environmental, social, and technical sustainability; economic feasibility; architectural identity; and self-construction.

Nevertheless, the project has undergone some major changes during the development and construction phases. The most relevant was the construction technology, since the local community, as well as the competent authorities, did not agree with the use of pallets, preferring more traditional limestone bricks and reinforced concrete to support the structure. Therefore, it was necessary to engage with a traditional construction company, undermining the initial goal of a community self-construction approach. The morphological layout has remained unchanged overall, as the functional layout, although the proportions of the central block have been modified during construction with a widening of its footprint on the ground to adapt them to those of the water storage tanks chosen by the builder.

This experience, in many ways positive, as it led to the realization of a structure useful to satisfy some of the primary needs of the local community, presented two critical issues in terms of consistency between design choices and actual implementation.

The first, and most significant, concerns the choice of the construction technology used for the realization. In the perspective of a critical analysis of the design and implementation process it may be useful to remember the reasons that led to this result. On the one hand, the local offices did not approve the proposed technological solution, as it was deemed not to comply with local construction standards. Unfortunately, the scientific literature and the selection of examples prepared at the time failed the objective of getting the proposed solution approved. On the other hand, even at the local level, the proposal did not meet the favor of the community, since the use of reuse material was perceived as a sort of debasement of the work, rather than as a virtuous application of circular economy.

Due to the negative perception by the local community and the resistance shown towards the technological proposal, despite the undoubted benefits that could have been derived from it, the constructive solution was reformulated according to a more traditional technology, which the community could feel like its own, without being passively subjected to it.

The change to the geometric proportions of the technological core of the building has had a negative impact on the project from the morphological and formal point of view, but not on the functional one. Although the geometric dimensions defined in the project were compatible with a plant solution (specifically rainwater storage tanks) available on the local market, the local builder in charge of the contract proceeded with a different choice, considering completely irrelevant the consequent morphological change.

The second case study refers to the General Hospital of Ayamè (Ivory Coast) and has been developed over several years in cooperation between the laboratory and an Italian NGO, the "Agenzia n.1 di Pavia per Ayamè," established some 30 years ago in Pavia.

Ayamé General Hospital has a catchment area that extends from Aboisso in the south to Songan in the north, covering a radius of about 100 km that includes several villages within plantations and forests. It covers an area of about two and a half hectares of which approximately eight thousand square meters were used for the construction of ten health pavilions and ten additional buildings.

The hospital area is bordered by boundary walls along the entire perimeter and is internally cut lengthwise by the river Anon Assouè; the two parts of the hospital are connected through a small concrete bridge at the west entrance.

Unfortunately, the initial choice of settlement of the structure, and the subsequent choices of territorial expansion, have led to the emergence of a currently critical situation for at least two reasons. On the one hand, the flows between one part of the hospital and the other are critical for the presence of a single point of promiscuous crossing (pedestrians, vehicles, litters, patients, visitors); on the other, the periodic flooding of the river, made more frequent in recent years also as a result of a progressive alteration of the traditional periods of precipitation, makes the structure particularly vulnerable and exposed to high risks.

The HGA has two separate entrances, both accessible from the main road, with different functions: the one to the north is intended for staff and for the storage of ambulances and staff vehicles; the one to the west is used by all types of users and is used as a main entrance. Both entrances are constantly supervised by employees located in special gatehouses located near the gates. In the space between the two entrances there is a pedestrian passage (not covered) that connects across the north area of the hospital and allows access to the pavilions and technical rooms that are located in this portion of the area.

In the area to the north are located the buildings that house the departments of pediatrics, maternity and obstetrics, the directorate, the pharmacy, maintenance, the mortuary chamber, the conference room, the storage of vehicles and several technical rooms. The area to the south accommodates different services, such as the laboratory analysis, radiology, medicine and surgery, the departments of dentistry and ophthalmology, the nutritional center, the public toilets, the kitchen and restaurant, the church and the mosque, and in the southernmost part of the hospital area the two incinerators have been built. The area of the hospital is strongly characterized by large green spaces, rows of trees, and gardens, some of which are also inside the pavilions or are very close to the entrances. The pavement is mostly in clay in some paved sections or with trotting lines that mark the connection paths between the various areas of the hospital.

The layout of the buildings is rather uneven, irregular, and inorganic; it is easy to attribute the cause of this difficulty to the fact that the buildings were built in different years and created in response to specific needs. There was no general planning that defined the rules and gave an order to the location of buildings in relation to possible connections and flows generated in the HGA.

There is in fact a constant movement along the connecting routes and rest areas. The traffic flows overlap with pedestrians, hospital staff, users and visitors who generate areas that are difficult to manage in terms of accessibility and, above all, security, because it is not possible to control their dynamics. One of the most critical areas is the west entrance, in front of the pharmacy, where many people stop daily, and near which the payment for drugs and tickets for services are received within the different departments. No less serious is the situation on the bridge that connects the two areas; pedestrian flows do not make it easy for ambulances and vehicles to pass, and there is also the overlapping of the paths of hospital materials with patients, which increases the possibility of contagion and infections.

The design and subsequent construction of the triage building have been developed as master thesis project in building engineering and architecture at the University of Pavia, by Ms. Barbara Braggion with Marco Morandotti and Daniela Besana (see Figs. 14.4 and 14.5), and are the result of several technical missions in Ayamé, which occurred in the years 2014 to 2016, when was clear the need for a

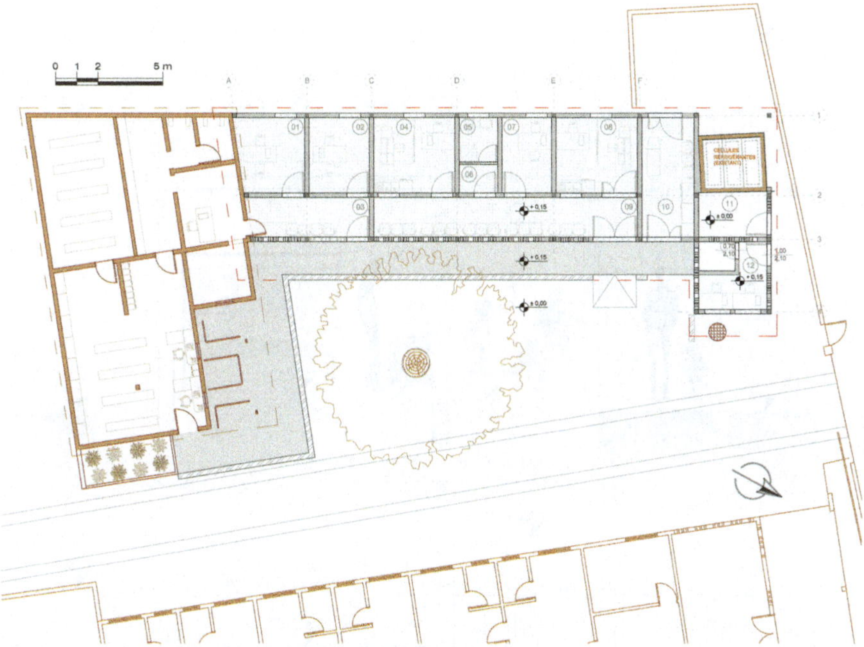

Fig. 14.4 Ayamè General Hospital. Plan of the new building of triage, next to the existing pharmacy

place near the entrance of the hospital for first aid services that were previously located in under-sized environments within the Pavilion of Medicine and Surgery. The functional requirements were quite clear: two to three rooms were required for dressing, an office/consultation room, a bathroom and the integration of the entrance hall with the possibility of expanding it to obtain a space for the payment for the tickets for the medical services received, and finally, resetting the external pavement that connects the entrance gate to the bridge, which helps with communication between the two sides of the hospital.

From the earliest stages of the project, it was essential to maintain a constant relationship with the client and local workers to have a common vision of the spaces to be defined and how to redistribute the flows into the hospital.

First, it was necessary to deal with the issue of the location within the HGA of the new department of triage: the main objective was to make it easily accessible and place it near the main entrances. Since the project was linked to an initial deployment of maintenance space in a new building, it was first decided to use the premises of the old maintenance space, whose location was very close to the public pedestrian entrance, and to work inside redesigning spaces. However, it was not possible to proceed in this way because the necessary spaces were greater than the available spaces of the old building in addition to the fact that the conditions in which the state of conservation concerned were not optimal. At this point it was decided to proceed with the design of a new entrance hall that included both the new triage and the functions of a mortuary and an access control point connected to it.

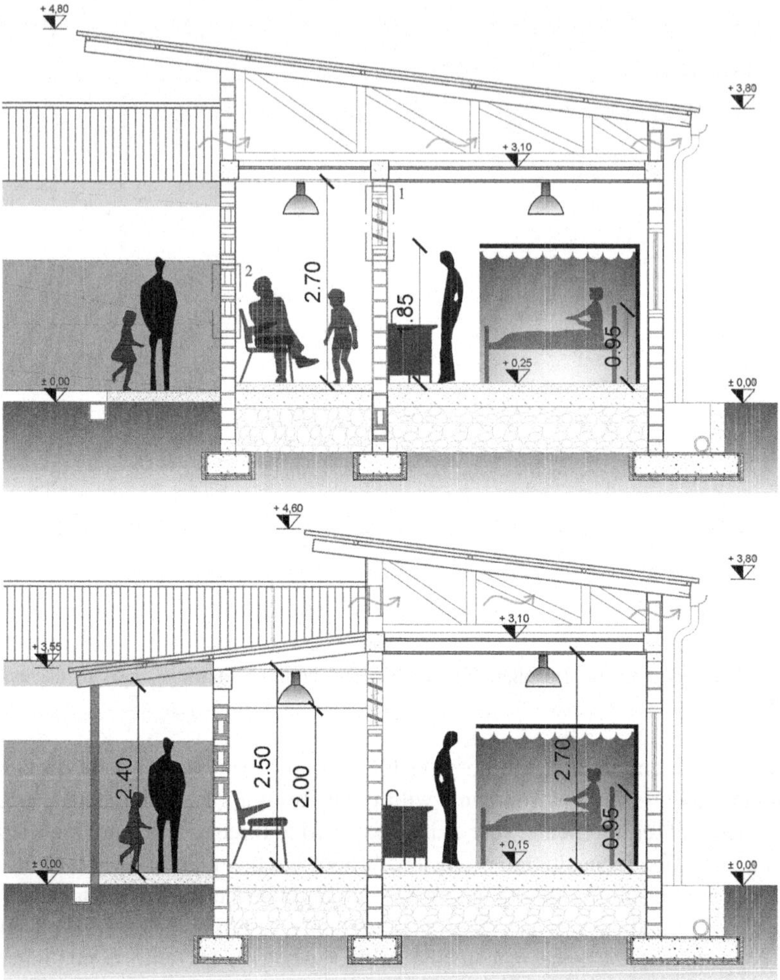

Fig. 14.5 Ayamè General Hospital. project section (top) and as built (bottom)

Another key point concerns the spatial constraints defined by the project area. Choosing the location of the new triage on the site of the old maintenance space collided with the constraints generated by the context: the presence of buildings in the area in front of the HGA, presence of cold stores that could not be relocated elsewhere in the complex, the proximity to the main entrance that had to be kept functioning, as well as the proximity to the premises of the pharmacy that inevitably would have been affected by the works, and finally, the presence of a tree in the entrance court.

The project area is in fact delimited to the north by the presence of the pedestrian and driveway entrance that must not be modified, to the south by the borders with the pharmacy, to the east by the boundary of the area, and to the west by the hospital main street.

The project defines a single block that contains all the required functions, all merged under a single cover in corrugated sheet. The structure provided a regular grid of pillars that followed the geometries of the area in order to harmonize the impact of the project in the context. The access to the building was settled in proximity to the entrance to the HGA and a raised pavement and a platform with an access ramp were provided to facilitate wheelchair transport for the patients and, at the same time, to protect the buildings from rain.

The triage building project includes a large waiting room that allows access to the three dressing rooms and the toilet. In the final part of the dressing block there is an additional space, including a filter area and a dressing room, dedicated to the treatment of infectious patients who should not come into contact with other users in order to reduce contamination. The last room, not connected to the rest of the pavilion, is designed as an extension of the adjacent pharmacy. For this reason, it has independent access from the outside with an internal door connected to the existing offices.

Building materials and technologies are traditional ones: the load-bearing structure in reinforced concrete, the concrete brick cladding manufactured at the foot of the building, a suspended wooden ceiling and two pitched roofs, one on the porch and one on the building, made by ventilated corrugated metal sheet. To ensure good indoor comfort, some openings in the upper part of the building were provided, to allow natural ventilation even under the roof, while in the main façade, some walls were designed with perforated blocks facing towards the inner patio.

In the following years the hospital was the subject of some other minor building interventions aimed at improving the general operating conditions, with reference to the radiology, pediatric and pharmacy pavilions.

The third case study differs significantly from the first two, being located in Colombia, in the city of Tunja, and concerns a large impact hospital project (see Figs. 14.6 and 14.7).

The University of Pavia and the Juan de Castellanos University Foundation of Tunja signed a framework cooperation agreement in April 2012. The agreement is focused on some specific fields of collaboration and exchange related to the topics of sustainable development, both on an architectural and territorial scale, and the definition and development of innovative healthcare buildings.

The Laboratory (project group: Marco Morandotti, Daniela Besana, Francesco Maccarone) has provided support for the morpho-typological development of the university campus and for the architectural design of some building interventions, including that of the new nursery clinic (Gomez Sierra et al., 2013), which was also the first building to be designed.

The project defines a new relationship with the area, which at the same time is: (i) symbolic and founded on the principle of an architectural settlement that gradually unravels while approaching, and which is always strongly connected to the land on which it rises, sometimes turning into a new soil; (ii) landscape-oriented as the building lies along the natural slope of the land, trying to integrate it as much as

Fig. 14.6 Tunja Nursery Clinic. Bioclimatic design section (top) and downstream facade (bottom)

possible with the side of the hill; (iii) morphological, because the building leans on some contour lines, naturally identifying an intermediate space between the volumes that compose it, which assumes the role of the pedestrian road connection on which are arranged some of the main functions.

The project is conceived on a modular scheme, both in regard to the patients' wing (where modularity is made evident by the chambers themselves, which constitute the basic element) and in regard to the surgical plate, articulated in three functional blocks per floor. The three hospital blocks and the plate are connected by two covered walkways which allow the flow of doctors and patients from one wing to another under controlled conditions. On the other hand, the patients' wing is accessible to visitors starting from the internal road, giving the latter the role of a real public distribution space, thus allowing a clear separation of flows.

The roofs are flat and are largely accessible and green-treated, like a real garden. The non-accessible parts can be integrated with photovoltaic and thermal panels for the production of hot water and electricity.

The project aims to introduce significant elements of technological and construction innovation, with particular reference to environmental sustainability and energy

Fig. 14.7 Tunja Nursery Clinic. Views of some rooms inside: type of room in hospital (top), inner gardens (center), children's playroom (bottom)

efficiency criteria in the design of the building envelope, as well as in the integration of innovative plant systems for the Colombian market, both in terms of energy production from renewable sources and in terms of comfort within the facilities.

The project also aims to pursue the general objectives of architectural and morphological innovation with regard to the criteria of spatial and functional organization of the structure, as well as the humanization aspects of hospitalization, applying (adequately so in the local context) the hospital design paradigms developed over the last 15 years at the STEP Laboratory (Morandotti & Besana, 2020).

An example of this approach is the design of the hospital rooms, which follows the idea of placing the two adjacent rooms' toilets aligned between the rooms. This solution involves an optimization of the plant networks thanks to the combination of two bathrooms through the provision of a common wall equipped between the two modules of the toilets. From a typological point of view, this solution allows a more formal cleanliness of the room, while ensuring the possibility of glazing the wall separating the room and the corridor without significantly compressing the view towards the outside of the room itself.

In this solution, the two bathrooms being aligned are one facing the corridor and the other towards the external front. Often, both to standardize the internal distributions and not to introduce distinctions between contiguous rooms, both toilets are kept ventilated and illuminated artificially. This solution inevitably leads to a greater use of surface area with the same number of rooms designed, a parameter not superfluous when it comes to hospital complexes of medium and large size. In this specific case, this solution has allowed a flexible structure to meet the needs of the client regarding the number of beds.

Considering a room module and thinking of the iteration of the module with two bedrooms and two bathrooms, it is possible to assume that it can replace the module of the room spaces with different functions, at the service of the patient themself. The module not occupied by the room, for example, leaves room for play areas for children divided by age group, rather than small greenhouses or outdoor terraces in which to insert the green project. Some of these spaces are eventually designed to be subsequently saturated in relation to a possible variation of users' requirements.

The project at the moment has not been developed at executive level due to a change in the direction of development of the Campus, which provided for the anticipation of other interventions, always defined under the same convention, but more oriented to the educational and research needs of the Foundation. Despite a slowdown in the implementation phase, also due to the great economic value of the construction, it was in any case an experience of great interest, for the intrinsic complexity of the work, the specificities of the Andean landscape and climate, and the interjections developed with the local community.

Figure 14.8 presents an evaluation map showing different outcomes within the case studies described in the paper.

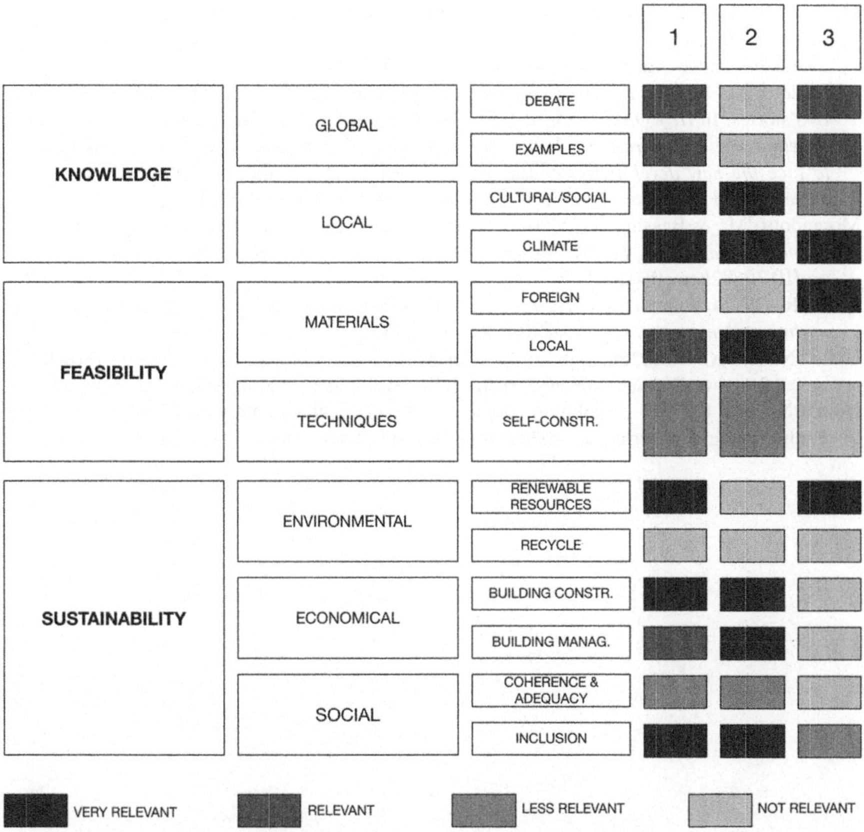

Fig. 14.8 Evaluation map of the relevance of each parameter, showing different outcomes within the case studies, already described in the paper

References

Angeon, V., & Bates, S. (2015). Reviewing composite vulnerability and resilience indexes: A sustainable approach and application. *World Development, 72*, 140–162.

Begashaw, B., Schmidt-Traub, G., & The Sustainable Development Goals Center for Africa and Sustainable Development Solutions Network. (2020). *Africa SDG Index and Dashboards Report 2020*. SDG Center for Africa and Sustainable Development Solutions Network.

Cutter, S. L., Boruff, B. J., & Shirley, W. L. (2003). Social vulnerability to environmental hazards. *Social Science Quarterly, 84*(2), 242–261.

Foti, M. (1991). *Progettare per l'autocostruzione*. CLUT.

Gomez Sierra, F. A., Morandotti, M., & Besana, D. (2013). Proyecto Campus y Clinica Materno-Infantil. In *Rivista "Cultura Cientifica", Tunja, Colombia* (Vol. 11, pp. 134–145) (ISSN 1657-463X - 9771657463005).

Lee, Y. J. (2014). Social vulnerability indicators as a sustainable planning tool. *Environmental Impact Assessment Review, 44*, 31–42.

May, J. (2010). *Architettura senza architetti*. Rizzoli.

Morandotti, M., & Besana, D. (2012). Architettura possibile. *Metodi, tecniche e materiali per il progetto nei paesi in via di sviluppo*, KMStudio Ed., Pavia, Retrieved from http://issuu.com/danielabesana/docs/architettura_possibile (ISBN 9788890750502).

Morandotti, M., Besana, D. (2014). Healthcare sustainable design in the Global South. A tale of two stories. In *III CONGRESSO CUCS, JUNCO—Journal of UNiversities and International development COoperation "Imagining cultures of cooperation: Universities networking to face the new development challenges III Congress of the Italian University Network for Development Cooperatio*n (CUCS)" (pp. 188–195) (ISBN 9788896894163).

Morandotti, M., & Besana, D. (2020). Hospital architecture and the challenge of humanisation. *Research and design, TEMA, Technologies, Engineering, Materials, Architecture.* https://doi.org/10.30682/tema0602q

UNHCR. (2016). Vulnerability screening tool. *Identifying and addressing vulnerability: A tool for asylum and migration systems*. ISBN 978-0-9924831-1-1.

United Nations. (2015a). Vulnerability-Resilience Country Profile (VRCP). A country-owned analytical framework for assessment of sustainable development in Small Island Developing States.

United Nations. (2015b). Resolution adopted by the General Assembly on 25 September 2015, Transforming our World: The 2030 Agenda for Sustainable Development.

Chapter 15
Coastal Vulnerability: Sustainable Settlements for Outdoor Tourism

Carlo Berizzi, Margherita Capotorto, Gaia Nerea Terlicher, and Luca Trabattoni

15.1 Italian Coasts in Transition

Today, coastal areas of the Italian peninsula have to deal with the impacts deriving from an intensive anthropization, despite the natural environment and landscape heritage. With a total length of 8300 km, 7500 km of coast remains natural (Isprambiente, 2019). This means that more than 9% of the Italian coast is now artificial[1] (Isprambiente 2019). The very nature of the coastal environment, which is dynamic, generates natural processes that, when added to those of anthropogenic origin, can modify its natural and biological characteristics, making the coasts extremely vulnerable territories. Despite the continuous natural evolution of this type of landscape, anthropic action is today the main cause of coastal erosion phenomena, significantly modifying the Italian coasts over the last 60 years. A significant part of these processes, particularly of the erosive type, is due to the anthropization resulting from the growth of tourism. A relationship between tourism and the landscape that has become increasingly conflicted since the 1950s. This

[1] This data is given by ISPRA (Istituto Superiore per la Protezione e la Ricerca Ambientale) that is an Italian public research body supervised by the minister of environment and of the protection of the territory and the sea. The institution carries out a number of tasks of national interest for the protection of the environment.

C. Berizzi · M. Capotorto · G. N. Terlicher
Department of Civil Engineering and Architecture, University of Pavia, Pavia, Italy
e-mail: Carlo.berizzi@unipv.it; Gaianerea.terlicher01@universitadipavia.it

L. Trabattoni (✉)
Department of Architecture and Planning, University of Technology of Opole, Opole, Poland
e-mail: L.trabattoni@po.edu.pl

E. Giorgi et al. (eds.), *Design for Vulnerable Communities*, The Urban Book Series,
https://doi.org/10.1007/978-3-030-96866-3_15

is when, in just a few years, the transition from low impact tourism to mass tourism took place, with an invasion of the coasts of the peninsula. The phenomenon led to the birth of the so-called "Med Wall" (EEA, 2006, p. 55): the overbearing urbanization along the Mediterranean in which a large part of the coast is dominated by concrete; mostly low-quality urbanization and mainly of a residential, tourist and seasonal nature that has largely forever changed the Italian coastal landscape. From that moment, not only did the seaside tourism demand have an exponential growth that is still significant today, but tourism in seaside resorts will become the dominant economic sector. In Italy, coastal tourism represents the main driving force of the national tourism economy; between 2016 and 2017 Italian tourists in seaside resorts accounted for 37.3% of the national total, while foreign tourists, 27% (Isprambiente, 2019).

Over the years, the mass phenomenon has led to a process of homogenization of the built landscape (as in the case of Rimini, Bibione, Lignano along the Adriatic coast of the peninsula), as well as its disfigurement. If we consider tourism in its various components of movement, stay and recreation, the vastness of the phenomenon emerges on different levels: landscape, environmental, territorial and socio-economic. The process of overbuilding the coasts in an attempt to respond to the unstoppable demand for seaside tourism has caused a significant exploitation of natural resources and the consumption and modification of coastal landscapes, weakening the coastal ecosystem. If in the years of the mass tourism boom there was a lack of sensitivity towards the coastal heritage, today there is awareness, and the consequences of that "urban tsunami"(Forman 2008, p. 265) are evident: erosion accelerates the mutability of the territory; soils and waters have become polluted and the ecosystem has lost its ecological balance, compromising biodiversity. In addition, coasts have to face the effects of climate change, which are causing a rise in sea levels, the perceived temperature on land and on the surface of the sea, as well as an increase in the frequency of extreme events that cause flooding. Uncontrolled and unconscious urbanization leading to these ongoing dynamics undoubtedly make these portions of the territory extremely vulnerable and unable to manage the "side effects" caused by the process itself. The consequences of this "trend" of uncontrolled landscape tourism are not just limited to the areas overlooking the coasts, but also include, even if in a different way, the innermost areas of these territories, which have experienced some effects of impoverishment. The tourist phenomenon has over the years in fact used a selective development approach that has almost exclusively included the development of the areas closest to the coast. This has led to a substantial fracture in the territorial dynamics in which a series of environmental and socio-economic disadvantages are reflected. Part of the territory that is today characterized by seasonal seaside tourism constitutes portions of territory defined as fragile: far from basic services (education, health and mobility) and from the main inhabited centres, and falling within the national strategy for inland areas subject to depopulation and environmental vulnerability.

15.1.1 Towards Sustainable Tourism: Impact and Potential

The dynamics taking place along the Italian coast caused by increased seaside tourism is inevitably worrying. Until 2019, forecasts had foretold an increase over the years of tourist arrivals in coastal regions; however, almost two years after the global health crisis due to the spread of the SARS-CoV-2 virus, future predictions are inevitably uncertain. In fact, tourism has been one of the most affected sectors in the current pandemic, with a considerable drop in international tourists. Despite this, data in Italy shows how the open-air tourism sector (i.e. structures that provide accommodation in their own settings, such as bungalows, tents, mobile homes owned by the structure, or camping pitches owned by tourists) has recorded less of a decrease when compared with other types of accommodation facilities. According to ISTAT[2] data, during the first year of the pandemic, hotel facilities (permanent structures) in Italy suffered a significant decrease in arrivals, verifying that between 2019 and 2020 there was a total decrease in tourists equal to 58%. Similar data has been verified for bed and breakfasts, which saw a 53% drop in guests. Despite this decrease, the open-air tourism sector saw a smaller decrease, equal to 40%, between 2019 and 2020. If there were approximately ten million tourists in 2019, in 2020 they were around six million. It also should be noted that before the pandemic, open-air accommodation facilities were largely oriented towards foreign tourism, while in 2020 the percentage of Italian tourists using these facilities has increased, expanding the market to new potential users. This data demonstrates how the pandemic has changed tourists' choices to preferring greener holidays and more contact with nature, which has become a necessity. The long quarantine period that families have had to face has certainly pushed many towards the search for a holiday within nature and has highlighted the need for safety from the spread of the virus even more, resulting in the choice of open-air accommodation. Open-air hospitality has therefore proved to be a valid alternative for living the holiday experience during the pandemic and this trend undoubtedly puts new challenges in front of us for a more sustainable tourism.

Sustainable tourism is, in fact, a theme that the International Community has been committed to promoting for years and constitutes one of the objectives within the UN's Agenda 2030. SDG 8 "Decent Work and Economic Growth" encourages to implement policies aimed at supporting sustainable tourism by 2030; SDG 12 "Responsible Consumption and Production" affirms that the development and application of tools to monitor the impact of sustainable tourism will translate into better economic, cultural and environmental results; SDG 14 "Life below water" promotes a sustainable use of the sea and oceans. It should also be highlighted that Europe is promoting the Green Deal, a series of policies to encourage a climate-neutral society through an action plan for a sustainable economy by 2050 in which all member states must commit to battle the effects of climate change and

[2] ISTAT (Italian National Institute of Statistics) is a public research body that produces official statistics, measuring and analysing collective phenomena in Italy.

environmental degradation by acting for a circular economy; these actions are intended to make significant changes also in the tourism industry. The direction of the European policies is clear, and initiatives such as "The European Destination of Excellence (EDEN)"[3] represent an accelerator for the transition towards sustainable tourism. Although the European Community has been promoting alternative tourism models, sensitive to the landscape and environmental impact for years, the pandemic phenomenon is certainly accelerating this type of development. What we are experiencing is a crucial moment for the future of the Italian coasts, for their conservation and undoubtedly for the relaunch of a sector of the economy that is still facing a serious crisis. A new design approach can promote the sustainable development of coastal areas by integrating the tourism-landscape combination as an opportunity to enhance the natural and cultural heritage and counteract land consumption.

Although open-air tourism is more sensitive to issues of sustainability and environmental impact, the open-air accommodation facilities located along the coastal areas of the Italian peninsula are almost always atopic and unable to act as elements capable to build the landscape combining nature and the built environment. The consequence of an approach that has often been purely functional until now led to the creation of low-quality and inaccessible open spaces, the absence of a relationship with the context, and an absence of a specific character for that place. In the awareness that tourism is often the main sector that supports the economy of maritime contexts, the right answer for sustainable open-air tourism consists of defining a new type of landscape that welcomes the demands of contemporaneity and guarantees the long-term care of the territory.

15.1.2 Different Issues for Different Italian Coastal Areas

Given the extent of the coastal area in the Italian peninsula, the research work has identified two different contexts of reference for specific areas characterized by different environmental conditions, in order to verify possible settlement solutions for sustainable outdoor tourism using an innovative approach. It follows the choice of two areas located along the east coast of the Italian peninsula overlooking the Adriatic Sea, in particular in the Veneto region and the Emilia Romagna region (see Fig. 15.1), the two most attractive regions of Italy for outdoor tourism, characterized by different geographical, economic and cultural situations, and particularly exposed to vulnerability.

The Emilia Romagna Region. Considerable erosion phenomena on the Emilia-Romagna coast already occurred during the 1960s, a period in which strong tourist development was already underway. The "Commissione De Marchi del 1970" study

[3] Eden is an initiative of the European Commission aimed at cities between 25,000 and 100,000 inhabitants belonging to EU countries or COSME countries that have adopted strategies to promote sustainable tourism in order to select pioneer destinations.

Fig. 15.1 Map of Italian Peninsula with Veneto region and Emilia Romagna region highlighted (Credit: AUDe Lab, University of Pavia)

proved that the phenomenon had been in progress for 10 years. The extraction activities of the quarries also contributed to the phenomenon, but the biggest contributor was the construction of "invasive" and rigid defence artefacts that had spread to be 77 km long by the end of the twentieth century (Gruppo Nazionale per la Ricerca sull'Ambiente Costiero 2006a, b). These interventions have altered the natural coastal system by degrading the dynamic nature of the coast. This has contributed to an acceleration of the regressive phenomenon of the coast, which we are trying to remedy today. Regional data referring to the year 2012 shows the presence of 44 km of coastline erosion (MATTM-Regioni, 2018). Between 2000 and 2012, the loss of sandy shore is estimated to be 540,000 square metres (MATTM-Regioni, 2018). Looking at the data over a longer period of time we can understand the urgency of the crisis: between 1960 and 2012, the Emilia Romagna region lost about 13 million square metres of beach, a loss equal to 37% of the total in the nation (Minambiente, 2017).

The Veneto Region. The entire coastal area of the region has undergone a strong development in the seaside tourism sector since the 1960s, which has resulted in a high consumption of land in various coastal portions. The artificialization of the coast has also destroyed many dune alignments of particular naturalistic importance. As in the Emilia Romagna region, in order to cope with the consequences due to the modification of the ecosystem, rigid systems of defence have been developed over the years. In the last 20 years the coastal situation has worsened, with a constant growth of the erosive phenomenon: in 2006 the erosion affected 25 km of the Veneto coast (Gruppo Nazionale per la Ricerca sull'Ambiente Costiero, 2006a, b); from 2007 to 2012 the presence of 52 km of coastline erosion has been recorded (MATTM-Regioni, 2018), and a loss of shore equal to 870,000 square metres.

In both regions, the coastal system is no longer in equilibrium and, among the fundamental causes that have triggered various phenomena, including erosion, there is the presence of rigid systems of defence (such as artificial cliffs). The current condition of the coasts necessarily leads us to reflect on interventions that do not compromise a natural system like the coast. The goal of the interventions must be to maintain the natural defence structure inherent in this type of landscape as much as possible.

The search for a new project proposal for the already consolidated sector of outdoor tourism wants, on one hand, to call into question the settlement methods and strategies adopted for the construction of open-air structures along the Italian coast until now, and on the other hand, to enhance the potential of this type of tourism, which, compared to the ordinary type of accommodation (such as hotels), certainly works with nature. Aware of the fact that tourism is necessary, both for the community and the local economy, as well as for the enhancement and management of the territory, the reflection starts from the idea that there is the possibility of intervening on the landscape sensitively and without altering its natural balance. The Global Code of Ethics for Tourism, in fact, recognizes tourism in nature, or ecotourism, as forms of enhancement and also of enrichment when responding to the request for hospitality; they respect the natural heritage and enrich it.

The presence of places with special characteristics, with singular landscapes between sand and sea, requires particular attention. Furthermore, the tourist propensity of these places, which leads them to be lived in only for certain seasons of the year, requires the search for new functions that can extend the period of stay. The coastal areas of Veneto and Emilia Romagna in fact undergo a high concentration of tourists during a few months of the year, a concentration that becomes a real phenomenon of congestion followed by the effect of "ghost town"for the rest of the year. The objective of the research reflects on possible settlement methods and on a functional programme that can allow coastal areas to be experienced for a prolonged period.

Can outdoor tourism reduce the vulnerability of coastal areas?

The starting point is the outdoor tourism sector; the problem is the high environmental, social and economic vulnerability of coastal areas; the point of arrival is sustainable tourism.

15.2 Methodology: Which Method? Which Goal?

At the foundation of the research on such a delicate and complex issue, which includes vastly different territories, there is the thought that to operate with a sustainable approach, it is necessary to know the context of the places and their nature. Although it is still a coastal territory, each portion of this landscape demonstrates its own peculiarities, characteristics and criticalities. A generic approach to the territory would be destructive as well as monotonous and would once again lead to a vulnerable landscape without character, as happened during the 1960s.

The need to contextualize the research is a priority, since only in this way can a contribution be made within a critical context that is sensitive to the characteristics of the specific place.

Therefore, the approach adopted for the search for sustainable tourism along the two coastal areas is that of research by design. The approach is certainly of an experimental type through which the final goal is a project. However it is, above all, in the preliminary phase of investigation and analysis of the territory and of the specific place, where the more theoretical study is addressed and which subsequently leads to the design phase. Research through design works in harmony with architectural practice and with the research process.

The research thus comprised the following phases (see Fig. 15.2):

- Theoretical research: research on the meaning of sustainable tourism and on the role of tourism in coastal areas.
- Investigation: analysis of coastal areas and the generic formulation of the problem in Italy; analysis of outdoor accommodation facilities and the mobile home product.
- Strategic research: identification of two areas of intervention, analysis of the specific places with detection of criticalities and problems, and subsequent definition of the objectives.
- Development of project proposals.

15.3 Open-Air Tourism: A Possible Answer

Outdoor tourism was born in England at the end of the nineteenth century (Luisi 2019) but it was after the Second World War, with the proliferation of the automobile, that it became a mass phenomenon. It was a response then, as it is today, to the

Fig. 15.2 Research by design process diagram (Credit: AUDe Lab, University of Pavia)

desire to travel at low cost and be immersed in nature, representing a real escape from the city (Battilani, 2001). Over the years, with the growth in the number of tourists, the structures that hosted campers began to organize the camping space by generating two different habitats: that of green pitches dedicated to tents and the more anthropomorphized one defined by pedestrian areas in which temporary housing units (bungalows) were distributed. Unlike the tents, the housing units had the peculiarity of being particularly comfortable and immersing the tourist into an extremely natural context at the same time. The paradox of comfort/luxury/safety and nature/adventure/wildness was expressed even more clearly in the following years and right up to the present day. The first tourist villages were born, and in addition to tent pitches and toilets offer a greater number of services for entertainment and recreational activities that assimilate tourist villages to theme parks and cruises (Luisi, 2019). Hence the desire of the owners of these structures to try to satisfy all the needs of tourists within the enclosure of the campsites transformed them into villages, breaking the relationship with nature. The ultimate goal became the maximization of space in order to place the largest number of housing units. In addition, the distance of the sector from the architectural world, which is an outcome of a regulatory apparatus that in Italy is not interested in any relationship with the landscape, has various consequences: the relationship with the context disappears, settlement models that correspond to the geography of the places are not proposed, the need to create a new model of inhabited landscape does not emerge.

Today, open-air accommodation facilities in Italy include campsites, glamping sites and tourist villages. These structures share the intention of placing the tourist in a natural setting through the arrangement of isolated and comfortable housing units for each family. If during the 70s the prevalent housing units were of a permanent nature (bungalow), in recent years the mobile home is the most popular type of housing unit. The ever-increasing national and regional regulations aimed at coping with the state of environmental and landscape degradation of the coastal territories have certainly influenced the spread of mobile housing units in these contexts. In fact, being equipped with wheels and therefore not permanently connected to the ground, mobile homes do not consume soil, a subject of strong debate today (Berizzi and Trabattoni 2019).

Compared to other countries, such as the United States, the mobile home in Italy is spreading exclusively in the world of outdoor tourism: these are minimal housing units on wheels produced on an industrial level and created specifically for the leisure sector. Precisely because it is a vehicle for leisure time, its use within the accommodation facilities is seasonal, and inside it has all the comforts: kitchen-living room, bedroom / s, toilets, veranda and a predisposition to have water connections, electricity and gas. Once placed inside the facility, it is connected to the various networks (water, gas, sewage, electricity) set up by the facility itself.

The great value of these housing units being removable, is that it gives the ability to give a new interpretation to those coastal areas that are mainly touristic. If until now the coastal settlements have been monotonous, atopic and decontextualized, contributing to the formation of mere enclaves of entertainment, the fragility and the vulnerabilities of these territories, as well as climate change, now gives us new

Fig. 15.3 Masterplan with mobile homes, water basins, phytodepuration systems and trees (Credit: AUDe Lab, University of Pavia)

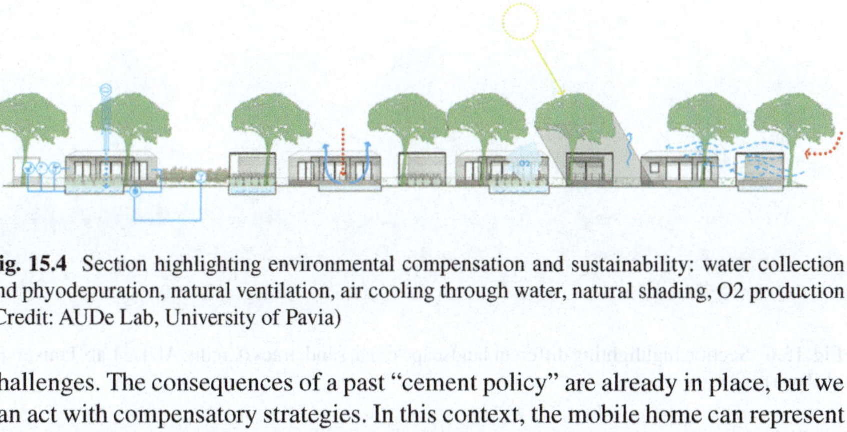

Fig. 15.4 Section highlighting environmental compensation and sustainability: water collection and phyodepuration, natural ventilation, air cooling through water, natural shading, O2 production (Credit: AUDe Lab, University of Pavia)

challenges. The consequences of a past "cement policy" are already in place, but we can act with compensatory strategies. In this context, the mobile home can represent an opportunity to build an integrated landscape in harmony with nature and with its own character identity. It is necessary to rethink the settlement system of these structures through an approach of integration between nature and the built environment. This means to work on one hand for a connotation of the built landscape capable of generating quality spaces, and on the other hand to work through mitigation strategies by providing shaded areas, rainwater recovery systems, phytodepuration systems and natural ventilation systems (see Figs. 15.3 and 15.4). The outcome must go in the direction of building a new landscape suitable for outdoor tourism but also capable of counteracting the effects of environmental impoverishment (see Figs. 15.5 and 15.6).

Inevitably, each intervention must be defined on the basis of the characteristics and morphology of the place, with its different peculiarities and criticalities.

Fig. 15.5 Masterplan with mobile homes settlement and natural landscape entering into the campsite (Credit: AUDe Lab, University of Pavia)

Fig. 15.6 Section highlighting different landscapes: sea, sand, trees (Credit: AUDe Lab, University of Pavia)

Precisely for this reason, the mobile home theme can represent a turning point. Its high versatility, in fact, allows incomparable settlement flexibility. The ability to move or remove the product as needed clearly responds to the urgency of not affecting the natural system, a system that has been suffering for too long and that is no longer able to "breathe". The soil can contain the climatic catastrophe, and in the case of the mobile home, the absence of permanent structures that anchor it to the soil allows for its high regenerative capacity.

Furthermore, a product such as that of the mobile home can not only avoid the disfigurement of a landscape such as the coastal one that characterizes the Italian peninsula and which constitutes an identity heritage, but can also constitute an ecological choice. The dry assembly system of the constituent elements of the mobile home allows a high recyclability and the ability to reuse its components in the future. In this way, the life cycle of the mobile home extends far beyond the tourist function, in which the individual components can be reused for new functions.

The reflection that led to the development of two settlement hypotheses in different coastal areas arises from the belief that it is necessary to rethink new forms of tourism through actions aimed at protecting the environment.

15.4 Designing Sustainable Tourism in Bibione

Bibione is a tourist city established around the 50s as a holiday resort. Located north of the Venice lagoon, it attracts thousands of tourists from all over the world every year, mainly from north-eastern Europe (Germany, Poland, Czechoslovakia). Data from Bibione Tourist information portal show haw, despite being a city of only about 2600 inhabitants, it sees about six million people passing through in the summertime.

This incredible numerical difference makes it difficult to read the urban and natural context.

If we try to define Bibione by ecological macrosystems linked to the tourist presence and anthropic use (Banham, 1971), it is possible to identify sets of spaces with a clear identity determined not only by their morphology but also by their use:

The first ecological system is undoubtedly that of the beach and the sea. With a stretch of beach more than 9 km long, this system represents the first tourist attraction of Bibione and is, therefore, its most "used" space.

Bibione beach is known to be one of the busiest in Italy. During the ongoing period of the present health emergency, the occupation system of the beach changed. By reducing the maximum number of people who could stay on the beach simultaneously, the occupation strategy has appeared more evident.

An orderly grid of equally spaced umbrellas completely covers the strip of sand next to the shoreline.

Behind the umbrellas, there are services for tourists, which change according to the activity to which they refer. The reflection on the use of public space (this stretch of coast is state-owned, and therefore public, but sold to exclusive private activities) which in part disappears as it is privatized to host exclusive services (because it is reserved only for tourists in the accommodation facility), it must be accompanied by a reflection on the ecological impact (on the natural system, in this case) of this situation.

Tourists use the beach as a massive gathering space. Except for the shoreline being still in public use, every centimetre of sand is subject to systematic use for various activities, from idleness to sporting commitment to orderly and free play.

The high concentration of people, the intensive use of space, and the massification of activities generate a very high environmental impact from CO2 emissions and waste production.

The second ecological system is that of the commercial city, or rather the built agglomeration. Bibione is characterized by buildings between 5 and 10 floors high, occupied mainly by hotels and holiday homes. The centre of Bibione has a commercial purpose, and even the places of historical identity (this being too short-lived) are almost entirely lacking.

This environment is also massively used by tourists, but mainly in the evening hours.

The holiday dimension, with the search for fun, is the basis of a reading of the urban space as a large holiday village which "dies" in the winter season.

In fact, during the so-called "out of season" months, the urban space remains empty, deprived of the presence of tourists. Many businesses remain closed for more than six months a year, making the urban space of Bibione a metaphysical space, as it has no interest and function, in addition to being a town also inhabited by the city's residents.

On the other hand, the urban commercial space represents the evening counterpart to the sea space in the summer months. The massive use of open space transports the "human tourist sector"from day to night. Together with the human mass and the opening of businesses, the environmental impact also remains constant, extending beyond the daytime hours.

The third ecological system is that of the wooded landscape. This system can be codified as the "Pineta" landscape, where the landscape of outdoor tourism is also located. The tree-lined area of Bibione represents a potential resource which can be used in two ways: one coded and one free.

The free space of the pine forest is a space overlooking the beach that generates a shady context for people's free use. The natural system of tall vegetation, characteristic of the maritime pines, which are the dominant vegetation of this context, suggests an interesting climatic variation to the warmth of the beach. For this reason, it is recognized as a resource by tourists and inhabitants.

The open-air tourism system represents the artificial image of the landscape, codified to generate living comfort. In fact, in the accommodation spaces, the pine forest is used or "built" to define a quality context that, although subsidiary to the "Bibione" system, represents a valid alternative.

The rational approach to managing the reception space, determined by commercial needs, means that the spaces within the enclosure are identified with a very orderly system of mobile homes or bungalows or open spaces. These are usually organized on orthogonal grids superimposed on the vegetation system, which is also ordered according to the same structure.

The campsite, therefore, wants to represent a housing ideal, which defines a hybrid between the other two ecologies mentioned so far: the natural element is integrated in a homogenous way with the anthropic one, but all are defined by an anthropized system linked to the management and their need to be used.

The subsidiarity of the enclosed space is linked both to a commercial and functional aspect, not representing a duplication of the urban space, neither for morphology nor for present activities. Concerning the ecological environment of the sea/sand, the space of the "pine forest"is not a substitute, presenting a potential for alternative and often supplementary use.

Finally, the high identity value of this ecological area must be recognized. The "pine forest"landscape is an iconic theme, representative of the Italian maritime tourism space. Therefore, this ecological environment is also supported by a marketable aspect that can suggest its rethinking according to its potential emphasis.

The expansion project of the Bibione International Campsite proposes a strategic intervention system in these areas.

The goal is to propose a vision of the landscape which takes into account various aspects:

1. Definition of a strategic anthropic landscape, using the mobile home system, which enhances the private space using the cluster theme.
2. Definition of a strategy to enhance the natural system, shade-trees-air, using water as an environmental mitigation system.

1. The project's starting point is the construction of an organized landscape with the accommodation system of mobile homes.

   As already explained, this accommodation system is characterized by an excellent sustainability value, being easily removable and therefore not affecting soil consumption.

   The first design choice is to define a housing system that integrates the private dimension with the collective one, using an ordered cluster system that maintains a possibility of "organic" perceptual use.

   The clusters that define the fundamental settlement matrix of the sector are developed with a double system of four units that are closed to the outside and meet in shared natural spaces.

   The clusters are organized with a linear flow system that distributes them homogeneously, guaranteeing easy technical and users' accessibility.

   A transversal flow system is then defined, which moves smoothly through the clusters and generates a transversal landscape perception determined by natural elements and visual variation.

   In this way, we want to generate a differentiated living environment, where the private dimension is never forced to interact but never loses the possibility.
2. The natural system present in the project area has already been mentioned, that of the Pineta. The project enhances existing trees (where it is compatible with the other design choices), integrating them with the proposed landscape idea.

   We choose to work with two elements for the construction of the landscape: the first is the plant elements, which can be rearranged according to better management of the light and shadow of the spaces. In this way, in fact, by alternating covered and open spaces, it is conceivable to generate pressure differences that favour convective air movements, which generates environmental comfort; the second element is water. In this context, the water topic is present in the lagoon and sea as characterizing elements of the landscape.

   However, we choose to face the water from a double point of view, aesthetically and technically. The technical point of view concerns the use of humid space as an air mitigation system. Combining the density strategy on the trees (shadow/light) using water tubs makes it possible to mitigate the air temperature, thus generating natural cooling ventilation.

We also work with the theme of water purification, integrating the landscape with a double system of phytodepuration tanks (dry and wet) positioned in the threshold spaces between the clusters to emphasize the organic movement of the transverse passages.

Furthermore, the landscape defined by the phytodepuration tanks also works on the visual density: the dry tanks are recognizable by the tall shrubs. The wet ones are water tanks with low essences, thus emphasizing the need for visual variation functionality to obviate the monotony of the landscape.

15.4.1 Outcomes

The project outcome identifies a strategic system that wants to work on the landscape, emphasizing its sustainable aspect. The approach to sustainability, already determined by the use of mobile homes as a living system, addresses the issue of water and natural ventilation in relation to the use of tree species.

The objective is the environmental compensation in the context of the "Bibione system", made vulnerable by the system of the three mentioned ecologies. With this in mind, the pilot project presented represents a plausible alternative in the redefinition of an extended system: that of outdoor tourism.

Open-air tourism, therefore, redefined according to the sustainability parameters defined in the project, could represent an effective compensation system in contexts, such as that of Bibione, which is characterized by a high anthropic impact.

The existing landscape, redefined according to the parameters of sustainability with particular attention given to water and air, becomes a recognizable element as a manifesto of a specific region and a strategy for safeguarding ecological vulnerabilities.

15.5 Designing a New Landscape in Cervia

Cervia is a tourist city in Emilia-Romagna and is part of the entire tourist sector of the Romagna Riviera. A long urban agglomeration, crossing various municipalities, defines the Adriatic coast for almost 50 km at this point that, starting from Cervia and reaching Cattolica without interruption.

The Pineta Family Village campsite is a typical example of the accommodation facilities in this region. It is spread over a longitudinal lot, with two short views, one facing directly onto the sea and the other onto the internal vehicular distribution ridge, via Matteotti, which defines the leading road network of the urban reference context.

The two adjacent lots on the long sides are also intended for hospitality activities, specifically for exceptional ones. To the north, the Montecatini colony represents a clear example of rationalist architecture, abandoned for years and therefore uninhabitable today.

On the south side is the logistic training base of the Italian army. It is followed by another abandoned maritime colony, the colony Varese or Costanzo Ciano.

In fact, the two colonies exemplify a drastic settlement attitude, which began in the 1930s but continued to the present day, determined by an exclusive relationship with the coastal landscape.

This sequence of accommodation activities is an example, in fact, of a settlement system that is repeated several times along the Adriatic coast of Romagna, which sees the accommodation business develop in the strip of land facing the beach up to the internal road.

In fact, the private structures, outside the historical centres, where the city/sea relationship is always balanced by the public space (seafront), are positioned in an exclusive way towards overlooking the portion of the beach, denying its relationship with the built context and also with the internal natural context.

This land occupation system generates a double effect concerning the use of the marine landscape: public access to the coastal area is solved with a system of secondary roads that develop alongside the accommodation activities (the same approach is repeated when the residential portion is before the historic core); the use of the beach is broken down according to activity, presenting itself as a sequence of private spaces.

The denial of the seafront as a public space is obviously also linked to the absence of a consolidated urban sector. The built-up space in which Camping Pineta is located, like most of the built-up area of the Adriatic coast of Romagna, is based on tourist activity.

The colonies are a historical reference example, almost all are abandoned because they are not used and are too inefficient to cater for the new tourist needs. Alongside the colonies, hotels and campsites have developed, and obviously holiday homes. The ensemble of these buildings defines an urban context of "transit", conveyed by the access to the sea (leading landscape) or historical centres (primary functions).

In summary, the relationship between the landscape and the coastal marine space (public for private use) is the qualifying element (near/far) of the accommodation activities of this anthropogenic macro-region. The qualification act that is proximity through an exclusive attitude binds the use and accessibility to the beach, implying a denial of the public promenade, and therefore denying the public dimension linked to the landscape to the urban context of reference.

Today the landscape system of the Italian coasts is put in crisis by general principles of erosion due to different causes (Archetti, 2019).

The risk of the disappearance of the beach, representing an actual ecological emergency, undermines the role of these spaces concerning the built and urban space. Both public contexts (when they exist) and private ones, fully defined in their relationship with the coastal space, undergo a significant crisis.

Camping Pineta Village is an example of a situation familiar to some accommodation sectors.

First of all, the uneven construction over time of dissimilar lots has generated a landscape discontinuity of the open space linked to the linguistic discontinuity of the build. The campsite alternates between several types of accommodation: mobile homes of different typology, shapes and heights; two-storey houses; hexagonal bungalows; rectangular bungalows; and collective services.

Alongside the construction discontinuity, there is also a very strong landscape discontinuity in the campsite, determined by the contingent natural conditions of the area: the proximity to the coastal landscape implies greater exposure to winds and the erosive effect of the sea on vegetation. It is still possible to distinguish two macro-landscapes inside the campsite:

- The first landscape, linked to the internal context, is defined by the arboreal presence of maritime pines along the Viale Matteotti road. The tall trees, typical of the Italian open-air tourism landscape, define a shaded and protected area.
- The second landscape begins at about three-quarters along the area. After a level change, a semi-desert landscape appears, influenced by the beach and the sea. The sandy soil and the strong winds mean that the vegetation in these areas struggles to grow, leaving large sunny areas.

Finally, the Camping Pineta Village is located, as mentioned, overlooking the sea. Access to the beach is exclusive and there is a portion for private use in the coastal space. However, the beach has been subjected to critical erosion that has led to its significant reduction, thus undermining the relationship with the accommodation space.

Therefore, the design reflection on the Pineta campsite starts from the need to find a coherent settlement solution that starts on the characteristics of the landscape. Taken in its objectivity, the landscape of the campsite is also now divided in its quality: the space of the pine forest is flourishing and in good condition; the space on the beach is clearly spoiled, where the plants fail to grow so close to the sea on sandy soil.

The settlement project, therefore, provides for a redefinition of the entire campsite, proposing a solution for two fundamental issues:

1. The confrontation with the sea, through the management of collective areas, in order to enhance the relationship with the landscape.
2. The definition of a landscape integrated with the existing ecological situation, taking into account the environmental difficulties mentioned.

Therefore, the complete redefinition of the organization of the campsite starts from the enhancement of the potential, trying to keep all the healthy vegetation intact and converting the damaged space into a sandy and sunny space.

The seafront, which starts tired from the winds, is converted into a coastal landscape, supported by the use of suitable mobile homes. We choose to propose a system of mobile homes facing the sea (first line) made of lodge tents, easy to place on the sand and slightly more spartan than traditional houses.

The second front, always on sandy soil, is proposed, with a system of houses with a terrace roof. This type of house allows the veranda to be developed partly above the house, thus enjoying a panoramic sea view (primary landscape).

The third front develops inside the pine forest and lives in an independent environment concerning the primary landscape, set up with "usual" mobile homes.

The design strategy proposes to define a landscape system starting from the concept of the seafront or of an urban space linked to the landscape, which instead of developing along the coastline (too short) becomes an internal ridge.

The campsite is interpreted as a longitudinal bridge space, which reaches the sea from via Matteotti. Despite its private status (exclusive for guests), this connective system wants to redefine the reading of the macro settlement context (Romagna Riviera) by explaining the change of landscape. The change of landscape, pine forest/sandy, emphasized by the landscape and settlement choices, is, therefore, a clear example of a critical situation converted into potential.

Finally, the internal ridge, or transversal seafront, is positioned in line with the current distribution of the campsite services (reception, restaurants, shops, swimming pools) and extended for the full length of the campsite.

Two hierarchically different strips are defined: one broad, mainly determined by accommodation functions (super-campers, mobile homes) and the other almost exclusively occupied by services.

Consistent with this organization, the campsite's waterfront becomes an opportunity to redevelop the coastal landscape, bringing a part of it into the campsite itself.

In this way, the present situation of the state-owned coast can be weakened and redefined with occasional and perhaps less destructive use.

15.5.1 Outcomes

The design of open-air tourism accommodation areas brings a potential reflection on the conservation and enhancement of the landscape. In the urban context of the Adriatic-Romagna coast, where the prevailing settlement system isolates the coastal landscape from the public context, the open-air receptive environment becomes an essential point of interaction.

Through reflection on the identity of the landscape and its vulnerabilities, the open space design can become an opportunity to redevelop an existing situation (that of the tourist space). Furthermore, the design action on the private space can start to lessen the difficulties of the public space (the coastal one).

Even if not weakened by anthropogenic presence, the massive weight of the industrial presence on the coast can be lightened. The weight of the environmental impact can be shared between the private and public landscape. By rewriting the internal campsite landscape as an extension of the sea landscape, it is possible to think about a redefinition of the human impact on the natural system.

15.6 Comparing Outcomes and Strategies

The two proposed design research experiences arise from the need to define a new sustainable landscape and a stable ecosystem in open-air accommodation through the definition of a light and reversible settlement system.

Efforts to counteract the vulnerability of the Italian Adriatic coasts can be activated starting with the tourism sector, which represents the main economic driving force and is responsible for the transformation of the territory. The coastal territory today presents vulnerabilities linked to two distinct conditions: on one hand, the action of man, and on the other, those natural actions which have evolved issues related to climate change. The first action, that of man, has evolved strongly over the last fifty years, transforming natural or man-made territories for agriculture and fishing into areas for the establishment of tourist activities. In the past, the coastal territory, through practices consolidated over time and closely linked to natural conditions, had created an ecological system that was in equilibrium. It was not always an approach of unconditional respect towards natural conditions but of a modification of controlled nature to make the territories habitable and usable for economic purposes. For example, in the area of the Delta del Po, the river was diverted and the geography changed to allow soil and water exploitation as productive elements. These actions, not always without disastrous consequences, for example with the Polesine flood, have often redetermined a balance between anthropization and nature. Man has thus learned to coexist with nature by modifying its landscapes and protecting them to safeguard environmental conditions and, therefore, productivity. The abandonment of these practices simultaneously led to the abandonment of the

OUTCOMES	BIBIONE	CERVIA
SOIL	redefined natural soil	active natural soil
STRATEGY	resilient sustainability	active and compensatory sustainability
GOAL	sustainable landscape, promote biodiversity, give character	sustainable landscape, promote biodiversity, give character, better climatic conditions
DESIGN ANSWER	beach extension	new environment creation

Fig. 15.7 Outcomes comparison between Bibione and Cervia projects in terms of soil definition, strategy for sustainability, goals and design answer (Credit: AUDe Lab, University of Pavia)

care of the territory, and it is for this reason that tourism, with its economic capacity, can represent a new hope for the protection and care of soils and landscapes. The care of the territory can arise where there is economic initiative and for this reason attention is placed on the accommodation facilities that characterize a substantial part of the Adriatic coast. From this consideration comes the first design response that combines the two projects of Bibione and Cervia; the construction of a new permanent landscape that combines the need to counteract the vulnerability of these places with the need to inhabit them. The soil thus becomes an element of definition and control of nature which takes on permanent aspects for these contexts. However, the two cases show a different attitude (see Fig. 15.7). In Cervia, the soil project works to counteract the effects induced by anthropization and climate change by preserving, as far as possible in a highly urbanized area, some natural features. The project extends the beach inside the open-air accommodation facility, allowing natural behaviour towards the sea by eliminating non-native or spontaneous vegetation, which had compromised the sandy strip. The project intervenes on an existing village that applied the same strategy throughout the area, namely that of camping in the pine forest, operating indiscriminately in the internal portion of the area and on the coastal strip. The beach enters the village not only in its being sandy soil but also in its collective functions which include some typical elements of accommodation facilities such as the swimming pool and the solarium. From the point of view of the construction of the character of the place, the theme of the beach that enters into the village coincides with the definition of a public "seafront" path that penetrates the area, defining hierarchies and references. The settlement system of the mobile units, with a temporary and possible variable character, supports the new landscape and soil system by adapting to it and strengthening it. Instead of counteracting the natural effect of winds and the sea with rigid elements, the strategy designed for Cervia defines the landscape according to specific geographical and climatic conditions. The identity of the landscape becomes an opportunity for the consolidation of a natural state and creates a new stable maritime landscape regardless of the presence or absence of human activity.

From this point of view, the strategy designed for Bibione is different. Also, in this case we have the definition of a new soil which, however, is conceived to be inhabited in a sustainable way and therefore conforms to the condition of an anthropized environment. The theme is that of an active soil sustainability. The open space, integrated with the settlement / housing system, is conceived as a system of resource compensation. The creation of phytodepuration tanks to clean rainwater and recover wastewater from bathrooms by reintroducing them into the water system, and with the possibility to use it for irrigation, as well as the introduction of water and plants to mitigate temperatures during the summer period and to contribute to energy saving, are all functional actions to create an environment suitable for being inhabited as well as inducing environmental and climatic conditions that promote biodiversity. This project assumes the pine forest as a starting condition and tries to make it even more habitable. The pine forest is extremely common along the Adriatic coast and represents a recent landscape built by man to ensure shade and naturalness near the sea. The maritime pine is a tree that is well suited to the coastal climate, allowing a large, shaded area compared to its trunk, thus leaving the soil little occupied by vegetation and therefore suitable for hosting tents, caravans or small homes, that could be fixed or mobile. It is also a tree that grows on soils that are partially sandy. For this reason, the proposal on Bibione takes on an easily exportable value along the Italian coast, representing a second feasible model. In this case, the natural environment responds to the built one by mitigating the vulnerabilities generated by the anthropogenic mass.

In both case studies, the construction of the touristic landscape of outdoor tourism is an opportunity to reflect on the possibility of building sustainable landscapes.

Bibione expresses an active and compensatory sustainability comparing to the resilient sustainability of Cervia. Reading the accommodation space as a necessary substrate integrated with mobile homes determines further reflection.

The vulnerabilities of the natural context, increased by strong anthropization, can be mitigated by the artificial system of outdoor tourism and can induce further research applied to public and non-accommodation contexts.

15.7 Discussion and Research Limitations

Sustainability applied to a more niche form of tourism is certainly more easily feasible than a form of mass tourism, such as the one that involves most of the Italian coastal areas, in which the application of sustainability principles on different levels is an intricate process. The difficulty is even greater when the intervention area is critical and vulnerable. However, it is necessary to understand the deeper meaning of sustainability applied to tourism and how, in a historical era such as the contemporary one, namely that of the Anthropocene, the trend of reversal can no longer be postponed. Designing for sustainability in many parts of the Italian coast is not just a question of protecting the environment and the landscape, but a question of "life or death"for the landscape itself. In the absence of immediate conscious actions, portions of the peninsula's territory could vanish forever, leading to disastrous consequences for the environment and for the health of its inhabitants. According to the World Tourism Organization, sustainable tourism must consider not only current social, economic and environmental impacts but also future ones. That is to say that today's actions will inevitably have a reverberation in the future. Unresponsive actions today could deprive future generations of benefiting from coastal areas.

The role of sustainability in the tourism sector inside vulnerable areas is of extreme importance to the point that its application can determine the fate of many coastal areas of the peninsula. The planning of accommodation facilities is one of the key issues, where the settlement strategy to be adopted plays a central role. Open-air tourism is certainly a form of tourism with a high potential to be able to define itself as sustainable. By adding determined measures to this way of using the landscape, it can be possible to preserve and define a new tourism landscape. Excluding the possibility of intervening with permanent structures, such as bungalows, can be the first step. Approaching to landscape with light and mobile architectures can reduce anthropization, allowing coasts to be more natural territories and less artificial. Leaving the landscape more natural allows natural processes undergo, strengthen coastal landscape. It means relating tourism to nature to create a harmonized landscape instead of a landscape of contrasts. Furthermore the high flexibility, not only of the settlement but also function of mobile homes, can allow for the management of the services of accommodation facilities through mobile units. In fact, there is no lack of examples of mobile units of minimal size and comparable to mobile homes, located in diversified contexts and with different functions. A clear example of the versatility of the product is demonstrated by the pandemic, that is still underway. The urgency of the spread of the virus and the need to allocate more spaces for healthcare in the cities has led to the placement of various mobile installations in many inhabited centres and in the vicinity of existing hospitals. Furthermore, the mobile home, if conceived as a combinable container, can allow articulation of the spaces to be used for the different services within the accommodation facility with different combinations, depending on the needs.

Even though the research outcomes demonstrate the high potential of mobile homes settlements for outdoor tourism, it demonstrates its limitations as well. As it stands, doesn't exist in Italy any strategy of control of these settlements, neither at a national level nor at a regional one. If there are regulations such as the distance of the settlement elements from the coast, there is currently no strategy aimed at encouraging the owners of the structures to adopt settlement measures and approaches that are sensitive to the natural landscape of the coast. In a practical way, this gap leaves strategic decisions to the owners of the structures, who can be free to approach the landscape in a more or less sensitive way. For this reason, the research that results in applied design has not yet reached the next stage, namely that of execution. Research limitations are reflected in the current lack of verification over time and the post-construction results of this type of settlement.

Thinking of a form of tourism that is sustainable in particularly vulnerable areas, leads inevitably to the need to measure the feasibility of the intervention as well as the practical efficiency of the planning intention. It is in fact the clearness in the intention/vision of planning that stands the quality of tourism landscapes. While intervening within purely tourist areas certainly raises issues that go beyond the mere design of the accommodation facility. One of the most common problems in these areas is in fact the social aspect. The lack of a unitary strategy of these places often results in a fragmentation of the man-made landscape: on the one hand, the existence of a pedestrian axis parallel to the coastline and on which the commercial services overlook the ground floor; around the axis a fabric consisting of condominiums mainly rented during the summer season; in the areas often furthest from the centre, the location along the coast and in the innermost areas of open-air accommodation facilities; and in the innermost areas of the territory the residential areas where the local population lives. The lack of homogeneity at the settlement level undoubtedly generates a divided space, and even more a series of spaces, between the low-quality building fabric that are hardly usable. Therefore, the legacy of a functional approach has led to an impracticable cohesion between the different areas of the urban fabric and an overcrowding phenomenon during the high season of a few specific areas for recreation. The contextual reflection on the landscape of tourism for the protection of the coastal landscape should go beyond the enclosure of individual accommodation facilities to embrace a wider landscape that contaminates the small, inhabited centres and other permanent accommodation facilities, reconfiguring the lost relationship between buildings and nature to find a new sustainable landscape in equilibrium.

References

Banham, R. (1971). *Los Angeles: The architecture of four ecologies*. Harper & Row.

Battilani, P. (2001). Vacanze di pochi, vacanze di tutti. L'evoluzione del turismo europeo.

Berizzi, C., & Trabattoni, L. (2019). *Mobile Home per il turismo all'aria aperta, storia evolutiva*. Vicolo del Pavone Editore.

European Environment Agency. (2006). *The changing faces of Europe's coastal areas* (EEA Report No 6/2006). Retrieved from June 1, 202, from https://www.eea.europa.eu/publications/eea_report_2006_6

Forman, R. T. T. (2008). *Urban regions: Ecology and planning beyond the city*. Cambridge University Press.

Gruppo Nazionale per la Ricerca sull'Ambiente Costiero. (2006a). Le spiagge della Emilia Romagna. *Studi Costieri, 10*, 83–88.

Gruppo Nazionale per la Ricerca sull'Ambiente Costiero. (2006b). Le spiagge del Veneto. *Studi Costieri, 10*, 89–94.

Isprambiente. (2019). *Annuario dei dati Ambientali* (89/20). Retrieved May 30, 2021, from https://www.isprambiente.gov.it/it/pubblicazioni/stato-dellambiente/annuario-dei-dati-ambientali-edizione-2019.

Luisi, S. (2019). Storia ed evoluzione dell'abitare mobile e del turismo all'aria aperta, pp 59–88, in Berizzi, C., & Trabattoni, L. *Mobile Home per il turismo all'aria aperta, storia evolutiva*. Vicolo del Pavone Editore.

MATTM-Regioni. (2018). *Linee Guida per la Difesa della Costa dai fenomeni di Erosione e dagli effetti dei Cambiamenti climatici*. Retrieved from http://www.erosionecostiera.isprambiente.it/linee-guida-nazionali

Minambiente. (2017). *L'erosione costiera in Italia. Le variazione della linea di costa del 1960 al 2012*. Retrieved June 2, 2021, from https://www.minambiente.it/biblioteca/le-variazioni-della-linea-di-costa-dal-1960-al-2012.

Renata Archetti (2019). Innovative Strategies, monitoring and analysis of the coastal erosion risk: The STIMARE project. In *The Proceedings of the 29th (2019) International Ocean and Polar Engineering Conference*. International Society of Offshore and Polar Engineering Conference.

Part IV
Social Engagement in Vulnerable Communities Between Digital and Humanist Visions

Chapter 16
Augmented Reality (AR) and Virtual Reality (VR) as Tools to Empower Vulnerable Communities: Opportunities and Challenges for Designers

Edgar Paul Martínez Ludert Muñoz de Cote, Alfredo Mauricio Flores Herrera, Emanuele Giorgi, and Tiziano Cattaneo

16.1 Introduction

Human imagination is one of the supreme tools that people have to confront the damaged systems in which the whole society is involved. In our Anthropocene complexity, where everyone is directly involved, information and technology play a transcendental role in how people relate to its own identity, context and otherness. Society must understand, now more than ever, how the "I" affects the "We", including those vulnerable communities and beings without a "voice" to defend themselves. All the new, evolving, and promising technological breakthroughs can have a significant impact on moral character and abilities (Rueda & Lara, 2020), in the behaviour of people, as cultural and social "animals". With a social drive and oriented to life, technology, with its interaction with people, can reach the possibilities of a new world, full of generosity, abundance, healthy information, empathy and in harmony with nature (Mumford, 1995).

E. P. M. Ludert Muñoz de Cote (✉)
Tecnologico de Monterrey, Escuela de Arquitectura, Arte y Diseño, Hermosillo, Mexico
e-mail: edgar.ludert@tec.mx

A. M. Flores Herrera · E. Giorgi
Tecnologico de Monterrey, Escuela de Arquitectura, Arte y Diseño, Chihuahua, Mexico

T. Cattaneo
Department of Civil Engineering and Architecture, University of Pavia, Italy

College of Design and Innovation, Tongji University, Shanghai, China

E. Giorgi et al. (eds.), *Design for Vulnerable Communities*, The Urban Book Series,
https://doi.org/10.1007/978-3-030-96866-3_16

16.1.1 AR/VR/AVR

In 1950, American science fiction writer Ray Bradbury wrote *The Veldt*, a science fiction essay first published in *The Saturday Evening Post* and later as an anthology in *The Illustrated Man*. This short story, based on Bradbury's imagination, evokes the first signs of a futuristic technology that had the ability to simulate hyper-realistic environments based on human desires. A decade after this dystopian text, in 1962, the inventor Morton Heilig created *Sensorama*, the first multisensory immersive machine that transported you to different experiences through three-dimensional colour cinematographic images experienced in first person (Xataka, 2018). The simulation was accompanied by stereo sound, movement, vibrations, smells and tactile stimuli such as the feeling of the wind, all adjusted according to the selection chosen by the user within a catalogue of experiences. A fundamental and innovative part of this machine was the cancellation of noise coming from the environment. The intention was to achieve a more focused and distraction-free immersion. Although this machine was not successful due to its high cost of production and maintenance, it remains as one of the first attempts to achieve a completely immersive experience that uses all the senses.

In the 1990s, the increase in the speed of processors, advances in computer graphics and the deployment of the web, brought the promise of virtual reality closer than ever (Riboldi, 2017). Nowadays, the initials AR and VR become familiar. It's common to hear these terms in all kinds of industry, from education and therapy to entertainment and arts. These technologies act from principles: play, teach, train, enhance, envision, educate, empathize, and interact with a non-physical reality. The differences between both platforms become important at the moment we call them for action.

Augmented Reality (AR) is a technology that involves the overlay of computer graphics on the real world (Silva, Oliveira y Giraldi, 2003): in other words, virtual objects merge with reality into a single image through the view of an apparatus.

On the other hand, *Virtual Reality* (VR) can be defined as a set of technologies that provides immersive experiences in computer generated worlds (Rueda & Lara, 2020). In virtual reality the immersion is complete: the outside vanishes and our body becomes a fundamental part of the equation. The most important distinction between Augmented and Virtual Reality is the ability of detaching us from the physical world (Rueda & Lara, 2020). Nowadays the best of both technologies are available.

Augmented Virtual Reality (AVR) uses a physical setting where the experiences take place. These installations become a safe and controlled space that enhances immersion through physical resources. The immersion of a virtual experience and the strength of the emotion to be transmitted can be expounded through the use of additional physical sensory stimuli (Harmsen & Mustar, 2019). Some relevant examples will be presented later.

16.1.2 Main Characteristics and Differences

For Silva, Oliviera and Giraldi (2003), AR needs three aspects to properly work: (1) the combination of real and virtual worlds; (2) interactivity in real time and (3) registration in 3D. VR technology also needs a set of three pillar elements to operate: (1) immersion as the technical capability of a system on its mission to evoque persuasive environments with which the participant can have an immediate—real time interaction (Rueda & Lara, 2020) with digital scenarios, objects and avatars; (2) presence, referred to the psychological experience of "being there" (Heeter, 1992), the person believes in their existence and inhabits that other world, possibly in another body for a lapse of time, this is the (3) third element and it is called embodiment, as the sense of experiencing the body or some body parts as "one's own" (Rueda & Lara, 2020). This is helped by motion tracking, enhancing the illusion by synchronizing movement and vision from a first-person perspective.

VR technology allows the creation of highly immersive-interactive virtual spaces where it is possible to play with the change of ego and the incarnation in other bodies and forms. Through these environments it is possible that the boundaries between one's own body and the perceived one become virtually blurred. Virtual Reality systems, unlike what a smartphone offers, are not necessarily going to become a day-to-day technology, but it will be much more relevant to call them when people want to get out of the ordinary to fantastic worlds and experiences in first person (Harmsen & Mustar, 2019). Although it could be said that this technology is still in its infancy according to all that it can become, artists, architects and designers are using it to amplify their experiences, services and products (Harmsen & Mustar, 2019). Jean Baudrillard, (1994) stated 17 years ago that virtual reality was going to become an aesthetic form, that computers will generate art. And he was not wrong.

VR becomes very relevant for journeys to places where few or no people can reach. This geographical or conceptual impossibility becomes feasible. Although there is much to improve in terms of the image or virtual representation, today other worlds can be explored, either in the eyes of another human being or from the perspective of an ant, a hawk, a toaster or a cell. Today it is possible to break gravity, accelerate or decelerate time and, in general, alter or eliminate any obstacle that in the physical world is technologically impossible today. The possibilities of virtual reality go far beyond gaming and entertainment applications. Many are betting on this as the ultimate empathy machine (Alsever, 2015; Rueda & Lara, 2020). For techno utopians, the promise and appeal of virtual reality is not just the reshaping of their own identities and potentials, but also the ability to shape those of others (Riboldi, 2017). Virtual reality has the potential to connect humans with other humans in a deep, intimate way, like no other means of communication can.

16.1.3 New Devices and Design Challenges

AR/VR devices, which are becoming more and more capable of penetrating everyday life, are strengthening their possibilities along with the quantitative development of the size of their market. The products related to virtual or augmented reality, in fact, have seen an exponential growth in their diffusion. It is even expected that AR/VR products will see their market grow from $ 30.7 billion in 2021 to $ 296.9 billion in 2024, thus transforming what the use of these products is today and paving the way for this technology is an almost daily and obvious presence in our society (Statista, 2021).

Certainly, time constraints such as the global semi-conductor shortages that recently affected various tech companies, such as Facebook, could impact the development of this trend. Anyway, despite this, the trend is clear, and the direction plotted. For this reason, it is essential for designers to reflect on how these technologies related to AR/VR can open new opportunities for the design of future territories, to understand how the potential of these technologies can be fully exploited and to imagine new methodologies and design strategies that will arise. Without any doubt, these new technological possibilities will change the way people imagine the practice of design in all social, urban and geographical contexts, but the design in vulnerable communities will be of particular interest, both because these communities will be the most affected by the design challenges of the future, and because in these communities the application of technological resources is at the same time the most limited but the most opportune.

16.1.4 Vulnerable Communities

In this chapter, the definition of "vulnerable community" is the one expressed in the *Design for Vulnerables* workshop: a group of people who share a generic state of limitation in being able to cope with contemporary or future needs. For this reason, the socio-economic or geographical context—the contextualization—is a fundamental element for defining the characteristics of a vulnerable community. While a "vulnerable group" lives a well-defined and constant state of vulnerability, a "vulnerable community" refers to those realities characterized by people with different personal needs, who they experience limitations in a discontinuous way over time. For example, while limitations of a group of elderly people can be easily defined (regardless of the geographic or socio-economic context to which they belong) and persist constantly over time, for a "vulnerable community", limiting characteristics depend entirely on the geographical or socio-economic context to which they belong and strongly depend on a precise moment of time (a moment of the day, a season particularly dry or cold, etc.). Elements that characterize a "vulnerable community" are the lack of sensitivity by the "formal" society, the limited awareness of their rights and basic needs by the community itself and a poor educational level, with the consequent limitation in being able to imagine future trends or the next environmental and social challenges that can violently hit the society in the coming years.

16.1.5 Research Question

Within this context, this chapter aims to understand which are the main contemporary or future advantages of applying AR/VR based design methodologies in vulnerable communities. To better understand AR/VR potentialities for "vulnerable communities", it was decided to analyse (1) case studies applied to "vulnerable groups" (characterized by specific problems of vulnerability or limitations) and (2) case studies applied to local communities (characterized by contextualized necessities). This, with the aim of defining the characteristics and potential of the application of AR/VR technologies in environments like those of "vulnerable communities". Therefore, a comparison was made between the results of AR/VR applied to "vulnerable groups" and "local communities".

16.2 Analysis of Existing AR/VR Projects Which Work with Vulnerable Groups

For the first analysis related to vulnerable groups, three case studies have been analysed. These cases share the intention of this technology in collision with cognitive sciences, cinema and neurodiversity and aim to transform viewers into a person with a different skin colour, a migrant and a person with autism.

The machine to be another The *Machine to be another* is a multidisciplinary AVR project by the Barcelona collective Beanotherlab that addresses the complex problem of racism and all those social phenomena with a negative impact linked to the tension between group stereotypes. From the combination of virtual reality, cognitive sciences and performance, this experience allows people to perceive themselves in a foreign body, position them in the eyes of another human being while interacting with tactile stimuli from a real person in a controlled space. The interaction becomes the most valuable element, because unlike other media, here there is a dialogue in real time. This unique methodology uses the temporary illusion of body transfer to its advantage. This occurs when the brain is tricked in terms of perception. Virtual reality has one goal: to convince the users that they are in another place and, often, that they are someone else (Presence - Embodiment). To do this, the human brain is fooled—in particular, the visual cortex and the parts of the brain that perceive movement (Parisi, 2015).

This collaborative open-source project relies on this technology to blur social conflicts that lead to marginalization, racism and any military combat. By putting ourselves in the place of the other, in their context, their life history, their daily life, often painful and vulnerable, users can understand, for a period of time, that other perspective of life. Although it would be naive to say that all people by experiencing these projects will change their social habits or behaviours 360 degrees, there is strong evidence from neurosciences that something changes when experiencing these realities. The *Machine to be another* addresses a complex problem and hopes over time to be part of the shift towards a future of more empathetic societies.

Carne y Arena In 2017, Alejandro Gonzales Inarritu imagined and created the immersive experience *Carne y Arena*, a 7-minute simulation of fragments of the dehumanizing and denigrating journey of all those migrants who leave their country trying to dignify their lives and that of their families (Fig. 16.1). This immersive installation in a controlled setting is the first one to appear at the Cannes Film Festival, so its relevance will impact massively and in a political level (Farago, 2017). The project relied on all the narrative resources that virtual reality can offer. This set of violent three-dimensional images, supported by real physical stimuli: the sand on the floor and the heat in the room that simulates the sun, make the spectators truly transport themselves to the desert, a place of desolation where they transform for a brief moment in migrants and feel the horror of experiencing the injustices, mistreatment and precariousness that society has imposed on these walkers.

It is interesting that filmmakers are beginning to look at the possibilities of this technology, as it offers another way of telling stories in first person, which completely changes the narrative dynamics, since it is the viewer who directs what they observe, making the experience unique and unrepeatable for each person. In the

Fig. 16.1 The immersive experience Carne y Arena by Alejandro Gonza-les Inarritu (2017) (credits of Pierre-Olivier Marinier Leseize)

words of Iñárritu "virtual reality, unlike cinema, is a medium that poses different narrative and theoretical challenges... you cannot change the frame, the characters have to be three-dimensional. It is almost a mixture of a video game with live theater". The theorist of design and technology Maldonado, 1999 affirms that theatricality is indispensable, becoming a basic component to construct narratives that are deeper and closer to reality.

Narratives are playful and authentic formats whose potential is vast for those who are interested in learning about people's life experiences. This form of information collection is constantly growing and evolving, over time it has managed to position itself and today it is recognized as a research tool that enriches the reading and projection of knowledge of social and cultural phenomena. Narratives have extraordinary potential in design research, being a means to understand events and actions of individuals and groups of people, particularly suitable in the study of human subjectivities and identities. Narratives serve to frame complex processes and tell them in a more accessible way, even being able to find connections and relationships that are not otherwise evident. Narratives, as a genre of action and verbal representation in everyday life, should be considered as instances of social action, as speech acts or events with common properties, recurring structures, cultural conventions and recognizable genres (Vasilachis, 2006). Narratives can present in different ways; films, stories, novels, poetry, and now virtual reality, they themselves delight and trap us with their depth and singular state, they help us to understand ourselves and others (Merriam, 2009), they make people feel the world and allow them to communicate the richness of their experiences. The personal dimension of the experience and its richly detailed description help others to become in a more real and human way what moves and inspires. The epistemological position of the lived experiences validates a basic way of constructing knowledge of the world.

Autism Empathy Tools Finally, in 2016, young designer Heeju Kim bet on inclusive and interactive design by creating the Autism Empathy Tools toolkit. The strongest argument is given through the experience acquired, because from this, a mutual understanding arises between different social groups, focusing on empathy as a tool for action. This set of empathy tools has the aim to understand the reality of people with autism and allows users to experiment through sensory immersion, a simulation of what it means to live with sensory processing disorder or SPD, a complex brain condition that affects the way sensations are experienced and information from the environment is processed. This problem can affect one or more senses directly influencing actions, reactions, and behaviour. People with autism can become more sensitive to the physical environment. It is difficult to precisely explain this disorder since each person's experience is completely different. The extrapolated sensory sensitivity of sounds, sights, smells, textures and flavours causes isolation and rejection of the physical and social environment; therefore this problem can become a challenge in the classroom, public spaces and even the home itself (Ludert, 2018). This kit has three main components: a virtual reality cardboard box with an application specially designed for smartphones that simulates vision altered by visual stimuli, headphones that cause hypersensitive sounds and a disposable candy that makes it difficult to speak. Although these tools cannot fully simulate SPD because their experiences are unique to each individual, they serve as a bridge to empathize and understand the difficulties that the environment causes in people with autism (Ludert, 2018).

Likewise, the National Autistic Society in 2018 launched its *Too Much Information* campaign, where in addition to generate updated and accessible information around autism, it was accompanied on that occasion by a 360 virtual reality video that simulated being in the body of a child with autism suffering a sensory collapse inside a shopping centre. This foundation has been characterized by doing research about autism and being a benchmark in terms of its awareness campaigns. The previous project and a peripheral investigation of SPD were the inspiration to create *My Mom's Favorite Restaurant* a AVR experience where people turn into a person with autism with sensory processing problems for a minute in a restaurant full of aggressive visual and sound stimuli. At the end of 2019, the Memory and Tolerance Museum hosted this experience, positively impacting more than 1000 people who were able to live for a moment what it means to have the sensory difficulties of this complex human condition. These 3 experiences are intended to simulate other realities. Jean Baudrillard affirms that simulation has the power of illusion (Baudrillard, 1994). To simulate is to pretend to have what users do not have, to supplant the signs or symptoms of an original reality (Lopez y Olivier, 2016). Both intend to create an illusory act where viewers live another reality and live it for a defined period of time.

16.2.1 Definition of Common Characteristics

Based on a background of knowledge and on these three experiences, some common characteristics can emerge, so to represent the main features and impacts of these technologies applied to vulnerable groups.

Empathy Empathy is a powerful force that informs and inspires design (Battarbee et al., 2014) architecture, the arts and any discipline that moves in the social field and is interested in approaching knowledge of others from the most human point of view. Empathy for others plays a key role in problem solving and the study of social phenomena; therefore, it is relevant to demand that creative disciplines attend diversity, including people with little standardization and in a vulnerable state.

Empathy consists of the ability to be aware of and sensitive to the feelings and thoughts of another person without having had the same experience (Battarbee et al., 2014). The philosopher Martin Buber (1985) defines empathy as that act of "entering with one's own feelings in the dynamic structure of an object, a column, a crystal, a branch, an animal or another human being, as if we wanted to contemplate it from inside to understand its formation and movement and perceive it with our own muscles". We stop being ourselves for a moment and become what we are trying to understand. It is immersion, not just interpretation or imitation. Taking the place of another person causes measurable changes in our thinking from the subconscious. American writer Willa Cather stated that empathy is the unique and wonderful experience of stepping into the skin of another human being, very similar to the words of the French writer Alphonse Daudet: "you have to get under your skin, see the world through your eyes and get to experience it through your senses" (quoted in Root-Bernstein & Root-Bernstein, 2002). The ability to experience empathy is a fundamental skill for proper performance in any helping profession, a human tool that provides knowledge unaffordable by any other means. Empathy requires deliberate acts that intentionally seek opportunities to connect with people in a meaningful way (Battarbee et al., 2014) and understand the other's vision from the depths. An amazing example, when it comes to working with child patients, is the virtual reality application *Nixi for Children*, this project arises to treat anxiety and fear in the run-up to an operation. This application prepares kids to experience an operation by interacting with an animated character in scenarios that simulate medical spaces. The designers thought about the emotions that patients go through before and after the operating room, an interactive children's narrative to help patients cope with a process that fills any human being with uncertainty.

Identity How do we act?, how do we think?, how do we look?, where we live?. Questions that respond to the construction of our identity. The “I” as a combination of emotional, cognitive, contextual and physical aspects that makes us unique in a community. These characteristics are possessed by an individual, through which it is known (Páramo, 2008). A good part of the identity we form it from the social interactions that begin with family, at school and with the people we know throughout life. Identity will influence the way we act in the world (Páramo, 2008).

Alterity–Otherness Understanding concepts such as otherness becomes necessary for disciplines that engage with human beings that come from different origins and contexts. The other is a different individual who is not part of their own community, who has a different history and identity. The term alterity derives from the Latin alter (the other) and it could be said that it is the philosophical principle of "alternating" one's own perspective (Krotz, 1994) or world view for that of the "other". Otherness becomes a mode of liberation that engenders new practices of freedom (Ruiz, 2009). For Karla Vega (2015), “otherness” is applied to the discovery that the “I” makes of the “other”, which gives rise to a wide range of images of the other, of “us”, as well as multiple visions of “him”. These representations are of human beings inhabiting different worlds within a shared world. We stop being ourselves to become for a moment in time what we are trying to understand.

16.3 Analysis of Existing AR/VR PROJECTS Which Work with Local Communities

After analysing projects for “vulnerable groups”, the authors expanded the research by extending to the study of AR/VR applications in local communities to observe the possible effects in a local and contextualized dimension.

In fact, the characteristics of a group cannot be mirrored with the characteristics of a community: the two populations experience their vulnerability in a completely different way and, therefore, the support that augmented or virtual reality technology can give is clearly different.

In fact, Virtual Reality and Augmented Reality have been used as tools to create projects in recent times, especially in areas like leisure, education, work and social relationships.

In the leisure field, games are the main goal for the use of VR and AR, for its capacity in reaching high realism and total immersion. Nevertheless, there are also other uses within leisure for VR or AR, such as Within. This is an application that has stored 360° videos from different sources. It presents TV shows, concerts, music videos and documentaries, and so it is an option to immerse yourself in popular culture. The Foo Show is another application that has an interview TV programme virtual format and what is particularly interesting about this application is that the host takes the user into gaming environments, where they can examine surroundings and interact with objects. Another app is Tripp, which encourages meditation and concentration with virtual videos and based on very simple instructions: just with the headset (and without hand controls) the visual and listening experience lead the users to a relaxation, which aims to help their well-being. With simple required tasks, users indicate how they feel during and after a short meditation exercise.

In the educational field, there is a great variety of applications, depending on what people are looking for: to discover places, applications as Google Earth RV offers aerial dive from any part of the world, even from space, with high quality 3D diurnal and nocturnal images. Such applications give the user access to know historical and/or distant places. Moreover, for history themes exists BBC Civilizations AR, Augmented Reality application that shows objects from different cultures with an audio explanation. These objects can be located on a plan, scaled, and rotated to suit the user for better exploration. For learning languages Mondly Languages VR is an application created to study idioms by interacting with virtual characters in a virtual place created in the user's real space. Immersion in different situations and in their virtual environments is very useful correcting pronunciation while developing socialization skills in users.

To be alerted about the current situations around the world exists NYT VR, which contains 360° video reports from the New York Times, which can be viewed with a smartphone or VR glasses. Such kind of applications lead the user to be immersed in various situations such as wars or investigations, receiving information about what people on field are really experiencing. Without any doubts, NYT VR is a useful tool to raise awareness about the different situations that are experienced in other places of the world. Spotcrimes is an application that provides geolocated information for some cities, about crime data. This application allows the users to filter its data, which are collected by the local Police Departments, the media or other sources. Such applications can represent very useful information to maintain user safety or to define communities where it is necessary to intervene.

Finally, Sunseeker is a very useful application for people who have to consider the sun path for their work (architects, designers, photographers, gardeners, etc.).

The application locates the user through the smartphone's GPS and shows a flat compass and a 3D compass in Augmented Reality with the path of the sun.

The AR/VR is reach also in applications that encourage teamwork in virtual environments: Spatial is a specialized application for work teams, which, through a realistic avatar made with a user's photo, colleagues can participate in virtual team meetings. Spatial creates a virtual work room where the attendees have the perception of a real meeting, discussing with colleagues' faces and handling virtual elements (notes, models, etc.) with their fingers. Spatial also allows interactions with smartphones to take notes, making the feeling of working in a real world even more real. Relying on virtual visualization tools and immersed in a virtual office with large screens helps to organize the work within Virtual Reality applications.

There are several other applications to help socialization in virtual environments. Fulldive, for example, is a VR based social network, where users can upload their own videos and images, allowing their contacts to follow their contents, commenting, and rating them, as in any social network. Altspace, a social network with avatars, allows several users to interact in different parts of the world, such as in a bar, another country or even a galaxy, offering meeting real people, socializing and even dating.

For information about places, animals, objects, texts, or products exists Google Lens, thanks to which, using smartphones, the user can take a photo of any object and the app will show results for their search, showing information on the internet about it.

16.3.1 Definition of Common Characteristics

Based on a background of knowledge and on these experiences, some common characteristics can emerge, so to represent the main features and impacts of these technologies applied to vulnerable groups.

Leisure Leisure is the pause in our daily life that leads people to relaxation and to enjoy some activity that produces pleasure. Leisure is considered important for life that its principles can be founded in the Universal Declaration of Human Rights. From practicing a sport, to simple contemplation, leisure is an activity that does not necessarily have a goal beyond producing satisfaction. Nowadays, with the use of technology, leisure has also changed, as says, by e-leisure we would consider the entire set of online leisure activities (virtual social networks, blogs, wikis, forums, virtual communities, etc.) and that has become in a basic activity in the life of new generations.

Work Job training is an important field where Virtual Reality has been gaining a place: future professionals face a dilemma when doing internships to strengthen their knowledge: understanding the positions' availability, the necessary tools to carry out their internships, as well living the situation that training is proposing. In addition, AR/VR technology has already widely shown its relevance in new social realities, by improving communication between people located in different places and giving them tools that make their work more efficient.

Education This area is where Virtual Reality has been used the most, since it has led to the generation of new teaching methodologies. According to Campos-Soto, there are results that reveal that in the last 20 years (1998–2018) the scientific production about this subject has grown exponentially, going from publishing 0.27% in 1998, to 14.48% in 2018. This trend represents how Virtual Reality has become a very useful tool in teaching, due to the use of graphics that motivate students to use it.

Social relationships Due to advances in technology, and a situation of social isolation due to the pandemic, for some people virtual social relationships have become almost the only alternative to socialize, offering several advantages over physical socialization, such as mentioned by. The most important benefit offered by a virtual social network is the possibility, without the necessity to move, to access to a virtual space where people can meet people who share similar interests and passions, no matter where they physically live.

16.4 Results and Conclusions

The most transcendental technologies do not come from reason, but from those forms of knowledge that promote the way of relating to each other more and more humanly. Technologies are nothing by themselves, there must be an intellect endowed with a purpose capable of integrating them into culture, that is, putting them at the service of life (Mumford, 1995). Tables 16.1 and 16.2 summarize the projects analysed in this chapter, in which VR technology becomes memorable, brings us closer to other realities, transforms us and cares for us. All these cases help us to visit the daily life, obstacles, context and the identity of other realities. The future possibilities of virtual reality and its peripheral tools lie directly in the collision and collaboration between disciplines. The research analysed case studies from art, cinema, medicine and design, all articulated from techno-social devices to generate a vision of empathy towards vulnerable groups and communities. In other categories of social impact, virtual reality can contribute to education by being an accessible learning resource to tell stories in first person, have a dialogue with characters that no longer exist on this planet, travel back in time and visit places that physically become impossible.

From the urban and the architecture enhances the possibility to visualize micro and macro spaces of the city restored, previously damaged, contaminated or forgotten. Bringing these tools closer to communities in order to insert an idea of caring for the nearby habitat can become interesting, that is, transplanting a vision of everything that could become a healthy and harmonious space. Virtual reality and virtual environments are fertile and growing fields of information technology. There is evidence that suggests that the use of this technology will improve the conceivable image or vision of a spatial proposal related to the design of the environment.

Table 16.1 Existing AR/VR projects which work with vulnerable groups

Project	Vulnerable group	Aim	Concepts	Technology
The machine to be another	Marginalized groups	Eradicate racism	Empathy Identity Otherness	AVR
Carne y sangre	Migrants	Visualize violence and acts of indignation towards the migrant community	Empathy Identity Otherness	AVR
Autism empathy tools and Journey towards empathy Nas	Autistic persons	Inform about sensory sensitivity	Empathy Identity Otherness	VR AVR 360 Video

Table 16.2 Existing AR/VR projects which work with local communities

Project	Leisure	Work	Education	Social relations
Within	Some of the content encourage enjoyment	Provided information can be useful for investigations	Provided information can be useful to increase knowledge	Content can be a topic in social gathering
Mondly Lenguages VR	Casual talks and environments help to relaxation	Learning different languages can improve work processes.	Provides access to specialized education	Helps to improve communication skills
Spatial	Can be used as a forum for different topics	Allows virtual work meetings with visual tools, realistic avatars and virtual elements easy to handle.	It has the possibility of lectures or presentations	Team work improve communications skills
Alspace VR	Casual talks and environments encourage relaxation and enjoyment	Work meetings can be virtually developed in any place	It has the possibility of lectures or presentations	Allows to socialize with different people from all over the world

The intention will be that through simulation, public officials and community members can make better decisions regarding proposed projects within their towns and cities (El Araby & Okiel, 2003). Although it is difficult to activate people to change habits and inappropriate social practices, designers can be assisted by cognitive sciences to build emotional and strategic narratives, using concepts such as nudges and the habit loop to modify human behaviour towards the benefit of the community and our planet.

16.4.1 Conclusions

As underlined by this research, applying AR/VR/AVR tools to design practices with vulnerable community could mean dealing with existing instruments (which in local communities mainly deal with leisure, work, education, social relations issues) to achieve benefits that usually are associated with application to vulnerable groups (empathy identity, and alterity - otherness issues). Anyway, this process would limit the positive results since it means using existing instrument that partially fits the necessities of describing a "vulnerable community". In fact, as demonstrated by this research, to apply AR/VR/AVR tools implies developing a new narrative about "vulnerable communities". Since the contextualization is so relevant while dealing with such communities, finding an appropriate way to interpretate and understand their needs, fears and hopes, developing appropriate AR/VR/AVR tools to support design processes, becomes a primary necessity.

While this research helped to clarify the framework of the relations between AR/VR/AVR tools—vulnerable communities—design opportunities, the next steps will be testing more appropriate design methodologies based on more appropriate AR/VR/AVR tools. So, following development of this research will move forward (1) definition of a narrative for vulnerable communities to be satisfactorily applied to AR/VR/AVR tools and (2) experiment application of AR/VR/AVR tools in design process with vulnerable communities, applied with research or didactic field activities.

Acknowledgements We thank the external reviewers for providing critical and helpful comments on earlier versions of this chapter and on the book proposal. This work was funded by Fondo de Investigación 2020–2021 del Observatorio de Ciudades del Tecnologico de Monterrey (Mexico), through the research project "Design for vulnerables", project duration from October 2020 to December 2021.

References

Alsever. (2015). *Is virtual reality the ultimate empathy machine*? Retrieved 3 June 2021, from https://www.wired.com/brandlab/2015/11/is-virtual-reality-the-ultimate-empathy-machine/.

Battarbee, Katja, Jane Fulton Suri, & Suzanne Gibbs Howard (2014). *Empathy on the edge. Scaling and sustaining a human-centered approach in the evolving practice of design*. Retrieved 9 de September, 2017, from http://5a5f89b8e10a225a44ac-ccbed124c38c4f7a3066210c073e7d55.r9.cf1.rackcdn.com/!les/pdfs/news/Empathy_on_the_Edge.pdf.

Baudrillard, J. (1994). *La simulación en el arte*. Centro Documental de la Sala Mendoza.

Buber, M. (1985). *Between man and man*. Macmillan.

Campos Soto, M. N., Ramos Navas-Parejo, M., & Moreno Guerrero, A.J. (2020). Realidad virtual
y motivación en el contexto educativo: Estudio bibliométrico de los últimos veinte años de Scopus. Alteridad, 15(1), 47-60. https://doi.org/10.17163/alt.v15n1.2020.04.

El Araby, M., & Okiel, A. (2003). *The use of virtual reality in urban design: Enhancing the image of Al-Ain City, UAE, CORP 2003, Vienna University of Technology, Austria*. Retrieved 3 June, 2021, from http://papers.cumincad.org/cgi-bin/works/paper/ea0f.

Farago, Jason. (2017). *Reseña: Alejandro González Iñárritu simula un viaje desgarrador en 'Carne y arena'*. Retrieved 6 June 2021, from https://www.nytimes.com/es/2017/05/20/espanol/cultura/inarritu-carne-arena-cannes.html

Harmsen, Lars & Alesa Mustar. (2019). *Parallel worlds – virtual reality in art and museum*. Retrieved 3 June 2021, from http://virtualspatialsystems.com/parallel-worlds-virtual-reality-in-art-and-museum/.

Heeter, C. (1992). Being there: The subjective experience of presence. *Presence, 1*, 262–271. https://doi.org/10.1162/pres.1992.1.2.262

Krotz, E. (1994). Alteridades y pregunta antropológica. *Alteridades, 4*(8), 5–11.

Lopez, Giraldo y Yamid Oliver. 2016. Aproximación filosófica a la realidad virtual y actual en el escenario de los videojuegos. Bogotá: Universidad Pedagógica Nacional Facultad de Humanidades, Departamento de Ciencias Sociales Licenciatura en Filosofía.

Ludert, E. (2018). *Diseñar para lo invisible, investigación-acción para hacer visible el autismo en la Ciudad de México*. Posgrado UNAM.

Maldonado, T. (1999). *Lo real y lo virtual*. Gedisa.

Merriam, S. (2009). *Qualitative research: A guide to design and implementation*. Jossey-Bass.

Mumford, L. (1995). *El mito de la máquina Técnica y evolución humana*. Pepitas de Calabaza.

Parisi, Tony. 2015. Learning Virtual Reality: Developing Immersive Experiences and Applications for Desktop, Web, and Mobile. USA: O'Reilly Media. Alsever, Jennifer. 2015. Is Virtual Reality the Ultimate Empathy Machine? Retrieved 3 June 2021, from https://www.wired.com/brandlab/2015/11/is-virtual-reality-the-ultimate-empathymachine/

Páramo, P. (2008). La Construcción Psicosocial De La Identidad Y Del Self. *Revista Latinoamericana de Psicología, 40*(3), 539–550. Bogotá: Fundación Universitaria Konrad Lorenz.

Riboldi, Mark. 2017. The Promise and Disappointment of Virtual Reality A Cultural History of VR—And its Repeated Failure to Catch On Retrieved 3 June 2021, from https://lithub.com/the-promise-and-disappointment-of-virtual-reality/

Root-Bernstein, R., & Root-Bernstein, M. (2002). *El secreto de la creatividad*. Kairós.

Rueda Jon & Lara Francisco. (2020). *Virtual reality and empathy enhancement: Ethical aspects*. FiloLab Scientific Unit of Excellence, Department of Philosophy, University of Granada, Granada, Spain.

Ruiz, Cesar. (2009). *La alteridad*. México: Revista Casa del tiempo Ed.Nov #25

Statista (2021). Retrieved 16 August 2021, from https://www.statista.com/statistics/591181/global-augmented-virtual-reality-market-size/.

Silva, Rodrigo, Jauvane C. de Oliveira y G.A.Giraldi. 2003. Introduction to augmented reality. Brazil. National Laboratory for Scientific Computation

Vasilachis, I. (2006). La investigación cualitativa. In E. de Investigación (Ed.), *Cualitativa, compilado por Irene Vasilachis* (pp. 107–149). Editorial Gedisa.

Vásquez Carbonell, M. y Silva-Ortega, J. 2020. Tendencias y características de la realidad virtual. Computer and Electronic Sciences: Theory and Applications. 1, 1 (dic. 2020), 36–70. https://doi.org/10.17981/cesta.01.01.2020.04

Vega, K. (2015). Construcción de la alteridad y el significado de la familia en las ventas ambulantes. In *Revista del Laboratorio de Etnología María Eugenia Bozzoli Vargas Escuela de Antropología*. Universidad de Costa Rica.

Xataka. (2018). *El primer simulador VR de la historia tenía forma de recreativa, y se inventó a finales de los 50*. Retrieved 3 June 2021, from https://www.xataka.com/historia-tecnologica/el-primer-simulador-vr-de-la-historia-tenia-forma-de-recreativa-y-se-invento-a-finales-de-los-50.

Chapter 17
Digital Participation for Inclusive Growth: A Case Study of Singapore's Collaborative Digital Governance Model

Minqing Ni

17.1 Digital Transformation Makes Strong Impact on Governance

Digitalization is omnipresent in today's social and urban life; it is changing human society with an irreversible trend. The COVID-19 outbreak has helped speed up the digital transformation. The world has brought a digital lifestyle under a state of "social distancing" and has also inspired new models of urban governance. Digital governance refers to the use of the ICTs[1] by governments to support public administration inner processes, provide services, and involve citizens in the policy-making process. Therefore, digital transition towards governance focus, the process which could support societal and urban transition and make strong impact on governance. Digital transition can be supported by specific tools to make governance more inclusive, participatory, and more efficient (Bonneau, 2020). Technologies always get modified in the context in which they emerge and are shaped by organizational, institutional, and different actors. As each digital governance initiative is embedded in its own organizational environment, it reveals different development levels of digital governance (Leosk, 2020). The issue of digital divide or digital exclusion has

[1] Information and communication technologies (ICTs) are considered the main driving force for economic growth and for improving the quality of governance. The Urban Agenda for the EU Digital Transition Partnership's action plan is explicit in the need to make use of the ICTs for more efficient administrative processes, for providing better quality and more user-friendly services and for improving the relationship between EU cities, citizens, and businesses. Retrieved from https://ec.europa.eu/futurium/en/system/files/ged/digital_transition_-_abc_-_final.pdf.

M. Ni (✉)
College of Design and Innovation, Tongji University, Shanghai, China
e-mail: niminqing@tongji.edu.cn

E. Giorgi et al. (eds.), *Design for Vulnerable Communities*, The Urban Book Series,
https://doi.org/10.1007/978-3-030-96866-3_17

been exposed in the process of digital transformation, and the promotion of digital inclusion cannot be ignored. Digital governance focus on changes in the urban physical and service environment supported by the digital technology and attach importance to the long-term and large-scale participation and cooperation between multiple stakeholders. As an experimental ground for urban public governance and social innovation, it is based on the joint participation of local administrative departments, enterprises, institutions, and citizens by public-private-people-partnership model. The government acts as the experiment initiator and enterprise providers of technology, products and services, universities and non-profit organizations as researchers, and citizens as creators are also the users of services and products. Digital governance is expected that the market and social actors participate in the inclusive growth of the city's economy and promote the interaction of interdisciplinary innovation.

17.1.1 From Public Engagement to Digital Participation

The research on public participation began in the 1960s, from the advocacy planning theory proposed by Paul Dvidoff, Ladder of Citizen Participation by Sherry R. Arnstein, which showed participation ranging from high to low, to John Thomas outlined the Effective Decision Model of Public Involvement that managers found an invaluable asset when making decisions about when and how to involve the public. Public participation is gradually accepted by city managers and urban planners. The origin of participatory design can be traced back to Scandinavia in Northern Europe in the 1970s, and the participatory design method is applied to computer system design experience. Later, the practice scope of participatory design expanded to urban planning and architectural design. The Charter of Machu Picchu signed in 1977 emphasized that urban planning must be based on professional designers, urban residents, and political, which forms participatory planning cooperate among different stakeholders.

From 2000 to 2015, the public participation movement starts in the initial attempts of digitalization, such as participatory learning and implementation, and participatory digital planning. In recent years, the application of digital participation has gradually promoted changes in urban governance, industrial production, and service models. Digital citizenship theory refer that participate online affects politics, economy and studied the use of technology and the openness of citizenship in the digital age (Mossberger et al., 2007). The European commission stated digital inclusion is their effort to ensure that everybody can contribute to and benefit from the digital economy and society (Digital Single Market, 2017). The MIT Future Citizen Media Center conducts a series of research and practice on the role of digital humanities in promoting citizen participation. In Germany, the Digital Skills Map[2] gathers and exchanges local digital knowledge training and innovation exchange platform, which helps participants to establish consensus and knowledge sharing on important digital skills and abilities in the future.

17.1.2 Digital Participation in Urban Governance

More and more studies and practices have promoted the role of collaborative digital governance in urban development. The three aspects of digital administration, digital services, and digital participation have constituted in the basic framework. The main digital participation in urban governance is to establish a one-stop public service that allows citizens to continuously update and give feedback for information. The emergence of new technologies has brought about tremendous changes among citizens, government, and other subjects, because these new tools and methods enable the government to explore new opportunities to help the public sector get rid of its hierarchical and rule-driven structure and bring about innovative governance methods (Goldsmith & Crawford, 2014).

The Boston City Government of the United States has launched a highly effective and low-cost solution: BOS:311[3] is an application designed by the Boston City Government for citizens to participate in urban governance digitally. Citizens can report to the government for repairs, complaints, and suggestions. The daily problems encountered in public spaces can also be tracked in real time to get the latest status at any time. Grassroots government personnel have also been empowered to stimulate action and self-efficacy, cultivate citizens' civic awareness, and enhance the urban environment. It also strengthens the responsiveness of government public services and enhances the trust relationship between government and society.

Digital participation also involves scoring systems for the walkability of urban streets, such as Walkonomics in the UK,[4] Rate My Street in the European Union,

---

[2] Berlin is part of the "Urban Agenda for the EU." Under its umbrella, a wide range of European players joined forces in 2016 to exchange views on life and work in cities. Together they want to promote growth and innovation in European urban regions. As part of the agenda partnership "Jobs and Skills in the Local Economy" Berlin is leading the implementation of the Digital Skills Map for Europe with its European partners. Retrieved from https://www.digiskillsmap.com/en.

[3] BOS:311 enables real-time collaboration with citizens, "deputizing" mobile users to become the city's eyes and ears. Citizens report potholes, graffiti, and other issues from anywhere in the city using their mobile phone. Retrieved from https://311.boston.gov.

[4] The Walkonomics project ran from 2010 to 2018 and provided an app to find the most tree-filled walking route anywhere in 9 cities around the world. The app and web-platform also allowed users to rate the walkability of over 700,000 streets around the world. Retrieved from https://walkonomics.com.

and Walkability in the Asian environmental organization.[5] These apps and websites collect citizens' evaluations on the safety, comfort, and convenience of various streets, so that citizens can find suitable routes and help local governments find sections of roads that need to be repaired. The Seoul Innovation Bureau (SIB) in South Korea uses online and offline methods to allow citizens to participate in the design of projects, such as rising rent prices, youth unemployment, and improving health care related social and collaborative economic projects (Digital Social Innovation, 2020). The exploratory application of digital participatory decision support tools in the field of environmental policy in Spain and the Netherlands helps support local-scale energy social innovation (Hewitt et al., 2020). In the current global epidemic period, Digital Social Innovation (DSI), as an information and communication technology, is increasingly playing an important role in mobilizing collective social and political actions to respond to outdated planning policies and practices (Mehmood & Imran, 2021). In China, the combination of data technology and community management has found a way to resolve the contradictions between the urban public space and the backward management. The concept of Science and Technology for Good and Participatory Design tools are used to evaluate the barrier-free environment of a community in Beijing based on the blind behavior observation and intelligent perception, it has been included in the first Accessibility Assessment Report and established it the first batch of barrier-free pilot community in Beijing (Urbanxyz, 2019).

The diverse participation tools in urban renewal have been widely used to form participating entities such as governments, developers, residents, experts, and NGOs, which are formed through exchanges, dialogues, consultations, and feedback. But there are still many problems such as low level of participation, small participation scope, lack of participation organization form and system, poor participation channels, and weak social foundation for participation. More public participation only stays at the stage of symbolic participation. While digitalization has brought convenience in urban governance, the COVID-19 epidemic has made the problem of digital exclusion even more prominent. The social distancing imposed by the epidemic has rapidly increased people's digital behavior in many areas, from food and services to education and office. When most people in society turn to online to maintain social connections, not everyone has this opportunity. Therefore, it is necessary to remove barriers to digital participation in order to create more favorable conditions for interventions designed to increase digital inclusion. Urban digital collaborative governance transforms urban governance from closed management to open governance, and one-way management to collaborative governance.

[5] Walkability is a mobile app developed to enable citizens to carry out walkability audits and convey the results to the authorities. The audit results are mapped from the GPS of the smart phone. Usage of mobile apps is becoming popular across the globe, largely due to the high growth of internet and smart phone penetration and people's willingness to contribute to improving the system albeit in a small way. This is a first of its kind app in the world and the idea is to promote and communicate that safe walking is everyone's right—irrespective of age, gender, education, and background. Retrieved from https://walkabilityasia.org/2012/10/03/walkability-mobile-app/#.

The case study from Singapore might have some valuable insights on how innovative models combined with digital tools can become a new channel for collaborative governance, as well as creative applications in urban community scenarios.

17.2 A Case Study of Singapore's Collaborative Digital Governance Innovation Model

In Singapore, the citizen-centered digital collaborative governance innovative model encourages and empowers social innovation, stimulates the subjective initiative of the public, and focuses on how to link the participation of multiple subjects: government, business, and citizens. The strategy has been created to improve digital capabilities to pursue innovation, by integrating the needs and produce digitally driven collaborative co-creation with citizens and companies. Over the years, the government set up innovation laboratories in the public sector and using design thinking to open the co-creation journey, discover problems, solve problems, and create value from the perspective of users, improving the supply of public services effectively and establishing an interaction mechanism for the collaborative governance of multiple subjects.

17.2.1 Smart Nation Plan

Singapore leans on the Malay Peninsula and the southeast port of the Strait of Malacca. Its limited land resources are the long-term and one of the biggest challenges facing Singapore's national development and urban construction. However, Singapore has established a unique modern urban governance model through far-sighted planning and practice. The Singapore Smart Nation Project has undergone about 40 years and a total of 6 stages of evolution. Throughout the process, the government has maintained a high value and vigorously promoted e-government and informatization, established clear development goals, and implemented it in a rolling and uninterrupted manner (Chang & Das, 2020). In all stages of the plan, they always implement the people-centered concept and adopt a decentralized execution and centralized guidance promotion mechanism. In particular, the "Smart Nation 2025" plan is the world's first smart nation concept. The government coordinates and builds a "smart nation platform" to build an operating system and infrastructure for data connection, and analysis which help to calculate the relevant needs of citizens. The innovative strategy of digital collaborative governance establishes a seamless, people-centered new service model, which is worth to explore and learn.

17.2.2 *Digital to the Core, Serve with Heart*

Singapore's collaborative innovation governance model not only focuses on accelerating urban digital transformation and innovation at the macro level, but also emphasizes on the innovation and development of governance tools at the micro level. Through the digital government blueprint *strategy*, it encourages and empowers social innovation, stimulates the subjective initiative of the public, triggers the social change, and focuses on how to link the participation of multiple subjects: the government, business, and citizens. The government is committed to building a more streamlined and stronger public institution, with digital to the core and serve with heart as the goal. The six strategies are embodied in: integrating services around citizen and business needs, strengthening integration between policy, operations, and technology, operating reliable, resilient, and secure systems, co-creating with citizens and businesses, facilitating adoption of technology, building common digital and data platforms, and raising the digital capabilities to pursue innovation (Govtech Singapore, 2020). These strategies all show the external drive and internal mechanism in the transformation process of multi-subject collaborative governance.

17.2.3 *3C: Connect, Collect, Comprehend*

The concept of Smart Nation is an important foundation for supporting Singapore's digital collaborative governance. 3C means Connect, Collect, and Comprehend. "Connect" and "collect" are the basic stages of the concept of a smart nation, and " comprehend " is its in-depth stage. The focus is on the analysis and integration of information, allowing the government to make better decisions with a long-term vision.

17.2.3.1 Connect

The "connect" concept of the Singapore Smart Nation refers to eliminating information and data isolated, providing a safe, economic, high-speed, and scalable national communication infrastructure to realize data sharing and instant connection. The project called "Wireless@Singapore"has implemented WiFi nationwide coverage measures, and the government has adopted more than 7500 hotspots to provide citizens with wireless Wi-Fi Internet services at a speed of up to 1Mbps in public spaces outside of their homes and offices. In addition, the government established the "Citizen Connect Centre (CCC)"to provide citizens with free Internet access tools, the vulnerable groups can also come to learn and easily access the services.

17.2.3.2 Collect

The "collect" concept refers to the use of sensor networks all over the country to obtain and aggregate more accurate and comprehensive real-time data, while at the same time anonymizing protection and management of the more important sensor data, and to a certain extent appropriate and share. The installation of sensors in the bathroom for families to remotely monitor whether the elderly is safe is also part of the Singapore Smart Nation Project. The Singapore government has installed unified and comprehensive sensors across the country, by changing the previous sensors configured through different agencies. In this situation, these sensors will be used to monitor various situations, from the public corridors, squares, toilets, and other public places. Communication technology is the most basic purpose of collecting information, and more importantly, it is better to use this information and serve the public, and make everything interconnected, allowing new technologies to spread more quickly, and connecting the entire country under a shared prosperity model.

17.2.3.3 Comprehend

"Comprehend "mainly refers to the rational use of acquired and collected data, and to establish a scientific and effective sharing mechanism for the public. Forecast the multi-dimensional scopes of urban management through algorithm deduction and model setting to better predict the needs of the people and provide more timely and humanized services. For example, in terms of transportation, data analysis is used to improve the passenger experience and the transportation system, making it easier for people to travel.

17.3 Design for Vulnerable: Design Thinking Empowers Public Service

Singapore's digital collaborative governance model for vulnerable groups is most worth learning from the use of design thinking to empower public services. The government has established the Public Service Division[6] with an agile and entrepreneurial public service system. The turning point was the Singapore Employment Service Center designed by the Ministry of Manpower of Singapore and IDEO Design Consulting company. In this project, the role of the design is not to renovate the physical space in the traditional sense, but to train public officials to experience

[6]The mission of Public Service Division in Singapore is: Developing strong leadership and engaged officers, building future-ready organizations, promoting good governance, which deliver excellent public services and enable effective government. Retrieved from https://www.psd.gov.sg.

the user service journey in an empathetic way, discover problems, solve problems, and create value from the user's perspective. After the project ended, positive feedback from citizens was obtained, which promoted confidence in applying design thinking to public policies. In 2012, the Prime Minister's Office of Singapore established The Human Experience Lab in the Ministry of Public Service as an informal design research department. In 2017, it was officially renamed Innovation Lab. Composed of an interdisciplinary professional team, the lab is committed to using design thinking to promote innovation in the public sector and to create creative, responsive, and people-oriented policies and services. The innovation lab combined with design thinking, behavioral insights, organization development, business process re-engineering, systems thinking, data analytics, futures thinking, and other methods is applied to various fields such as policy, housing, employment, welfare, public services, and urban operations. In this process of open co-creation, the Singapore government has also created an environment to encourage and support self-organizing teams to create bottom-up innovation, and various financing mechanisms to release resources and cost sharing among different agencies in public services.

The government has also established a platform for citizen participation seed funds and public service innovation challenges program to empower citizens and social organizations and promote problem-solving. The government has also established Our Singapore Conversation (OSC) project to gain public opinion and public concerns, and five core aspirations have emerged: opportunities, purpose, assurance, spirit, and trust.[7] These have formed the role of empowering public service innovation subjects to participate. At the same time, a clear reform goal has been formulated from the individual to the entire government: at the individual level, it focuses on cultivating the innovation ability of each public official, making him an active innovator with innovative thinking and skills; at the institutional level, it aims to reshape the organization to create an institutional community with innovative thinking based on the operation mode and management model of the entire government; at the level of the entire government, it is committed to establishing a clear governance structure, supporting and promoting cross-institutional communication and collaboration, and fostering an innovative culture of the entire government. During this co-creation process, Singapore government create an environment to encourage and support self-organizing teams to create bottom-up innovation, and various financing mechanisms to release resources and sharing among different agencies in public services, the following two case studies are used to analyze the innovation laboratory through five roles: moderator, facilitator, developer, collaborator and mentor, to promote the public sector from progressive to subversive, from individuals, institutions to the entire government system innovation process.

[7] https://www.reach.gov.sg/Read/Our-SG-Conversation.

17.3.1 User-Centered Design: Design for Disables

In 2016, National Council of Social Services in Singapore did survey about disabled people; 62% of the respondents showed they were not tolerated by society and did not have the opportunity to exert their own value. At the same time, it is found that it is difficult for disabled people to obtain public infrastructure, receive proper care, use assistive technology equipment, etc. At first, the relevant government departments collected feedback and suggestions for improvement by widely distributing questionnaires, but they ended in failure. The main reason is that the survey questionnaire relies on the participants' self-awareness, and the limited field of vision hinders potential solutions. In order to explore more problems of disabled group, the Joint Committee of Innovation Laboratories adopted an empathy-based design method to conduct in-depth research on the daily lives of 25 disabled persons and recorded what they saw, heard, and touched from their perspectives, and feeling, needs, desires, and obstacles that even they can hardly name. After collecting the key touch points, the lab organized some collaborative design workshop, inviting as many stakeholders as possible to participate in it, including government officials, professional medical and nursing staffs, the disabled people and their families together more than 140 people to explore possible solutions. Finally, a social service ecological map of Singapore with more than 30 proposals, and 4 of them are currently being implemented. For example, Box of Joy project is a set of personalized surprise toolkits to help individuals with cognitive impairments or in need to learn and explore new things. Social Fusion project is an accelerated incubator that cultivates business thinking; it helped the disabled people who with ideas can also integrate into social work and entrepreneurship. The project has improved the quality and social inclusiveness of the living environment of disadvantaged groups. It won the iF Service Design Award in 2017 and it is still ongoing.

17.3.2 Collaborative Service Design: LifeSG

LifeSG[8] is a comprehensive life service app based on feedback through six months of citizen participation; it aims to identify and better understand the needs of users. It upgraded from the Moment of life project. Since October 2018, the Smart Nation Project and People Everywhere (SCOPE) have organized numerous workshops to help better understand how the LifeSG program can improved people's needs. LifeSG is not just a manifestation of connecting government services, it also represents how actively collecting citizen feedback can help enhance the service delivery and design of applications.

[8] LifeSG offers a simplified experience across government services so citizens can save time and focus on the things that matter. Retrieved from https://www.life.gov.sg.

LifeSG is for new parents for whom the most frustrating thing is to go to different institutions to handle all kinds of affairs and fill in the forms repeatedly, such as applying for household registration, seeing a doctor, finding a suitable nursery, etc. It makes young parents who are at a loss feel pressured. In 2017, Singapore's Ministry of Social and Family Development intends to develop a one-stop service application for parents of children aged 6 and below. The Innovation Lab was invited to serve as a project consultant to guide the design process and promote cross-departmental collaboration. In the stage of exploring and defining problems, the project team adopted design methods such as role-playing, storyboards, user journeys, and service blueprints, and worked with young parents to outline important touchpoints for children aged 0 to 6 in different stages and scenarios. In the solution phase, due to the communication and collaboration of multiple departments, the laboratory organized a temporary working group, seconded public officials from 15 relevant departments, and assigned the leaders in charge to be directly responsible for the project to promote collaborative optimization of services, improve processes, and adjust policies. Today's upgraded application program integrates 70 comprehensive services such as pension benefits, job hunting, taxation, housing, medical care, and education.

17.4 Singapore's Collaborative Digital Governance Innovation Model

Singapore's Collaborative Digital Governance Innovation Model could be summarized as four points: From "government for you" to "government with you"; Cross-regional, cross-departmental, and cross-level data interconnection and information sharing; Encourage citizens to transform from consumers to co-creators; Bridging the digital gap, let everyone benefits from a smart nation. The innovative model plays a different important role in each project, and flexibly embeds the design method into the whole process of the project, from exploring the problem, defining the problem to finding the appropriate solution and implementing the plan. Through joint collaboration, on one hand, the foundation of democratic legitimacy is created and the basis for decision-making is formed; on the other hand, it is conducive to finding solutions that are generally satisfactory, holistic, comprehensive, and inclusive.

17.4.1 From "Government for You" to "Government with You"

With the promulgation of the Smart Nation Plan, the Singapore government has made a very important change in its government service strategy: transforming from "government for you" into "government with you," changing the past strong government's thinking of doing everything, emphasizing on the government and the public, scientific research institutions, and the private sector to get better cooperation. Singapore government mainly starts with the integration of multiple entities, working together to co-create more value, and encourages the private sector, the public and the government to negotiate and cooperate, work together, innovate, and create more convenient, more dimensional, and better public services. The Singapore government has strengthened cooperation with multinational companies, scientific research institutes, private companies, and the citizen, and has actively established and supported large-scale non-profit backbone research institutions, and adopted measures such as formulating overall strategic plans, optimizing funding support policies, and establishing basic research platforms. The government uses the concept of openness and innovation to attract all parties in the society to participate in the construction of a smart country and realize the integration of the government and the multi-party forces of the society.

17.4.2 Cross-Regional, Cross-Departmental, and Cross-Level Data Interconnection and Information Sharing

The Singapore government uses advanced and creative technology and internal integration and cooperation to improve the efficiency of the public sector and transform to an integrated service management model. In 2015, the Singapore government created the digital twin technology city information model called Virtual Singapore. As a digital collaboration platform for data-driven governance, it accurately maps and reproduces Singapore in the real physical world, including the five core elements of connection, data, model, interaction, and application. It has the key characteristics of dynamic, accurate, real-time, and two-way interaction. Throughout the entire life cycle of the real-world system, it provides functions such as monitoring, analysis, diagnosis, prediction, decision-making, and control. The system opens to the public departments, enterprises, citizens, and research institutions to cope with Singapore's increasingly complex social issues. Virtual Singapore contributes a lot of value to collaborative governance innovation, which is mainly manifested in the following aspects:

- **Extreme visualization enhances the participation experience.** The 3D simulation model of virtual Singapore[9] has extremely fine granularity, including topography, buildings, vegetation, public facilities, etc. It creates an immersive participation experience for users, the real scenes and instant feedback mobilize citizens' motivation and enthusiasm for participation.

[9]Virtual Singapore is a dynamic three-dimensional (3D) city model and collaborative data platform, including the 3D maps of Singapore. When completed, Virtual Singapore will be the authoritative 3D digital platform intended for use by the public, private, people, and research sectors. Retrieved from https://www.nrf.gov.sg/programmes/virtual-singapore.

- **Scientific decision-making with full-scale and all-element data.** Virtual Singapore integrates all the data of all elements, such as basic city data, dynamic real-time data, government affairs data, business data, etc. Through data analysis, it can diagnose current problems and establish a consistent, intuitive, and easy-to-understand cognition for different stakeholders system by reducing communication costs caused by information asymmetry, and build a scientific data foundation for collaborative governance. For example, through data analysis, the best roof to install solar panels can be obtained to assist in discussion and decision-making.
- **Open platform and technology empower social innovation.** Virtual Singapore integrates data, models, algorithms, visualization, simulation, and other capabilities. Users can make proposals on the virtual Singapore platform to simulate the operation of experimental prototypes in the real world. While obtaining evaluation and feedback, they can also interact with other citizens and stakeholders, and continuously optimize and promote solutions that are mutually agreed. This not only reduces the cost of trial and error and the threshold of innovation for the collaborative creation of various social forces and various market entities, but also forms an inclusive and open experimental environment and creates a positive social innovation atmosphere.

17.4.3 Encourage Citizens to Transform from Consumers to Co-creators

The Singapore government encourages citizens to transform from consumers to active co-creators and contributors. The government cannot and should not solve all the problems. In many ways, individuals can actively use digital tools to shape their communities. For example, many informal WhatsApp and Facebook groups are united around a common vision and needs. Citizens and community groups can use open data (data.gov.sg) and APIs to jointly create new solutions and cooperate with the government to meet demand and promote development.

Singapore also attaches great importance to the training of ICT talents, who can help companies and governments seize opportunities when digital opportunities arise. At the same time, as the use of digital technology becomes more and more common, people need to re-master their skills to prepare for work in the digital age. The core of this work is the SkillsFuture movement; the tripartite partners have established TechSkills Accelerator (TeSA), which provides various programs to cultivate digital leaders, upgrade ICT skills, and support the transition to Non-ICT professionals in the ICT industry. Another plan is SkillsFuture for Digital Workplace, which aims to equip all Singaporeans with the mindset and basic functional skills of the future economy. Companies need to prioritize improving the skills of their employees and encourage individuals to develop the habit of lifelong learning. The government will also implement a plan to improve the digital literacy of public officials. The goal is to train 20,000 public officials in data analysis and data science by 2023.

17.4.4 Bridging the Digital Gap, Let Everyone Benefits from a Smart Nation

The Singapore government ensures that everyone benefits from a smart country. They believe that technology could become a balancer in society, including vulnerable groups such as the elderly, low-income people, and the disabled can seize the opportunities provided by digital technology. The government will work closely with the people and private companies on how to strengthen support for disadvantaged groups, regardless of age or ability, to get the support of the Home Access Programme and Enable IT Programme in the Smart Nation Program. The government cooperates with enterprises to improve the standards of digital services and establish similar digital concierge services and digital clinics in the community to assist those who are not comfortable with the use of digital services. Businesses and individuals can voluntarily participate in Silver Infocomm Junctions which are learning hubs located island-wide for seniors to start their digital learning journey. It helps to bridge the digital divide between seniors (50 years and older) and enable seniors to live a more convenient digital lifestyle and connect with the community.

IMDA also launched a basic digital skills course, which clarifies the basic digital skills that everyone needs to enjoy greater convenience and participate meaningfully in the digital society. The government will also uphold the principle of inclusive design, ensure that all products and services comply with international accessibility standards, and encourage enterprises to do the same. The government will continuously review its digital service standards to ensure that its digital standards and design principles are inclusive. In addition, the government also strongly supports digital projects, creates opportunities for community participation, and encourages more Singaporeans to adapt to the changes brought about by technology. Comparing with the previous failure of Google's sidewalk experiment in Toronto, Canada, Singapore was able to implement this strategy how to use the collected private information to serve the people smoothly. This is also reflecting to a certain extent the people's understanding of technology and the sense of trust in technology. Singaporeans have already enjoyed the new services behind the information. This is the best endorsement.

17.5 Singapore Lesson Learnt

In the past 40 years, Singapore has always adhered to the "citizen-centered" collaborative governance concept of digital government, and it has achieved world-leading development results. Its collaborative governance model will bring the following lessons to the city who are under digital transformation:

First, **digital technology empowers diverse social governance.** Usually, urban governance has pain points such as participation and difficulty in coordination. Singapore's collaborative governance model combines top-down and bottom-up to strengthen digital empowerment and diversified social governance, stimulate public initiative and enthusiasm, to play participate in the operation of digital transformation scenarios. It creates a scientific, refined, and intelligent new paradigm of "digital governance" for mega cities, and form a digital city innovation ecosystem of co-construction, co-governance, and sharing.

Second, **use design thinking to drive innovation in the public sector.** From exploring problems, defining problems, to proposing innovative ideas and solutions, and to conceiving prototypes, testing, verifying and constantly reviewing and reflecting, the whole process try to establish a pragmatic "people-oriented" design framework, methods, tools, and responded in a gradually complex social needs and public issues. It will help to increase public participation, transparency and inclusiveness, the prospects will respond to global challenges and achieve a new "digital normal" of sustainable development. This dynamic process attaches importance to the relationship between the subject and the object in the social ecology and institutional environment. The citizens and even vulnerable groups take process of social learning, participation in decision-making, and collective effectiveness transformation, which will lay a solid foundation for the digital transformation of urban governance and bring great value to the promotion of sustainable community economic construction.

Third, **governance has changed from "for the people" to "by the people."** Human behavior cannot be designed, but it can create conditions that are conducive to the occurrence of specific behaviors. How to create such conditions and environment to make people willing to participate, digital social innovative design interventions can provide support for various independent and related subjects in this complex system. At the level of policy implementation, the public sector assists in completing the coordination work among the government, enterprises, and citizens. This model of tripartite cooperation aims to fully mobilize the enthusiasm, initiative, and creativity of citizens and enterprises to participate in the development of wisdom and enhance the dominant position and participation of citizens and enterprises. Singapore is the absolute dominant force in the Smart Nation Plan; it brings strong execution efficiency, its transparent and efficient government strategy, and attaches great importance to the development, collection, and aggregation of data resources, and vigorously promotes it.

Singapore will be well placed to succeed in this new digital world, continue to open new chapters, accelerate the construction of digital industries, complete the integration of the ecosystem, and transform into the digital industry. It really needs to acknowledge to the digital participation strategy which created by Singapore public sectors, who bring all the stakeholders work together for the exciting digital future. Digital innovation empowers the promotion of public service innovation, which is an important content that affects the modernization of the national governance system and its capabilities. Singapore's collaborative digital participation model promotes value co-creation and opens up the strategy of public service innovation in the context of digital transformation from the theoretical and practical levels. The essence of digital transformation is the process of deeply embedding digital technology in products and services and reshaping value creation by using digital capabilities. Singapore attaches great importance to digital technology to empower public services from citizen participation, data sharing, business collaboration, and system collaboration, and enabled the rise of new business model, such as the sharing economy and open innovation, which have allowed societies and individuals to redefine the way people work and interact (Infocomm Media Development Authority, 2019). At the new normal period, the collaborative digital participation method creates new initiatives combining top-down and bottom-up; it will improve the implementation of digital life and the construction of an innovative model of digital governance through governments, enterprises, and citizens' efforts. It also stimulates the initiative and enthusiasm of the public, guides market entities to participate in digital scenarios, with the expectation that the market and social actors contribute fairly across society and creates opportunity for all by accelerating inclusive growth in the digital transformation.

Acknowledgements This work was supported by the Foundational Research Funds for the Central Universities of Tongji University, project duration from May 2021 to April 2024.

References

Bonneau, M. (2020, March 4). Digital Transition in cities – how can it benefit citizens? https://urbact.eu/digital-transition-cities-benefit-citizens

Chang, F., Das, D. (2020). Smart Nation Singapore: Developing Policies for a Citizen-Oriented Smart City Initiative. Developing National Urban Policies, pp. 425-440. https://doi.org/https://doi.org/10.1007/978-981-15-3738-7_18.

Digital Single Market. (2017). *August 16*. Digital Inclusion for a better EU society. https://wayback.archive-it.org/12090/20171006212139/https://ec.europa.eu/digital-single-market/en/digital-inclusion-better-eu-society

Digital Social Innovation. (2020). April. *Seoul Innovation Bureau.*. https://digitalsocial.eu/case-study/78/seoul-innovation-bureau

Goldsmith, S., & Crawford, S. (2014). *The responsive city: Engaging communities through data-smart governance*. John Wiley & Sons.

Govtech Singapore. (2020, March 15), Digital Government Blueprint.. https://www.tech.gov.sg/digital-government-blueprint/.

Hewitt, J., De Boer, C., & Flacke, J. (2020). Participatory development of digital support tools for local-scale energy transitions: Lessons from two European case studies. *Global Transitions, 2*, 138–149.

Mehmood,A., Imran, M. (2021). Digital social innovation and civic participation: toward responsible and inclusive transport planning. European Planning Studies, 29(10), 1870-1885.

Infocomm Media Development Authority. (2019). Digital Economy Framework for Action.. https://www.imda.gov.sg/-/media/Imda/Files/SG-Digital/SGD-Framework-For-Action.pdf.

Leosk, N. (2020). Digital Transition ABC - Main principles, components and best practices for the development of local digital governance.. https://ec.europa.eu/futurium/en/system/files/ged/digital_transition_-_abc_-_final.pdf.

Mossberger, K., Tolbert, J., & Mcneal, S. (2007). *Digital Citizenship: The Internet, Society, and Participation*. The MIT Press.

Urbanxyz. (2019, January).. Urban Governance Laboratory of Shuangjing Street Based on New Urban Science. http://www.urbanxyz.com/sj/1-zhuye/xc46d4a7b.html

Chapter 18
The "Rebuilding the World" Post-graduation Programme at ENSAP Bordeaux: Responding to Contemporary Urban and Environmental Issues, Through an Innovative Pedagogy with a Humanist Vision

Carlos Gotlieb and Hocine Aliouane-Shaw

18.1 Introduction

"Rebuilding the World" (RBW) is a post-master's level programme that was set up in the Architecture department of the Bordeaux National School of Architecture and Landscape (ENSAP Bordeaux) in October 2016.

It is interested in new postures and new approaches to urban planning and architecture and its teaching at the present time and is part of a specific multidisciplinary educational and research space at ENSAP Bordeaux.

The programme adheres to the new paradigms currently put forward by urban planning to deal with the global environmental crisis, strongly highlighted over the last few decades with the rise of globalisation and metropolisation.

The RBW training is nurtured by a series of experiences of international cooperation carried out by ENSAP Bordeaux with foreign universities in emerging countries. The aim of this cooperation is to study in a comparative view the phenomena linked to the critical evolution of territories as a consequence of globalisation, in order to find ways of acting locally through project practice.

Today, the project practice, which is the central subject of the teaching programme, is faced with new global challenges and the urgent need to find ways of acting locally. Therefore, the project practice is invested with a new cultural role at the service of populations that are anchored in specific territories and have a capacity for concrete action.

C. Gotlieb (✉) · H. Aliouane-Shaw
ENSAP Bordeaux, Talence, France
e-mail: carlos.gotlieb@bordeaux.archi.fr

E. Giorgi et al. (eds.), *Design for Vulnerable Communities*, The Urban Book Series,
https://doi.org/10.1007/978-3-030-96866-3_18

This article has a twofold objective: to present the key concepts which are at the origin of RBW and to describe the precise content of the programme as well as a review 5 years after its establishment.

18.2 Key Concepts Addressed in the RBW Programme

The RBW post-master programme stems from the recognition of three fundamental axes for understanding and acting in the contemporary territory, in order to effectively address the new challenges that urbanism is currently facing:

1. The study of phenomena that act on the whole planet with direct consequences on the evolution of contemporary human settlements, regardless of their size and level of development:
   - The Anthropocene.
   - Global changes.
2. The knowledge of various new operational concepts to be favoured when intervening on human settlements faced with the above-mentioned phenomena:
   - Resilience.
   - Degrowth.
   - Ecological transition.
   - Environment and identity.
   - Empowerment.
3. Taking into account new ways of approaching interventions in the living environment that propose alternative and innovative methods of action:
   - Territorialists and the local project.
   - Transitional and tactical urban planning.

18.2.1 New Phenomena That Act on the Whole Planet

Once considered as the goal to be achieved by all societies wanting to be modern, economic development and growth have been looked at critically for the past few decades because of the negative effects that these dynamics of evolution have caused on the planet, endangering the future of humanity. Human action is now considered to be at the origin of the Anthropocene, a new geological era which has caused considerable environmental changes. In the absence of awareness of the necessity to modify the behaviours that are at the root of these alterations, societies risk creating irreversible harm to artificial or natural environments, as well as to the human communities that inhabit them.

The Anthropocene: This term refers to the new geological era that follows the Holocene, the geological era that emerged after the Ice Age and lasted for a period of about 10,000 years. The Holocene was characterised by great stability in the biosphere, particularly in terms of CO_2, which generated a very favourable environment for the development of agriculture, enabling without which the emergence of human civilisation would not have been possible.

The notion of the Anthropocene was first introduced in February 2000 in Cuernavaca, during an international programme on the geosphere and biosphere by Paul Crutzen, Nobel Prize winner for his work on the ozone layer, who put forward the idea that humanity had entered a new geological era that replaced the Holocene. The latter was based on the observation that the acceleration of human activity following the years of strong development after the Second World War had led to an unprecedented degradation of the biosphere. Man's actions to propel social and economic development in different parts of the world, thanks to technological progress, had allowed an unprecedented intervention on the natural environment, considered as an inexhaustible or at least easily renewable resource. This approach was now questioned as it appeared to have produced a boomerang effect that endangered the very continuation of humankind. It is therefore human activity that has produced a new geological era, the Anthropocene. Unlike previous eras which were generated by nature's exclusive dynamic of intervention, in the Anthropocene the evolution of human history and natural history are intertwined. In this respect, climate change, which is one of the phenomena that characterise the Anthropocene, is the result of human activity. It is therefore clear that this activity has direct consequences on the collapse of biodiversity, leading to a deterioration of the biosphere, which must be addressed as a matter of urgency for the survival of life on Earth.

Paradoxically, the environmental crisis, which is the result of an uncontrolled emphasis on man's action on the Earth, has led to an awareness that the solution to this crisis lies in a radical change in this mode of action. This requires a review of modern notions of development, which, even when subsumed under the concept of sustainable development, continue to be present in the growth models of many countries.

Global changes: This dual approach to developments on a global scale, a vision that integrates both natural phenomena and cultural and societal phenomena in an articulated manner, is now the subject of a specific study theme, that of global change.

The basic idea, as in the case of the Anthropocene concept, is that human action has a direct impact on the evolution of the environment on a planetary scale and a concrete impact on the critical state of the biosphere, hence the need for an approach that crosses these two types of natural and socio-cultural issues. This has a direct influence on the new way of approaching its object of study by environmental geography, where the examination of social and cultural issues now becomes a priority. With the impact of human activity on biodiversity now widely highlighted as a direct factor in climate change, it is important to take a systemic and multidisciplinary approach to studying environmental change. Indicators such as the ecological footprint clearly show that human activity has been directly damaging the environmental balance of the planet for several decades, irreversibly destroying its natural capital without a return, in the absence unless of a change in societal and cultural behaviour. The reaction to this very negative framework is however present, either in the approaches of intellectuals and various protesting citizens movements, or in the progressive awareness of these themes in the world summits on the environment and the ensuing climate change negotiations and policies that are increasingly adopted by the countries involved.

18.2.2 The Knowledge of New Operational Concepts to Deal with the Crisis and Design Human Settlements According to New Paradigms

The environmental crisis has also called into question certain modes of action specific to the modern vision, such as that of the development and growth of societies that considered nature as an entity at the service of human action that had to be constantly mastered through technical progress.

This dichotomy between the natural and the artificial environment has inexorably led to a brutal divorce between man and his environment, as can be seen in the recent phenomena of financialisation and the relocation of certain activities. In addition to the consequences in the devaluation of production methods rooted in local cultures, these phenomena contribute to the progressive degradation of the environment.

However, this situation has led to a series of reactions from certain intellectuals and activists who advocate the establishment of a new relationship between populations and their environment. Accordingly, several new operational concepts have emerged in recent years, followed in some cases by new strategies for action put in place by political stakeholders.

Resilience: Resilience is a notion borrowed from physics. It has been studied in psychology as the ability of people to overcome very traumatic crisis situations. When applied to cities and territory, the concept of resilience can be defined as the capacity of an urban system to absorb a crisis which produces disturbances in its environmental balance, its social and economic organisation and its infrastructures.

This understanding of the notion of resilience highlights the idea that, even in situations of extreme vulnerability, it is possible to activate existing resources within a society and to re-use them to reconstruct organisational systems that have been severely disrupted during a crisis. This applies in particular to territories that are isolated or not well situated in global networks and do not benefit from the advantages of globalisation and metropolisation.

Degrowth: The concept of degrowth is based on an awareness of the urgent need to abandon the idea of growth to which contemporary society has gradually become accustomed.

This concept is based on the recognition that our society must move away from a productivist process that began with the advent of the industrial era because of the damage that this process has caused to the planet and because of the gradual scarcity of available resources it has induced, particularly fossil fuels, which are necessary to maintain this mode of development.

The notion of degrowth appeared in March 2002 in the context of a symposium held at UNESCO entitled "Undoing Development, Remaking the World". It is a provocative concept, a political slogan, aimed at launching debates to imagine new non-consumerist modes of production in contrast to the behaviour that has become widespread thanks to new advances in global communication networks. The basic idea is that the growth model of industrial capitalism, which has accelerated in recent years with the rise of globalisation, has generated a race to produce goods and services. This increasingly excessive production, imposed by an economic system that depends on it for its survival, considerably harms the human and natural environment without being inherently necessary for the real fulfilment of human life. Hence the emergence in different parts of the world of proposals that imagine frugal modes of production, situated at the antipodes of the productivist vision that has created an exacerbation of needs, fuelled in recent years by a logic of accumulation of capital rent.

The notion of degrowth also proclaims itself as a protest mode of action, opposing the notion of sustainable development adopted in 1987 by the UN. According to the proponents of degrowth, the notion of sustainable development, born of the objective of favouring economic development to satisfy the needs of current populations without compromising those of future generations, is an ambiguous idea and an oxymoron, because both terms have an opposite meaning and are assembled with a strictly opportunistic purpose. It would have been adopted so that the actors who hold power in the capitalist system could give themselves a clear conscience in the face of nature protection issues and thus continue the race towards development without hindrance. The proponents of degrowth, on the other hand, propose to create a new frugal relationship between man and his environment, putting forward concepts such as re-use and recycling in opposition to the frantic race towards consumerist production.

Ecological transition: The observation that climate change is caused by industrial development and the desire to promote frugal behaviour in the face of the foreseeable scarcity of non-renewable resources has led some authors to propose experiments in new resilient development strategies based on the principles of degrowth. This is the case of Rob Hopkins, who defines transition (which has been extended to include the ecological transition) as a new concept for action in the face of the environmental crisis. This concept refers to the implementation of new and locally designed initiatives aiming to reorganise energy production to meet the needs of human activities in a frugal mode, to develop low-energy transport, and above all, to encourage the re-learning of know-how and skills linked to the natural environment that are still rooted in local tradition but forgotten at the level of neighbourhoods or villages. These initiatives have enabled the emergence of various citizens movements such as the "Cities in Transition" movement, which aim to provide communities with an educational tool for implementing innovative actions to reduce energy consumption. The main idea is to be able to face two phenomena whose advent is to be foreseen in a joint and ineluctable way in the medium term: global warming and Peak Oil (the moment when the demand for oil will be greater than the available resources).

In line with these citizen actions and following the recommendations of international conferences on climate and environment, new policies with concrete measures have been put in place by different countries. In France, the awareness to fight against climate change and promote environmental preservation, as well as the will to reinforce the country's energy independence by offering access to energy at a competitive cost, are at the origin of a specific law approved in 2015, the law on energy transition for green growth. This law is part of a broader strategy put in place by the government from 2012 to address the issues of ecological transition. This strategy includes, among other proposals, several measures to promote the resilient development of territories, encourage the implementation of circular and low-carbon economies, reduce environmental, social and territorial inequalities, and invent new economic and financial models.

Environment and identity: The awareness of the need to establish a new mode of action with regard to the environment, different from that resulting from productivist doctrines, has also been accompanied by an evolution in the worldview of geographers. This has enabled them to enrich their view of their discipline by integrating a multidisciplinary vision, with contributions from philosophy and anthropology.

This is the case with Augustin Berque and his interest in reconstructing the relationship between man and his environment, questioning the certainties in terms of apprehension of the world, present since the Renaissance in Western culture and its adherence to the Cartesian rationalist vision. According to this vision, man and his environment are two dissociated entities and the relationship between the individual and the Earth is something abstract: nature is understood as an entity distinct from man, at the service of the improvement of his conditions, thanks to the manipulation of the environment through technical progress. This vision which persists to the present day and has spread everywhere in the context of industrialisation and, more recently, globalisation. It has led to the rise of a feeling that man does not belong to a particular environment, causing a significant break between the individual and his environment. Yet, on the contrary, what characterises a culture is precisely the relationship of identity between man and his environment which, in the case of very ancient cultures, is very visible through myths. It is also the relationship between man and his environment that is at the origin of language, because language is a mode of representation of the reality that surrounds us that was constructed to be able to communicate by creating abstract concepts from concrete facts. In the field of architecture, this desire on the part of certain geographers to re-establish a new cultural relationship between man and his environment, with a view to reconstituting an identity that has vanished in the context of the globalised world, finds its equivalent in the neo-vernacular postures advocated by certain architect. The latter propose to revisit and reclaim vernacular architecture as a frugal approach that could provide appropriate responses to contemporary environmental issues.

Empowerment: Another concept that has gained importance in recent years in order to deal effectively with global changes is empowerment. As we have already pointed out, the damage to the biosphere is basically the result of human actions which, in the context of a productivist vision, have distanced us considerably from our environment. In order to repair the damage caused to the planet by this productivist attitude, there is a present need to recreate a solid link between man and his environment by changing the ways in which we relate to it. This implies considering the environment not as an abstract resource to be mechanically exploited to satisfy the needs of humanity, but rather as an instance with which we must reinvent, in a very localised manner, modes of action that integrate the logic and rhythms of nature instead of cutting ourselves off from it radically.This approach makes it possible to create a new cultural link between man and his environment, in order to respond effectively to the environmental issues that are leading to the fragility of the planet. It gives fundamental importance to the action of human groups on their environment, whatever means they have to act. Empowerment is thus understood as a liberating process that allows communities, understood as human groups sharing interests and identities in the same territory, to increase control over their environment. It is about fostering individual and collective achievement within a community while enabling the community to maintain a balanced relationship with the environment. This objective of promoting the empowerment of social groups opens up a new field for urban and architectural design. This new understanding is no longer focused on promoting performance and technical progress, as was the case in the productivist vision, but now aims to promote the creation of new, more responsible relationships between man and his environment.

18.2.3 New Thoughts and Intervention Approaches in the Territory That Offer Alternative and Innovative Methods of Action

Faced with the extent of the questions that call into question the productivist approach and given the damage that the latter has caused to the environmental balance, new stances are emerging within the disciplines that deal with territorial planning and urbanism at different scales. These stances are the antithesis of a scientific approach, characterised by a compartmentalised vision of reality. The aim of this compartmentalising study was to isolate reality into several specific objects of study, so as to be able to effectively develop rational methods for acting on a case-by-case basis. This method, which has proved to be very effective within the framework of the productivist vision, is no longer valid for dealing with global changes which, as mentioned above, require a multidisciplinary approach to deal with the contemporary environmental crisis.

In the case of urban planning, it is no longer a question of designing ideal but ultimately harmful models to create efficient spatial entities, in line with the productivist vision. Rather, it is a matter of rediscovering a new culture of action anchored in the territory, capable of acting locally and in stages, to restore disturbed balances. The aim is to be able to provide, through the project, effective responses that are both complex and modest, and that can create a new culture of action in the face of the environmental and social crisis that is spreading throughout the world.

Territorialists and the local project: Faced with the understanding of the territory as an abstract space where the life cycles of individuals develop without valorising the identity characteristics of the site, seen as a simple means of economic activity, the territorialists, a group of intellectuals who militate for an alternative way of acting in space, led by the Italian architect-urban planner Alberto Magnaghi, propose to return to the local territory understood as a common good for the inhabitants and as a producer of elements for the reproduction of biological and social life.

Two types of commitment have been made by territorialists to respond to these issues. The first is to identify long-term strategies to reverse the long-lasting trends that have caused the current environmental crisis. The second is to put in place immediate actions to protect cities from the consequences of climate change.

These short- and long-term strategies are based on the idea that the territory is the result of a long-standing sedimentary process that has produced anthropogenic actions on a natural environment. As a result, the territory is a common good that cannot be privatised or sold, but which requires re-signification through active citizen action and by revalorising the knowledge that has led to its constitution throughout history. This requires a recomposition of knowledge, which is currently hyperspecialised and sectorialised on the one hand, and on the other hand, the construction of systems of relations between disciplines in order to bring about the emergence of new professions that will enable us to act as closely as possible to the territory.

Under the leadership of Alberto Magnaghi, since 1991 the territorialists have set up an international network to promote this new posture of action on the environment, first in Italy and then more widely in Europe.

Transitional and tactical town planning: The empowerment approach applied to territorial management shows the need to revisit certain principles of action of modern urban planning, characterised by a top-down mode of action generally established by local authorities. This mode of action results from a will to control territorial development on the part of public authorities in order to avoid drifts due to excessive intervention of market forces. As it has been invested with a productivist vision, modern urbanism has favoured the implementation of a technocratic vision causing the progressive distancing of populations from their environment.

In view of the new environmental challenges that require awareness and new modes of action on the part of inhabitants who are closer to local realities, new bottom-up participatory approaches have been tried out in recent years. In this regard, transitional urbanism and tactical urbanism are two new approaches which favour citizen participation as a prerequisite for the definition of spatial projects. Since these projects are by nature intended for communities, these approaches aim to make these communities more responsible in defining objectives and actions to be implemented, as well as to act as indicators to measure the effectiveness of the actions carried out.

Furthermore, the will to involve the population in a participatory dynamic is the basis for the development of more playful approaches understood as simple and effective tools to involve the community in its integration in the process of designing a project.

18.3 The Rebuilding the World (RBW) Programme

18.3.1 Postulates and Objectives of the RBW Programme

The RBW programme is based on the premise that it is necessary to create a new modality of action on the territory to face the current environmental crisis. To this end, the programme uses a project-based action research approach to teach a new culture of intervention in the environment, taking an active and "decompartmentalised" interest in the new paradigms of designing space for humans.

This programme is in line with a humanist stance of urbanism fully claimed by the teachers of the training, adhering to the postulates of degrowth in opposition to those of productivism.

Faced with the environmental crisis, the notion of project promoted by RBW implies an action to improve the condition of the environment, moving away from the one conveyed by modern thinking which privileges the performance dimension of the project. Invested with an ethical dimension, the project is conceived as an instance of collective creation and sense sharing prior to any intervention in space, in other words as a process of close and dynamic interaction between man and his environment.

The RBW training is thus particularly aimed at developing the capacity of designers to induce and manage projects adapted to the constraints of their environment, and to encourage the emancipation of local populations through collaborative processes of design and production of spatial structures. The ambition is to support the transition towards a resilient mode of management of the territory, relying on bottom-up dynamics that mobilise energies and resources closest to the material and cultural realities of the territory.

In emerging countries, experimental projects are systematically taking place in sectors that are economically and socially in a "stall" condition and where the application of urban planning regulations comes up against many obstacles. By re-establishing a direct link with the inhabitants according to a logic based on participation and economy of means, new ways of thinking and making projects are emerging in these sectors. And if the designers' objective is to induce a process of effective improvement of the inhabitants' living environment, they quickly become aware of the fact that the dynamics set in motion on the scale of a specific project require the reworking of the planning procedures on a larger scale in order to last.

This implies that the professionals responsible for steering and building the projects are also capable of calling on the political representatives and technicians in charge of territorial development to propose an evolution towards more transversal and multidisciplinary modes of spatial management, open to the various scales of the territory, in order to respond effectively to the current environmental crisis.

18.3.2 Background of the Programme

The methods and pedagogical contents developed in the RBW training are based on several experiments such as workshops and research projects, carried out in France and abroad by the teachers who designed this programme (see Figs. 18.1 and 18.2). These experiments have made it possible to analyse the question of the "ecumene" (the vision of the world constructed by individuals or human groups) as approached by Augustin Berque, that is to say as an interaction between issues on a global scale that are posed in a generalised manner in the territory and local responses that produce a specific culture of the environment. These experiences have created an intellectual base that has made it possible to theorise the new issues that arise on a planetary scale in the face of the environmental crisis, through action-research that takes place in very contrasting environments, particularly in emerging territories where the crisis is quite violent. The RBW programme thus poses the idea of the project as an instance of research/action that questions and at the same time actively intervenes in the making of the environment.

Based on the previous experiences of ENSAP Bordeaux, the objective of the RBW programme is to give students and professionals the ability to situate and position themselves, through project practice, in contrasting cultural contexts and disciplinary specialities in order to address in a very open manner the current questions concerning the space making. This ability is also an asset for understanding the complexity of project situations and processes of transformation through projects in the European context. If the objective is to allow the understanding of the stakes and the modalities of intervention in other cultures, it is also the occasion to re-interrogate the practices of the project and the methodologies of its culture of origin.

It is a question of offering professionals already trained in the disciplines of space the possibility of apprehending other contexts of production of the city and of bringing them to mobilise the skills and know-how acquired in their culture of origin to collaborate in the design, realisation and evaluation of projects with professionals and researchers from other cultures.

Fig. 18.1 Example of conference of the first sequence: Basurama, Intervention in public space by reusing waste

Fig. 18.2 Example of conference of the first sequence: Play the City

18.3.3 Experimental Character of the Programme's Pedagogy

Bringing the designer to approach the process of intervention in space from a renewed angle in terms of interaction between man and his environment requires the implementation of experimental pedagogical approaches, in a concomitant and "decompartmentalised" manner.

On the one hand, from the "objective" understanding of global phenomena that take place in different contexts as well as the characteristics of their receptivity in the corresponding environments (migratory movements, fractures between communities caused by wars, desertification of the countryside, mercantilisation of territories by the arrival of mass tourism...). This is studied by deepening the understanding of global problems that impact different environments, based on multidisciplinary approaches focusing in particular on socio-economic and environmental issues. In addition, there is a knowledge of new resilient approaches developed locally by various groups to deal with crisis situations caused by global phenomena. In particular, it is a question of studying field experiences implementing innovative bottom-up approaches that favour collaborative modes of action with regard to users.

On the other hand, in order to create a new culture of action in the territory, different from the top-down productivist approaches, it is very important to implement pedagogical devices favouring a "subjective" involvement in the territory. The programme is based on the assumption that this "subjective" involvement favours a process of collective appropriation of the environment by designers and inhabitants. In terms of pedagogical devices, this translates into an interest in investigative approaches that encourage a sensitive, intuitive and close encounter with the environment (walks, urban drifts, playful experiences, etc.). This is also reflected in the in-depth experimentation of the transitory phases of collective project construction (tactical or transitory urbanism), which are seen as knowledge-producing bodies that allow the effective anchoring of actions in the territory.

RBW training also focuses on placing the students in a professional situation, through an internship period in a high-level host structure located in emerging countries, in a close relationship with a partner university of the RBW programme. This real-life practice, which takes place in the middle of the programme, aims to develop students' capacities to act creatively and effectively on the complexity of urban systems, whatever their context. This specific offer in the field of transcultural projects, open to the understanding of environments and cultures in their diversity, represents a unique opportunity for students to train in a creative and reflective professional practice enriched by a confrontation with a real situation in these environments.

18.3.4 Programme Sequences

The programme consists of three sequences spread over 1 year.

The first sequence aims to progressively accompany the student in the definition of a personal project on which he/she will work throughout the programme. During this sequence, the student is expected to work on questions concerning the deconstruction and reconstruction of project situations (in the broadest sense), through the understanding and questioning of the processes of constitution of contemporary environments. The objective is to present theoretical tools and case studies to question productivist postures in the making of the contemporary environment and to present alternative approaches by calling upon a diversified body of knowledge and specific know-how. In order to achieve this, the sequence addresses three lines of thought aimed at understanding and acting in the territory according to a non-productivist vision.

Firstly, the dynamics operating in territories in transition, particularly those resulting from the phenomena of globalisation and metropolisation, as well as alternative economic approaches conceived locally. It is important in this axis to address the issues related to the experiences of circular and solidarity economy which are quite evident as an alternative process of development closer and more engaged with local realities.

Secondly, the preservation of environmental balances in connection with the updating of knowledge on the local environment. In this area, it is particularly important to follow the recommendations of the territorialists, that is to revive modes of action in the territory that are rooted in geography and history and that have become blurred by the progressive establishment at a global level of decontextualised modes of production that are part of the dynamics of delocalisation.

Thirdly, the cultural emancipation of local populations to promote resilient processes of territorial development. The objective of this pedagogical axis is to present different methods of involving the population at the local level in the design of their environment, with the support of participative experiences developed by collectives such as "Architects Without Borders" (Architectes Sans Frontières), or playful experiences or the elaboration of collaborative cartographies.

At the end of the first phase, enriched by all these contributions of deconstruction-reconstruction of a project that questions a territory under the prism of contemporary environmental issues, the student must be able to define a personal research project that has incorporated the decreasing approach promoted by RBW.

The second sequence refers to a practice of immersion in a "laboratory" territory, conceived as a "side step" in the learning process (see Figs. 18.3 and 18.4). The theoretical elaboration of the student's personal project, initiated during the first pedagogical sequence in terms of methodology and research tools, begins at this stage with the simulation of one or more action scenarios that involve the student in a research-action process. This takes place in a "laboratory" territory—preferably located in a foreign cultural context—in order to confront the student with an effective action dynamic "in situ", in relation to the issues addressed in his or her personal work. This territory used for practical immersion is different from the one in which he or she develops his or her personal work. Thus, the student is involved in the real-time dimensioning of the effects of an action in the transformation of the environment and at the same time takes a reflexive "side step ". These practices are the result of close collaboration with universities and local authorities, either within the framework of decentralised cooperation associations initiated by ENSAP

Fig. 18.3 Study of the evolution of rural areas near Guanajuato, Mexico, as part of the research practice immersion (Magali Risler)

Bordeaux, or with teaching and research networks that collaborate with the RBW programme. The challenge for the student is to understand the exchanges that take place between the designers, the project and the cultural environment of the project and to identify how and at what stages of the project the interaction between the experimental actions and their contribution to the project occurs.

The third sequence is developed after the return of practical immersion experiences in laboratory territories and has as its aim the finalisation of the student's personal work. This last sequence therefore refers to a return on experience and a repositioning in a framework of greater theoretical generality. This phase involves a critical reflection based on a combination of field experiences and the completion of a final project. This project is elaborated in the framework of a collective final seminar held every week, allowing for the results, the points of view and a comparison between the different experiences. This contributes to the collective construction of a critical vision of the problems that may arise in different ways depending on the sites studied. These themes will be used to feed the action-research mechanisms and to develop the network of laboratory territories linked to the RBW training. Several thematic families have emerged since the creation of the programme in order to

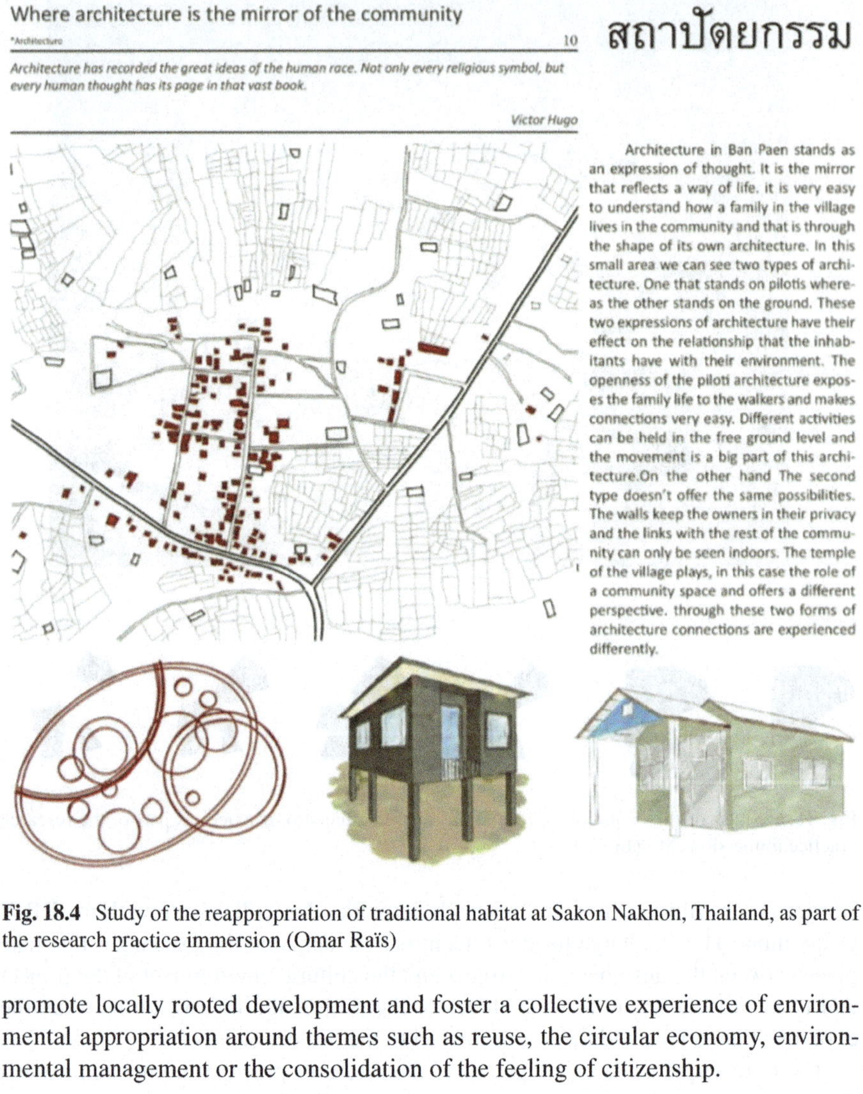

Fig. 18.4 Study of the reappropriation of traditional habitat at Sakon Nakhon, Thailand, as part of the research practice immersion (Omar Raïs)

promote locally rooted development and foster a collective experience of environmental appropriation around themes such as reuse, the circular economy, environmental management or the consolidation of the feeling of citizenship.

18.3.5 Types of Action Research Developed After Five Years Since the Programme's Inception

As RBW is a programme widely open to young professionals from different countries, the themes of personal action research are very diverse, usually related to issues in the students' home countries. This work is enriched by the practical immersion and research experiences developed in a foreign territory in the middle of the programme, which allow students to take a step back and look at their study territory from a different angle.

Several types of action research have been carried out over the 5 years since the programme was set up. They can be classified into three main families:

1. The project as an instance of resilient interaction in the territory, favouring the emergence of new identity relationships in the constitution of the environment. This theme is mainly aimed at territories in a « stall » situation within large metropolises and are subject to marginal use that damages the environment, particularly through illegal occupations resulting from migratory processes. However, it is possible to set up circular and solidarity economy mechanisms in these types of territories, with the aim of socially integrating migrants, replacing rural production methods and setting up resilient activities that create a virtuous productive circle with regard to the environment (as an example, we can cite the work of Magali Risler: Process of development of urban agriculture in Kinshasa. See Fig. 18.5).
2. The project as a tool for mediation and the creation of citizenship to recreate links in fractured territories.

   This type of action research has been applied in particular to territories subject to conflict. This is the case in Lebanon, where students from this country have proposed the implementation of actions very close to the territory (participatory approaches, walks, etc.). In some cases, the objective was to promote a change in the perception by the different actors of sites that create a border between communities of different religious confessions in Beirut (Beirut, towards a Curative Urban Dynamics by Saïd El Helou). In another case, it was a question of promoting the reinstatement, according to innovative mobility modalities to be imagined, of the railway line that crosses the Lyban from East to West (A participatory journey on the rails of history by Christelle El Hage).
3. The project as a new approach to the creation of commons in very precarious spaces.

   This type of action research has often characterised the work of students from Latin American countries who have been interested in new modes of intervention and community appropriation of public spaces in very precarious neighbourhoods (this is the case of Marcos Colina's work: Interfaces. The creation of the urban common in self-produced neighbourhoods, The case of Valle del Pino, Venezuela).

   At the end of the RBW programme, the students are able to continue their work in a real context by involving the actors concerned in the implementation of the proposed actions or to continue their reflection in the framework of a doctorate.

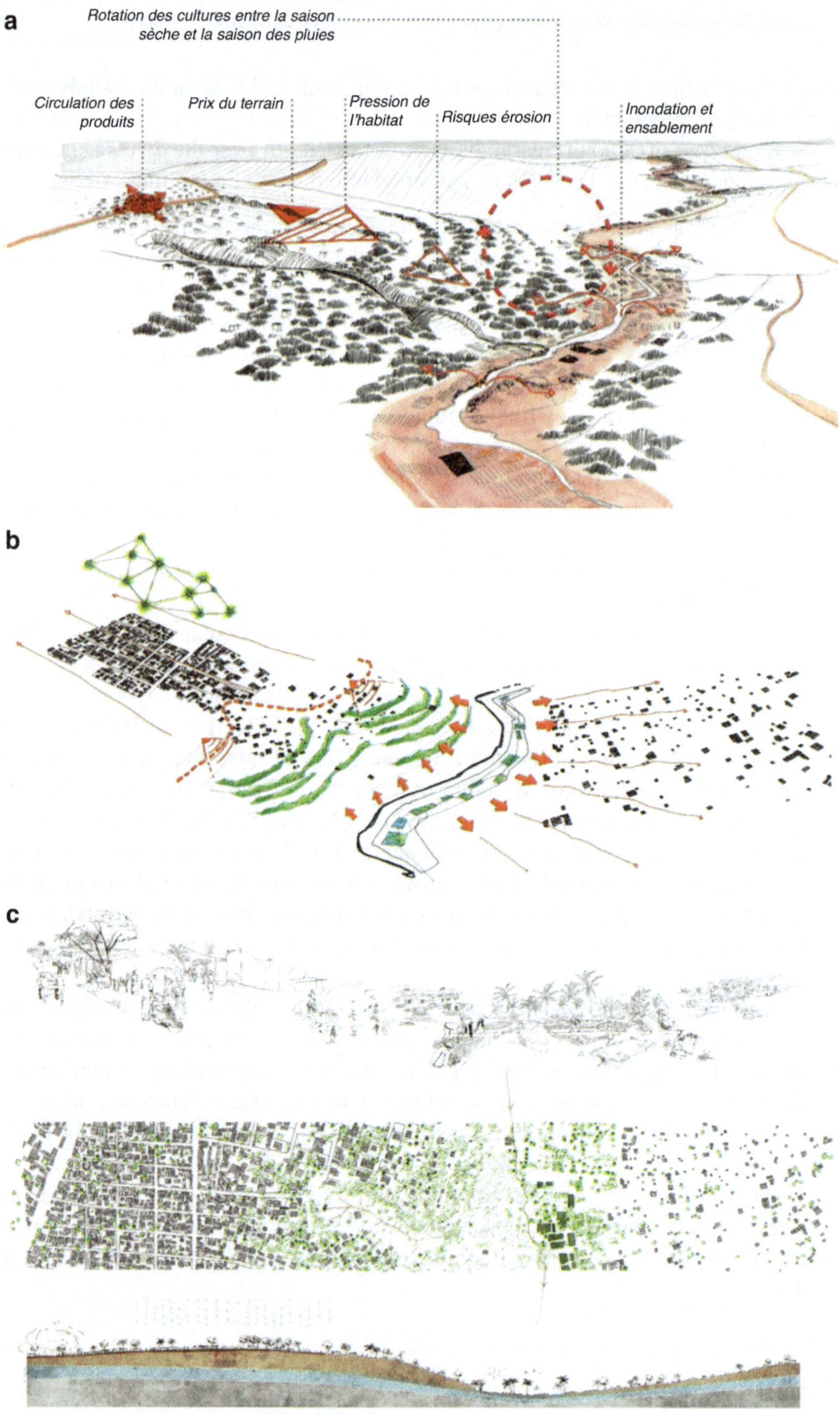

Fig. 18.5 Example of student final work: Proposal for a circular farm economy cycle to integrate rural migrants and clean up polluted territories in Kinshasa, Congo (Magali Risler), dynamics and proposed territorial strategy (Fig.18.5a, 18.5b); plan and section of the studied area (Fig. 18.5c); detail of habitat unit with proposed aquaponics production system (Fig. 18.5d)

d

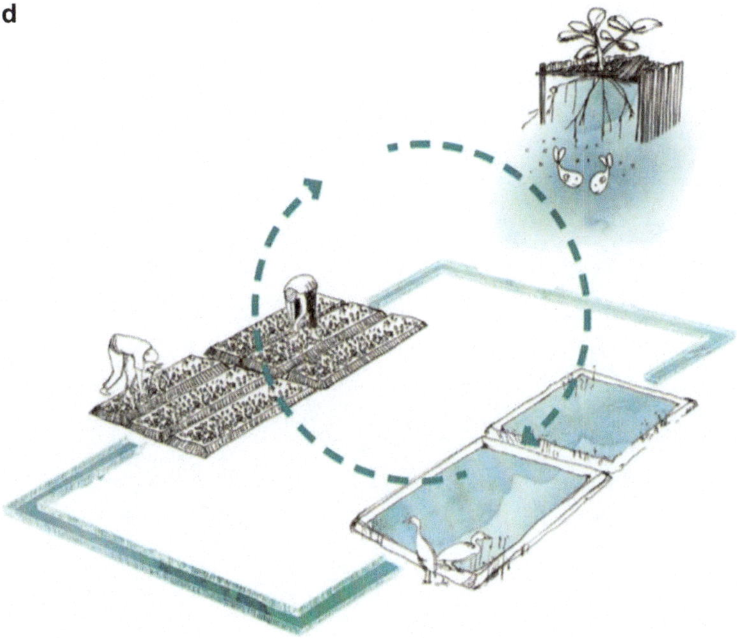

Fig. 18.5 (continued)

Bibliography

Beau, R., Goupil, C., Koenig, C., & Renouard, C. (2020). Manuel de la grande transition. *Former pour transformer, Campus de la transition, Paris, Éd. Les Liens qui libèrent*

Berque, A. (1996). *Etre humains sur la terre*. Gallimard.

Berque, A. (2010). *Milieu et identité urbaine. Notes pour un dépassement de la modernité.* Editions Donner Lieu.

Bonneuil, C., & Fressoz, J. B. (2013). *L'évenement Anthropocène. La terre, l'histoire et nous.* Edition du Seuil.

Careri, F. (2020). *Walkscapes. La marche comme pratique esthétique.* Babel.

Hopkins, R. (2018). *Manuel de transition. De la dépendance au pétrole à la résilience locale.* Editions écosociété.

Lecoq, M. (2018). *De l'habitant à la ville. L'exercice du droit à la ville.* MetissPresses.

Letouche, S. (2019). *La decroissance* . Que sais je? Humensis.

Magnaghi, A. (2002). *Le projet local.* Mardaga.

Tan, E. (2017). *Play the City. Games informing the Urban Development.* Jap Sam Books.

Part V
Vulnerabilities in Context: Analysis and Projects in the U.S.–Mexico Border Region

Chapter 19
Climate Resilient Development Pathways in the US-Mexico Border Region: The Case of the El Paso del Norte Metropolitan Area

Simone Lucatello

19.1 Introduction

Climate change impacts and effects at the regional level is an established field of study. The downscaling of the impacts of climate change at the local level is based on different techniques, methodologies and projections that provide well-informed climate risk maps and possible trends (Smid & Costa, 2013). However, the frequency of the impacts, as much as the severity of damage caused, particularly across borders at the urban and city level, remain an ongoing discussion. Extreme weather events and their impacts do not respect geographical boundaries; rather, they offer a complicated scenario when they become transnational. For the purposes of this article, we consider that climate change creates uneven and asymmetric impacts on populations in the border region of the USA and Mexico, due to cultural, economic and environmental differences as well as the diverse governance setting. The complex and cascading impacts of climate change can be summarized in the following broad areas:

1. Wetland and terrestrial ecosystems, which make up most of the region's areas of biodiversity, will be strongly affected by increasing droughts and reduced precipitation. This may have impacts on agricultural cycles and crop production, posing threats to food security systems among the border communities.
2. Climate change projections available for the region point to a severe reduction and stress on water resources that may increase existing vulnerabilities linked to the water cycle, from supply to distribution. As we recently experienced during the winter storms in Texas, compound effects on water systems may stress the

S. Lucatello (✉)
Instituto Mora, CONACYT, Mexico City, Mexico
e-mail: slucatello@institutomora.edu.mx

E. Giorgi et al. (eds.), *Design for Vulnerable Communities*, The Urban Book Series,
https://doi.org/10.1007/978-3-030-96866-3_19

supply of energy and the infrastructure used for farming, industry and the indigenous communities or pastoralist communities prevalent in the border region.
3. Transborder governance needs to create adaptive capacity systems to combat climate change through border cooperation and the implementing of diverse governance approaches—as well as strengthening environmental border institutions. This is mostly for water management and adaptation during heat waves and other climate hazards.
4. Cities and settlements will face higher temperatures and extreme weather conditions: urban planning in the area must consider important changes in its inception and development if the vulnerabilities of the population and its exposure to climate change impacts are to be reduced over the coming decades.

We begin this chapter by describing some of the key characteristics of climate change impacts in northern Mexico based on recent and available projections. Secondly, under the theoretical framework of the double exposure (DE) and Climate Resilient Development Pathways (CRDPs) from the United Nations Intergovernmental Panel on Climate Change (UN-IPCC, 2018), we set the context for a discussion about the complexity of climate change variables, adaptation and resilience in border areas. We then take a closer look at the case of El Paso del Norte Metropolitan Area (PNMA) by analysing specific climate impacts and development actions in the area. We finally provide a set of possible adaptation scenarios under the CRDPs framework for the case study and we offer some recommendations for implementing these pathways under the current circumstances.

19.2 Theoretical Framework: Double Exposure, Climate Change Adaptation and Climate Resilient Development Pathway Concepts

In this section we describe and propose a combined theoretical framework, with three different but interlinked domains, to provide the grounds for the case study analysis: Double Exposure (DE), Climate Change Adaptation (CCA) and Climate Resilient Development Pathways (CRDPs) concepts (see Fig. 19.1).

Starting with DE, this is part of a recent set of qualitative methodologies applied to the complex field of environmental and socio-economic studies and it involves the idea that climate change, together with economic dynamics and the interaction with vulnerable societies, provokes severe stressed conditions to socio-ecosystemic relations in a given context (Leichenko et al., 2010). The DE approach applies simultaneously and sequentially to communities and groups who are highly exposed to climate hazards and socio-economic vulnerabilities originating from different sources like poverty and lack of urban planning, among others (Bolin et al., 2013).

As analysed by Grineski et al. (2015), the double exposure framework has applicability for a case study of social and climatic drivers in the US-Mexico border metropolitan area of El Paso del Norte, which is made up of the city of Ciudad

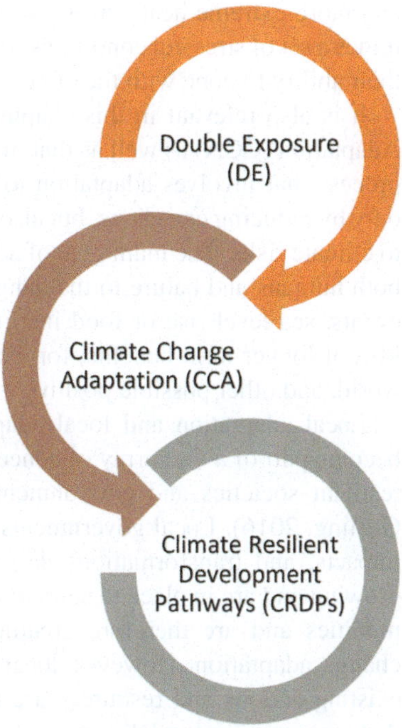

Fig. 19.1 Theoretical Framework and Interlinkages among DE, CCA and CRDPs

Juarez (Mexico) and El Paso (Texas). As explained in the next section, this area is a recognized case of DE because of a mix of social drivers such as rapid urbanization, industrialization, migration and violence, alongside general climate prediction models pointing to an increase in temperatures and extreme events that may severely affect the population in the area. Therefore, the identification of linkages between environmental, economic and social components (sustainable development) provides a useful example of the key facets of the double exposure framework. Once these linkages are identified, solutions and actions towards resilient development pathways can follow.

Before discussing the CRDPs it is also important to mention that the DE concept goes hand in hand with the *Climate Gap,* another framework that encompasses environmental justice considerations associated with climate change impacts. This climate gap concept is concerned with preparing evidence about the conditions of already vulnerable communities (mostly ethnic minorities and low-income populations) and their relations with climate impacts (Morello-Frosch et al., 2012). As an example, low-income populations and ethnic minorities such as Latinos and Afro-American populations can suffer disproportionately from increasing heat waves in urban areas, where households and general infrastructure are not resilient to climate change. Previous studies in Ciudad Juárez (Mexico) have also shown this trend in the identification of Double Exposure as well as the Climate Gap (Grineski et al., 2012). Different spatial models adopted under this framework proved that higher population density, together with lower economic incomes, social class differences and more extreme heat waves (for example, in the surroundings of Juárez), determines a set of stressful conditions for socially marginalized populations that reduces their ability to cope with the effects of climate change.

It is also relevant in this chapter to introduce the concept of Climate Change Adaptation (CCA) as well as that of mitigation, for this is now widely accepted as a process that involves adaptation to actual or expected future climate impacts not only by reducing emissions, but also to promote concrete actions to reduce exposure to climate risks. The main goal of adaptation is in fact to reduce the vulnerability of both humans and nature to the dangerous impacts of climate change (e.g. extreme events, sea-level rise or food insecurity, among others), and it also envisages the idea of longer term benefits, for example, increased yields in some regions of the world, and other possible positive impacts (Arnell et al., 2019).

Local adaptation and local adaptation solutions to climate change have also become part of a vast array of concepts and practices used by communities to build resilient societies and environments against climate change impacts (Arnell & Gosling, 2016). Local governments are more and more addressing climate-related impacts, and transformations derived from it. Global and local movements are growing and are in place to support climate adaptation initiatives among local communities and are therefore creating different solutions which promote climate change adaptation. However, local actors involved in the process recognize that existing actions and resources are insufficient for achieving local adaptation and that more accurate and diverse tools for understanding local vulnerability, as well as for subsequently providing adaptation plans, are needed. There is also a strong and growing recognition that streamlining climate adaptation resources, and providing services to communities, can help to comply with the Paris Agreement and other international commitments such as the Agenda 2030 and the Sustainable Development Goals (SDGs) (Nordgren & Meerow, 2016).

A second important framework for this chapter is the adoption of the concept of Climate Resilient Development Pathways (IPCC, 2014). The 5th Assessment Report of IPCC presented the idea of "*climate-resilient development pathways*" (CRDPs) as crucial development trajectories to envision the possible future under different climate scenarios as well as to foresee complex processes of deliberation and implementation to solve the climate crisis in different scales and contexts (Roy et al., 2018). The CRDP also tries to respond and addresses crucial questions by offering future scenarios. However, given the ongoing global discussion around the outreach of these pathways, and how and which of them are really resilient, how do we distinguish them from sustainable development actions? For the purposes of this article we use CRDPs as a proxy to sustainable development efforts that include resilience perspectives in social and urban planning to reduce climate change impacts. In particular, in order for development decisions to reach sustainability in a changing climate (be they economic, environmental or socio-political) they need to include a set of actions, patterns and societal models that will endure over time. Climate resilient development pathways are, then, new decision-focused approaches which are to include flexibility and account for future uncertainties to accomplish a better sustainable development (Bosomworth et al., 2016).

CRDPs also allow for a progressive implementation of development planning, actions and collective adaptative decisions by several stakeholders, as well as offering inputs and priorities to decision makers. These pathways can be seen, therefore, as complementary activities in an overall vision of pursuing sustainable development and climate change resilience. According to Chesterman (2020), resilience is also a system that can have both high or low relations with environmental crises, disease outbreaks (such as the COVID-19 pandemic), extreme events or armed conflicts, amongst others. It means that the level of resilience within an affected community or natural system has different degrees of influence depending on the capacity of the community, city or country to recover and quickly adapt to climate impacts. According to a widely used definition from Walker and Salt (2012), climate resilience is also recognized as the *"ability to cope with actual or expected climate-related stresses and shocks and keep functioning in much the same way"*.

19.3 Key Characteristics of Climate Change Impacts in Northern Mexico

The multidimensional challenges that the northern Mexico region faces with climate change are also a combination of a high degree of economic integration among the states along the border, and their insertion into regional and local processes of economic and environmental development. As Leichenko and others pointed out, the border region is characterized by the above-mentioned "Double Exposure" (Leichenko et al., 2010), meaning that both economic and environmental processes determine the regional landscape in terms of urbanizations, planning and other dynamics such as labour conditions and migration—all entwined with climate change impacts. It is therefore of great importance to identify the drivers of climate vulnerability and adaptation capacities in the U.S-Mexico Border region in the context of multidimensional contemporary challenges. This can be helpful not only to determine transboundary public policies to climate impacts, but also to understand future climate pathways and scenarios for the northern part of Mexico.

Before we describe climate drivers and impacts in northern Mexico, we need to clearly define the geographical area we are studying in this chapter. Literature, available both in Mexico and the USA, speaks of the *Border Region* in general, or the US-Mexico border, or the southwest of the USA, or the northern part of Mexico. Given the heterogeneity of the definitions and documents reviewed for this article, the analysis in this chapter encompasses the Metropolitan area of El Paso del Norte, which includes the city of Ciudad Juárez (Mexico) and El Paso (Texas). The northern part of Mexico and the border region also stretches from San Diego-Tijuana on the Pacific coast to the Matamoros-Tamaulipas in the Gulf of Mexico, a distance of about 3.200 km.

It is also important to clarify that this analysed geographical part of the border matches with the definition used in the US National Climate Assessment (NCA) for the Southwest region and also from the 6th Mexican National Communication to the United Nations Framework for Climate Change (UNFCCC), which includes the U.S. border states of California, Arizona, and New Mexico and Texas, while on the Mexican side it includes Baja California, Sonora and Chihuahua. The Paso del Norte corridor is considered here due to its key importance for commercial and economic drivers of the border area. An important variable to include in this chapter is that of demographics and, in particular, the population concentration along the border and the system of city-pairs that constitute an integrated and dynamic bilateral urban system (Wilder et al., 2013). Border counties and *municipios* are very important when discussing climate change implications, given their multidimensional vulnerability and exposure to climate change.

In terms of clear climate hazards and climate change impacts across the area, there are still few comprehensive studies that encompass the case study analysed in this article. The vulnerabilities highlighted and presented in the abstract come from scattered empirical evidence and from multiple contexts within the region. It is worth mentioning though, that even if we rely upon qualitative and quantitative studies since the early 1960s, the evidence of vulnerability, impact and consequences related to climate change, may not be calibrated enough to clearly address the exposure of border populations to climate risks and extreme events (Wilder et al., 2013). In this regard, robust information is mainly concentrated on urban, agriculture, water, health and other socio-economic impacts, and we try to assess key vulnerabilities in this context.

Broadly speaking, the border region presents high aridity and high temperatures for most of the year: rains fall in the summer months, due mainly to the North American monsoon, while precipitations in winter are mostly associated with both Californias. The region is also affected by variability in precipitation due to the El Niño phenomenon or ENSO, which has dramatically changed the way droughts and floods, as well as the hydrological cycle, are affecting the population and the agricultural processes. This has clear impacts on planning sustainability efforts along the border and in the cities (see Fig. 19.2).

Further challenges, in climatic terms, for the regions are due to the lack of meteorological stations, particularly on the high mountains in northern Mexico, which does not allow for continuity in records of precipitations and aridity or other indicators that complicate the comprehension of the border region's climate (Arnell et al., 2019). Lack of data and model projections, methods and downscaling may add uncertainty to the projections available, which are mostly based on the IPCC and the US National climate assessments.

Current Climate models available, including RCP, CMXC, Cordex and others, converge on the common point that temperatures for the border region will increase in the near and long-term future, at least to the year 2080 (IPCC, 2018). Average annual temperatures may increase from a range of 1 °C to 3.5 °C in the next 50 years (Magaña et al., 2012). Temperatures will be high during the summer and significant increases should be of concern to the central Sonoran Desert and the northwest

Fig. 19.2 Climatic Characteristics of Northern Mexico (Chihuahua). Source: personal elaboration from INEGI 2020

Chihuahua State with a major increase in the frequency of heat wave episodes. At the same time, being desert zones, high cooling temperatures overnight during winter could be experienced. With respect to precipitations, models show that they will decrease greatly along the Pacific coasts and parts of the Arizona-Sonora border. The western Chihuahua Desert will likely experience the same trend. The lack of precipitations may strongly increase during the period 2030–2080 causing the High Colorado River stream flows to diminish. These changes may cause strong water stress for the region and the population dependent on this vital liquid (see Fig. 19.3).

Type of Climate Hazard	Likelihood of Impact	Direction of Change
Annual number of days with maximum temperatures > 38 degrees Celsius	High	Increasing
Heat waves duration	High	Increasing
Cooling temperatures days	High	Increasing
Annual precipitation	Medium-high	Decreasing
Droughts	High	Increasing
Summer Precipitations	Medium-high	Decreasing

Fig. 19.3 Projected Climate Change Impacts for the Border Region. Source: Personal elaboration based on Wilder (2013)

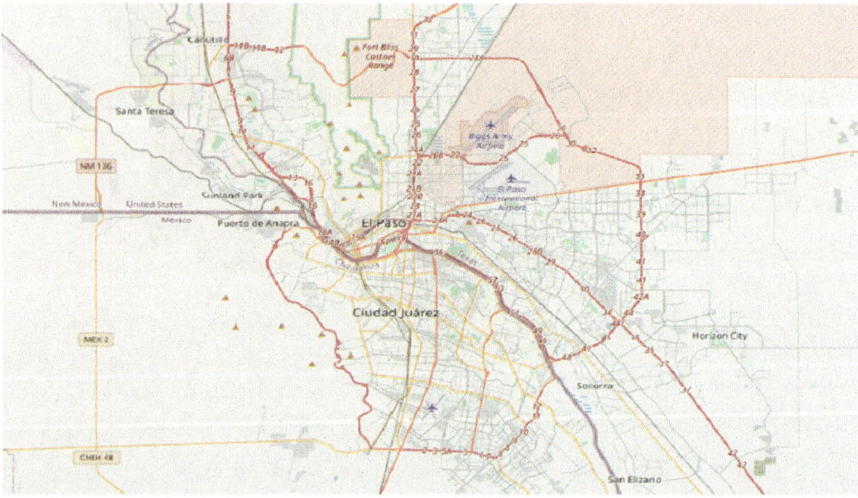

Fig. 19.4 The El Paso del Norte Metropolitan Area (Texas-Chihuahua). Source: personal elaboration in QGIS maps

19.4 El Paso del Norte Metropolitan Area

The metropolitan area of Paso del Norte is among the most complex international cross border regions in the world and the largest metropolitan area along the entire US-Mexico border. It currently hosts almost 3 million people, and it is considered a major economic centre, with the conjunction of important urban areas such as El Paso in Texas and Ciudad Juárez in Chihuahua (OECD, 2010) (see Fig. 19.4). In some literature, the city of Las Cruces (New Mexico) is also included in this area, but will not be included in the case study used for this article.

The transboundary nature of the region offers a picture of what the Double Exposure effect may look like in the future: the population growth is increasing at a high rate, by approximately 30% by 2050, but it also presents intra-regional social, economic and environmental differences in its main features. Ciudad Juarez is a complex example of a city where a mix of economic determinants, violence and harsh environmental conditions create the conditions for a climate change emergency. Currently, more than 70% of the Ciudad Juárez workforce is employed within the so-called *maquiladora industry*, which is one of the most important manufacturing exporting systems in the region (Delgado Ramos et al., 2015).

The area is also known for providing a vast array of services, including business services, trade, health and transportation, all supported by a strong link with the transnational US-Mexico manufacturing sector. The interconnection of the cities offers a very broad and deep linkage with the territory and is constantly shaping the environment of the border. Since the implementation of the North American Free Trade Agreement (NAFTA), currently named the USMCA after the recent changes to the main agreement, the Paso del Norte Region is experiencing a deep integration process. The area is also affected by a growth in population migration and dynamics, given the search for employment in maquiladora industries and other services, mostly by unskilled workers. This is also why the area is characterized by asymmetric conditions and long-standing weaknesses primarily associated with low incomes (OECD, 2010).

Under the DE framework, poverty is therefore a clear driver for vulnerability on both sides of the border, together with deficient health services, environmental degradation, and continuing violence. Even though it is not part of the main content of this chapter, the narcotrafficking issue and the battle for drug markets along the border is a constant threat to the creation of a more cohesive and homogeneous metropolitan zone.

Figure 19.5 shows, from the most recent data available from Texas governmental data, the number of heat waves that occurred in 2016, one the of worst years on record for this climate hazard. The spatial distribution of heat waves shows a pattern where el Paso del Norte is highly affected by the phenomena. Previous records available and accessed by the author confirm that this record high year is much more significant compared to the average of two heat waves during 2012–2016 (see Fig. 19.6).

19.5 Climate Change Adaptation and Climate Resilient Development Pathways: Scenarios for the Case Study

The border region in the case study analysed for this chapter shows that under the DE framework, the issues of poverty, water stress, lack of decent housing and urban planning create multiple asymmetries between the US and Mexican sides of the border. The climate change component and continuously rising temperatures, mostly heat waves, create even more daunting challenges for local governance, including effective health prevention programmes related to climate change adaptation strategies. In this section we will briefly discuss border vulnerability per sector in the area of the case study, and offer some possible implications for adaptation policies in the border region and the adoption of CRDPs.

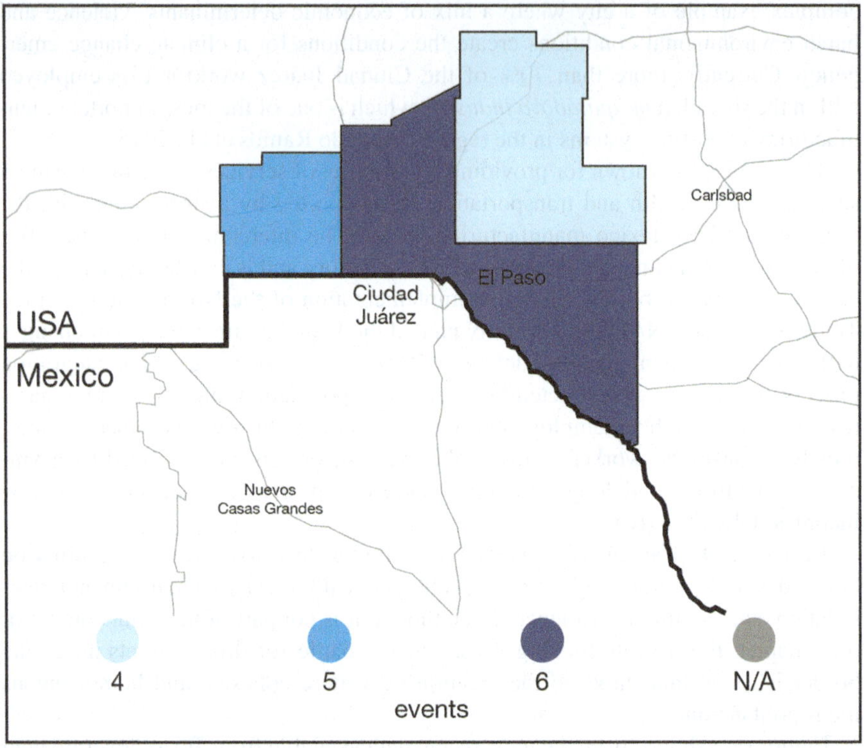

Fig. 19.5 Spatial distribution of extreme heat waves recorded in El Paso County (2016). Source: personal elaboration Texas EL Paso County data

1. **Water supply.** The entire water system of the metropolitan region has been under stress for the last 30 years. Changes in the water supply for the area are due to heavily exploited aquifers, which are also affected by pollutants and salt intrusion, as well as agricultural and industrial intensive extraction processes. The water system will therefore experience a significant decrease in storage reservoirs for urban use and irrigation. The provision of water may also be affected by energy use during winter times, as experienced at the end of 2020 during the winter storm in Texas.

   Ecosystems and human activities are experiencing severe deterioration, and the region of the main transboundary river, the Rio Grande, is having an increasing number of conflicts over water distribution. Ageing infrastructure, including dams and water storages, are becoming insufficient on both sides of the border and water governance and management need to be revisited.

   CRDP1:

   It is extremely important to keep preparing for an increase in water reservoirs and to deal with emergencies caused by climate change, because the drought periods are now longer and recurring more frequently and on a bigger scale

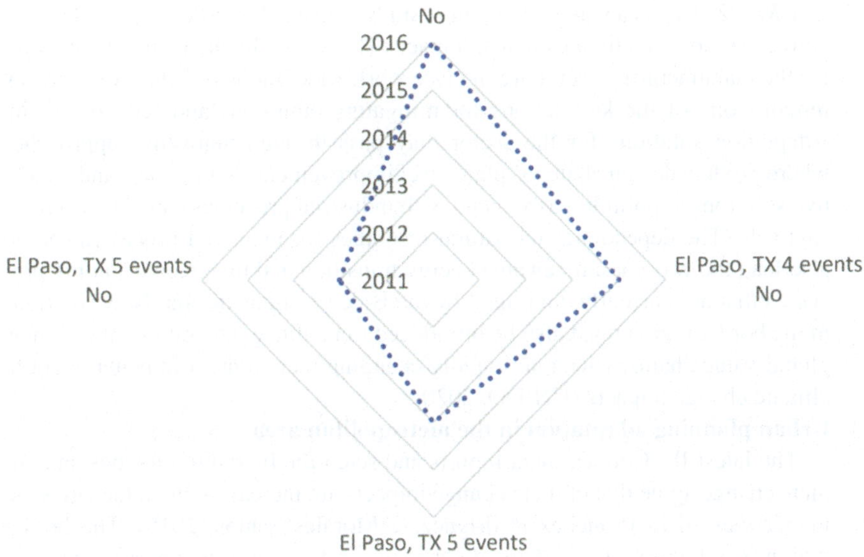

Fig. 19.6 Number of extreme heat waves recorded in El Paso County during 2011–2016

(IPCC, 2018). Water reserves are jointly managed by the U.S. and Mexican sections of the International Boundary and Water Commission (Cila and IBWC), set up in 1944, and its main role is dealing with the distribution of water and border limits. Since water is becoming scarce, the demand for the liquid and its management requires strong cooperative effort from both countries, as well as common planning in the longer term. Recent developments in the metropolitan area are using new recycling process systems to avoid the water-waste in flushing and other civilian uses, so are paving the way for resilient pathways. Given the projections that the Hueco Bolson will run dry by 2025, water adaptation measures will apply a circular system, where a drink-expel and drink again will be implemented. Together with other technological solutions, a potable water reuse plant system will be offered (Schlanger, 2018).

2. **Agriculture, farming and manufacturing industries**

Water is vital for many natural and human made processes. Linked to it is the implementation of adaptation systems to agriculture and manufacturing processes in the border region. Agriculture and manufacturing play a critical role in the regional identity, landscape, food and culture as well as in other spheres. Around 80% of the employment in the case study area is in the agricultural and maquiladora sector (Hurd, 2012). Higher temperatures are leading to more frequent droughts, which translate into a loss of crops, livestock, etc. Both small and large ranches can be affected by the changing climate conditions, and this can be even more severe if farmers' practices still follow customary processes without taking into account climate variability models.

CRDP2: This is an area where most study is needed in order to provide alternative solutions to climate change impacts. However, climate resilient pathways for the manufacturing sector are in place worldwide. Industrial energy efficiency remains one of the key actions for mitigating emissions and reducing them. Adaptation solutions for the sector may pass through innovative approaches where product design changes, plant level improvements in materials and collective solutions to pollution associated with industrial processes can play a significant role. The dependence on natural resources for industrial processes should also lead the way for a renewable energy transition and the green economy processes that are currently envisaged by the Biden Administration. New solutions in the border region must also be introduced to facilitate integration into the new global value chains, which are rapidly changing their content to better adapt to climate change impacts (UNIDO, 2020).

3. **Urban planning adaptation in the metropolitan area**

The latest IPCC assessment reports and scientific literature on cities and climate change agree that climate change impacts are increasing in all the cities, no matter their sizes (Sánchez Rodriguez & Morales Santos, 2018). The border region is not an exception. Climate change can have impacts on infrastructure, energy and transportation in the cities, and can shift the behavioural patterns of citizens in many ways. A general framework for the city of Ciudad Juarezin in climate resilience, for example, is provided by the State Plan for Climate Change (Programa Estatal de Cambio Climático, 2019) of 2019, and the Chihuahua climate change law (2013), where mitigation and adaptation goals are set for urban, ecological and terrestrial ecosystems. In the case of El Paso, TX, no state or local plan is available though resilient initiatives are pushed forward by local councils. The main problem is that Texas is among the US major emitters of CO_2, and state legislature on climate change has not had sufficient impact. Moreover, the state has demonstrated hostility towards municipalities who regulate themselves in terms of how to address climate change impacts and issues. The partial information available offers a patchy scenario for the metropolitan area, where Mexico and the Chihuahua State, in particular, is setting a governance framework for CCA and Resilient pathways (at least in normative terms) while the city of El Paso is far from providing a plan to reach resilient scenarios to climate change.

In the case of Ciudad Juarez, there has been a rapid growth of the population over the past 30 years that has caused a process of disorderly and irregular urbanization, affecting the original environment where settlements grew. Therefore, a large population and poorly planned urbanization make the urban centre little prepared for adapting to the effects of climate change, mostly in the form of heat waves and rising temperatures. It is crucial, then, to implement actions to prevent risks and reduce the exposure of the population to rapidly shifting climate impacts. Some of the actions necessary to be able to strengthen the resilience of the urban area of Cd. Juarez include the renewal of "Atlas of Risks in the Urban Development Programs" in order to better map critical points in the city to prevent the creation of new formal or informal settlements (PECC, 2019). The risk mapping and planning must consider the functionality of the watersheds and landscape units, to provide better conservation and restoration actions.

CRDP3:

Another important adaptation action points to the governance process of the bilateral basin councils to incorporate ecological zoning criteria as an instrument to prevent water overuse on both sides of the border. The metropolitan city must also include emergency plans that consider the relocation of the most vulnerable human settlements when facing an increasing number of droughts. Finally, CAA in the metropolitan area of El Paso Norte should look to road infrastructure, which under the stress of high temperatures faces challenges of maintenance and new materials that can support the increasing weight of trucks and commercial traffic in the border area.

Concerning housing and settlements, ventilation and cooling systems should be designed to use less fossil fuel energy and include renewable energies as power sources. Recycling water systems to reduce severe environmental impacts, as mentioned in the section above in this article, should also be implemented. Another important adaptation scheme is the improvement of green areas on the roofs of buildings, through the placement of "gardens" on top of buildings that, in addition to absorbing CO_2, would help the air conditioning of these buildings and reduce water runoff in urban areas.

19.6 Concluding Remarks

In this chapter we offered a proxy to the understanding of the Double Exposure and CRDPs frameworks for the Metropolitan area of el Paso del Norte. Given the complexity of the case study and variables included, the chapter only looks at some specific issues such as water and heat waves in relation to city and urban planning for future scenarios.

Border zones are characterized by unevenly distributed patterns, where cooperation and bilateral processes are needed to improve satisfactory environmental and social standards for the USA and Mexico. Scientific and academic research is needed to improve solutions, for example, for water distribution and water management, but also to look for new solutions to prevent health impacts due to increasing heat waves and other climate hazards. International and transboundary cooperation is also needed to create agreements that favour joint and mutual decisions for the benefit of local communities.

Concerning CRDPs, they can represent development trajectories that follow existing and new practices to transform and change conditions from negative to positive scenarios, or simply for improving current conditions. While there are several ways to generate development pathways, these can also help to complement local development agendas, or any other form of development goals, in order to achieve greater climate resilience. The important thing is to keep on the horizon the resilient component of development and its relation to climate change impacts.

However, it remains to be seen how those pathways complement or differ from sustainable development actions towards climate change adaptation and resilience efforts already being undertaken at international, national and local level.

References

Arnell, N. A., & Gosling, S. (2016). A global assessment of the impact of climate change on water scarcity. *Climatic Change, 134*(3), 371–385. Springer.

Arnell, N. W., Lowe, J. A., Challinor, A. J., & Osborn, T. J. (2019). Global and regional impacts of climate change at different levels of global temperature increase. *Climatic Change, 155*, 377–391.

Bolin, B., Barreto, J. D., Hegmon, M., Meierotto, L., & York, A. (2013). Double exposure in the sunbelt: The sociospatial distribution of vulnerability in Phoenix, Arizona. In C. Boone & M. Fragkias (Eds.), *Linking Urban ecology, environmental justice and global environmental change: A framework for urban sustainability* (pp. 159–178). Springer.

Bosomworth, K., Owen, C., & Curnin, S. (2016). Addressing challenges for future strategic-level emergency management: reframing, networking, and capacity-building. *Disaster*. https://doi.org/10.1111/disa.12196

Chesterman, S., Neely, C. N., Thornton, P., & Pereira, L. (2020). Climate-resilient development pathways. In *SADC futures: Developing Foresight Capacity for Climate Resilient Agricultural Development Knowledge Series. CCAFS Report*. CGIAR Research Program on Climate Change, Agriculture and Food Security (CCAFS).

Delgado Ramos, G., De Luca, A., Vázquez, Z. V. (2015). *Adaptación y mitigación urbana del cambio climático en México*. PINCC-UNAM, CDMX.

Grineski, S., Collins, T., Ford, P., Fitzgerald, R. M., Aldouri, R., Gilberto, V.-A., Aguilar Maria de Lourdes, R., & Lu, D. (2012). Climate change and environmental injustice in a bi-national context. *Applied Geography, 33*, 25–35.

Grineski, S., Collins, T., McDonald, Y., Aldouri, R., Aboargob, F., Eldeb, A., Lourdes Romo Aguilar, M., & Velázquez-Angulo, J. G. (2015). Double exposure and the climate gap: Changing demographics and extreme heat in Ciudad Juárez, Mexico. *Local Environ, 20*(2), 180–201. https://doi.org/10.1080/13549839.2013.839644

Hurd, B. (2012). Climate vulnerability and adaptive strategies along the Rio Grande/Rio Bravo border of Mexico and the US. *Journal of Contemporary Water Research & Education, 149*(1), 56–63.

IPCC (2014). Assessment Report 5, Chapter 20, WG2.

IPCC (2018). 1.5 Degrees special report.

Leichenko, R. M., O'Brien, K. L., & Solecki, W. D. (2010). Climate change and the global financial crisis: A case of double exposure. *Annals of the Association of American Geographers, 100*(4), 963–972.

Magaña, V., Zermeño, D., & Neri, C. (2012). Climate change scenarios and potential impacts on water availability in northern Mexico. *Climatic Research, 51*, 171–184. https://doi.org/10.3354/cr01080

Morello-Frosch, R., Pastor, M., Sadd, J., Shonkoff, S. B. (2012). *The climate gap: Inequalities in how climate change hurts Americans & how to close the gap*. Program for Environmental and Regional Equity, editor. University of Southern California (pp. 1–32).

Nordgren, M., & Meerow, S. (2016). Supporting local climate change adaptation: Where we are and where we need to go. *Environmental Science & Policy, 66*, 344–352.

OECD. (2010). *Higher education in regional and city development*. The Paso del Norte Region, Mexico and the United States.

Programa Estatal de Cambio Climático (2019), Chihuahua.

Roy J, Tschakert P, Waisman H, Abdul Halim S, Antwi-Agyei P, Dasgupta P, Hayward B, Kanninen M, Liverman D, Okereke C, Pinho PF, Riahi K, Suarez Rodriguez AG. (2018). Sustainable development, Poverty eradication and reducing inequalities. In: *Global warming of 1.5°C. An IPCC special report on the impacts of global warming of 1.5°C.*

Sánchez Rodriguez, A., & Morales Santos, A. (2018). Vulnerability assessment to climate variability and climate change in Tijuana. *Mexico. Sustainability Journal MDPI, 10*, 2352.

Schlanger, Z. (2018). *Quartz report, A Major US City will start drinking its own sewage*. Retrieved from https://qz.com/1353825/a-major-us-city-will-start-drinking-its-own-sewage-others-need-to-follow/.

Smid, M., & Costa, A. C. (2013). Climate projections and downscaling techniques. *International Journal of Urban Sciences, 22*(3), 277–307.

UNIDO. (2020). *World report*: *Promoting resilient climate industry*.

Walker, B., & Salt, D. (2012). *Resilience practice: Building capacity to absorb disturbance and maintain function*. Island Press.

Wilder, M., Garfin, G., Ganster, P., Eakin, H., Romero-Lankao, P., Lara-Valencia, F., Cortez-Lara, A. A., Mumme, S., Neri, C., & Muñoz-Arriola, F. (2013). Climate change and U.S.- Mexico Border Communities. In G. Garfin, A. Jardine, R. Merideth, M. Black, & S. LeRoy (Eds.), *Assessment of climate change in the Southwest United States*: A report prepared for the national climate assessment (pp. 340–384). A report by the Southwest Climate Alliance. Washington, DC: Island Press.

Chapter 20
Nutritious Landscapes: The Food Environments in the Metropolitan Peripheries of Mexico

Aleksandra Krstikj, Greta Lukoseviciute, Christina Boyes, and Moisés Gerardo Contreras Ruiz Esparza

20.1 Introduction

The global trend of increasing overweight and obesity in the population, resulting from an upsurge in processed food intake and the reduction of physical activities, has gradually shifted focus from developed countries toward the urban poor in emerging regions (Popkin, 2006). In Latin America, the absolute annual percentage point change in weighted prevalence of overweight and obesity for women in urban areas is among the highest in the world (Popkin et al., 2012). Specifically in Mexico, the age-standardized mean for girls' body mass index (BMI) from 1975 to 2016 was almost four times higher than in other parts of the world (Abarca-Gómez et al., 2017). Obesity and overweight are leading causes of several other preventable and noncommunicable diseases, such as diabetes (Felber & Golay, 2002), hypertension (Seravalle & Grassi, 2017), and depression (Preiss et al., 2013). Even though some countries are already exploring programmatic and policy measures to face this global health crisis, very few engage in serious efforts to mitigate dietary risks (Popkin et al., 2012).

A. Krstikj (✉)
Tecnologico de Monterrey, School of Architecture, Art and Design, López Mateos, State of Mexico, Mexico
e-mail: sandra.krstik@tec.mx

G. Lukoseviciute
Tecnologico de Monterrey, School of Architecture, Art and Design, Puebla, Mexico

C. Boyes
Centro de Investigacion y Docencia Economicas, A. C., DEI, Ciudad de México, Mexico

M. G. Contreras Ruiz Esparza
Universidad Nacional Autónoma de México, Instituto de Ingeniería, Ciudad de México, Mexico

E. Giorgi et al. (eds.), *Design for Vulnerable Communities*, The Urban Book Series,
https://doi.org/10.1007/978-3-030-96866-3_20

There is mounting evidence that the neighborhood food environment, understood as "the interface that mediates one's food acquisition and consumption with the wider food system" (Holdsworth & Landais, 2019), strongly influences dietary behavior and obesity (Lytle & Sokol, 2017; Crawford et al., 2014; Giskes et al., 2011). The World Health Organization identified food environment interventions as strategies for creating improvements in dietary patterns and weight status (Vandevijvere et al., 2015). The rapid urbanization, economic development, and demographic change in emerging regions are reshaping food environments for the majority of the world urban poor, nevertheless, very few studies focus on how social and physical environments in those regions condition dietary habits. In a recent systematic literature review focused on the influence of local food environments on food behavior, Pitt et al. (2017) found nineteen articles from the USA, seven from the UK, and one each from Mexico, Canada, Australia, and the Netherlands. There is strong evidence of inequality in food access only from the USA, but trends are less apparent in other countries (Black et al., 2014).

Although geographical patterns of food production and distribution undoubtedly clarify some aspects of the food security problem, wider integration of contextual socioeconomic factors is necessary, especially when researching food environments of the most vulnerable populations living on the urban fringe of emerging regions. For example, Holdsworth and Landais (2019) and Gálvez-Espinoza et al. (2018) studied food environments in Africa and Chile, respectively, and recognized the importance of lifestyle, social and community networks, living and working conditions, and general socioeconomic and cultural determinants in eating healthy.

This study explores food environments in the periphery of Mexican metropolises, focusing on availability, accessibility, and acceptability of food that embody inequities and determine health outcomes. Availability is related to the physical presence and walkable proximity (1 km or 15-min walk) to food sources that includes foods offered/served in various settings. Accessibility on the other hand describes food's availability in a form and location that facilitate its consumption, for example, the price of food products in relation to income and the presence and quality of the pedestrian infrastructure to reach food sources. Acceptability is used as an indicator of food quality, people's education related to nutrition and cultural habits that can affect food choices and thus determine health outcomes. In the specific context of the Mexican periphery, we set out to explore dietary modernity and shed light on possible factors or underlying mechanisms that influence changes in dietary behaviors. This pilot study, based on a preliminary survey in various metropolitan zones, probes the role of factors outside of transportation infrastructure and the prevalence of fast food and convenience stores—measures which are commonly used to address accessibility and availability of food sources, to examine the peri-urban food environment in Mexico and set the course for future research.

The paper is divided into four main sections: Section 20.2 provides a review of the literature and the theory driving our research, Sect. 20.3 describes the method for the study and the survey metadata, Sect. 20.4 presents the findings of the pilot survey, and in Sect. 20.5 we discuss the implications of the findings and summarize the take-away messages.

20.2 Literature Review and Theory

Today, 55% of the world's population lives in urban areas, a proportion that is projected to escalate to 68% by 2050 (UN, 2018). Latin America and the Caribbean are one of the most urbanized regions globally, with 81% of people living in urban areas (UN, 2018). As urbanization and income increase, the prevalence of diabetes, heart disease, obesity, mental health problems, alcohol and drug abuse, and violence increase as well (Ezzati et al., 2005; Mendis & Banerjee, 2007). Obesity has become progressively more widespread among the socially disadvantaged (Friel et al., 2007) and is often accompanied by underweight among the urban poor (Mendez et al., 2005). Ezzati et al. (2005) predicted that cardiovascular disease risks are going to systematically shift to low-income and middle-income countries and, together with the persistent burden of infectious diseases, further increase global health inequalities. Thus, preventing obesity should be a priority from early stages of economic development.

Health inequality is not a natural phenomenon but a result of societal action, planning, and public policy (Commission on Social Determinants of Health, 2008). The existence of health inequality illustrates the unjust and inequitable development of food environments in different settings. Mounting research focused on the relationship between the food environment and health outcomes (Caspi et al., 2012; Feng et al., 2010; Kelly et al., 2011; Kirkpatrick et al., 2014; Glanz et al., 2016) attests to the fact that the social and urban environment, and the lifestyles they promote, result in differences in health outcomes. While urban living has undoubtedly improved many aspects of health, such as availability of services and technologies (Bloom & Canning, 2000), conditions of housing and sanitation (WHO, 2009), and life expectancy (Roser et al., 2013), the disadvantages of urban poverty can be worse than rural poverty. In 2007, Van de Poel et al. compared child health outcomes between urban and rural areas in 47 developing countries. They found that while health outcomes were better in urban than in rural areas of developing countries, in nine out of 47 countries children from lower socioeconomic households in urban areas had higher rates of mortality than their rural counterparts.

Factors such as the availability, accessibility, affordability, food quality, and store characteristics are consistently identified as key determinants of store choice and purchasing behaviors that often result in less healthy food choices within community nutrition environments (Pitt et al., 2017). Poorer neighborhoods, even in developed contexts, often suffer from decreased access to health products and lower quality fruits and vegetables (e.g., Black et al., 2012). Thus, inequalities in access may disproportionately affect the poor. Likewise, infrastructure, market access, percentage-working women in urban areas, and norms and institutions have significant influence on food consumption in developing contexts (e.g., Pandey et al., 2020).

A more complex understanding of the urbanization processes, which goes beyond demographic shifts, is necessary to better understand dietary changes (Pandey et al., 2020). Several of these factors are inherently geographic, as are many of the studies of the urban food environment, but attention to reliability, validity, robustness, and sensitivity and specificity to change are lacking in many existing studies (Lytle & Sokol, 2017). Spatial or not, more robust measures of food environments in different urban development contexts are necessary to improve the effects of community-level food environments on individual dietary behaviors and to inform public policy aimed at reducing obesity. Built-up patterns, individual, social, and cultural factors play a crucial role in determining how individuals interact with their food environment (Lucan, 2015). This suggests that researching only geographical distribution of food resources based on datasets like commercial data lists may often misrepresent the real food environment. Surveys on dietary habits and perceived conditions of food environments could be very useful to complement such geographical measurements.

The global change to urban lifestyle has contributed to the unequal distribution of health both within the national urban–rural divide and within rich and poor cities (Friel et al., 2011). While the research discussed above attest to the interest that the urban area has attracted in the context of healthy food environments, the peri-urban area is less researched. The peri-urban area does not have a clear definition yet, although there have been attempts to clarify its meaning and boundary. For example, the UN often refers to peri-urban development as "zones of transition from rural to urban land uses located between the outer limits of urban and regional centers and the rural environment. The boundaries of peri-urban areas are porous and transitory as urban development extends into rural and industrial land" (https://en.unesco.org/events/peri-urban-landscapes-water-food-and-environmental-security).

Peri-urban areas, sometimes called "edge-cities" (Garreau, 1991), "edgeless cities" (Lang, 2003), or "post-suburbs" (Phelps & Wood, 2011), are located at the urban periphery but contiguity itself cannot explain the phenomenon (Murgante et al., 2007). Key features of peri-urban areas are their mediating role in migration between rural and urban and their partially developed infrastructure and services (Iaquinta & Drescher, 2000). Although researchers and governments have long acknowledged the potential role of peri-urban areas in alleviating food insecurity of urban areas, few studies probe into the food environment of peri-urban dwellers. A recent study by Murphy et al. (2018) used quantitative, spatial, and qualitative data to clarify the understudied relationships between local food environments and health outcomes across various suburban developments in Australia. They found that interrelated challenges of car dependency, poor public transport, and low-density development hampered healthy food access in the suburbs.

20.2.1 The Food Environment in Mexico's Peri-Urban Zones

Barrera-Cruz et al. (2013) reported that Mexico has the second highest global prevalence of obesity in the adult population (30%). In 2012, 26 million Mexican adults were overweight and an additional 22 million were obese. Thus, concern over diet-related health issues, health inequality, and the food environment are of particular importance in Mexico.

Mexico has recently adopted new policies regarding food labeling according to the INFORMAS Initiative (White & Barquera, 2020). Moreover, studies have been performed on Mexican food environments in the various levels of consumption, such as household, school, and community (Rosales Chavez et al., 2020). The study found that while both nutrient-dense and energy-dense foods were present in all levels, individuals were more likely to consume foods with added sugars, fats, and salt options. In this review, most studies on the community food environment were done in rural areas, whereas most studies on the school food environment were conducted in urban settings. The food environments of the Mexican peri-urban areas have been largely absent from studies.

In a recent study by Pineda et al. (2021) on the relationship between the retail food environment and BMI in Mexico, food outlet density was used as a measure. The study found that in the retail food environments, the convenience store that does not have a variety of healthy food options was the most accessible food source, especially in second lowest, middle, and second highest income households. Pineda et al. (2021) argue that this is likely a result of zoning laws in the highest income neighborhoods and lack of services in the poorest. However, the authors point out that "as households increase their available income there tends to be a cross-over to a higher availability of food outlets—particularly convenience stores."

Mexico's urbanization and economic growth is dependent on extractive industries, which in turn have triggered the accelerated growth of cities and de-peasantation (Rincon et al., 2019; Araghi, 1995). Mexico City is the fifth-largest urban agglomeration in the world, and its influence is observed throughout the whole country, but especially in the central region (Rincon et al., 2019). The central region is a dense patchwork of urban networks where some of the largest peri-urban zones of Mexico City, Toluca, Cuernavaca, Puebla, Pachuca, and Querétaro, converge (Ávila-Sánchez, 2011). The peri-urbanization in Mexico goes beyond the passive relocation of population and involves employment activities and luxury residential developments, as well as civic functions and amenities. The peri-urbanization has taken place mainly on agricultural lands, 60% of which were expropriated from collectively owned lands or *ejidos* (Ávila-Sánchez, 2011).

Mexico's peri-urban inhabitants face degradation of their livelihoods while conflicts for space with other social groups increase. This results in unequal access and use of urban-rural peripheral spaces, which increases marginalization and spurs the redefinition of lifestyles and dietary habits (Ávila-Sánchez, 2011). Local resources are not included in policies addressing the mitigation of urbanization's negative impacts. For example, agriculture in peri-urban areas has not been included in or encouraged as part of policy despite its potential to encourage the consumption of healthier, locally produced foods while improving local livelihoods (Ávila-Sánchez, 2011). Research on food environments and dietary habits in peri-urban areas of Mexican metropolises is necessary to support policy aimed at community health in these zones.

The Mexican street market, also known as "*tianguis*," is a common source of fresh, unprocessed and often locally grown and sourced food products for the residents of the country's urban areas. However, as Fig. 20.1 illustrates, peri-urban

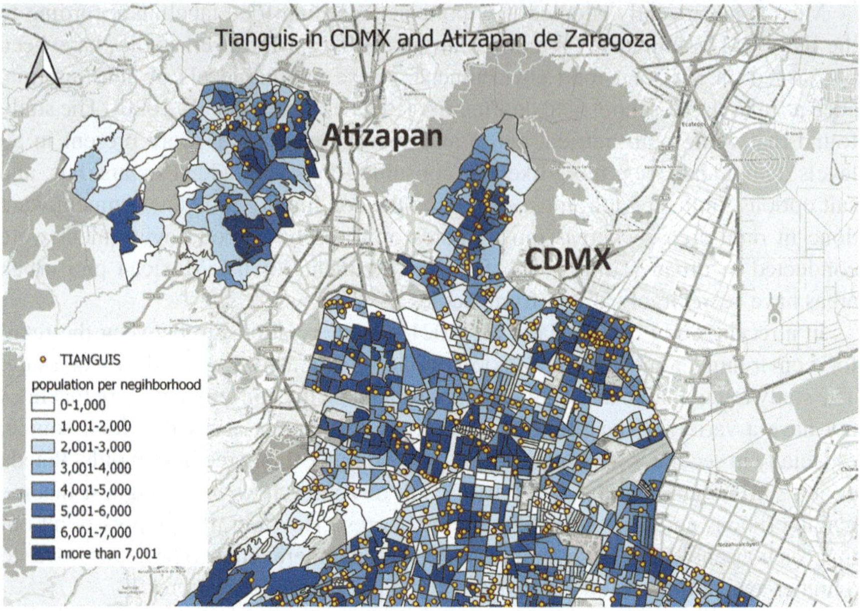

Fig. 20.1 Tianguis in the north of Mexico City (CDMX) and the municipality of Atizapán de Zaragoza (data from INEGI, 2020 for CDMX, and from field survey by authors for Atizapán de Zaragoza)

areas have a significantly lower density of *tianguis* than their urban neighbors do. In Mexico City, there are 1327 *tianguis* for approximately nine million people, while in Atizapán de Zaragoza, the industrial north periphery of Mexico City Metropolitan where half a million of people live, we found only 27.

In this preliminary study, we argue that the peri-urban areas with their specific character of fragmented landscape, partial infrastructure, and insufficient services influence food environments and dietary habits. Drawing on the insights of Pineda et al. (2021) and Murphy et al. (2018), we specifically argue that peri-urban residents differ from urban and rural residents in their attempts to eat healthy and their consumption of a modern diet characterized largely by their consumption of store-bought vs. homemade meals and processed foods. We define healthy eating as "higher in plant-based foods, including fresh fruits and vegetables, whole grains, legumes, seeds, and nuts and lower in animal-based foods, particularly fatty and processed meats" (Cena & Calder, 2020). Relative consensus in the literature on nutrition suggests that diets which prevent disease and encourage better health share the characteristics of being low in processed foods, higher in whole grains, fruits, and vegetables, and containing less saturated fats, trans fats, animal proteins, and refined or added sugars (Pistolatto et al., 2018). While regional dietary patterns can influence the exact components of a healthy diet, regional diet differences need not impact the inclusion or absence of features considered "healthy," although targeted nutritional initiatives may be adaptable to local tastes and preferences for specific flavors or ingredients.

Peri-urban residents have access to fewer healthy food sources, as illustrated by the reduced density of *tianguis* in Atizapán de Zaragoza in comparison to Mexico City. Likewise, peri-urban residents are frequently commuters to the larger urban centers their cities border. From the surrounding area of Mexico City, for example, 2.25 million daily trips were registered in 2017 (INEGI, 2017). As a result, and contrary to healthy diet and food infrastructure and services, transportation infrastructure is often well developed in peri-urban areas, to reduce the traffic congestion between peri-urban and urban zones. Long commutes and decreased access to unprocessed food sources may encourage the consumption of meal outside the home, as well as meals that are processed and easy to prepare, even when peri-urban residents make active efforts to eat a healthy diet. Thus, our argument is that, despite attempts to eat healthy and driving for access to better food choices, peri-urban residents face inherent structural disadvantages that make them more vulnerable to the consumption of highly processed and easily prepared foods. We further argue that there is no reason to expect peri-urban residents to have less desire to eat a healthy diet, or to put less effort into doing so, and when they have access to healthy foods, they will make efforts to eat healthy.

We draw three hypotheses based on the preceding work done in this area, our knowledge of Mexico's urban and peri-urban zones, and the theory detailed above. Namely:

Hypothesis 1: Peri-urban residents who live in more marginalized areas will be more likely to consume a modern diet than urban residents.

Hypothesis 2: Peri-urban residents who report driving for groceries will be less likely to attempt to eat a healthy diet than peri-urban residents who do not drive for groceries.

Hypothesis 3: Peri-urban residents who consume a modern diet will be likely to put effort into eating healthy.

20.3 Methods

This is a preliminary study. The sample is a non-random convenience sample which was gathered by distributing a voluntary response survey in the communities of origin of 40 student volunteers from 17 states in Mexico (see Table 20.1). Data collection began in Fall 2020 and was completed in Spring 2021.

While the non-random nature of the sample does present significant limitations, the resulting sample has some interesting characteristics which we believe demonstrate it to be a tough case in which to find evidence for our theory. Our sample is largely homogenous in socioeconomic status, age, and education level. If support is demonstrated in this sample, whose results we caution the reader to view as purely descriptive rather than unbiased and generalizable, we would expect support to be much stronger in a truly representative sample of the Mexican population, as the impacts of marginalization, access to vehicle transportation, and availability of local healthy food sources such as *tianguis* should have stronger effects as marginalization increases.

Table 20.1 Select demographics of Mexico vs. sample population

	Mexico (total)	Sample
% Age 15–64	66	99
% Female	51	57
% Male	49	43

Select demographic information comparing the Mexican population (2018) and the pilot survey population. Data on the Mexican population sourced from https://www.gob.mx/conapo/documentos/la-situacion-demografica-de-mexico-2018. Percentages for the sample population are rounded.

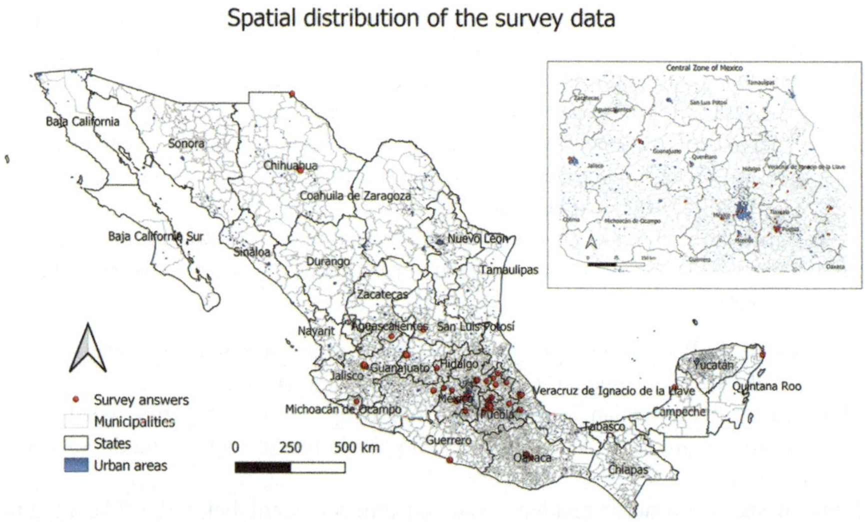

Fig. 20.2 Spatial distribution of the collected survey data

20.3.1 Descriptive Statistical Analyses

The data for this study comes from a voluntary response-based, non-random survey of 453 participants from more than 70 neighborhoods in 17 Mexican States (Fig. 20.2).

Respondents answered a battery of 57 questions related to their dietary habits, transportation, and shopping habits. Prior to our analyses, we discarded all respondents who left more than five questions blank. We also discarded all answers that were nonsensical (e.g., one respondent replied that they bought food for 41 individuals each week, another indicated eating more than 80 meals per day at home, another self-identified as a dolphin). The final sample comprised 349 individuals, 199 who self-identified as females, 150 as male, and one member of the sample identified as bisexual. As shown in Fig. 20.3, the vast majority of the sample is under 40 years of age. The youngest respondent was 15 years old, and the oldest respondent was 75 years old. A range of lifestyles and diets are represented in the sample, including vegetarian, vegan, allergy, and protein-based diets.

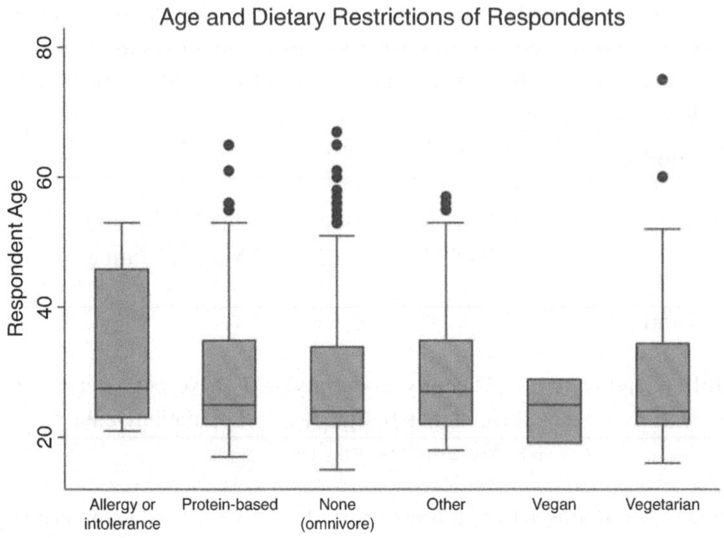

Fig. 20.3 Boxplot of respondent age and dietary restrictions. As the plot demonstrates, respondents were predominantly young adults, and young adults were also more likely to have specialized diets of any kind

Most respondents reported at least some university education, either in progress or already completed. Furthermore, most respondents live in areas characterized by very low to medium marginalization. Notably, the sample is not representative of the general population of Mexico. Despite this shortcoming, it is important to note that the relative homogeneity of this group of respondents leads us to believe that the variations seen in this sample are greater in the general population. Thus, although the results should be seen as descriptive, the data presented can be viewed as a tough case, within which we should see less variation than we would in a representative sample and within which it should be harder to observe evidence for our hypotheses.

The dependent variables for this chapter are continuous indices dietary modernity and attempted healthy eating, described in detail below. As a result, we rely on Ordinary Least Squares regression due to its properties as the best, linear, unbiased estimator. To ensure our model choice was correct, we used Maximum Likelihood to identify the appropriate link function for the model. The resulting link function was Gaussian, and the estimation results demonstrated no substantive or statistical difference in significance.

The independent variables for the first models in the study are driving for groceries, gender, age, population density, and marginalization. The second set of models incorporate the same set of independent variables together with dietary modernity. These controls allow us to address social, transportation infrastructure, and socioeconomic factors, and basic demographic characteristics which may impact consumption practices.

To assess the hypotheses, we settled on two base models which allow us to examine the roles of specific factors on dietary modernity and efforts to eat a healthy diet. We tested each model with and without the inclusion of standard errors clustered by neighborhood.

Models 1 and 2:

$$\begin{aligned}\text{DietaryModernity}_i = &\ \alpha + \beta_1\text{Drive for groceries}_1 + \beta_2\text{Gender}_2 + \beta_3\text{Age}_3 \\ &+ \beta_4\text{Populationdensity}_4 + \beta_5\text{Marginilization}_5 + \varepsilon\end{aligned}$$

Models 3 and 4:

$$\begin{aligned}\text{EfforttoEatHealthy}_i = &\ \alpha + \beta_1\text{Dietary modernity1} + \beta_2\text{Driveforgroceries}_2 \\ &+ \beta_3\text{Gender}_3 + \beta_4\text{Age}_4 + \beta_5 + \text{PopulationDensity}_5 + \\ &+ \beta_6\text{Marginalization}_6 + \varepsilon\end{aligned}$$

The indices used in this set of preliminary studies were created using primary component factor (PCF) analysis. For the index of dietary modernity, the components include the consumption of processed fruits, vegetables, legumes, meat, poultry, fish, and milk substitutes. Additionally, the index includes having food delivered to one's home, eating out, making purchases based on marketing, publicity, and brand, and purchasing foods based on ease of preparation and shelf life. Ordering food to go and purchasing food in the supermarket were also included in the index. PCF identified four retainable factors, of which the first had an eigenvalue of 3.48677 and all items were positively signed and matched theoretical expectations in their relative contributions. Thus, factor 1 was retained as the index of dietary modernity. The Cronbach's alpha coefficient for the index is 0.7519, placing it well within the acceptable range.

The second index used in these pilot studies is that of effort made to eat healthy. The components of the index are including unprocessed fruits, vegetables, legumes, fish, meat, and poultry, as well as whole grain cereals, milk products, and organic products in their diet, as well as paying attention to health, calories, organic labeling, and the presence of GMOs and preservatives in purchasing decisions. PCF identified four retainable factors, of which the first had an eigenvalue of 3.83687 and all items were positively signed and matched theoretical expectations in their relative contributions. Thus, factor 1 was retained as the index of healthy eating. The Cronbach's alpha coefficient for the index is 0.7697, placing this index, like the first, well within the acceptable range.

20.4 Results

As we emphasized earlier in this chapter, the results presented here are from a pilot study on a population of volunteer respondents that skews heavily toward the working-age population in Mexico. Models 1 and 2 examine the role of specific factors on individuals' consumption of a modern diet. Models 3 and 4 focus on the consumption of and attention paid to a healthy diet. Notably, the differentiation between urban, peri-urban, and rural zones is not reflected in the initial estimates. To identify the impact of respondents' food environments, we graph the results by respondents' zone of urbanization in Figs. 20.4 and 20.5.

The preliminary results, which do not account for different zones of urbanization, show that driving for groceries increases the likelihood that respondents will eat a modern diet, as does marginalization when neighborhood clusters are not included. Furthermore, older respondents were less likely to consume a modern diet. Where healthy eating is concerned, older respondents were more likely to make an effort to eat healthy or to buy healthy food products, as were individuals who ate a modern diet (see Table 20.2).

Figure 20.4 shows the results of the first set of models by zone of urbanization. Looking at the figure, although the confidence intervals of rural respondents are too large to allow us to speak to the discrete influences of specific factors on their level of dietary modernity, it is clear that marginalization has different impacts on urban

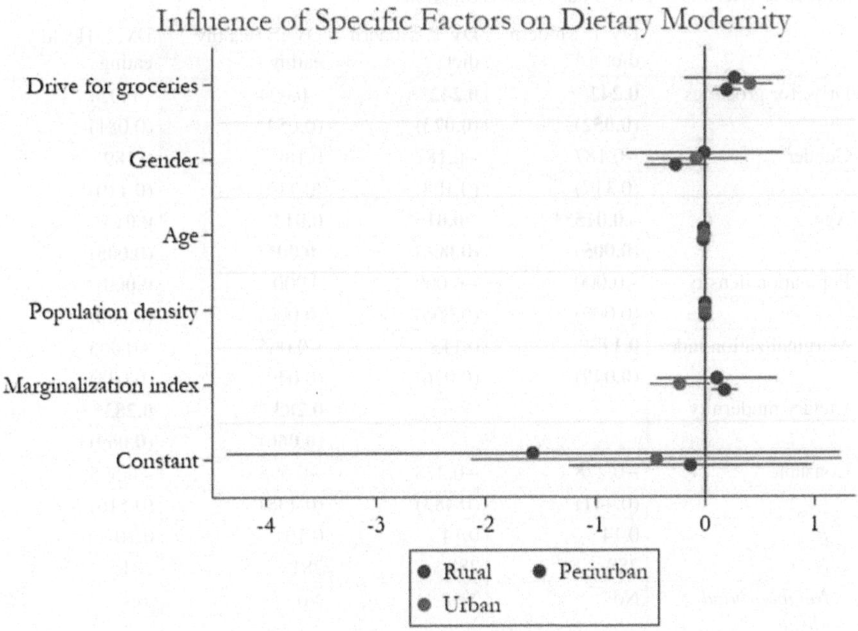

Fig. 20.4 Influence of Specific Factors on Dietary Modernity

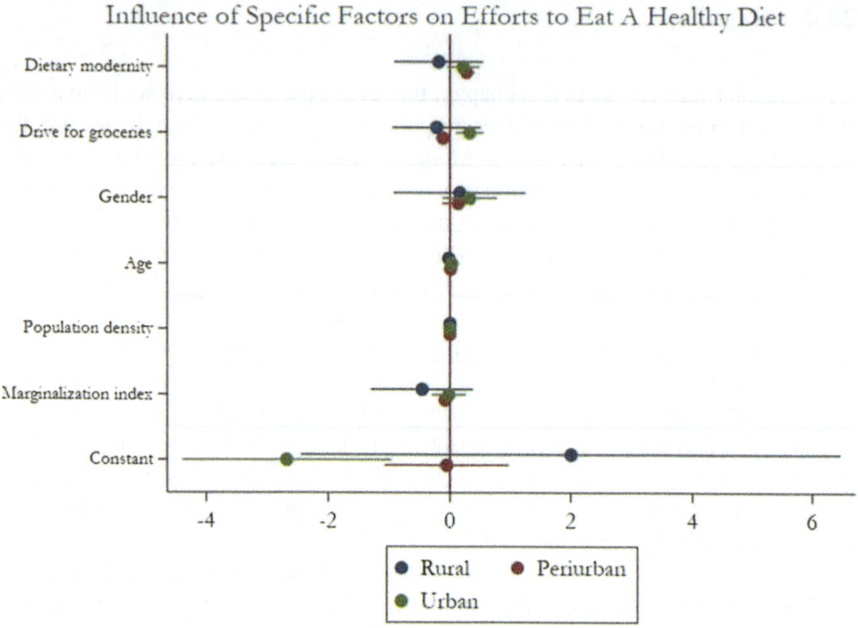

Fig. 20.5 Influence of specific factors on efforts to eat a healthy diet

Table 20.2 Results of preliminary OLS estimations

	DV 1: Modern diet	DV 1: Modern diet	DV 2: Healthy eating	DV 2: Healthy eating
Drive for groceries	**0.242****	**0.242****	−0.024	−0.024
	(0.052)	**(0.073)**	(0.054)	(0.081)
Gender	−0.187	−0.187	0.189	0.189
	(0.117)	(0.108)	(0.117)	(0.110)
Age	**−0.015****	**−0.015***	**0.013****	**0.013***
	(0.005)	**(0.006)**	**(0.005)**	**(0.005)**
Population density	−0.000	−0.000	0.000	**0.000****
	(0.000)	(0.000)	(0.000)	**(0.000)**
Marginalization index	**0.113***	0.113	−0.065	−0.065
	(0.049)	(0.076)	(0.049)	(0.038)
Dietary modernity			**0.283****	**0.283****
			(0.060)	**(0.069)**
Constant	−0.278	−0.278	−0.595	−0.595
	(0.441)	(0.483)	(0.438)	(0.516)
R^2	0.14	0.14	0.10	0.10
N	288	288	281	281
Neighborhood cluster	No	Yes	No	Yes

*$p < 0.05$
**$p < 0.01$

versus peri-urban residents. Specifically, peri-urban residents who live in areas characterized by greater marginalization are more likely to eat a modern diet. Like their urban peers, peri-urban residents who drive for groceries are more likely to eat a modern diet, as well.

As Fig. 20.5 illustrates, the impact of specific factors on the effort of respondents to eat a healthy diet is nearly uniform across zones of urbanization with one key exception. When we control for dietary modernity, the remaining peri-urban respondents who drive for their groceries are less likely to pay attention to health factors or attempt to eat a healthy diet. However, peri-urban respondents who report eating a modern diet are also more likely to report that they attempt to eat a healthy diet.

Based on the graphed results of this pilot study, we find tentative initial support for all of our hypotheses. We stress, however, that the results of this study are an invitation to further and more systematic examination of the impact of peri-urban food environments on individual consumption behaviors and, consequently, health.

20.5 Discussion

The preliminary results of the survey analysis showed that reliance on automobile transport to procure food and marginalization are positively related to shifts toward a modern diet. In this context, the older generation and people eating modern diets are more likely to pay more attention to healthy food intake. Therefore, the people most at risk of eating unhealthily are peri-urban residents from marginalized neighborhoods who drive for groceries. Yet, many of these individuals also report trying to eat healthy. These analyses demonstrate tentative initial support for all three of the study's hypotheses.

The results of the preliminary analyses have significant theoretical and practical implications. From a theoretical perspective, our study adds to the existing literature on urban food systems and resilience by highlighting the need for an integrated context-specific approach for adaptive urban planning that can support community health and sustainable development. We highlight the importance of considering development patterns of peri-urban zones not only as a potential or constrain in providing healthy food for the nearby city, but also as important factors in providing healthy food environments for peri-urban dwellers. The practical significance of these results can be found in clarifying that automobile infrastructure and marginalization should be crucial aspects for city planners and governors interested in developing nutritious landscapes, especially in the urban periphery.

Pineda et al. (2021) associated the density of convenience stores in neighborhoods in Mexico with higher BMI and associated higher risk of unhealthy diet to lower- and middle-income neighborhoods. Our study supports the fact that the biggest shifts in dietary modernity can be found in lower- to middle-income neighborhoods. What our study adds to the discourse of food environments in Mexico is the spatial aspect of this phenomena linked to urban development patterns and a push to evaluate not only what exists in the urban periphery, but what does not. In specific, respondents in the urban periphery did not express a decreased desire or decreased effort to eat healthily. One potential intervention to improve health outcomes in peri-urban zones might, therefore, be the creation of additional *tianguis* in such areas.

Previous studies, such as Ávila-Sánchez (2011) and Chavez et al. (2020) touched upon the potential of peri-urban zones to provide fresh food to urban zones but did not evaluate the food environment in these zones in relation to urban conditions. The preliminary results of our study point to peri-urban areas as critical, yet under-studied, zones where the developing infrastructure increases access to processed food, thus changing dietary habits. Murphy (2018) found similar patterns in the suburbs in Australia and proposed that building up density and quality public transport in peripheral areas might reduce the risk of unhealthy diets in those zones. We agree with this assessment, but note that there is nuance to the development of infrastructure that should not be overlooked. While transportation infrastructure may allow easier egress from peri-urban zones for commuters, it may accompany greater reliance on processed foods and dining out. To avoid such impacts, simultaneous investment in local markets and agriculture could potentially be beneficial by offering residents nearby access and availability of healthy food sources and supporting local agricultural producers whose livelihoods vanish as urbanization increases. Previous research has argued that investments in automobile infrastructure nowadays must be contextual (Siemiatycki, 2005) and complementary to greater strategical mobility and economical schemes (Enright, 2013). We argue that it also must be aligned with public programs focused on nutrition and healthy food environments. Only through integrated planning approaches focused on both mobility and nutrition would be possible to form urban nutritious landscapes.

Pandey et al. (2020) found that additional to infrastructure, percent of working women and policies and institutions are also significant in food consumption in India. At this preliminary phase, our study did not include the indicator of working women, but did consider marginalization. The percentage of working women could be significant in future studies to assess the gender gap in development of food environments, and we suggest that future studies undertake a careful examination of gender's influence in peri-urban zones of differing cultural contexts.

The limitations of this study are the non-random convenience sampling and the cross-sectional data used, which does not permit causal inferences to be made. Nevertheless, the similarity of the sample's socio-demographic characteristics only emphasizes the differences that most likely exist in the general population. For example, if the peri-urban development context is significant for food environments in a sample with similar socioeconomic characteristics, it can be expected that the significance will increase in a larger general sample that would integrate more varied groups. As a first probe into the food environments in different urbanization contexts, this study sets the base for future research focused on peri-urban environments that is largely missing from the current discourse. We suggest the use of a random and representative panel survey design to clarify the dynamic changes in the food environments.

Finally, our study merely investigates the shifts in dietary modernity linked to urbanization and tries to reveal significant underlying mechanisms that affect those changes. However, public policy and governance is a crucial part of any effort to improve equity of food environments that affect community health. Thus, the take-away message of the study is that peri-urban space plays a significant role in determining efforts to eat healthy. Future planning and public policy should take into account the link between the produced urban space, socioeconomic processes, and the produced food environments and find mechanisms to integrate more local markets or *tianguis* in peri-urban areas since they can determine health outcomes in the long run.

Acknowledgments This research received no external funding.

References

Abarca-Gómez, L., et al. (2017). Worldwide trends in body-mass index, underweight, overweight, and obesity from 1975 to 2016: A pooled analysis of 2416 population-based measurement studies in 128·9 million children, adolescents, and adults. *The Lancet, 390*(10113), 2627–2642. https://doi.org/10.1016/S0140-6736(17)32129-3

Araghi, F. A. (1995). Global depeasantization, 1945–1990. *The Sociological Quarterly, 36*, 337–368.

Ávila-Sánchez, H. (2011). Socio-territorial changes in peri-urban food production spaces in Central Mexico. *Norois, 221*, 39–51.

Barrera-Cruz, A., Rodríguez-González, A., & Molina-Ayalaet, M. A. (2013). Escenario actual de la obesidad en México. *Revista Médica del Instituto Mexicano del Seguro Social, 51*(3), 292–299.

Black, C., Moon, G., & Baird, J. (2014). Dietary inequalities: What is the evidence for the effect of the neighbourhood food environment? *Health Place, 27*, 229–242. https://doi.org/10.1016/j.healthplace.2013.09.015. Epub 2013 Oct 17. PMID: 24200470; PMCID: PMC4948665.

Black, C., Ntani, G., Kenny, R., Tinati, T., Jarman, M., Lawrence, W., Barker, M., Inskip, H., Cooper, C., Moon, G., & Baird, J. (2012). Variety and quality of healthy foods differ according to neighbourhood deprivation. *Health Place, 18*(6), 1292–1299. https://doi.org/10.1016/j.healthplace.2012.09.003. Epub 2012 Sep 18. PMID: 23085202; PMCID: PMC3679513.

Bloom, D., & Canning, D. (2000). The health and wealth of nations. *Science, 287*(5456), 1207–1209.

Caspi, C. E., Sorensen, G., Subramanian, S. V., & Kawachi, I. (2012). The local food environment and diet: A systematic review. *Health & Place, 18*(5), 1172–1187.

Cena, H., & Calder, P. C. (2020). Defining a healthy diet: Evidence for the role of contemporary dietary patterns in health and disease. *Nutrients, 12*(2), 334. https://doi.org/10.3390/nu12020334

Commission on Social Determinants of Health. (2008). Closing the gap in a generation: Health equity through action on the social determinants of health. In *Final Report of the Commission on Social Determinants of Health*. World Health Organization.

Crawford, T. W., Jilcott Pitts, S. B., McGuirt, J. T., Keyserling, T. C., & Ammerman, A. S. (2014). Conceptualizing and comparing neighborhood and activity space measures for food environment research. *Health Place, 30*, 215–225. https://doi.org/10.1016/j.healthplace.2014.09.007. Epub 2014 Oct 10. PMID: 25306420; PMCID: PMC4253906.

Enright, T. E. (2013). Mass transportation in the Neoliberal City: The mobilizing myths of the Grand Paris Express. *Environment and Planning A: Economy and Space, 45*(4), 797–813. https://doi.org/10.1068/a459

Ezzati, M., Hoorn, S. V., Lawes, C., et al. (2005). Rethinking the "Diseases of Affluence" paradigm: Global patterns of nutritional risks in relation to economic development. *PLoS Medicine, 2*(5), e148.

Felber, J. P., & Golay, A. (2002). Pathways from obesity to diabetes. *International Journal of Obesity, 26*, S39–S45. https://doi.org/10.1038/sj.ijo.0802126

Feng, J., Glass, T. A., Curriero, F. C., Stewart, W. F., & Schwartz, B. S. (2010). The built environment and obesity: A systematic review of the epidemiologic evidence. *Health & Place, 16*(2), 175–190.

Friel, S., Akerman, M., Hancock, T., Kumaresan, J., Marmot, M., Melin, T., Vlahov, D., & GRNUHE Members. (2011). Addressing the social and environmental determinants of urban health equity: Evidence for action and a research agenda. *Journal of Urban Health, 88*(5), 860–874. https://doi.org/10.1007/s11524-011-9606-1. PMID: 21877255; PMCID: PMC3191214.

Friel, S., Chopra, M., & Satcher, D. (2007). Unequal weight: Equity oriented policy responses to the global obesity epidemic. *BMJ, 335*(7632), 1241–1243.

Gálvez-Espinoza, P., Iglesias-Vejar, L., Vizcarra-Catalán, M., Palomino, A. M., Valencia, A., & Schwingel, A. (2018). "Comemos así porque…". Entendiendo las conductas alimentarias de mujeres chilenas de bajos recursos. *Revista Médica de Chile, 146*(8), 882–889. https://doi.org/10.4067/s0034-98872018000800882

Garreau, J. (1991). *Edge city: Life on the new Frontier*. Doubleday.

Giskes, K., vanLenthe, F., Avendano-Pabon, M., & Brug, J. (2011). A systematic review of environmental factors and obesogenic dietary intakes among adults: are we getting closer to understanding obesogenic environments? *Obesity Reviews, 12*(501), e95–e106.

Glanz, K., Johnson, L., Yaroch, A. L., Phillips, M., Ayala, G. X., & Davis, E. L. (2016). Measures of retail food Store environments and sales: Review and implications for healthy eating initiatives. *Journal of Nutrition Education and Behavior, 48*(4), 280–288. (e281).

Holdsworth, M., & Landais, E. (2019). Urban food environments in Africa: Implications for policy and research. *Proceedings of the Nutrition Society, 78*(4), 513–525. https://doi.org/10.1017/S0029665118002938

Iaquinta, D. I., & Drescher, A. W. (2000.) *Defining Periurban: Rural-urban linkages and institutional connections*. A Paper Published by the Partnership Programme of the Food and Agriculture Organization of the United Nations (FAO). Retrieved 3 July 2021, from http://www.fao.org/3/x8050t/x8050t02.htm.

INEGI (Instituto Nacional de Estadística, Geografía e Informática, México). (2017). Encuesta Origen Destino en Hogares de la Zona Metropolitana del Valle de México (EOD). Retrieved 4 July 2021, from https://www.inegi.org.mx/programas/eod/2017/.

Kelly, B., Flood, V. M., & Yeatman, H. (2011). Measuring local food environments: an overview of available methods and measures. *Health & Place, 17*(6), 1284–1293.

Kirkpatrick, S. I., Reedy, J., Butler, E. N., et al. (2014). Dietary assessment in food environment research: A systematic review. *American Journal of Preventive Medicine, 46*(1), 94–102.

Lang, E. (2003). *Edgeless cities: Exploring the elusive metropolis*. Brookings Institution Press.

Lucan, S. C. (2015). Concerning limitations of food-environment research: A narrative review and commentary framed around obesity and diet-related diseases in youth. *Journal of the Academy of Nutrition and Dietetics, 115*(2), 205–212. https://doi.org/10.1016/j.jand.2014.08.019. Epub 2014 Oct 14.

Lytle, L. A., & Sokol, R. L. (2017). Measures of the food environment: A systematic review of the field, 2007–2015. *Health & Place, 44*, 18–34. https://doi.org/10.1016/j.healthplace.2016.12.007. Epub 2017 Jan 27.

Mendez, M., Monteiro, C., & Popkin, B. (2005). Overweight exceeds underweight among women in most developing countries. *The American Journal of Clinical Nutrition, 81*, 714–721.

Mendis, S., & Banerjee, A. (2007). Social inequities in cardiovascular health and cardiovascular disease outcomes. In *Background Paper of the Priority Public Health Conditions Knowledge Network of the WHO Commission on Social Determinants of Health*. World Health Organization.

Murgante, B., Las Casas, G. B., & Danese, M. (2007). The periurban city: Geostatistical methods for its definition. In M. Rumor, V. Coors, E. M. Fendel, & S. Zlatanova (Eds.), *Urban and regional data management: 2007*. CRC Press, Taylor & Francis.

Murphy, M., Badland, H., Jordan, H., Koohsari, M. J., & Giles-Corti, B. (2018). Local food environments, suburban development, and BMI: A mixed methods study. *International Journal of Environmental Research and Public Health, 15*(7), 1392. https://doi.org/10.3390/ijerph15071392

Pandey, B., Reba, M., Joshi, P. K., et al. (2020). Urbanization and food consumption in India. *Scientific Reports, 10*, 17241. https://doi.org/10.1038/s41598-020-73313-8

Phelps, N. A., & Wood, A. M. (2011). The new post-suburban politics? *Urban Studies, 48*(12), 2591–2610.

Pineda, E., Brunner, E. J., Llewellyn, C. H., Mindell, J. S. (2021). The retail food environment and its association with body mass index in Mexico. *International Journal of Obesity 45*(6), 1215–1228. https://doi.org/10.1038/s41366-021-00760-2

Pistollato, F., Iglesias, R. C., Ruiz, R., Aparicio, S., Crespo, J., Lopez, L. D., Manna, P. P., Giampieri, F., & Battino, M. (2018). Nutritional patterns associated with the maintenance of neurocognitive functions and the risk of dementia and Alzheimer's disease: A focus on human studies. *Pharmacological Research, 131*, 32–43. https://doi.org/10.1016/j.phrs.2018.03.012

Pitt, E., Gallegos, D., Comans, T., Cameron, C., & Thornton, L. (2017). Exploring the influence of local food environments on food behaviours: A systematic review of qualitative literature. *Public Health Nutrition, 20*(13), 2393–2405. https://doi.org/10.1017/S1368980017001069. Epub 2017 Jun 7.

Popkin, B. M. (2006). Global nutrition dynamics: the world is shifting rapidly toward a diet linked with noncommunicable diseases. *The American Journal of Clinical Nutrition, 84*(2), 289–298. https://doi.org/10.1093/ajcn/84.1.289

Popkin, B. M., Adair, L. S., & Ng, S. W. (2012). Global nutrition transition and the pandemic of obesity in developing countries. *Nutrition Reviews, 70*(1), 3–21. https://doi.org/10.1111/j.1753-4887.2011.00456.x. PMID: 22221213; PMCID: PMC3257829.

Preiss, K., Brennan, L., & Clarke, D. (2013). Obesity and depression. *Obesity Reviews, 14*, 906–918. https://doi.org/10.1111/obr.12052

Rosales Chavez, J., Garcia, L., Jehn, M., Pereira, M., & Bruening, M. (2020). Relationship between different levels of the Mexican food environment and dietary intake: A qualitative systematic review. *Public Health Nutrition, 23*(11), 1877–1888. https://doi.org/10.1017/S1368980019004294

Roser, M., Ortiz-Ospina, E., & Ritchie, H. (2013). Life expectancy. *Published online at OurWorldInData.org*. Retrieved from: https://ourworldindata.org/life-expectancy [Online Resource].

Ruiz Rincón, V., Martínez-Alier, J., & Mingorria, S. (2019). environmental conflicts related to urban expansion involving agrarian communities in central Mexico. *Sustainability, 11*(23), 6545. https://doi.org/10.3390/su11236545

Seravalle, G., & Grassi, G. (2017). Obesity and hypertension. *Pharmacological Research, 122*, 1–7. ISSN 1043-6618. https://doi.org/10.1016/j.phrs.2017.05.013

Siemiatycki, M. (2005). Beyond moving people: Excavating the motivations for investing in urban public transit infrastructure in Bilbao Spain. *European Planning Studies, 13*(1), 23–44. https://doi.org/10.1080/0965431042000312398

UN's Department of Economic and Social Affairs. (2018). 68% of the *world population projected to live in urban areas by 2050, says UN*, published online 16 May 2018. Retrieved July 1, 2021, from https://www.un.org/development/desa/en/news/population/2018-revision-of-world-urbanization-prospects.html..

Van de Poel, E., O'Donnell, O., & Doorslaer, E. V. (2007). Are urban children really healthier? Evidence from 47 developing countries. *Social Science & Medicine, 65*(10), 1986–2003.
Vandevijvere, S., Dominick, C., Devi, A., Swinburn, B., & International Network for Food and Obesity/Non-communicable Diseases Research, Monitoring and Action Support. (2015). The healthy food environment policy index: findings of an expert panel in New Zealand. *Bulletin of the World Health Organization, 93*(5), 294–302. https://doi.org/10.2471/BLT.14.145540
White, M., & Barquera, S. (2020). Mexico adopts food warning labels, why now? *Health Systems and Reform, 6*(1), e1752063.
World Health Organization. (2009). *Cities and public health crises: Report of the International Consultation 29–30 October 2008 Lyon, France*. World Health Organization.

Chapter 21
Drawn Across Borders

Stephen Mueller and Ersela Kripa

21.1 Background

21.1.1 An Invisible Risk

Dust is an unseen threat. Exposure to high levels of airborne particulate, or *dust*, contributes to increased incidence of many respiratory diseases and other adverse health impacts. In most situations—except the most extreme dust storms—it is difficult for individuals to determine the presence of hazardous levels of airborne dust without sophisticated monitoring equipment. To protect themselves from dangerous levels of exposure, inhabitants of arid and dusty regions must rely on official monitoring equipment to detect and report when air quality reaches hazardous levels. In most contexts, local governance and weather stations are capable of issuing timely and accurate air quality warnings to local populations. However, in multijurisdictional contexts—like those throughout the US–Mexico borderland—accurate monitoring and timely detection is problematized by the fragmentation of environmental air quality detection and reporting structures (Mueller, in press). In the borderland, hazardous levels of airborne dust elude detection, as countless particles travel on the wind, crossing multiple territories and boundaries.

The Paso del Norte border region faces particularly heightened challenges related to transboundary dust transfer. The highly populated region, consisting of the El Paso–Ciudad Juárez binational metroplex, is located within the transnational Chihuahuan Desert. This arid and urbanized landscape is characterized by low annual rainfall and a high degree of water scarcity. With these dry conditions, large desert areas and dry lake beds in the region routinely create dusty and sandy ground conditions. The geological composition of the region exhibits a particularly high

S. Mueller (✉) · E. Kripa
Texas Tech University, El Paso, TX, USA
e-mail: S.Mueller@ttu.edu; Ersela.Kripa@ttu.edu

E. Giorgi et al. (eds.), *Design for Vulnerable Communities*, The Urban Book Series,
https://doi.org/10.1007/978-3-030-96866-3_21

concentration of fine silica, a very light and fine material that can be easily carried by the wind. One major origin site for airborne particulate in the region, the Samalayuca Dune fields south of Ciudad Juárez, contains sand composed almost entirely of pure quartz—a crystalline mineral composed of silica. These climatological and geological factors conspire to create elevated levels of suspended airborne particulate throughout the year, and especially in the spring season, when large dust storms commonly sweep through densely urbanized population centers.

Anthropogenic impacts in the Paso del Norte have amplified the negative health impacts of dust transfer. The region's rapid industrialization since the 1965 establishment of duty-free zones in response to the North American Free Trade Agreement (NAFTA) has contributed to the pollution and contamination of the ground's surface. As these contaminants attach to dust particles and become airborne during dust storms, they create particularly dangerous air quality. Urbanization and infrastructural neglect further compound this issue, as unpaved roads and desertifying urban peripheries contribute to airborne particulate.

Social and economic characteristics of the region indicate the population is particularly vulnerable to the impacts of transboundary dust transfer. Low wages and lack of access to health care increases the likelihood that industry workers will contract and not receive the proper care for various respiratory ailments and other diseases related to dust exposure.

The cities of the region share similar concerns. El Paso has been ranked in the top twenty polluted cities for both ozone and particulate matter by the American Lung Association (ALA) (Price, 2016). Studies have shown that low air quality in the region contributes to increased levels of respiratory disease, including asthma, bronchitis, fibrosis, etc. A 2003 study published by the Commission for Environmental Cooperation of North America showed a correlation between increases in particulate matter (PM_{10}) in the air and ozone pollution due to commercial traffic and respiratory disease and death in infants in Ciudad Juárez (Romieu et al., 2003).

As climate change continues to impact the borderland, the presence of airborne dust in the region will continue to increase, causing significant health risks.

21.1.2 Problematics of Measuring Atmospheric Conditions

Measuring atmospheric conditions poses many statistical and operational challenges. The air quality monitoring stations on both sides of the border region cover large geographical areas and their calculations show regional averages, producing a single Air Quality Index (AQI) that can obscure significant differences in air quality. By averaging regional data, the readings might conceal hazardous air quality in high pollution locations. The regional average thus erroneously invents and perpetuates the idea of a seemingly unified regional airshed, despite the reality of distinct and sometimes wildly different local atmospheric phenomena within the Paso del Norte.

There are currently six air quality monitoring stations in the El Paso–Ciudad Juárez binational metroplex, which cover areas ranging between four and sixteen square miles. However, the urban density and topographic diversity of both cities produces hyperlocal conditions. The most underserved neighborhoods coincide with the highest amounts of unpaved roads, which contribute to dust pollution in the air.

Instead of increasing the spatial resolution of their sensor network to better capture these local differences, US federal regulating agencies are instead shuttering some stations. The U.S. Environmental Protection Agency (EPA) has new requirements for the collocation of multiple sensors on site, to assist in quality control, and is reducing the number of sites and thereby increasing the portion of the airshed each site represents (TCEQ, 2016).[1] The generalization of air quality exposes border communities to respiratory health risks and provides a false sense of security in the highly fluctuating air pollution densities.

21.1.3 Cross-Jurisdictional Data

Airborne particulate matter knows no boundaries. Airsheds cross national and state borders, demanding atmospheric jurisdictions that do not align with land-based jurisdictions. Transboundary and transnational constituencies share airsheds yet are underserved by fragmented regulatory atmospheric assessments (Kripa & Mueller, 2022, in press). Fragmented transboundary atmospheric jurisdictions, along with relatively low investment in urban-, neighborhood- and micro-scale atmospheric sensors in the US–Mexico border region, create blind spots and asymmetric data assessments that disproportionately impact border communities (Mueller, in press).

Air quality monitoring data is subject to the impacts of what we have termed the "data border," abrupt disruptions to transboundary spatial data, and barriers to data access (Kripa & Mueller, 2021). Access to complete and equivalent geospatial data across political borders is hampered by an array of challenges including: *informational challenges* (gathering relevant, complete, and current data at adequate resolution and comparable scales), *geophysical challenges* (locating data in continuous and comparable cross-border geographies), and *administrative challenges* (maintaining affordable and up-to-date data).

The divided atmosphere in the Paso del Norte region relies on selective monitoring and piecemeal reporting across its multiple atmospheric jurisdictions, effectively disenfranchising a number of significant microenvironments from contributing to air quality assessments. The monitoring network is sparse (Eades, 2018)[2] and the districts each monitor represents may not in fact be indicative of the full range of microenvironments impacting populations in the region. Loopholes in monitoring standards will render detrimental differences in air quality near pollution sources all but invisible. As long as polluters are adequately spatially distributed their effects

---

[1] In 2016, the Ojo de Agua station was "collocated to support exceptional events."

[2] "Air quality monitors on both sides of the border are spread thinly throughout the border region."

are absorbed and averaged out in the region's emissions reports. Unless polluters are spatially clustered, federal regulations do not require monitoring organizations to locate sensors nearby TCEQ (2015).

Further, Mexico and the USA measure and assess air quality and air pollution differently. Mexico does not supply data on fine particulate, while the USA does not supply data on Total Suspended Particulate Matter (TSP). The limited resources and political initiative in the region have plagued the network with technical challenges. Sensors in Mexico have failed to meet EPA standards in a series of audits, and reports can no longer be generated from a cross-border air quality website.

The diversity of monitoring technologies, languages, and databases for air quality monitoring make cross-jurisdictional analysis difficult in the region. Despite these challenges, a means for measuring and visualizing cross-border particulate flows is necessary.

21.1.4 Proprietary Data

Visualizations of particulate flow often require expensive, proprietary, or specialized equipment, software, and sensing technology. Data from monitoring stations and research facilities is fragmented—sequestered in private, academic, and governmental databases. This geopolitical *atomization of the atmosphere* suspends binary geographic divisions—between cities, states, regions, and nations—in a complex, volumetric, and intersecting assemblage of proprietary institutional information. Environmental and atmospheric management in the Paso Del Norte region must contend with mineralized administrative and physical barriers between two countries (the USA and Mexico), three states (New Mexico, Texas, and Chihuahua), and two cities (El Paso and Ciudad Juárez). Asymmetries in reporting frequencies, environmental standards, protocols, funding, and systems of measurement in the region yield divided and discontinuous datasets. Neighboring jurisdictions are rendered atmospherically independent, myopic, and protectionist.

21.1.5 Nephelometry

Nephelometry is a term borrowed from atmospheric science, which describes a method of measuring the concentration of airborne particulate by detecting the amount of light scattered within a volume. The method is an important component in conducting air quality and air pollution assessments. *Nephelometers*—the instruments that conduct this measurement—employ a light source and a light detector to assess the amount of scatter, but vary widely in their cost and the resulting level of resolution and utility of the data collected. Research-grade nephelometers capable of fine calibration to assess multiple sizes and types of known particulates and pollutants come at high costs, which limits the deployment of robust monitors to a small number of sites, typically operated and maintained by specialists in environmental monitoring agencies within state or federal government, or by academic institutions.

With technological advances in recent years, a number of *low-cost dust sensors* are now widely available and can provide some level of reliable particle concentration detection. Some low-cost sensors, for instance, are capable of distinguishing between different common particle sizes ($PM_{2.5}$ and PM_{10}) with known impacts on public health. Using a light scattering method similar to their high-cost equivalents, these sensors can detect and report particulate matter in the air, even at very low densities.

While any one low-cost sensor may not provide the accuracy and reliability of the cost-prohibitive stations in more official monitoring networks, data from a multitude of low-cost sensors can begin to demonstrate statistically significant patterns. A deployment of multiple low-cost sensors is also more likely to capture anomalous local conditions, registering areas of high exposure that larger networks with limited sites would not detect. By utilizing many small-scale sensors, we can collect data at a resolution that directly contradicts the perceived monolithic continuity of the airshed, revealing the airshed instead as a highly differentiated assemblage of wildly different dust exposure conditions. Atmosphere is radically more discontinuous than current representations allow. By deconstructing atmosphere into discrete observations, we can uncover the intense—and sometimes fatal—discrepancies and inequalities elided by prevalent techniques.

21.2 Pilot Project: Delta Fabrics

Building on these potentials, we set out to augment the perception of transboundary atmospheres on discrete sites by developing low-cost sensor arrays and near real-time visualizations which privilege the local condition and the singular observer over the statistical averaging of environmental data at the territorial scale.

The project was first developed in a pilot project in 2017 in a different border landscape—the Hong Kong–Shenzhen border in the Pearl River Delta region. The pilot project, entitled DELTA FABRICS, exposed dangerous differences in official, reported air quality measurements and the realities on-the-ground in one of the world's most rapidly advancing transboundary metropolitan regions.

In that borderland—as in the US–Mexico borderland—airborne dust is a transnational vector which crosses otherwise fixed economic and political boundaries. But with mainland China's notorious air pollution problems focusing attention on Beijing and Shanghai, serious air quality issues in the Pearl River Delta are often overlooked. The Chinese government limits access to air quality data and adheres to a less stringent standard for reporting on conditions of human health than other international bodies. The pilot project sought to initiate an open-source data harvesting and reporting mechanism to supply more complete and reliable air quality data to the citizens of the region.

To complete these goals, AGENCY (an architecture, urbanism, and advocacy practice directed by the authors) built an air particulate sensor and traveled to Hong Kong and Shenzhen to conduct on-the-ground measurements of fine airborne particulate ($PM_{2.5}$), known to be hazardous to human health and life. The project resulted in a series of visualizations and materializations transforming thousands of data points to expose the "deltas"—the changes or differences— between reported and measured data.

The prototype dust sensor used an Arduino microprocessor programmed to calculate air quality measurements from a small dust sensor component. The battery-operated assembly fit within a small nylon pouch, and collected data continuously, every 5 s, then stored the data on a microSD card for later retrieval (see Fig. 21.1).

Traveling with the sensor, we conducted daily "border walks" along the land border between Hong Kong and Shenzhen, and along the waterfront in Shenzhen (see Fig. 21.2). The walks traced the boundaries at the limits of each jurisdiction, in order to gather air quality information far from the official monitoring stations. While the dust sensor was activated, we could measure particulate matter density in air pollution.

Compared to the use of data from static monitoring stations, the walking method allowed us to reach a broader sample of urban environments, passing through many different types of neighborhoods and land uses. As we walked, we witnessed and recorded some perceptible differences in air quality, especially around vacant land, construction sites, and industrial zones prevalent in this peripheral urban landscape.

At the end of each walk, the sensor's data was downloaded and visualized, detailing the fluctuations in air pollution through the various neighborhoods and urban environments we had visited. The data was then geolocated by distributing the readings along sample points following the path of each day's walk, and then partitioned into categories (e.g., "good," "moderate," "unhealthy for sensitive groups," "unhealthy") according to the standards set by both US and Chinese sources (see Fig. 21.3). Since each country has different standards for these categories, two different sets of maps were constructed, revealing which neighborhoods would be included or excluded as sites of concern according to the different standards.

The data collected by our sensor was then compared to data reported by US and Chinese monitoring stations the same day, to better understand the disparities between local air quality experienced in each neighborhood and the regional averages captured at the official monitoring stations off-site. Many discrepancies were uncovered, with our sensor recording many values much higher than the reported values, especially in areas of the border walks where airborne dust and higher levels of atmospheric pollution were visible. These areas, typically near large construction sites, roadways, or industry, routinely recorded as "spikes" in the sensor data, indicating that intermittent bursts of high concentrations of atmospheric pollution were common.

Numeric air pollution data collected from the sensor was then translated into a topographic surface model and described in 2D animated drawings showing the contours of this newly imagined topography. We describe the topographic surface as "isopulvic" (from the Latin: "iso" meaning *same*, "pulvis" meaning *dust*), where

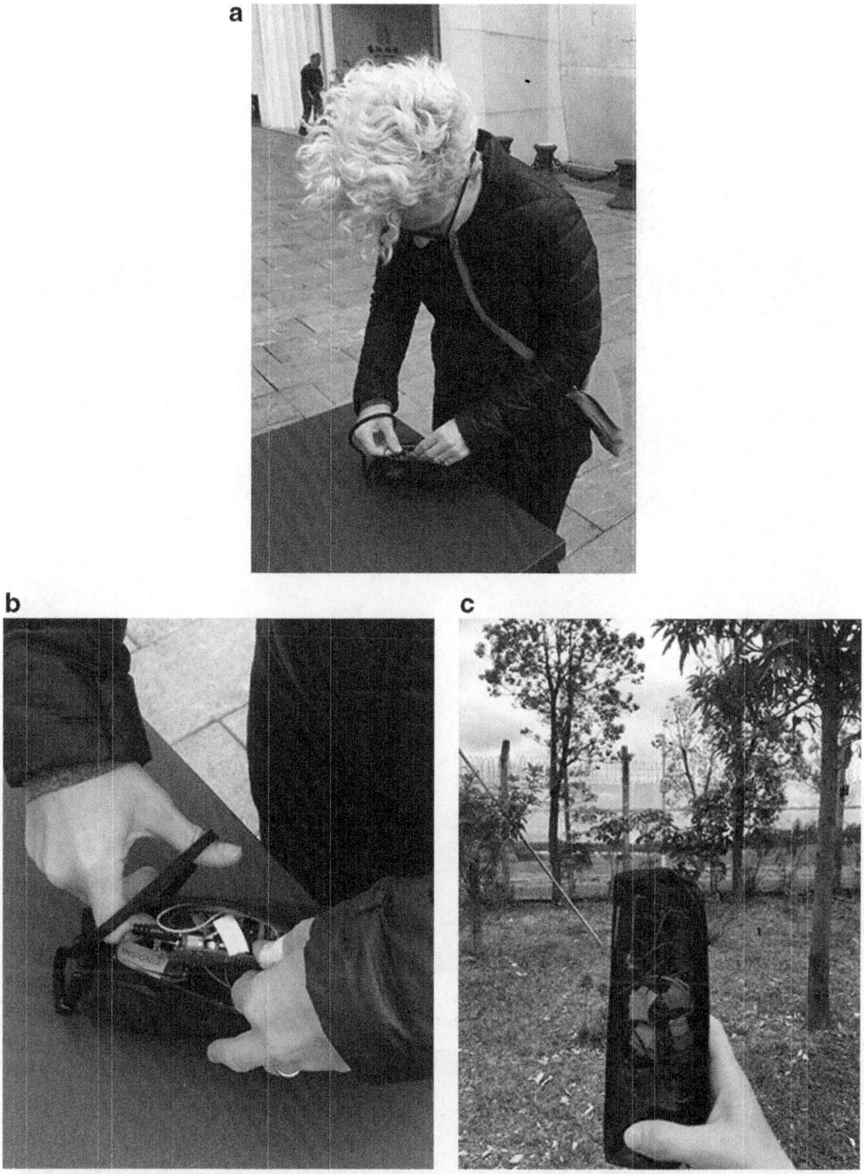

Fig. 21.1 Delta Fabrics—Prototype Dust Sensor. (**a**) Powering on the low-cost dust sensor at the start of a "border walk". (**b**) Detail view of low-cost dust sensor. (**c**) Low-cost dust sensor collective data on airborne particulate levels near Hong Kong–Shenzhen land border (Photos Courtesy AGENCY Architecture LLC)

each height in the model represents a consistent level of exposure to atmospheric particulates (see Fig. 21.4). By reading the *peaks* and *valleys* of this model, viewers can assess areas of high and low exposure, respectively. Steep *slopes* indicate

Fig. 21.2 Border Walks. (**a**) Dust origin site–vacant lot. (**b**)Dust origin site–cleared lot with construction activity(Photos Courtesy AGENCY Architecture LLC)

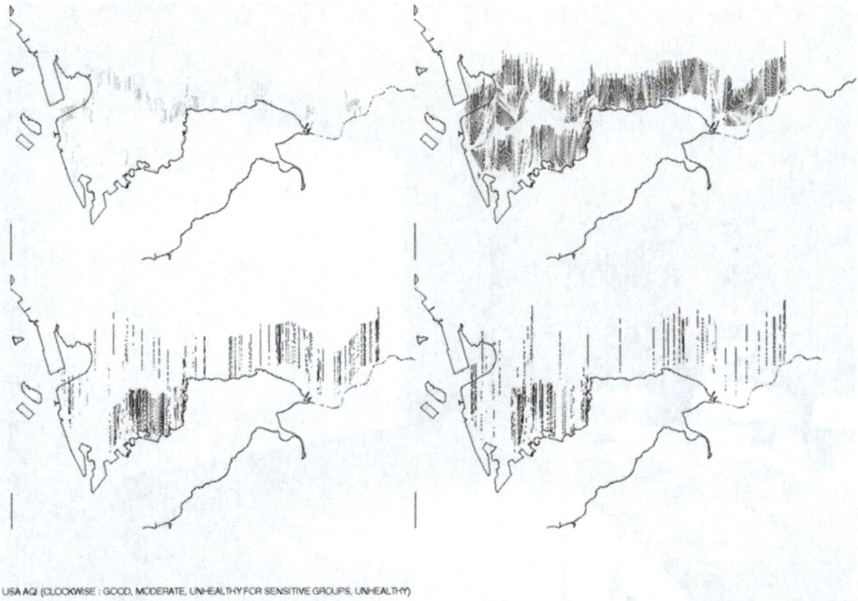

Fig. 21.3 Measured Air Quality Index (AQI) by U.S. Standards (Drawings Courtesy AGENCY Architecture LLC)

Fig. 21.4 Delta Fabrics—Isopulvic Topography 3D-Printed model describing intensities of air pollution along the Hong Kong–Shenzhen land border (Photo Courtesy AGENCY Architecture LLC)

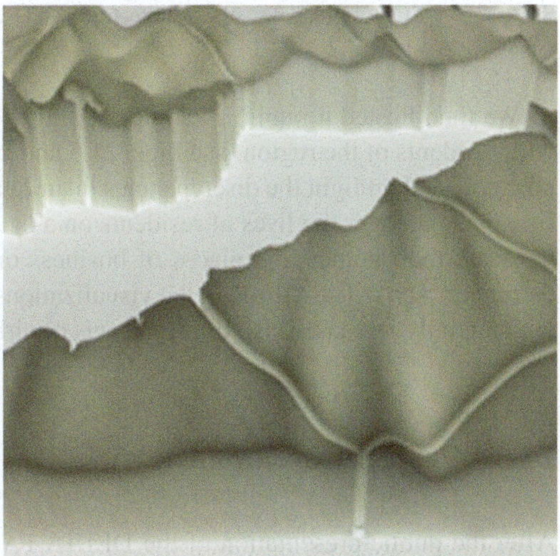

rapidly changing values between neighborhoods, signaling areas of asymmetric exposure and distribution of infrastructural resources. Broad areas of high slope indicate the neighborhoods most likely to be subject to prolonged and increased risk. In the isopulvic model, the land border dissolves, and new boundaries of atmospheric exposure emerge, registering a wholly different set of border conditions woven throughout the Hong Kong/Shenzhen region.

Fig. 21.5 Delta Fabrics—Animated Pollution Discrepancy Maps Presented at Hong Kong–Shenzhen Bi-City Biennale of Architecture and Urbanism (Photo Courtesy AGENCY Architecture LLC)

We then hosted a public awareness event to discuss the findings of the project with residents of the region (see Figs. 21.5 and 21.6). We used the models and visualizations to highlight the discrepancies in air quality reporting and air quality exposure that endanger the lives of residents on a daily basis. Many attendees were able to locate their homes and places of business on the maps and models, and some were surprised to learn through the visualizations that their daily environments were at potentially increased risk of high levels of airborne particulate.

21.3 Project Development

After the public presentation of the DELTA FABRICS pilot project, we sought to continue our development of the sensor and visualization design for application in our own environment—the El Paso–Ciudad Juárez metroplex. Without the time and travel constraints we faced during the pilot project, and with new partnerships with communities in the US–Mexico border region developing, we could develop broader and more inclusive, participatory strategies engaging multiple sites and stakeholders.

We were invited by the New Cities Future Ruins curatorial initiative in 2018 to develop a project for a group exhibit including other artists addressing

a
b

Fig. 21.6 (**a, b**) Delta Fabrics—Public Awareness Event Presented at Hong Kong–Shenzhen Bi-City Biennale of Architecture and Urbanism (Photo Courtesy AGENCY Architecture LLC)

environmental concerns in the US–Mexico borderland. The exhibition, entitled "Sections: New Cities Future Ruins at the Border," included work that addressed shared environmental concerns across three conceptual strata—earth, air, and water. With funding available for the exhibit from the National Endowment for the Arts (NEA), we were able to develop the functionality of our sensor prototype, create multiple sensor deployments, and develop more impactful means of communicating the results to a broader audience. We used the opportunity to develop a new project, increasing the capacity of the sensor design in both the geographic and temporal resolution of the data, as well in the immediacy of the feedback between the data collection and a geographically distributed public audience. Considering the results of the pilot project and these new potentials, we outlined a series of objectives for the next stage of development.

21.3.1 Increasing Spatial Resolution

First and foremost, the project was designed to increase the spatial resolution of air quality monitoring data available to the population of the El Paso–Ciudad Juárez metroplex. The pilot project had shown that local and microclimatic site conditions can significantly impact air quality, resulting in divergences between reported regional characteristics and the realities of particular sites. "Neighborhood" and "block-level" data is an invaluable tool to assess the asymmetrical distribution of air quality in the region, especially in neighborhoods at or near the international boundary.

21.3.2 Increasing Temporal Resolution

The project also endeavored to increase the frequency, or temporal resolution, of air quality data available to residents. Dust events and other airborne hazards can emerge in minutes. Air quality in the region can change drastically throughout the day, most notably in sudden "dust storms" or "haboobs" that pose great risk to respiratory health. Residents in the region might have access to daily air quality assessments via weather reports and apps. This AQI is often calculated as a daily (24-h) average of hourly readings. Expert analysts can typically access data averaged every hour or every 30 min from weather stations. A sensing system capable of recording granular temporal data—for example, by documenting changing conditions every minute or every second—is better equipped to detect changing conditions and sudden threats.

21.3.3 Providing Real-Time Visualization

In tandem with an increased tempo data gathering, the project sought to increase the rate and speed of delivery of data visualizations capable of communicating rapidly changing conditions to a wide public.

21.3.4 Increased Participation

Combatting the high-cost and exclusivity of governmental and institutional air quality monitoring services, the project proposed to increase public access to and participation in air quality monitoring efforts through the wide distribution of low-cost sensors, designed from readily available Arduino/Raspberry Pi components. The design of the sensor as affordable, low-maintenance, and easily programmable ensured a broader representation of the air quality of the region through increased participation of a diverse range of sites and stewards for the monitors.

21.3.5 Developing Toolsets for Desert Cities

The project was a design exercise to develop a toolset that renders visible specific urban and infrastructural issues shared by similar contexts globally. The process mined hyperlocal data towards a forensic material and geospatial investigation of the densifying desert cities on the US–Mexico border, and index shifting urban and geological territories, to forecast sites and strategies for improvement in this growing crisis.

With these goals in mind, we set out to prototype a low-cost, open-access, and cross-jurisdictional distributed air quality monitor sensing system capable of providing real-time data and fine-scale visualizations of transnational particulate flows.

21.4 Engaging a Binational Airshed

21.4.1 Paso del Norte (PdN) Airshed

The project was sited in, and sought to reconstitute an imaginary for, the Paso del Norte (PdN) airshed, a binational airshed including El Paso County (Texas), Doña Ana County (New Mexico), and Ciudad Juárez (Chihuahua, Mexico) (see Fig. 21.7). The airshed circumscribes a vast desert region with a history tied to the travel of

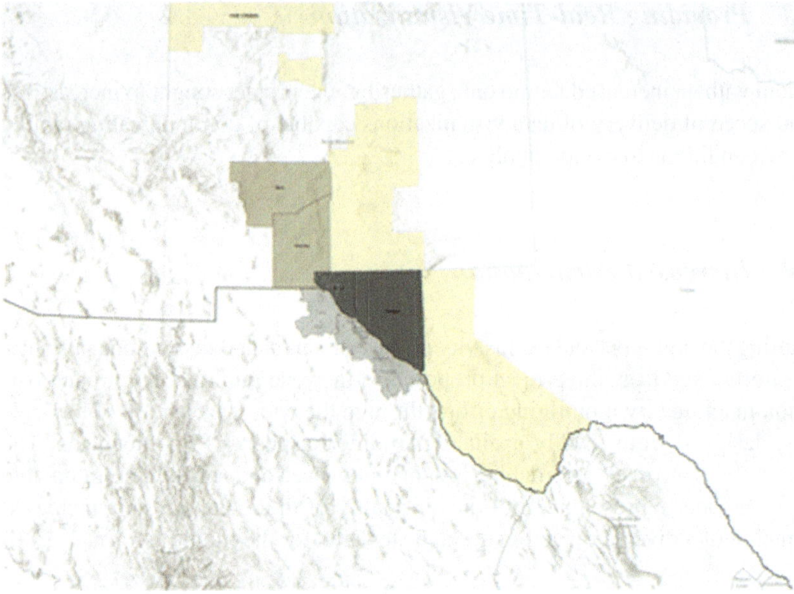

Fig. 21.7 Paso del Norte Airshed Showing Multiple Counties and Environmental Regulatory Jurisdictions (Courtesy POST-Project for Operative Spatial Technologies)

airborne particulate.[3,4] Within the Chihuahuan Desert—a vast binational geology and ecosystem—the airshed includes several "origin sites" for dust transfer throughout the region as well as several large metropolitan areas, including the populations of El Paso, Ciudad Juárez, and Las Cruces—a total over 2.5 million inhabitants share the air basin. The cities of El Paso and Ciudad Juárez are divided by the US–Mexico border, but share this transnational atmosphere, which conditions public health, national security, and international relations.

In this region of high volumes of dust transfer, dust clouds and hazy conditions often become trapped in the Rio Grande/Rio Bravo valley, hanging ominously between the Franklin Mountains in the USA and the Juárez Sierras in Mexico, a thickened atmosphere spanning the two cities across the border. This particulate-laden air is shared between the inhabitants of the two cities and the two countries. Particulate matter is often trapped close to the ground due to temperature inversions, which occur nocturnally. In the dry desert, at a considerable altitude of 3700 feet above sea level, the diurnal temperature swings in the summer can reach up to 70 °F, which create the stratification of cold and warm air producing the inversions. As

[3] The nearby Samalayuca Dunes have long been characterized as an atmospheric liability, its shifting dunes contributing to hazards on the ground and a major source for dust flow through the nearby inhabited regions. Samalayuca was noted as an especially dangerous area to cross along the historic Camino Real, resulting in many deaths.

[4] Oñate was reportedly the first to cross the dunes establishing a trail in 1598. Doniphan struggled to move through in 1847.

warm air traps cold air near the ground, humidity, dust and other particulates are trapped in a swirling cloud without a possibility of releasing into the atmosphere above. Patterns of urbanization (construction and demolition) and related ecological transformations (desertification, erosion) throughout the USA and Mexico can be mapped in this region by identifying the changing frequencies, constitutions, and flows of regional dust events.[5]

The Paso del Norte atmospheric sensing regime, like in many border regions, is underequipped to serve its population and fragmented due to the multiple jurisdictions the airshed spans. The World Air Quality Index Project (AQICN) shows six monitoring stations along the border gather and report hourly air quality data for the region. Texas and New Mexico environmental agencies (Texas Commission on Environmental Quality, TCEQ; and New Mexico Environment Department, NMED, respectively) operate stations on the US side. While it is typical for an environmental monitoring site to monitor regularly for $PM_{2.5}$ and PM_{10}, many of these stations do not. Six monitoring sites in Ciudad Juárez apparently share reporting regularly with TCEQ and five of these are managed by Mexican Secretariat of the Environment (SEMARNAP). Navigating this patchwork remains difficult. Many of the stations at the time of this writing showed missing or malfunctioning equipment or were missing current or complete data on their publicly available web services.

21.5 Nephelometry Project Overview

NEPHELOMETRY is a distributed network of dust sensors in Ciudad Juárez and El Paso, transcribing a shared aerial environment across the national boundary. The project documents the scattered territory of airborne particulate through a series of experiments in collective *drawing*. The procedural visualizations *draw* information gathered by individuals on both sides of the border in real time, while the lungs of the citizens *draw* in the dust of the territory. The project provides a model for transboundary environmental assessment and action capable of addressing issues of environmental justice in the borderland. The project engages the atomized atmosphere of the Paso del Norte region to forge new binational atmospheric jurisdictions and test representational strategies which can bridge administrative divides.

[5] Decades of urbanization have transformed the airshed, contributing new pollutants to its already substantial natural sources. A 2010–2013 study found that dust transmission in Ciudad Juárez is more a result of vehicles traveling on unpaved roads than wind-blown dust on open land, although both sources are significant (EPA, 2017).

21.5.1 Real-Time Visualization, Spatial and Temporal Resolution

The designed sensors provide real-time airborne particulate ($PM_{2.5}$) measurements and Air Quality Index (AQI) calculations to an online geospatial visualization hosted on a project website, allowing users and the public to gauge particulate flow and possible impact in real-time. Data from the sensors is uploaded, via the onboard Wi-Fi connection, to an online cloud server. From there, a series of project-specific algorithmic operations converts the received numeric and spatial data to updated graphic visualizations and mappings. Summary graphics and regional maps are updated near-real-time on the project website using an interactive GIS database service. The project enables a density of sensors capable of visualizing particulate flows at the neighborhood scale. The sensors are capable of logging and uploading data every second.

21.5.2 Public Sensing

The project proposes a low-cost sensor, designed from readily available Arduino and Raspberry Pi components. The measurements generated by the sensors are meant to be publicly available for use by researchers, advocacy organizations, and the public at large, who are encouraged to develop their own uses and applications of the platform. A series of discrete, low-cost dust sensors uncover the intense—and sometimes fatal—discrepancies and inequalities elided by the atmospheric administrative fragmentation and blind spots of current federal and binational protocols. Data from the sensors are used to make a number of static drawings and animated visualizations, rendering visible the spread and density of cross-border dust flows.

21.6 Nephelometry Project Sites

Sensors were installed in several urban communities and urban environments vulnerable to exposure to hazardous air quality on either side of the US–Mexico International Border Fence. Several sites were selected in and around El Paso (see Fig. 21.8) and Ciudad Juárez (see Fig. 21.9), representing a diversity of urban, geologic, and environmental conditions. The specific locations were selected to represent some of the environments most underserved or under-represented by current sensing infrastructure, including informally settled environments, where a combination of environmental neglect and rapid desertification produces high level of pollution and high variation in measured concentration of airborne particulate glossed over by regional sensing networks.

Fig. 21.8 Project Site in El Paso, Texas (Photo Courtesy AGENCY Architecture LLC)

Fig. 21.9 Project Sites in Ciudad Juárez, Chihuahua (Photos Courtesy AGENCY Architecture LLC)

21.6.1 El Paso

El Paso sites included a site near the downtown rail depot, steps from international border fence; a site near Smeltertown Cemetery, the site of decades of environmental pollution from the nearby Asarco smelter; a residential site near a dry riverbed (arroyo) below the foothills of the Franklin Mountains; a residential site on a mesa at the foothills of the Franklin Mountains, and a residential site in the Upper Valley, the floodplain of the Rio Grande.

21.6.2 Ciudad Juárez

Ciudad Juárez sites included residential sites near the international border and urban core, residential sites along major and minor trafficways, and residential sites near areas of development.

21.6.3 Anapra + Sunland Park

One project site was located in Anapra (Chihuahua, Mexico), a largely self-built community on the periphery of Ciudad Juárez characterized by unpaved roads. One project site was located near Sunland Park (New Mexico, USA), a low-income city on the opposite side of the Rio Bravo/Rio Grande from Anapra. The two communities are bound by a shared history of binational pollution and jurisdictional divides, near the site of the former Asarco plant, responsible for polluting both communities with airborne contaminants for decades (see Fig. 21.10).

21.6.4 Rio Bosque Wetlands Park

One project site was located in Rio Bosque Wetland Park, a riparian landscape just north of the international border to the extreme southeast of El Paso (see Fig. 21.11).

Sensors were placed at each site for continuous operation for 1 week. The success of the project relied on the participation of several community volunteers. Volunteer stewards living at or with direct access to the sites periodically assessed the equipment for proper functioning, verified connection to Wi-Fi, and communicated with our project team if any issues arose.

Fig. 21.10 Anapra (Chihuahua, Mexico) left and Sunland Park (New Mexico, USA) right crossed by the International Border Fence center right (Photo Courtesy AGENCY Architecture LLC)

Fig. 21.11 Nephelometry in Rio Bosque Wetlands Park (Photo Courtesy AGENCY Architecture LLC)

21.7 Nephelometry Project Details

21.7.1 Sensor Design

The project used a low-cost Grove Dust Sensor, which incorporates a photo diode to measure the relative opacity of environmental air passing through the sensor housing. The unit records the Low Pulse Occupancy (LPO) time, a value correlated to particle concentrations in the air—determining factors in determining Air Quality Index (AQI) values. The sensor is capable of differentiating two measurements, including particles greater than 1 μm ($PM_{1.0}$) and particles greater than 2.5 μm ($PM_{2.5}$), important benchmarks for air quality levels with different impacts on human health.

21.7.2 Sensor Bundle Design

The sensor passes its data to a Seeeduino Wio Tracker board, where it is both stored in an onboard microSD card and passed to a Grove Uart WiFi serial transceiver module. The transceiver communicates with local Wi-Fi, with additional capability to engage Bluetooth and cellular signals. The sensor bundle is powered by a small photovoltaic solar panel. Power from the solar panel passes through a LiPo Rider, which is capable of directing the charge directly to the bundle components in daylight, while also charging an onboard lithium polymer (LiPo) battery pack for continued operation in cloudy or dark conditions.

21.7.3 Sensor Housing Design

The sensors were housed within ubiquitous objects that inhabit the borderlands, effectively becoming invisible, in order to limit bystander curiosity and tampering. Two such objects were selected, which were hollowed and "hacked": security cameras and small electrical junction boxes.

21.7.3.1 Security Camera Housings

Empty security camera housings were "hacked" in order to hold the sensor circuit bundle. In order to fit the interior packaging precisely within the "found object," the security camera housings were first precisely measured and digitally modeled. Within the digital model, custom, flexible, packaging was designed as a series of interlocking parts, each one formed specifically to hold elements of the six sensor components. The interlocking modules, when assembled, tightly fit into the camera shell (see Fig. 21.12).

The parts were 3D-printed using flexible filament, allowing for slight compression against the sides of the housing, and cushioning against vibrational damage. The modules serve an aerodynamic function, facilitating air movement in from the front opening of the housing past the dust sensor. Several prototypes of the modular design were made to test and refine the proper airflow and functioning of the sensor, directing air and particulate matter efficiently in high- and low-wind scenarios. The layout of the modules and location of the sensor within the housing was also calibrated to reduce the impact of sunlight on sensor readings. The housing design navigates the relationship between the required form of the infrastructure of surveillance and the functionality of a measuring device, in order to reverse the direction of surveillance and monitoring in the border region. Once the design was finalized, the 3D-printed components were mass-produced. Multiple housings were emptied and outfitted with the custom-designed flexible inserts and sensor bundles (see Fig. 21.13).

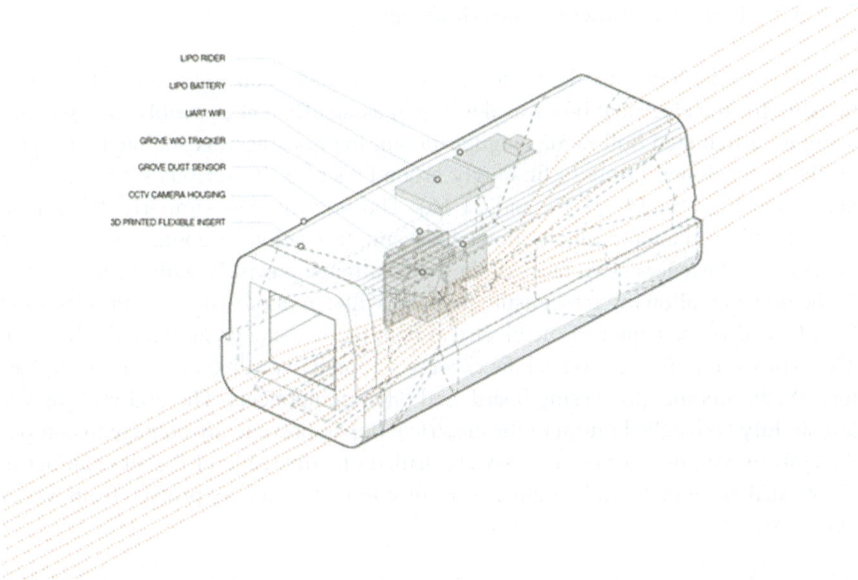

Fig. 21.12 Nephelometry—Sensor Bundle in Camera Housing (Drawing Courtesy AGENCY Architecture LLC)

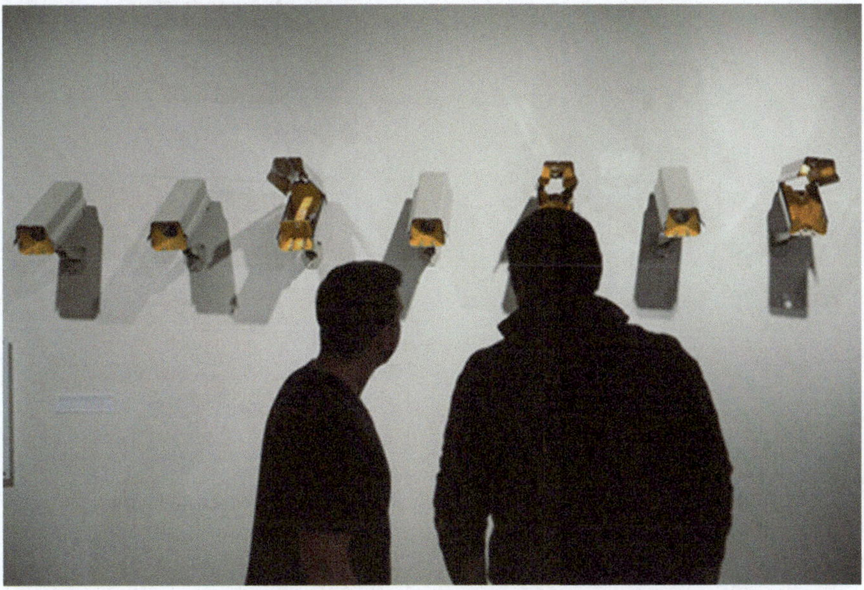

Fig. 21.13 Security Sensor Housing Exhibited at Rubin Center for the Visual Arts (Photo courtesy Rubin Center for the Visual Arts)

21.7.3.2 Electrical Junction Box Housings

Building on the logics explored in the security camera housing design, the sensor housing for the electrical box was developed as an efficient assembly. Designed as a single, articulated hollow rectangular prism, the housing was sculpted with precisely dimensioned insets to fit snugly around each of the various sensor components (see Figs. 21.14, 21.15, and 21.16). The housing was similarly 3D-printed using flexible filament, which allowed for simple friction-fit connections without the need for fasteners. Two faces of the prism aligned precisely with the inside face of the housing, allowing the housing to insert with slight pressure into the electrical box. Placed in the center of the electrical box, the housing created an air chamber. The exposed top, front, and back faces were sculpted to receive the dust sensor, battery, Wi-Fi module, processing board, and other components. The bottom face was left slightly above the bottom of the electrical box to prevent any damage to components from windblown rain. Holes were drilled on either side of the air chamber to allow airflow, and flexible tubing was inserted in order to protect from water exposure.

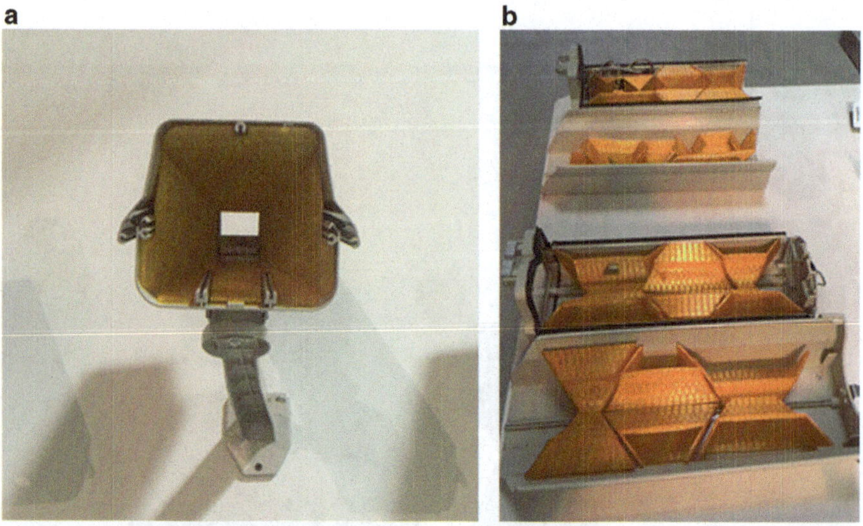

Fig. 21.14 (**a**) Front View, assembled (**b**) Top View showing 3D-printed sensor housing and placement of sensors

Fig. 21.15 Electrical Junction Box Housing (Prototype)(Photos Courtesy AGENCY Architecture LLC)

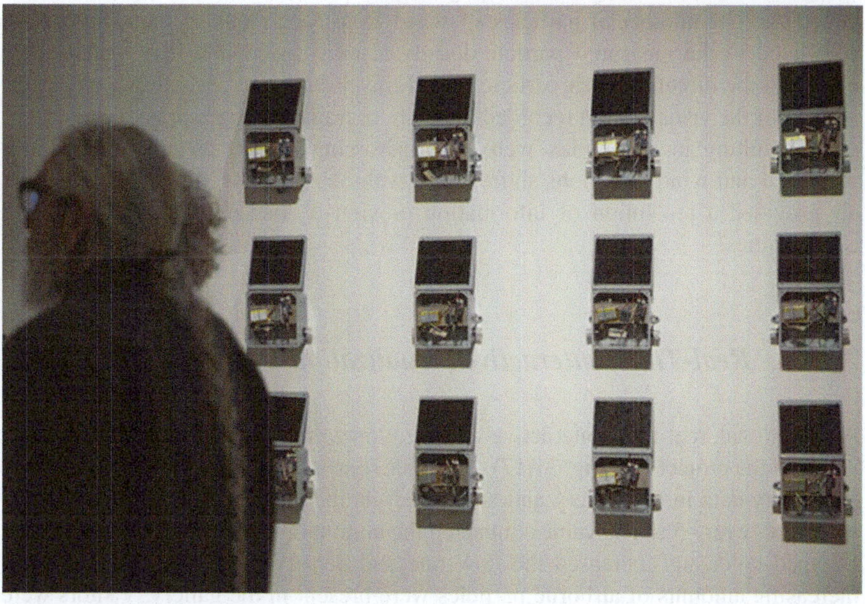

Fig. 21.16 Electrical Junction Box Housings Exhibited at Rubin Center for the Visual Arts (Photo courtesy Rubin Center for the Visual Arts)

By inhabiting the peripheral manifestations of surveillance and measuring infrastructures, the project quietly takes over and redirects information flows in understated but democratic ways—towards the affected individuals who inhale the dust.

21.7.4 Sensor Network Design

The project enlists individuals to measure and report the air quality of their backyards, consequently participating in a larger collective drawing activity where each hyperlocal environment forms part of the larger atmospheric measurement.

21.7.5 Online Visualization Design

The combined array of sensors provided real-time airborne particulate ($PM_{2.5}$) measurements and Air Quality Index (AQI) calculations to an online geospatial visualization webpage hosted on a project website, allowing users and the public to gauge the intensity and direction of particulate flow in real-time. Data from the sensors was uploaded, via the onboard Wi-Fi connection, to an online cloud server, and used as inputs for a series of particle cloud simulations. Each site of the sensor network was geolocated on a navigable 3D topography, and hosted a simulated particle cloud generator. The number of particles spawned from each cloud in the visualization correlated to the measured particle density at each site, while live weather data impacted the direction each particle would travel during its virtual life span. One version of the visualization recorded the path, or trail of each particle during its life span, resulting in a complex web of crisscrossing vectors as particle densities increased and wind directions shifted. The web-based live-feed map of hazardous air provided a resolution of information previously inaccessible to the affected communities.

21.7.6 Real-Time, Interactive Visualization Design

An additional, real-time, interactive visualization was designed for the gallery exhibition of the project (see Fig. 21.17). A sample sensor bundle in the gallery collected air quality data in the gallery and connected via the gallery Wi-Fi to upload measurements every 5 s. The values impacted the animation of a particle stream, which changed color and increased the flow rate and density of a particle cloud when increasing amounts of airborne particles were present in the gallery. Visitors were invited to interact with the sensor by disturbing a nearby soil sample, blowing into the sensor, or lighting a match nearby.

Fig. 21.17 Nephelometry Online Visualization Exhibited at Rubin Center for the Visual Arts (Photo courtesy Rubin Center for the Visual Arts)

21.7.7 Transboundary Drawings

A series of final transboundary drawings registered the ways in which each of the sensor sites was connected to both origin sites for dust transfer, and other locations throughout the binational metroplex. Interpolating and then drawing the path of a particle emitted from a single sensor, using agent-based modeling and atmospheric data, we were able to project the likely travel of contaminants registered by the sensors. This connected each sensor site visually with other areas and sites throughout the city it was "drawn" to, describing the interconnectivity of several sites forged by the routes of these unseen particles. Similarly, we were able to project backwards, to identify the likely origins of the contaminants we sent. Sensor stewards gained a new perspective on the likely sources and composure of the airborne pollution entering their homes and bodies.

Insisting on the autonomy of each sensor, each home, and each measurement to remap the territory, without "averaging" or smoothing the results, this "automatic" drawing process revealed emergent atmospheric phenomena elided by regional statistical models and summaries.

The collected data from each sensor was layered with wind pattern information and mapped onto a digital topographic model of El Paso and Ciudad Juárez. The information produced a live drawing that tracked dust travel patterns over time, much like a weather map, highlighting binational air quality shifts as a moving front between two nations across the territory (see Figs. 21.18 and 21.19). A thick and layered collection of dust trails swirled in the valley of the Juárez and Franklin

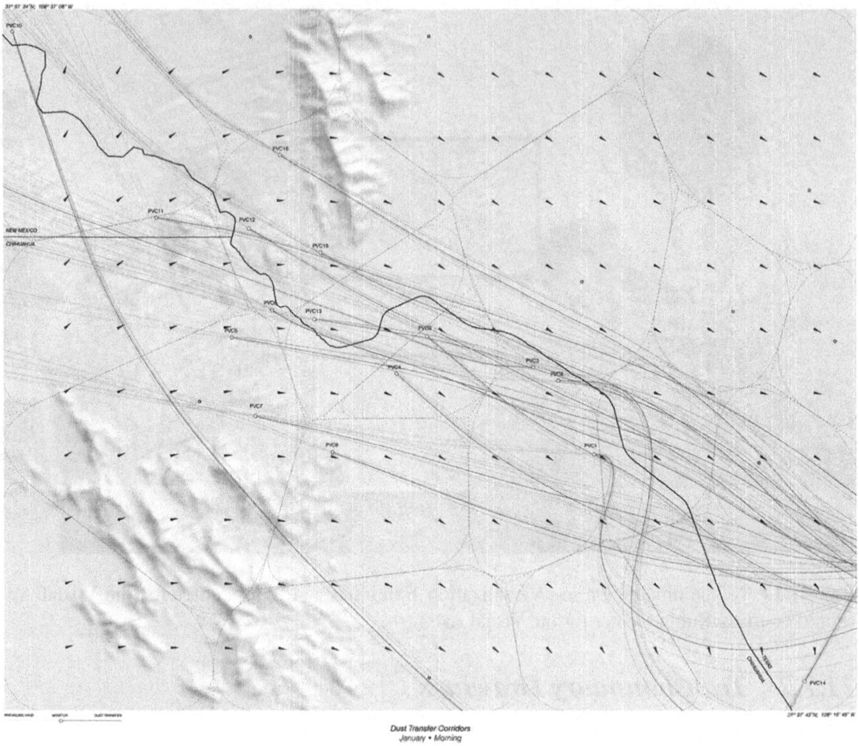

Fig. 21.18 Nephelometry Transboundary Drawing Dust Transfer Corridors, January, Morning (Copyright © AGENCY Architecture LLC)

mountains, leaving traces of multiple lines drawn in the air of the region, ignoring the international border, and creating a collective atmospheric marking of individual voices from each sensor location. The collective drawing highlights the need for a re-orientation of information from regional averages to hyperlocal measurements and empowers the singular observer's perspective. Through the drawing, each sensor's data contributes to a collective of individuals with common needs, goals, and desires.

21.8 Results and Discussion

The drawings render visible the spread and density of airborne particulate by recording its concentration hourly and uploading it to an accessible online platform. *Nephelometry* operates at local and global scales by tracking microscopic matter and highlighting its trajectories as indices of urban desertification and infrastructural neglect. It operates at the individual and collective scales, by imaging unique data within a larger tapestry of atmospheric multitude. The project seeks to combat

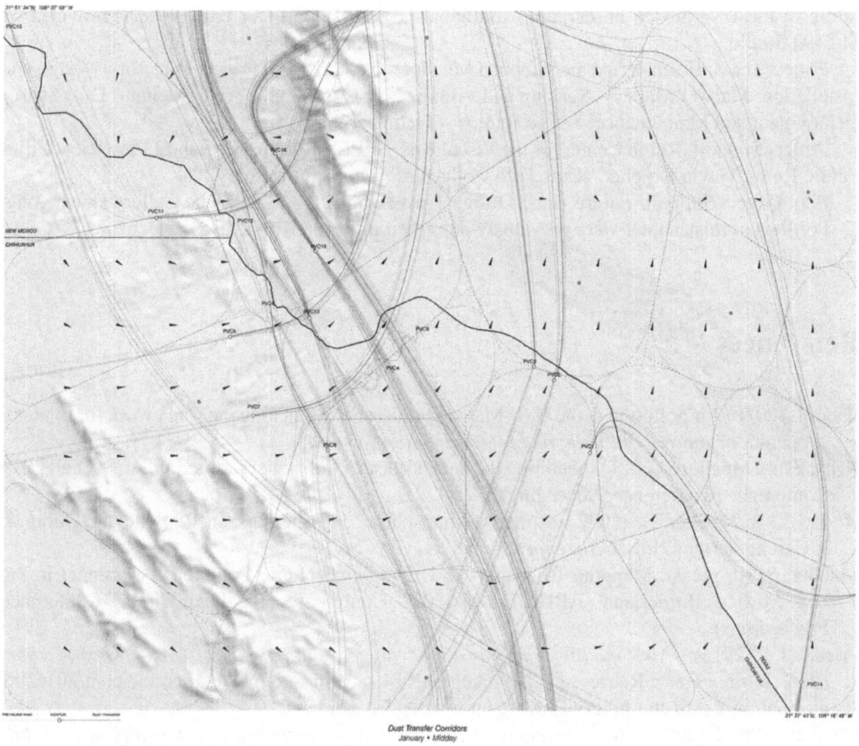

Fig. 21.19 Nephelometry Transboundary Drawing Dust Transfer Corridors, January, Midday (Copyright © AGENCY Architecture LLC)

the "coarse spatial and temporal resolution" of aeolian (wind-blown) dust data which limits the efforts to model, predict, and combat health risks of airborne particulate in the region. Instead of relying on networks that produce regional averages by intentionally avoiding pollution sources, the project conducted targeted sampling, intentionally seeking out sources of pollution to deploy sensors. While dust affects human health uniquely, the overall image of its trajectory emerges as a shared entity throughout the binational territory at large. The project imagines and images cross-jurisdictional airsheds extrapolating readings within microenvironments to understand the airspace across the US–Mexico border.

Acknowledgments The pilot project, DELTA FABRICS, was supported through a residency for New Cities Future Ruins, with Future+ Aformal Academy and Handshake 302, supported by Design Trust (an initiative of the Hong Kong Ambassadors of Design), as part of the Hong Kong-Shenzhen Bi-City Biennale of Urbanism/Architecture (UABB), 2018, and supported in part by Texas Tech University.

NEPHELOMETRY was funded in part through a National Endowment for the Arts (NEA) grant awarded to the Stanlee and Gerald Rubin Center for Visual Arts at the University of Texas - El Paso. The curator, Kerry Doyle, commissioned Ersela Kripa and Stephen Mueller (AGENCY), along with three other artists, to develop and install projects concerning the shared ecological future of the US–Mexico borderland. Additional project support (for pilot project) provided by SEEED Studio.

Project Team: Ersela Kripa and Stephen Mueller—Research and Project Directors, Design and Installation; Mary O'Malley—Sensing and Visualization Consultant; Irving Cuellar—Dust Sensor Deployment and Maintenance; Sergio Ramos—Technical Support.

Project Support: Rubin Center for the Visual Arts, SEEED Studio, Fab Lab El Paso, New Cities Future Ruins, Gavin Kroeber, Texas Tech College of Architecture (CoA).

Pilot Project Support: Future + and Design Trust Hong Kong, Handshake Gallery Hong Kong.

Portions of this chapter were previously published in Ersela Kripa, "Desert Densities" ACSA.

References

Eades L (2018) Air pollution at the U.S.-Meχico border: Strengthening the framework for bilateral cooperation. *Journal of Public and International Affairs 64.*

Kripa E and Mueller S (2021) Inhabiting the data border. In Less talk more action, ACSA Fall 2019 Conference [Conference Proceedings].

Kripa, E., & Mueller, S. (2022, in press). Atmospheric Jurisdictions. In R. Beneytez-Duran & J. Garcia-German (Eds.), *Airscapes*. Actar.

Mueller, S. (in press). Mapping Blindspots in Urban Atmospheric Pollution Assessment in the U.S.-Mexico Borderland. ARCC/EAAE 2022 Conference: Resilient Cities [Conference Proceedings].

Price, J. (2016) *Air quality one of the biggest threats on U.S., Mexico border. Borderzine.* Retrieved 29 Aug 2021, from https://borderzine.com/2016/11/air-quality-one-of-the-biggest-threats-on-u-s-mexico-border/.

Romieu, I., et al. (2003). *Health impacts of air pollution on morbidity and mortality among children of Ciudad Juárez, Chihuahua, Mexico.* Commission for Environmental Cooperation of North America, 10 Nov 2003.

Texas Commission on Environmental Quality (TCEQ). (2015). *Texas five-year ambient monitoring network assessment.* Retrieved 29 Aug 2021, from https://www.tceq.texas.gov/assets/public/compliance/monops/air/annual_review/historical/2015-5yrAAMNA.pdf.

Texas Commission on Environmental Quality (TCEQ). (2016). *2016 Annual Monitoring Network Plan.* Retrieved 29 Aug 2021, from https://www.tceq.texas.gov/assets/public/compliance/monops/air/annual_review/historical/2016-AMNP.pdf.

United States Environmental Protection Agency (EPA). (2017). *Highlights Report August 2016.* Retrieved 29 Aug 2021, from https://www.epa.gov/sites/production/files/2017-01/documents/b2020_highlights_report_2016_hq.pdf.

Chapter 22
Sustainable Energy Through Design: An Approach to Alleviate Energy Poverty in Vulnerable Communities on the US–Mexico Border Region

Viviana Barquero Diaz-Barriga and Ardeth Barnhart

22.1 Vulnerable Communities in the US–Mexico Border Region

The Intergovernmental Panel on Climate Change Chair's Vision Paper towards preparation of the 6th report (IPCC, 2017) has mandated a greater use of social science disciplines for gaining insight into issues related to our ability to adapt to climate change. This vision therefore acknowledges the importance of studying vulnerable communities from a broader perspective as a way to ensure the best evidence to inform policy-making processes. This can be done through the assessment of livelihood patterns and its influence toward capacity building. Large uncertainties on how regions will develop in the future, and how they can prioritize local actions, investments, and policy responses that are locally relevant, are true gaps in the sustainable development available knowledge (Seto et al., 2014).

The analysis offered in this chapter focuses on the Western portion of the US–Mexico border defined by Wilder et al. (2013), particularly in the Chihuahua Desert. This region is characterized by high stress and severe degradation of resources due to migration, desertification, high temperatures, and toxicity of land, water, and air resources (Wilder et al., 2013). This is mainly due to the social and economic development presented in recent years. Modern industrialization and the intensification of migration have pushed the formation and growth of communities which do not meet the necessary elements to satisfy the well-being of their inhabitants, provoking

V. B. Diaz-Barriga (✉)
Tecnologico de Monterrey, Escuela de Arquitectura, Arte y Diseño, Querétaro, Mexico
e-mail: viviana.barquero@tec.mx

A. Barnhart
Arlington, MA, USA

E. Giorgi et al. (eds.), *Design for Vulnerable Communities*, The Urban Book Series,
https://doi.org/10.1007/978-3-030-96866-3_22

societal vulnerability in both sides of the border, but increasingly in the Mexican side (Kopinak, 2004; Wilder et al., 2010; Lara-Valencia & Giner, 2013).

Vulnerability, as a concept, is widely linked to disaster and hazard studies since early 1940s and until the 1970s, dominating a hazard-centered perspective (Wisner, 2016). It was until the 1980s when some empirical studies showed that political and social conditions of hazard-affected communities contributed to the occurrence of disasters. The finding of societal causes was crucial to claim that environmental problems are socially constructed, where vulnerability is perceived as social weakness, considered the main cause of natural disasters (Dunlap, 2011). This approach to vulnerability allows to focus not only on the hazard perspective but on the related social phenomena as well (Quarantelli, 2000a, 2000b). This conception is important in the narrative of this chapter. We assume that vulnerability arises from the lack of environmental, social, and economic assets in order to build resilience and adaptive capacity to cope with climate change-related risks.

In this regard, we dispute that the lack of urban design projects, planning processes, and its relationship to promote efficient and sustainable use and production of energy within the region have direct consequences in its ability to build capacity. A number of works have explored vulnerabilities to climate change in urban settings through several approaches in recent years (Michael et al., 2019; Weber et al., 2015; Castan-Broto & Bulkeley, 2013; Ellena et al., 2020) but few have considered analyzing vulnerable conditions in the border from the lack of cohesive urban planning and energy access approaches. We therefore explore throughout this document whether sustainable urban energy design might respond efficiently towards adaptation in this particular region.

22.1.1 Prevalence of Vulnerability and Its Relation to Energy Poverty in Communities Across the US–Mexico Border

Communities located in the Western portion of the border (from San Diego-Tijuana to Paso del Norte Area) are located in the Sonoran and Chihuahua Deserts, considered drylands. Arid regions often lag far behind in well-being rates compared to the rest of the world basically due to their socioeconomic condition which, in the majority of cases, is related to its natural capital (Safriel & Adeel, 2005). Natural resources in drylands are severely affected by desertification, which is the result of many socioecological factors, often triggered by poor resource management operations (Reynolds et al., 2007). Electricity demand is likely to increase as a result of future cooling needs due to higher temperatures as well as higher water pumping demand due to less availability of shallow and surface water resources. Moreover, this increased electricity demand will peak more frequently, putting energy transmission and distribution infrastructure at risk (Frisvold et al., 2013; Tidwell et al., 2013; IPCC, 2014). Consequently, energy poverty emerges as a very important issue to consider not only in the context of basic energy access, but also in the economic and social cost that energy access represents to border communities.

As context, it is important to highlight that globally, energy poverty affects about 789 million people, from which 84% are in rural areas (IEA, 2020). In Mexico, 43.9% of the population is considered poor (CONEVAL, 2020) and 36.7% of households are considered below the energy poverty line (García-Ochoa & Graizbord, 2016). One characteristic of the US–Mexico border region is the high rate of poverty suffered by urban and rural inhabitants. A California "climate gap" study found that households in the lowest income bracket use more than twice the proportion of their total income on electricity than households in the highest income bracket (Shonkoff et al., 2009). In Texas, a study found that almost half of the families with extreme (0–30% Area Median Income) and very low (30–50% Area Median Income) income faced difficulties in paying electricity bills. Participants sacrificed mostly clothing and food to pay electricity bills, and in extreme circumstances, they cut transportation, medicine, and housing needs. In border counties, the energy burden rate ranges from 8% to 18% (Harmon & Moss, 2019). On the Mexican side, a comprehensive study analyzing the compliance of absolute energy needs per household (García-Ochoa & Graizbord, 2016) shows that US–Mexico border states have an average of 25% of households considered energy poor, being Chihuahua the border state with the highest index, with 35.8%. A significant part of households in the border region suffer from deprivation of the thermal comfort assets, considering essential due to its extreme weather conditions.

22.1.1.1 Energy Poverty

One significant issue regarding energy poverty is the lack of a consistent definition and thus a method for measuring it. Research on energy poverty started in the United Kingdom and other countries of the European Union representing a variety of issues and themes, including fuel poverty, access to electricity, clean energy financing for rural areas, household income expenditure in energy, household appliances, among others. Thus, the definitions surrounding energy poverty literature are constantly suited to fulfill the objectives of a specific study.

While today the most used definition is the one confirmed by the International Energy Agency which states that energy poverty is comprised of the lack of access to electricity and reliance on traditional biomass fuels for cooking (IEA, 2020; Jones, 2010). United Nations has a more specific definition arguing that a household is energy poor when it does not have modern cooking fuels and a bare minimum of electric lighting to read or for other household and productive activities (Gaye, 2007). In the UK, Dr. Boardman (1991) established that a household that spends more than 10% of the total household income on fuel should be considered in the fuel poverty range. This definition became the most widely used, mainly due to the introduction of a threshold that is relatively simple to measure. These definitions are aligned with the concept of absolute poverty (Sen, 1983), where a cutoff line (in this case, the access to modern energy) determines poverty. However, most European countries have agreed that energy poverty is more than just simple access to modern energy sources, but the capacity of adequate heat, or provide other energy services in their homes, at affordable cost (Pye et al., 2015). Healy (2004) developed an index, following the relative deprivation concept (Townsend, 1979) where he measured fuel poverty through objective and subjective indicators. The objective indicators had to do with household-specific conditions and equipment, while subjective indicators measured the feelings and concerns of energy needs and the satisfiers of individuals in the household.

The main objective of studying energy poverty is to notice that energy services (e.g., lighting, heating and cooling energy, cooking, energy for telecommunication and entertainment), although taken for granted in many societies, remain unavailable to a large share of the world's population. This is due to not only the lack of the energy resource but also the lack of households' spending capacity. This represents a major barrier to human development and the achievement of objectives set forth by the international community.

22.1.1.2 The Social Dimension of Energy

The relationship between energy and vulnerability is becoming an important issue to consider in the national and international agendas on development and public policy. Clean and affordable energy sources are important considerations for increasing quality of life, as well as reducing poverty, since energy is related to almost every single activity in daily life. Thus, the United Nations (UN) through the Sustainable Development Goals is encouraging nations "to ensure access to affordable, reliable, sustainable and modern energy for all" by 2030 (UN, 2015, p. 19).

The social dimension of energy can be understood from several perspectives: at the household and community level, seen from the individual consumer point of view; at the community or county level in which the utility company plays a big role in the decisions taken, although affecting directly to the end user; and finally from the federal government point of view, where in many regions subsidies and apparently cheap energy production are still the solution towards providing energy access to vulnerable regions. The cost of energy to maintain basic needs and a healthy lifestyle at the household level generates a significant economic burden (Harmon et al., 2017) and this burden is mostly carried by the end user. As stated before, some policies accept subsidize electricity and fuel as a solution towards increasing energy prices; however the poorest communities remain unable to pay full cost. Another important aspect regarding energy poverty is that a poor household is often unable to afford basic assets to fulfill their energy needs such refrigerator, electric fans, and in some extreme climate regions, cooling and heating systems (García-Ochoa & Graizbord, 2016; GNEB, 2016) and also comply with basic weathering for standard housing conditions.

In this regard, the work of Kozulj (2009) highlights three main findings that are applicable to the border region: (i) access to electricity to the poor is not a priority in public policy and planning programs; (ii) access to energy is focused on rural communities leaving aside the fact that 80% of the population lives in urban areas, so that poverty polygons have increased in recent years where access to energy is an issue; and (iii) poor populations spend a larger proportion of their income in energy services compared to other social classes.

22.1.1.3 The Justice Dimension of Energy

Energy Justice research aims to apply justice principles to energy policy, energy production and systems, energy consumption, energy activism, energy security, and climate change (Jenkins et al., 2016). Energy security assesses the security of supply and production, and emergent insecurities, such as availability of pricing, with a view to promoting the safeguarding of energy supply and indigenous production capabilities (Ang et al., 2015; Månsson et al., 2014). Energy justice evaluates where injustices emerge and which affected sections of society are ignored, as well as which processes exist for their recommendation in order to reveal and reduce such injustices (Jenkins et al., 2016). People receiving less than others portrays the unequal distribution and access to energy through both financial and physical terms. Justice has to do with the distribution of benefits as well as burdens.

There are at least three forms to analyze energy justice. According to Jenkins et al. (2016) we can find the distributional energy justice, which not only raises concern of the location of production facilities and its relationship with nearby communities, but also questions the distributional burden of energy prices and subsidies. Distributional justice can be manifested in the actual sitting of energy infrastructure unveiling the existing imbalances between supply and demand, and the lack of planning between energy intensity and localized production across communities and cities. Another approach is the recognition-based justice, which acknowledges misrepresentations or ignored sections of the society. This usually happens due to unfair location of power plants, or when certain communities struggle more to achieve energy needs, or people that is under the energy poverty line. A third energy justice outlook is the procedural justice, where decision-makers have sought to engage with communities without success. In this viewpoint, it is necessary to promote local knowledge mobilization, greater information disclosure, and better institutional representation in order to foster a sense of community ownership and empowerment.

Addressing energy justice in the border region is particularly challenging. Many of the human-rights abuses associated with extractive industries and the renewable energy supply chain are currently occurring within this particular region, causing conflict and representing a further obstacle to the energy transition. The socioeconomic context of the region is also an important aspect to consider in addressing energy justice, as it is one of the regions with higher levels of inequality. The adaptation of tailor-made energy transition policies that overcome this particular issue is one of the most relevant challenges this region faces. The 2013 energy reform in Mexico set in motion a model for massive implementation of renewable energy projects, mainly private and in large-scale. This made impossible to overcome the wide gaps in exclusion for socioeconomic and territorial implementation. The generation of energy from fossil fuels has violated human rights, provoking scenarios of exclusion and serious implications for Mexico's indigenous and rural communities. Likewise, many renewable energy infrastructure projects have generated socio-environmental conflicts (Acuña & Serrano, 2017). These relate to the lack of transparency and participation, violation of the rights of native peoples, lack of knowledge of traditional ownership and uses of land, obstacles to access natural resources, and environmental degradation (Wilder et al., 2016).

22.2 Community Physical Design and Its Role on Energy Access

Urban settings are normally comprised of a myriad of communities and neighborhoods where systems are connecting in all kinds of ways. Several studies have encountered issues and plausible solutions regarding sustainable access and uses of energy at the major urban scale (Arnfield, 2003; Seto et al., 2014; Nichols & Kockelman, 2015; Bai, 2016; Güneralp et al., 2017; Carreon & Worrel, 2018; Abbasabadi & Ashayeri, 2019). However, there is a significant lack of information when scaling these issues at the community or neighborhood range.

The 5th IPCC report in its chapter of human settlements (Seto et al., 2014) acknowledges that key drivers towards energy transition are density, land use mix, connectivity, and accessibility as interrelated factors. However, the majority of Urban Climate Action Plans are mainly focused on energy efficiency measures and not in specifics of land-use or other cross-sectoral measures that imply urban design strategies, either in the large urban scale or at the community-neighborhood level.

The physical form of a community has been analyzed from several perspectives, being public health one of the major assessment subjects (Frank et al., 2004; Handy et al., 2002), passing through its effects on land surface (Zhang et al., 2013; Guo et al., 2020) up to the optimal benefits on well-being and quality of life (Smith et al., 1997; Feng et al., 2018; Kwon et al., 2019). However, there are few studies that correlate physical or morphological design to energy access and use at the community level (Ratti et al., 2005; Rickwood et al., 2008; Ko, 2013; Roselt et al., 2015; Yang & Quan, 2016; Güneralp et al., 2017), among others, where the focus is on major urban form and density strategies and its relationship on the energy use in buildings.

As the literature review unveils, there is little pondering on the issues regarding energy use, energy access, and its relationship to the urban form. We address the fact that the growth-oriented resource and energy-intensive models promoted for urban development have created communities more vulnerable than ever. Neighborhoods and communities are being pushed towards a non-back situation where the excessive intention of using energy is segregating the communities that can afford access to energy from the ones that cannot.

Since 2002 the World Energy Outlook has devoted attention to the topic of energy access, analyzing and presenting a detailed status, particularly in developing countries. Modern energy access, although there is no single internationally accepted definition, can be described as "reliable and affordable access to both clean cooking facilities and to electricity, which is enough to supply a basic bundle of energy services initially, and then an increasing level of electricity over time to reach the regional average" (IEA, 2020). This definition, although can be used as benchmark, does exclude the more urban categories regarding operational, transport or other embodied energy constraints, that are crucial to economic and social development at the community level.

Increasing arguments about the importance of not only access to energy but also to sustainable energy sources are posing governments and planning officials to a myriad of scenarios. These scenarios imply the necessity of developing cleaner and more efficient sources in order to meet basic energy needs. This is relevant, particularly when thinking at the neighborhood or community level, as the transmission and distribution system is generally quite inefficient. The challenges for supply energy demand and secure it over the long-term, while being efficient and reducing energy-related pollution, are immense. Communities and cities must decide how much of their energy needs they should produce for themselves, how much to rely on imported energy, and how much energy can be saved through conservation and greater energy efficiency.

The planning, development, and deployment of communities must foster universal access to more sustainable energy forms. Designing for energy access is important for minimizing environmental impacts maintaining community health and safety, as well as reducing the access gap among inhabitants, empowering a vibrant community. Because most communities rely on energy from large production sources often far away from the end consumer, knowing how energy is accessed and used at the local level can be an effective way to promote greater energy security, increasing on-site or near-site capacity.

Framing our argument in terms of environmental justice, and more accurately of energy justice (Sovacool & Dworkin, 2014), it is fair to say that urban design and urban planning should focus on the future for our next generations, away from nuclear waste, depletion of fossil fuels, and the pollution of the atmosphere and climate. Therefore, our approach towards understanding how the physical form of communities might be related to sustainable energy access is seen in Fig. 22.1 as an analytical framework.

We acknowledge that urban forms are complex socioecological systems; however this framework focuses only at the physical dimensions of these forms. We understand that urban scale is always intertwined with the physical form at the community scale, and that these dimensions are pushing back and forth its relationship with the urban forms of energy, directly affecting system behaviors.

Understanding how the physical form of a community plays a significant role on the way its inhabitants have access to energy and use it, is crucial for building capacity towards adaptation and decarbonization planning processes.

For the purposes of this chapter, we defined three dimensions at the urban physical form: the natural environment, the built environment, and the socioeconomic environment. The natural environment relates to surroundings and conditions that are not human made, where living and non-living things exist on Earth. It is composed of ecological systems such as soils and vegetation, and of natural resources

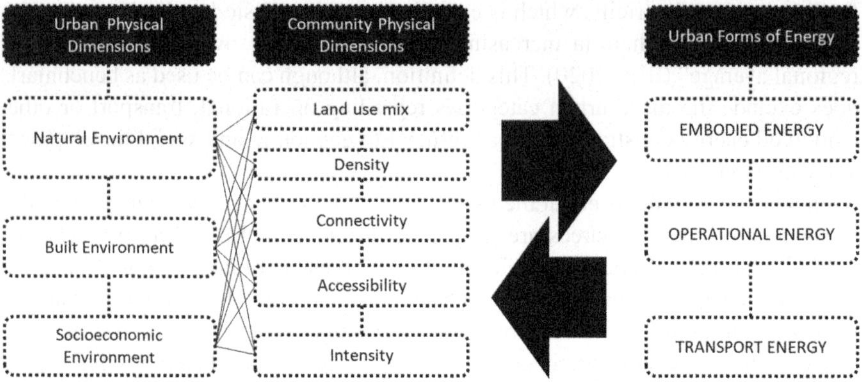

Fig. 22.1 Analytical Framework

such as air and water. The built environment is seen as areas that have been fundamentally transformed and influenced by human activity, such as cities, towns, and infrastructure. The socioeconomic environment is the layer of actions, services, and activities that humans perform either on the natural or on the built environment.

In our framework, the physical dimensions at the community level are characterized with different layers of information, being land use mix the first one as it helps us understand how is the community divided and functioning. The second layer is density, which refers to the number of people inhabiting in the same plot of land. Then, we consider connectivity as a way to recognize how direct and permeable the community is, and how efficient transportation network functions. Accessibility is the layer that includes all the features related with universal design, incorporating all members of the community at equal peace. Finally, we consider the intensity layer, referring to the concentration of commercial, services and other activities happening on the streets; the volume of spatial interactions that the community has to offer. Figure 22.2 shows a depicted analysis of main characteristics shared by vulnerable peri-urban border communities. This analysis is based on communities in the Chihuahua Desert, which share particular characteristics in terms of land use, density, connectivity, accessibility, and intensity.

The way we use energy in the urban environments is divided into three main categories: Embodied Energy, which is the one consumed in the manufacture and distribution of materials used in the construction and their associated infrastructure, as well as in operation and maintenance of structures and infrastructure; Operational Energy which refers to the energy consumed in the heating and cooling of the buildings, and in running the appliances used within them; and Transport Energy (both private and public) being the energy consumed in travels undertaken for both work and domestic or other recreational purposes. A series of parameters were proposed to measure in order to understand the use of urban energy in communities (Fig. 22.3).

These parameters were selected after reviewing several studies about the measurement of energy use (IAEA, 2020; Chen & Chen, 2015; IEA, 2020; among others). This approach is necessary for understanding how these factors relate to the

LAND USE MIXTURE	DENSITY	CONNECTIVITY	ACCESIBILITY	INTENSITY

LAND USE MIXTURE	DENSITY	CONNECTIVITY	ACCESIBILITY	INTENSITY
- Single land use - One-storey households - Near agricultural land - Near livestock farms	- Very Low density (~65 square meters per person) -Medium land use (~50% of land use) -Medium land cover (~70% of land cover)	- Permeability through the community with grid system -Connectivity to larger urban settlements only by car -Public transport routes that pass through the community, where existing, are insufficient	- Limited Access to primary education (up to 6th grade) - Access to local stores - Limited Access to workplace (in farming) - Access to public space below the WHO standards - Access to governance through municipality facilities available - Limited Access to walkable streets - No universal access capacity	- Limited access to formal recreational and sport activities - High interaction at street level and front porch, particularly among women and children

Fig. 22.2 Physical Dimensions of Typical West US-Mexico Border Region communities

OPERATIONAL ENERGY	TRANSPORT ENERGY	EMBODIED ENERGY
Annual amount of energy (fuel and electricity) consumed in heating and cooling buildings	Annual amount of energy consumed (fuel) traveling within the community ---- or ---- Annual amount of cash invested in public transportation for traveling within the community	Annual amount of energy consumed in the manufacturing of materials used in construction and maintenance of houses and other buildings within the community over that year.
Annual amount of energy (fuel and electricity) consumed in cooking	Annual amount of energy consumed (fuel) for traveling outside the community to get basic domestic needs (food, water, health, education, jobs,) ---- or ---- Annual amount of cash invested in public transportation for traveling outside the community to get basic domestic needs	Annual amount of energy consumed in the manufacture of materials used in paving streets and other infrastructure services provided to the community over that year.
Annual amount of energy(fuel and electricity) consumed in running appliances used in buildings to maintain average quality of life	Annual amount of energy consumed (fuel) for traveling outside the community for recreational purposes ---- or ---- Annual amount of cash invested in public transportation for traveling outside the community for recreational purposes	Annual amount of energy consumed in the distribution of materials used in construction and maintenance of houses and other buildings within the community over that year.
	Annual amount of energy consumed (fuel) for the suppliers that bring needs to the community (e.g. supply of goods and groceries to the local store, if any)	Annual amount of energy consumed in the distribution of materials used in paving streets and other infrastructure services provided to the community over that year.

Fig. 22.3 Proposed parameters for measuring Urban Energy Variables

way communities and neighborhoods are physically designed and built. The parameters for operational, transport, and embodied energy aim to inform the energy use within the community of study. It is important to highlight that all parameters remain general, as are not intended to become indicators at this stage, but rather an approximation for first empirical observations.

22.2.1 Energy Use and Community Physical Design

The physical form of a community directly affects how energy is consumed throughout the categories explained above. Vulnerable communities in arid regions, particularly in the US–Mexico West border region, have a historic harsh relationship with resources and its related infrastructure (Frisvold et al., 2013; Tidwell et al., 2013; IPCC, 2014). Physical characteristics of vulnerable communities in border region can be linked to the use of energy, directly affecting access to resources, services, materials, and infrastructure. To put this clear, vulnerable communities, particularly in the Mexican side, are likely suffering for energy poverty in great sense due to the way their communities are designed and built. Using as case study three peri-urban communities from Chihuahua Mexico, we were able to identify certain physical characteristics that push towards a greater use of energy, provoking difficulties for livelihoods to access to affordable energy resources, appliances, transportation, and to materials and other important infrastructure services that are usually provided by the local government or civil associations (refer to Box 22.1).

The most vulnerable communities are those who do not have access to quality energy infrastructure. We know that energy infrastructure is also dependent on road infrastructure, buildings, planning, and governance capacity, so it is important to understand these factors in order to generate design projects that are relevant and allow access to energy.

In the same way that poorly constructed or designed buildings hinder an adequate use of energy, reduce comfort levels, and increase operating costs, a lack of urban planning strategy related to energy use, efficiency, and clean energy affect a whole community. There are many publications available about different sustainable criteria that meet with sustainable physical planning and design, and the vast majority agree on particular carbon reduction strategies, which in many cases also help in creating a more energy efficient community.

Nevertheless, communities within the border remained absent of basic urban characteristics that are well known for better climate and energy performance, as seen in Fig. 22.2. One common strategy, known for millennia, is to take into consideration environmental conditions and build as a response to high temperatures, with regional materials. The microclimate that can be created in small communities by using vegetation, considering the building height, the street width, and the wall and roof orientation and color brightness, among other passive strategies, can be the most typical solutions towards a more efficient use and conservation of energy at the community level. These strategies must be considered for plans and regulations, in order to create a better, more responsive urban design.

Box 22.1: Facts About Energy Poverty in Three Vulnerable Communities of Chihuahua, Mexico

In 2015, three communities in Chihuahua were visited and studied in terms of energy poverty, being Miguel Hidalgo, Cuatro Vientos, and La Quemada, as are considered by Mexican Government evaluations to be living below the poverty line (CONEVAL, 2020).These communities share particular characteristics, from economic, social, and energy access perspectives. In order to understand the degree upon these communities were in terms of energy poverty, the Satisfaction of Absolute Energy Needs Methodology (García-Ochoa, 2014) was implemented. Results show that 86.5% of households are considered to be living in energy poverty, being the ones with the least diversified livelihoods more prone to experience higher energy poverty. In the majority of cases the households more dependent on local economic sources were considered the more energy poor.Findings show that in these communities, energy poverty is correlated to chronic illness in livelihoods, as a fail to achieve recommended comfort temperatures due to a lack of cooling or heating devices. Migraine and diabetes-related metabolic derangements were the most significant ailments across households, provoking a significant health issue for the community. Furthermore, it is considered that consistent absence to work (about 1.5 times per week compared to 0 in non-energy poor households) is related to illness, representing a severe inconvenience to both employer and employee (Barquero, 2016).

55% of energy poor households within these communities are lacking for efficient roofing (Barquero, 2016). Metal roofing is the most common material for these dwellings, and often does not provide a conditioned space, particularly if it is not insulated or weathered. An interesting observation among these communities and similar in Chihuahua, is that, because roofing is in so bad quality, local government almost every year generate campaigns to donate zinc sheets (which have a high thermal conductivity rate), among other provisional construction-type materials, as part of the Solidarity Support State Program (Chihuahua, 2021).

According to García-Ochoa (2014), the percentage of energy poverty in urban and peri-urban households of Mexico is higher than the one presented in rural zones. Although energy poverty in rural households is higher when compared to the total stratum, when compared to the total population sampled there are more urban households considered energy poor than there are rural. This is evident as 78% of total Mexican population is living in urban areas.

It is clear, therefore, that it is necessary to develop strategic planning approaches that eventually will transition communities towards climate change adaptation, social inclusion, energy autonomy, and economic development in order to approach energy transition in Mexico.

22.2.2 *Energy Poverty and Community Physical Design*

The authors assume that physical design plays a significant role in the way energy is accessed, consumed, and conserved. Communities within the border region have particular characteristics that are shared in both US and Mexico sides. As discussed throughout the chapter, communities in the border have particular physical aspects as depicted briefly in Fig. 22.3. The community physical dimensions presented in Fig. 22.1 can be evaluated with different parameters, and the assumption is that, the higher the parameter, the less energy poverty will be present in a community.

This theoretical approach on the relationship between the physical aspects of a community and its role towards decreasing energy poverty should encourage an energy planning strategy. A specific set of parameters needs to be developed so we can create particular settings for each form of energy used in these communities. It is important to consider that not every form of urban energy is evenly found or can be equitable measured on each physical dimension. We have considered the assumptions showed in Fig. 22.4, for further research and development of parameters and indicators, in order to prove our theory.

Figure 22.5 tries to identify where to find the elements of embodied, operational and transportation forms of energy according to each physical dimension. The parameters of measurement for each form of energy were presented on Fig. 22.3.

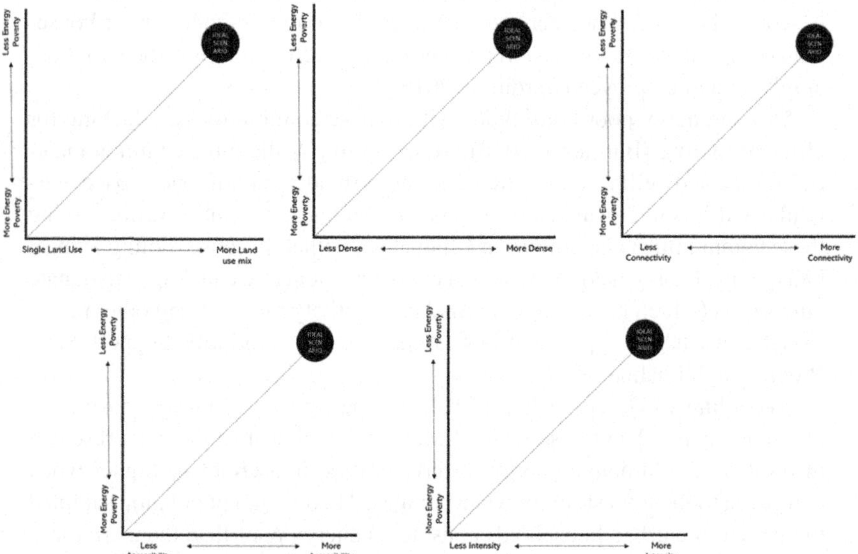

Fig. 22.4 Ideal scenarios for physical dimensions and their relationship with energy poverty

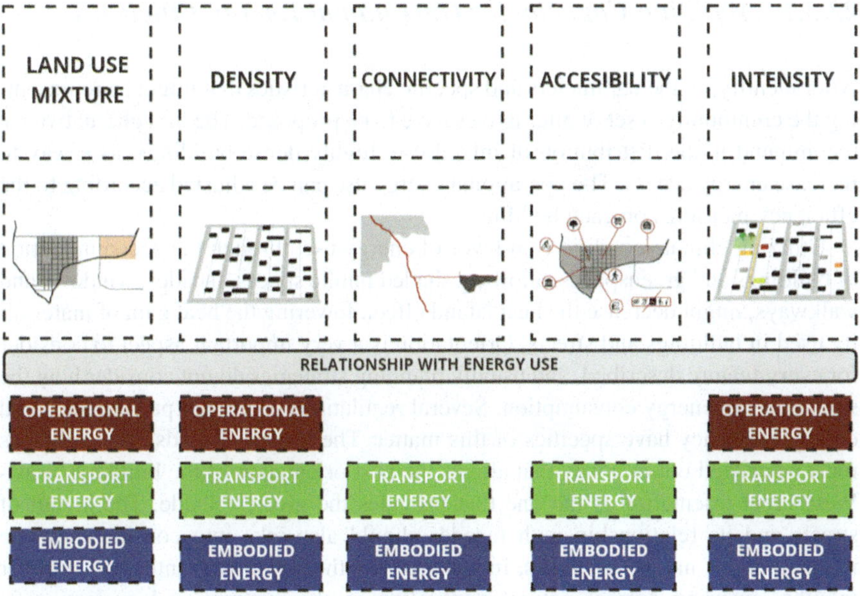

Fig. 22.5 Urban forms of energy and its theoretical relation with physical dimensions

22.3 Urban Capacity Building for Sustainable Energy Access as Adaptation to Climate Change Effects

Building capacity is one of the most challenging responsibilities any institution or organization can have. According to United Nations (UN, 2020) building capacity is "the process of developing and strengthening the skills, instincts, abilities, processes and resources that organizations and communities need to survive, adapt, and thrive in a fast-changing world." Taking into consideration this definition, the idea of building capacity for sustainable energy in communities across the US–Mexico border region is huge. Although, energy is considered a resource, the ability to have sustainable access to this resource forces to consider the skills, abilities, and processes that need to be found or created. Diverse organizations that have generated certain criteria for sustainable urban development might serve for our purpose, such as de Green Building Council (USGBC, 2021), Living Building Challenge (LBC, 2020), or even Mexican regulations (DGN, 2012). In this sense, we believe it is necessary to consider some specific strategies, focusing on capacity building for adaptation to climate change, promoting sustainable energy access.

The first thing to consider when building capacity is to establish minimum environmental and energy objectives specific for each community. A diagnosis must be developed, taking into consideration what we have discussed throughout this document, regarding the three types of urban energy use and its relationship with physical aspects of community design.

22.3.1 Building Capacity for Operational Energy Efficiency

After identifying the parameters and specific characteristics that might be representing the community, a set of alternatives have to be proposed. The first alternative we recommend is the distribution of mixed use, highly dense buildings, as a way to increase energy access. The operational energy use can be adjusted according to the efficiency measures on each building.

Utilizing climate conditions in favor of energy use, through the accommodation of buildings and greenspaces, allowing shaded public spaces, facades, windows and walkways, might decrease the heat island effect, lowering the heat gain of materiality used in buildings and streets. Orientation is a very important aspect to consider for every feature described, and usually planning strategies do not consider how the sun affects in energy consumption. Several regulations on energy performance and energy efficiency have specifics of this matter. The layout of roads and walkways must be carried out in a way that allows the best orientation of the buildings. East-West street orientation is the one that provides the greatest shade. The layout of streets and its relationship with facades should also take into consideration the maintaining of natural air flows, in order to use them in both winter and summer months, ensuring correct natural ventilation in the summer and protection in the winter.

Building Energy Performance is currently understood as dependent upon urban geometry, building design, systems efficiency, and occupant behavior. There are different actors controlling each factor, being urban planners and designers the ones that interact mainly at the urban geometry; architects in building design; systems engineers in the system efficiency; and users remain the responsible of occupant behaviors.

Plot configuration for future construction should always consider south facing facades not only considering non-regular patterns—this helps for solar heat gain and passive air conditioning systems. These particular features directly affect the operational energy use. To verify compliance, it is advisable to use a graphic study of shadows in floorplans. This might be a skill that can be taught to community leaders so they comply with their own operational energy use plan.

22.3.2 Building Capacity for Transport Energy Efficiency

This type of urban energy is related to the amount of energy used for transportation of people within the community and outside the community, but also implies transportation of goods and services in and out the community. In order to increase capacity for this aspect, it is necessary to understand that the majority of these communities are dormitory towns forcing a high commuting to larger cities or industries within the area. It is necessary to provide quality public transportation, shading roads for walking and biking distances, and interaction with the rural landscape that often exists in-between. Public transportation ideally should be hybrid or electric, and should be connected with other transportation sources. Community members are often forced to buy old trucks that are very contaminant, because there is no other transportation source. This brings other set of problems related to operation and maintenance of the car or truck, and the energy and time consumed per household.

Furthermore, many communities do not have markets or grocery stores to consume within the town. People have to go to bigger cities to provide basic needs in terms of food and goods, and other services. Traveling across towns is very common and generate a considerably expenditure to every household. The solution is to create urban alternatives that foment connectivity.

22.3.3 Building Capacity for Embodied Energy Efficiency

Reduce the heat island effect with building height, street width, and vegetation, maintaining natural air flows, greenways, and water elements to contribute to evaporative cooling. Use deciduous trees and wide crowns to protect the lower floors of homes in summer and allow access from the sun in winter. Incorporate heat absorbing surfaces, bright surfaces, and permeable surfaces. Consider maintain as pristine and natural the surfaces as possible. It is important to consider the amount of energy consumed in the manufacture and construction of paving streets. The use of traditional building methods, such as adobe and structural woodwork, should remain feasible, as it is less expensive and less energy intensive.

Strategies should incorporate local, nature-based solutions, which can serve as climate control elements. The institutions that help bringing construction materials and that develop streets and other urban infrastructure must consider a planning strategy that promotes the use of local materials and knowledge. Other type of embodied energy is related to manufacturing and distribution of materials for construction and maintenance. A possibility of generating a market of local materials that serve to produce their own homes, but also to explore the possibility of generating a business model based on green, sustainable materials can add value and build capacity not only into the energy conservation and efficiency perspective, but also in other dimensions.

Street lighting is considered into the embodied energy capacity building strategy, as it is from the public realm. It is important to consider energy saving technologies, but also a strategy based on small-scale renewable energy that might incorporate more efficient public lighting, allowing to have more and better lighting within the community. This also helps building capacity in terms of security measures. A cogeneration system for heating and cooling might be interesting to explore. Communities in the border in general have cold winters and hot summers, and few have the economic possibility to pay for artificial climatization. Not only having good physical design aspects outside their homes but also creating a system that can be shared by blocks or even larger amount of homes could be a solution for increasing capacity for the embodied energy uses.

22.3.4 Sustainable Energy Access as Adaptation to Climate Change Effects

It is clearer than ever the agreement among academics and policy makers in the fact that energy and climate change adaptation are deeply interconnected. The debate now focuses on how the energy sector would adapt to climate change impacts.

Sustainable energy can be achieved through design and planning processes focusing in three main steps: the conservation, the efficiency, and the production of energy at the community scale. Conservation efforts relate to passive strategies that allow to use as less energy as possible, promoting consumption reduction in both fuel and electricity sources. The transmission and distribution lines often waste electricity when transmitting from the source to the end point; therefore it is highly recommended to develop distributed forms of energy that are nearer to the end user. For the efficiency process, it is also recommended to manage community forms of energy, developing strategies that promote technology infrastructure that reduces considerably the consumption of kilowatt hours. Typical strategies are focused on LED lighting, day-control sensors, and efficient appliances for both public and private spaces. Finally, getting on-site clean energy production, is the ideal for sustainable communities to become more resilient. By promoting a suitable distributed energy portfolio mix, these communities are increasing their adaptation to climate change effects.

Despite the increasing demand for the energy sector to address climate change impacts and propose mitigation efforts, the energy sector can also facilitate and build climate resilience. Energy is a basic human necessity, and energy supply is an indispensable element in all economic activities. As the impacts of climate change grow, energy demand is expected to increase accordingly. It is necessary to understand the role of reliable, affordable, and sustainable modern energy services in strengthening societies' capacities to better adapt to climate change impacts (Sharma, 2019). Climate change is expected to increase energy demand to support adaptation measures, such as space cooling, water supply, and cold chains.

Furthermore, communities located in arid regions often lag far behind in well-being rates compared to the rest of the world basically due to their socioeconomic condition which, in the majority of cases, is related to natural resources (Safriel & Adeel, 2005). Natural resources in drylands are being affected severely by desertification, which is the result of many socioecological factors, often triggered by poor resource management operations (Reynolds et al., 2007). The importance of studying drylands is based on the fact that they comprise 41% of the total land area in the world (Safriel & Adeel, 2005), and that, as a result of increasing greenhouse gases in the atmosphere, many parts of the Earth are expected to become drier (IPCC, 2014).

22.3.5 *The Community Energy Model as a Capacity Building Strategy*

A community energy model can be an alternative to the current climate change scenarios, as it incorporates the participation of locals and, hence, offers a system concurrent with local needs, values, and resources. It encourages local food production while generating electricity from a clean energy source. With a localized system, local level governments become independent producers of energy and food which might also help respond faster to natural and social threats to the community, or to the larger economy on which it depends. A well implemented community energy model might discourage urban migration, or better said, could encourage migration to the rural living. This has been seen particularly in industrialized nations, where people from urban environments, tired of unhealthy habits, have begun to migrate to rural settings. However, this remains a challenge in emerging economies where the rural economy is seen as the poorest sector of the society.

A community energy model could be implemented as part of a whole sustainable community plan, promoting adaptability and absorptive capacity. Adaptability, according to Walker et al., is "the capacity of actors in a system to influence resilience" (Walker & Devine-Wright, 2008). Resilience could be defined as the capacity of a system that ensures shocks and stressors do not have long-lasting effects and consequences in a socioecological system (Frankenberger et al., 2014). A community energy model could help in building resilience using an Absorptive Capacity approach, coping with the ability of the communities to minimize exposure to shocks and stresses, and recover quickly when exposed (Frankenberger et al., 2012). Communities currently adapting localized energy and food production are able to maintain a more sustainable livelihood, improve their overall well-being, and promote economic, environmental, and social benefits.

22.3.5.1 Community Integrated Sustainable Energy (CISE) Strategy

Food and energy production are the two most important economic activities in the modern world. Current research on climate change argues that both activities will become highly vulnerable in the near future. In the case of agriculture, the lack of resources, the stress in water availability, and a significant increase in pests and crop disease will severely affect the whole food production system, particularly in arid regions (Frisvold et al., 2013; IPCC, 2014). Due to an increase in temperatures, electricity demand will peak as a result of future cooling needs and water pumping, putting transmission and distribution lines at risk, affecting overall electricity availability (Tidwell et al., 2013; IPCC, 2014). Localized production systems are becoming both an adaptation and mitigation response towards potential climate change risks.

In recent years, there have been some efforts to localize food and energy production; however, most communities in the world are becoming highly dependent on third parties for their basic food and energy consumption needs. In this context, this research introduces a Community Integrated Sustainable Energy (CISE) strategy as a localized effort to promote self-sustained agriculture and energy generation. This strategy has two goals: first, to generate enough electricity to fulfill basic household electricity needs; and second, to produce enough vegetables to cover the recommended daily intake per person. The purpose of this paper is to model a Community Integrated Sustainable Energy strategy based on data of three rural communities from northern Mexico. In these communities, although having access to the electrical grid, more than 80% of their population are living below the energy poverty line, meaning they do not fulfill their basic energy needs. Furthermore, their vegetable daily intake is about one third of the recommended by the World Health Organization (WHO, 2016; INEGI, 2014).

A community integrated system that produces electricity and food on the same land is being considered as a result of a literature review on community energy topics (Bogdanski et al., 2010; Coughlin et al., 2010; Hess, 2009; Hoffman & High-Pippert, 2005; Walker & Devine-Wright, 2008; St. Denis & Parker, 2009), sustainable energy literature (Rogall, 2009; Prandeki, 2014), as well as on PV and agriculture co-location research (Macknick et al., 2013; Ravi et al., 2016).

A Sustainable Energy strategy is defined, for the purposes of this paper, as a strategy that should improve overall livelihood conditions through the adoption and implementation of renewable energy projects. This strategy can be an alternative to the current energy and food system production as it incorporates participation of locals and hence offers a system concurrent with local needs, values, and resources. It encourages local food production while generating electricity from a clean energy source. With a localized system, local level governments and communities can become independent producers of energy and food and might help respond faster to natural and social threats to the community, or to the larger economy upon which the community is dependent (Hess, 2009). Furthermore, a CISE strategy might discourage urban migration, or better said, could encourage migration to rural living, enforcing the back-to-land culture. This has been seen particularly in industrialized countries, where people from urban environments, tired of unhea lthy habits and the adoption of a sustainability conscience, have begun to migrate to rural lifestyles (Halfacree, 2001, 2007; Sanders, 2004; Guthman, 2004).

The CISE strategy could be implemented as part of a whole sustainable community plan, promoting adaptability and absorptive capacity. Adaptability, in the words of Walker and others (2004), is "the capacity of actors in a system to influence resilience." Resilience could be defined as the capacity of a system that ensures shocks and stressors do not have long-lasting effects and consequences in a socio-ecological system (Frankenberger et al., 2014). A CISE strategy could help in building resilience acting as an Absorptive Capacity approach, coping with the ability of the communities to minimize exposure to shocks and stresses, and recover quickly when exposed (Frankenberger et al., 2012). Communities currently adapting localized energy and food production are able to maintain a more sustainable livelihood, improving their overall well-being, promoting economic, environmental, and social benefits.

The Community Integrated Sustainable Energy (CISE) strategy proposed here is expected to perform at the community level, as a distributed generation type system. Figure 22.6 depicts how this system will provide enough energy and food to fulfill community requirements. If any remaining surplus, this will be sold first to the community market and other community-owned businesses, and then to nearby industry and farming facilities. The idea is to start in phases, by first investing in the required energy and vegetable provisions for the interested communities, and then to promote further investment options for aggregating both energy and food production in a larger scheme.

The proposed business model that enables the CISE Strategy to happen is that of a cooperative. In Mexico, there are several types of cooperatives, with the production cooperative as the focus of this research (DOF, 2009). A production cooperative is the one in which cooperative partners work together for the production of goods and services, benefiting from mutual work while sharing equal responsibility. The adoption of a cooperative business model is important for sustainable development, as it generates self-employment in situ and potentially improves livelihood conditions without the necessity to migrate to the city. Cooperative partners will be in charge of managing both agriculture and energy production.

Investment in the cooperative will be an individual decision; however, this research assumes that each household will participate in the cooperative with an initial investment equivalent to the amount of energy required to fulfill their current energy needs. The cooperative will return the investment in a maximum period of 25 years. Both energy and agriculture produce are shared with the cooperative partners according to the level of involvement for which they have subscribed. The energy produced is shared through a renewable energy policy called virtual net metering (see Fig. 22.7).

Virtual net metering allows the users to allocate the energy produced to one or multiple clients, no matter where they are in the country. Electricity goes through the grid and a credit is provided to the different owners of the renewable energy system; this credit is granted in their electricity bill despite being away of the renewable system location (DOF, 2014; CRE, 2016).

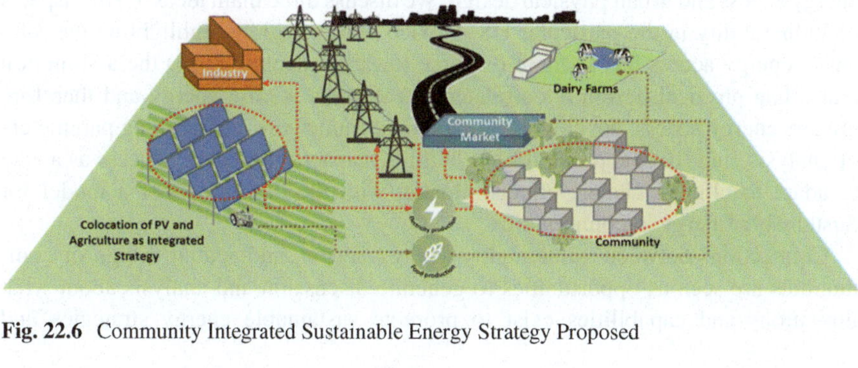

Fig. 22.6 Community Integrated Sustainable Energy Strategy Proposed

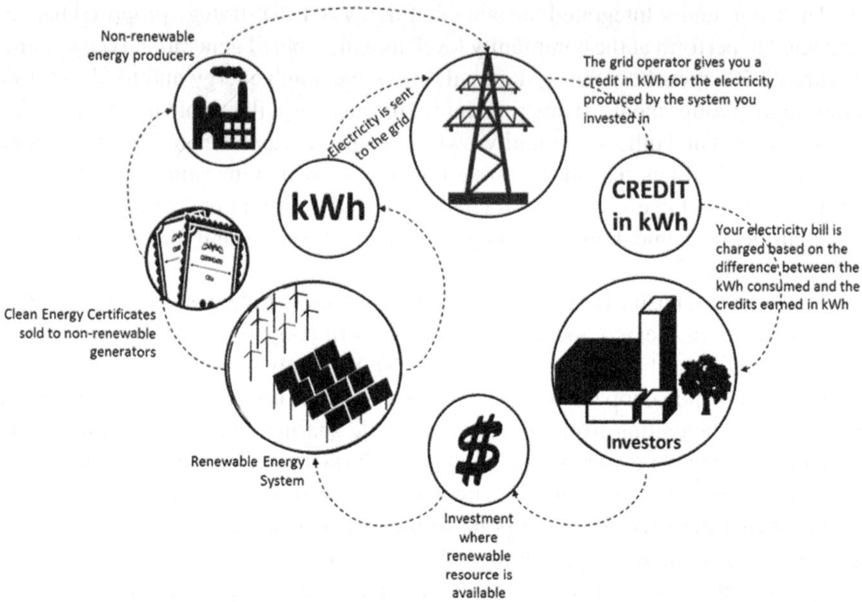

Fig. 22.7 Application of Virtual Net Metering Program (Source: author, based on CFE documents)

A Community Integrated Sustainable Energy (CISE) strategy can be an alternative to the current energy and food system production as it incorporates participation of locals and hence offers a system that is concurrent with local needs, values, and resources. It encourages local food production while generating electricity from a clean energy source. It promotes the three steps towards Sustainable Energy and is a viable option for vulnerable communities aiming to adapt to climate change.

22.4 Conclusions and Next Steps

Throughout this chapter we were able to go through several ideas that are related to energy access and urban physical design. We discuss three main ideas: (i) the aspects of Vulnerability in the particular US–Mexico border region, highlighting the relevance energy access has as main resource towards adaptation; (ii) the assumption that urban physical design is correlated to the ways we use energy and therefore hinders energy access, developing a theoretical framework and defining parameters of analysis, and (iii) the possibility of building capacity for energy access as a way to adapt to climate change effects, closing with the description of a model for Sustainable Energy in communities.

Connections between vulnerability, energy access, and spatial design of communities are seen as opportunities to generate discussion and analysis about what limitations and capabilities exist to promote sustainable energy strategies and implement specific community urban design aspects that will help alleviate energy poverty, creating a framework for understanding energy use in urban and semi-urban environments, particularly in the west side of the US–Mexico border region, that is considered among the most vulnerable due to its harsh climate conditions.

Communities in this area struggle to get access to better energy conditions, not only due to the energy transmission and distribution systems, but due to the urban configuration, its physical design, that forces livelihoods to be unsustainable. Characteristics such as the land use mixture within the urban setting, its density, the connectivity and accessibility these communities have to basic services, as well as the intensity of activities happening within border communities, create an important gap on how energy is used and distributed. This is an important consideration for an energy justice standpoint. We acknowledge that urban configuration of border communities may be responsible of how energy is used at three forms: the operational form that is related to the use of energy within buildings and homes; the transportation form which responds to the use of energy for transportation purposes, not only for the commuting within the community and outside, but also how to transport primary goods and services that have to be provided to the community; and the embodied form which helps understand the energy consumed for the manufacturing and distribution of materials used for community infrastructure.

A better, more sustainable access to energy is necessary not only for providing better livelihood conditions to border communities in the short run, but also as a way to build capacity towards climate change effects. By creating a Sustainable Energy Strategy that localizes energy generation, empowering community members by getting their own energy sources combined with better physical urban design, households should be able first, to reduce energy needs in terms of cooling and heating dwellings, and commuting to work or recreational activities, as physical design will provide bioclimatic, nature-based solutions towards those issues.

Energy poverty diminishes quality of life, increases disease, provokes a gap in communication and education for household members, relates to insecurity incidents, among other preoccupant situations. Developing a strategy that faces energy poverty not from the individual or household perspective only, but as a common problem, enables us to frame it as a social problem, a community problem, that has to be solved focusing on several perspectives. This chapter aims to highlight the fact that good urban design and minimum physical aspects of community configuration might help increase household's resilience and adaptation. Energy is one of the most important resources for modern living, as it provides economic wealth and human wellbeing. Energy enables capabilities that otherwise could not be granted to the world we live in. We need to take actions that help develop better communities, in its design and in its configuration. It is a matter of policy, but most of all, it is a matter of design.

We need to pursue the adoption and deployment of sustainable energy strategies in order to empower border communities. It is important to start generating models to advance knowledge in terms of energy use and consumption and urban physical design. Energy justice should be embraced through adaptation to the socioeconomic context, energy generation at the necessary scale overcoming existing barriers to exclusion implementing community energy projects and designing participatory approaches. Without a doubt, a just energy transition is necessary and urgent in Mexico. The country now has the opportunity to undertake a progressive and timely transition that allows for better scenarios of social, environmental and climate justice, and that responds to current social demands. Further research is necessary to develop ideal scenarios, indicators, and other parameters that allow to better understand this relationship and inform policy makers. The adoption of a sustainable energy strategy such as a community energy system has to be further explored.

References

Abbasabadi, N., & Ashayeri, J. (2019). *Urban energy use modeling methods and tools: A review and an outlook.* Elsevier Science.

Acuña, G., & Serrano, R. (2017). Los conflictos socio-ambientales energéticos en America Latina: A propósito de las energías renovables en la agenda 2030. *Revista Internacional de Direito Ambiental, 6*(16), 153–196.

Ang, B. W., Choong, W. L., & Ng, T. S. (2015). Energy security: Definitions, dimensions and indexes. *Renewable and Sustainable Energy Reviews, 42*, 1077–1093. https://doi.org/10.1016/j.rser.2014.10.064

Arnfield, A. (2003). Two decades of urban climate research: A review of turbulences, exchanges of energy and water, and the urban heat island. *International Journal of Climatology, 23*(1), 1–26.

Bai, X. (2016). Eight energy and material flow characteristics of urban ecosystems. *Ambio, 45*(7), 819–830.

Barquero, V. (2016). *Livelihood assessment for rural Delicias Chihuahua as means for developing a community energy model.* The University of Arizona. ProQuest Dissertations Publishing, 10158147.

Boardman, B. (1991). *Fuel poverty: From cold homes to affordable warmth.* Belhaven Press.

Dubois, O., Jamieson, C., & Krell, R. (2010). *Making integrated food-energy systems work for people and climate: An overview.* Food and Agriculture Administration FAO.

Carreon, J. R., & Worrel, E. (2018). Urban energy systems within the transition to sustainable development. A research agenda for urban metabolism. *Resources Conservation and Recycling, 132*, 258–266.

Castán-Broto, V., & Bulkeley, H. (2013). A survey of urban climate change experiments in 100 cities. *Global Environmental Change, 23*(1), 92–102.

Chen, S., & Chen, B. (2015). Urban energy consumption: Different insights from energy flow analysis, input–output analysis and ecological network analysis. *Applied Energy, 138*, 99–107. https://doi.org/10.1016/j.apenergy.2014.10.055

Chihuahua, G. D. (2021). *Gobierno de Chihuahua.* Retrieved 2021, from chihuahua.gob.mx: http://www.chihuahua.gob.mx/contenidos/entrega-desarrollo-social-mas-de-450-toneladas-de-alimento.

CONEVAL. (2020). *Informe de Pobreza en Mexico 2020: el país, sus estados y sus municipios.* Consejo Nacional de Evaluación de la Política de Desarrollo Social. Retrieved September 30, 2021, from https://www.coneval.org.mx/coordinacion/entidades/Paginas/Informes_Pobreza_Evaluacion_2020.aspx

Coughlin, J., Grove, J., Irvine, L., Jacobs, J., Phillips, S., Moynihan, L., & Wiedman, J. (2010). *A guide to community solar: Utility, private and non-profit project development.* National Renewable Energy Laboratory.

CRE. (2016). Comisión Reguladora de Energía: *Regulación de Electricos.* Retrieved May 12, 2021, from http://www.cre.gob.mx/documento/faq-regulacion-electricos.pdf.

DGN (2012) *Dirección General de Normas Mexicanas*. Retrieved September 30, 2021, from http://www.2006-2012.economia.gob.mx/conoce-la-se/atencion-ciudadana/procesos-administrativos/dgn.

DOF. (2009). *Ley General de Sociedades Cooperativas*. Diario Oficial de la Federación.

DOF. (2014). *Reglamento de la Ley de la Industria Electrica*. Diario Oficial de la Federación.

Dunlap, R. (2011). Environmental sociology. In G. Ritzer & J. M. Ryan (Eds.), *The concise Encyclopedia of sociology* (pp. 189–191). Wiley-Blackwell.

Ellena, M., Breil, M., & Soriani, S. (2020). The heat-health nexus in the urban context: A systematic literature review exploring the socio-economic vulnerabilities and built environment characteristics. *Urban Climate, 34*, 100676.

Feng, J., Tang, S., & Chuai, X. (2018). The impact of neighbourhood environments on quality of life of elderly people: Evidence from Nanjing, China. *Urban Studies, 55*(9), 2020–2039.

Frank, L., Andresen, M., & Schmid, T. (2004). Obesity relationship with community design, physical activity, and time spent in cars. *American Journal of preventive medicine, 27*(2), 87–96.

Frankenberger, T., Constas, M., Nelson, S., & Starr, L. (2014). *Current approaches to resilience programming among non-governmental organizations*. s.l., s.n.

Frankenberger, T., Spangler, T., Nelson, S., & Langworthy, M. (2012). *Enhancing Resilience to Food Insecurity amid protracted crisis*. Food and Agriculture Organization.

Frisvold, G., Jackson, L., Pritchett, G., & Ritten, P. (2013). Agriculture and ranching. In G. Garfin, A. Jardine, R. Merideth, M. Black, & S. LeRoy (Eds.), *Assessment of climate change in the Southwest United States: A report prepared for the National Climate Assessment* (pp. 218–239). Island Press.

García-Ochoa, R. (2014). *Pobreza Energética en América Latina*. Comisión Económica para América Latina y el Caribe CEPAL, Naciones Unidas.

García-Ochoa, R., & Graizbord, B. (2016). Caracterización espacial de la pobreza energética en México. Un análisis a escala subnacional. *Economía, sociedad y territorio, 16*(51), 289–337. Recuperado en 30 de septiembre de 2021, de http://www.scielo.org.mx/scielo.php?script=sci_arttext&pid=S1405-84212016000200289&lng=es&tlng=es

Gaye, A., (2007). *Access to energy and human development. Human Development Report 2007/2008*, s.l.: United Nations Development Program.

GNEB (Good Neighbor Environmental Board). (2016). Climate change and resilient communities along the U.S.-Mexico Border: The Role of Federal Agencies: *Seventeenth Report of the Good Neighbor Environmental Board to the President and Congress of the United States*. EPA 202-R-16-001. Washington, D.C.: U.S. Environmental Protection Agency. nepis.epa.gov/Exe/ZyPDF.cgi/P100QFGF.PDF?Dockey=P100QFGF.PDF.

Güneralp, B., Zhou, Y., Ürge-Vorsatz, D., Gupta, M., Yu, S., Patel, P. L., Fragkias, M., Li, X., & Seto, K. C. (2017). Global scenarios of Urban density and its impacts on building energy use through 2050. *Proceedings of the National Academy of Sciences of the United States of America, 114*(34), 8945–8950.

Guo, A., Yang, J., Xiao, X., Xia, J., Jin, C., & Li, X. (2020). Influences of urban spatial form on urban heat island effects at the community level in China. *Sustainable Cities and Society, 53*, 101972.

Guthman, J. (2004). Back to the land: The paradox of organic food standards. *Environment and Planning A: Economy and Space, 36*(3), 511–528. https://doi.org/10.1068/a36104

Halfacree, K. (2001). Going "back-to-the-land" again: Extending the scope of counterurbanization. *Espace, Populations, Societes, 1*(2), 161–170.

Halfacree, K. (2007). Back-to-the-land in the twenty-first century—Making connections with rurality. *Tijdschrift Voor Economische en Sociale Geogragie, 98*(1), 3–8.

Handy, S., Boarnet, M., Ewing, R., & Killingsworth, R. (2002). How the built environment affects physical activity—Views from urban planning. *American Journal of Preventive Medicine, 23*(2), -64, 73.

Harmon, D., Haley, M., & Funkhouser, E. (2017). *Energy poverty research landscape analysis: Working paper*. Texas Energy Poverty Research Institute. Retrieved from www.

txenergypoverty.org/wp-content/uploads/2017/05/TEPRI-Energy-Poverty-Research-Landscape-AnalysisMay-2017.pdf
Harmon, D., & Moss, J. (2019). *Low-income community profile series. Part 1: Texas overview*. Texas Energy Poverty Research Institute. Retrieved from www.txenergypoverty.org/wp-content/uploads/2019/06/TEPRI_LICPTexasOverview_August2019_REVISION_v3.pdf
Healy, J. (2004). *Housing, fuel poverty, and health: A pan-European analysis. s.l.* Routledge.
Hess, D. (2009). *Localist movements in a global economy: Sustainability, justice, and urban development in the United States*. The MIT Press.
Hoffman, S. M., & High-Pippert, A. (2005). Community energy: A social architecture for an alternative energy future. *Bulletin of Science, Technology, and Society, 25*, 387–401.
IAEA. (2020). *Energy, electricity and nuclear power estimates for the period up to 2050*. International Atomic Energy Agency Vienna International Centre.
IEA. (2020). *World Energy Outlook 2020*. International Energy Agency. Retrieved from https://www.iea.org/reports/world-energy-outlook-2020
INEGI. (2014). *Encuesta Nacional de Ingresos y Gastos de los Hogares*. [Online] Retrieved from https://www.inegi.org.mx/rnm/index.php/catalog/685/related-materials.
IPCC. (2014). *Impacts, adaptation and vulnerability observed*. Intergovernmental Panel on Climate Change, Working Group II. United Nations.
IPCC. (2017). *Chair's vision paper*. Addis Ababa, Ethiopia: AR6-SCOP/Doc. 2.
Jenkins, K., Mc Cauley, D., Heffron, R., Stephan, H., & Rehner, R. (2016). Energy justice: A conceptual review. *Energy Research and Social Science, 11*, 174–182.
Jones, R. (2010). *Energy poverty: Now to make modern energy access universal?* World Energy Outlook, International Energy Agency/OECD.
Ko, Y. (2013). Urban form and residential energy use: A review of design principles and research findings. *Journal of Planning Literature, 28*(4), 327–351.
Kopinak, K. (2004). *The social costs of industrial growth in Northern Mexico*. CUMS ASIN: B008YSU1EK.
Kozulj, R. (2009). *Contribución de los servicios energéticos a los Objetivos de Desarrollo del Milenio y a la mitigación de la pobreza en America Latina y el Caribe*. Programa de las Naciones Unidas para el Desarrollo PNUD.
Kwon, M., Pickett, A., Lee, Y., & Lee, S. (2019). Neighborhood physical environments, recreational wellbeing, and psychological health. *Applied Research in Quality of Life, 14*(1), 253–271.
Lara-Valencia, F., & Giner, M. E. (2013). Local responses to climate change vulnerability along the Western reach of the US-Mexico border. *Journal of Borderlands Studies, 28*(2), 191–204.
LBC. (2020). *Living building challenge*. Retrieved from https://living-future.org/lbc/.
Macknick, J., Beatty, B., & Hill, G. (2013). *Overview of opportunities for co-location of solar energy technologies and vegetation*. National Renewable Energy Laboratory.
Månsson, A., Johansson, B., & Nilsson, L. J. (2014). Assessing energy security: An overview of commonly used methodologies. *Energy, 73*, 1–14. https://doi.org/10.1016/j.energy.2014.06.073
Michael, K., Deshpande, T., & Zievogel, G. (2019). Examining vulnerability in a dynamic urban setting: The case of Bangalore's interstate migrant waste pickers. *Climate and Development, 11*(8), 667–668.
Nichols, B., & Kockelman, K. (2015). Urban form and lifecycle energy consumption: Case studies at the city scale. *Journal of Transport and Land Use, 8*(3), 115–128.
Prandeki, K. (2014). Theoretical aspects of sustainable energy. *Energy and Environmental Engineering, 2*(4), 83–90.
Pye, S., et al. (2015). *Energy Poverty and vulnerable consumers in the energy sector across the EU: Analysis of policies and measures*. European Commission.
Quarantelli, E. L. (2000a). Disaster research: Sociohistory of the field. In E. F. Borgatta & R. J. V. Montgomery (Eds.), *Encyclopedia of sociology* (pp. 681–688). Macmillan Reference USA.
Quarantelli, E. L. (2000b). Disaster research: Sociohistory of the field. In E. F. Borgatta & R. J. V. Montgomery (Eds.), *Encyclopedia of sociology* (pp. 681–688). New York, NY.

Ratti, C., Baker, N., & Steemers, K. (2005). Energy consumption and urban texture. *Energy and Buildings, 37*, 762–776.

Ravi, S. et al., (2016). Colocation opportunities for large solar infrastructures and agriculture in drylands. *Applied Energy, 165*, 383–392.

Reynolds, J., et al. (2007). Global desertification: Building a science for dryland development. *Science, 316*, 847–851.

Rickwood, P., Glazebrook, G., & Searle, G. (2008). Urban structure and energy: A review. *Urban Policy, 26*(1), 57–81.

Rogall, H. (2009). *Sustainable economy: Economical theory and practice of sustainable development*. Metropolis Verlag.

Roselt, K., Quaas, I., Genske, D., Klawonn, U., Männel, L., Reich, A., Ruff, A., & Schwarze, M. (2015). 'Effort' (energy efficiency on-site)—A new method for planning and realization of energy-efficient neighborhoods under the aspects of sustainability. *Proceedia Engineering, 118*, 1288–1295.

Safriel, A., & Adeel, Z. (2005). Chapter 22: Dryland systems. In *Ecosystems and human well-being: Current state and trends. s.l.: Millenium ecosystem assessment series* (Vol. 1, pp. 623–662).

Sanders, J. (2004). Back to the land: The Enduring dream of self-sufficiency in modern America. *Journal of American History, 101*(1), 305–306.

Sen, A. (1983). Poor, relatively speaking. *Oxford Economic Papers, 35*, 153–169.

Seto, K., Dhakal, S., Bigio, A., Blanco, H., Delgado, C., Dewar, D., Huang, L., Inaba, A., Kansal, A., Lwasa, S., McMahon, J. E., Müller, D. B., Murakami, J., & Nagendra, H. (2014). Human settlements, infrastructure and spatial planning. In O. Edenhofer, R. Pichs-Madruga, Y. Sokona, E. Farahani, S. Kadner, K. Seyboth, & J. Minx (Eds.), *Climate change 2014: Mitigation of climate change. Contribution of Working Group III to the Fifth Assessment Report of the Intergovernmental Panel on Climate Change* (pp. 923–1000). Cambrige University Press.

Sharma, V. (2019). Access for adaptation? Reviewing the linkages between energy, disasters, and development in India, *Energy Research and Social Science, 52*, 10–19.

Shonkoff, S. B., Morello-Frosch, R., Pastor, M., & Sadd, J. (2009). Minding the climate gap: Environmental health and equity implications of climate change mitigation policies in California. *Environmental Justice, 2*(4), 173–177. https://doi.org/10.1089/env.2009.0030

Smith, T., Nelischer, M., & Perkins, N. (1997). Quality of an urban community: A framework for understanding the relationship between quality and physical form. *Landscape and Urban Planning, 39*(2–3), 229–241.

Sovacool, B. K., & Dworkin, M. H. (2014). *Global energy justice: Problems, principles and practices*. Cambridge Univ. Press.

St. Denis, G., & Parker, P. (2009). Community energy planning in Canada: The role of renewable energy. *Renewable and Sustainable Reviews, 13*, 2088–2095.

Tidwell, V., Dale, L., Franco, G., Averyt, K., Wei, M., Kammen, D., & Nelson, J. (2013). Energy: Supply, demand, and impacts. In G. Garfin, A. Jardine, R. Merideth, M. Black, & S. LeRoy (Eds.), *Assessment of climate change in the Southwest United States: A report prepared for the national climate assessment* (pp. 240–266). Island Press.

Townsend, P. (1979). *Poverty in the United Kingdom: A survey of housheold resources and Standards of Living* (1st ed.). Penguin Books, Ltd..

UN. (2015). *Draft Outcome document of the United Nations summit for the adoption of the post-2015 development agenda*. United Nations.

UN (2020). *Capacity-building*. United Nations Academic Impact https://www.un.org/en/academic-impact/capacity-building.

USGBC (2021). *United States Green Building Council*. Retrieved from https://www.usgbc.org/.

Walker, G., & Devine-Wright, P. (2008). Community renewable energy: What should it mean? *Energy Policy, 36*, 497–500.

Weber, S., Sadodd, N., Zell, E., & Sherbinin, A. (2015). Policy-relevant indicators for mapping the vulnerability of urban populations to extreme heat events: A case study of Philadelphia. *Journal of Applied Geography, 63*, 231–243.

WHO. (2016). *Global strategy on diet, physical activity and health*. World Health Organization. Retrieved 12 May 2021, from http://www.who.int/dietphysicalactivity/fruit/en/

Wilder, M., Aguilar-Barajas, I., Pineda-Pablos, N., Varady, R. G., Megdal, S. B., McEvoy, J., Merideth, R., Zúñiga-Terán, A. A., & Scott, C. A. (2016). Desalination and water security in the US–Mexico border region: Assessing the social, environmental and political impacts. *Water International, 41*(5), 756–775. https://doi.org/10.1080/02508060.2016.1166416

Wilder, M., Scott, C. A., Pablos, N. P., Varady, R. G., Garfin, G. M., & McEvoy, J. (2010). Adapting across boundaries: Climate change, social learning, and resilience in the U.S.–Mexico border region. *Annals of the Association of American Geographers, 100*(4), 917–928. https://doi.org/10.1080/00045608.2010.500235

Wilder, M., et al. (2013). Climate change and U.S.-Mexico border communities. In G. Garfin, A. Jardine, R. Merideth, M. Black, & S. LeRoy (Eds.), *Assessment of climate change in the southwest United States: A report prepared for the national climate assessment. A report by the Southwest Climate Alliance* (pp. 340–384). Island Press.

Wisner, B. (2016) Vulnerability as concept, model, metric, and tool. In *Oxford research Encyclopedia of natural hazard science*. Retrieved 30 Sep. 2021, from https://oxfordre.com/naturalhazardscience/view/10.1093/acrefore/9780199389407.001.0001/acrefore-9780199389407-e-25.

Yang, P. P., & Quan, S. J. (2016). Urban form and energy resilient strategies: A case study of the Manhattan Grid. In *Urban resilience: A transformative approach* (pp. 153–172). Springer.

Zhang, Q., Schaaf, C., & Seto, K. (2013). The vegetation adjusted NTL urban index: A new approach to reduce saturation and increase variation in nighttime luminosity. *Remote Sensing of Environment, 129*, 32–41.

Chapter 23
Design with Vulnerable Communities

Emanuele Giorgi, Tiziano Cattaneo, Alfredo Mauricio Flores Herrera, and Virginia del Socorro Aceves Tarango

23.1 Introduction

Funded by the Observatory de Ciudades—Tecnologico de Monterrey (Mexico), the *Design for Vulnerables* project began in October 2020, with the aim of researching how urban-architectural design must change in the coming decades to face the challenges in vulnerable communities. In fact, there is a widespread perception that architectural and urban design must evolve rapidly to face the challenges of contemporaneity (Nast, 2019; Giorgi et al., 2020; Cattaneo et al., 2018; European Association for Architectural Education, n.d.; Giorgi, 2020).

Design for Vulnerables is part of a broader research activity that has been implemented in recent years within an international research network, mainly between Mexico, Italy, and China, whose focus is on understanding the role of design architecture and urban development in sustainable territories, ready to face the challenges of the coming decades. Understanding the phenomena that characterize our territories is a priority: contemporary socio-environmental crises are putting the whole of humanity to a severe test, in particular exposing the most vulnerable populations to uncertainty.

E. Giorgi (✉) · A. M. Flores Herrera
Tecnologico de Monterrey, Escuela de Arquitectura, Arte y Diseño, Chihuahua, Mexico
e-mail: egiorgi@tec.mx

T. Cattaneo
Department of Civil Engineering and Architecture, University of Pavia, Italy

College of Design and Innovation, Tongji University, Shanghai, China

V. S. Aceves Tarango
Tecnologico de Monterrey, Escuela de Humanidades, Chihuahua, Mexico

E. Giorgi et al. (eds.), *Design for Vulnerable Communities*, The Urban Book Series,
https://doi.org/10.1007/978-3-030-96866-3_23

As these social realities become increasingly complex, the question that "Design for Vulnerables" wants to answer is "how can design contribute to the empowerment of vulnerable communities in the coming years"?

Committed to sustainable territorial development, the School of Architecture, Art and Design of the Tecnológico de Monterrey on the Chihuahua campus began working for years with vulnerable local communities to be able to offer proposals and solutions to their needs.

Design for Vulnerables is organized by the School of Architecture, Art and Design and the School of Humanities of Tecnologico de Monterrey. Strategic partners of the project are the China Lab of the University of Pavia (Italy), the Environmental Futures Lab and the Desis Lab (Design for Social Innovation and Sustainability) of the Tongji University (China), Accionética Association which has been working for years with the Paso Del Norte community and the Association of Italian Researchers in Mexico (ARIM).

Vulnerability is the result of specific spatial, socio-economic–demographic, cultural, and institutional contexts (Kuhlicke et al., 2011). Thus, since the concept of vulnerability is closely linked to the context to which it refers, the authors have decided to approach the issue from two fronts. On the one hand, contextualizing the theme as much as possible with an interpretative analysis and design works in the specific context of Paso Del Norte, a vulnerable colony in the city of Chihuahua (Mexico), working with residents of the community. On the other hand, expanding the disciplinary vision thanks to the contribution of experts from various disciplines that allow highlighting what challenges will strike vulnerable communities in the coming decades.

Thus, in January and February 2021, an interdisciplinary workshop focused on producing design solutions for Paso Del Norte, attended by students and professors of the Tecnologico de Monterrey, professionals active in the city of Chihuahua, civil associations, PhD students of the University of Pavia (Italy) and a Spanish architect and (main actors) residents of the Paso Del Norte Community. The workshop took place over 6 weeks, through virtual meetings to address the limitations imposed by the pandemic. For this reason, before the workshop, the residents of the community were organized so that everyone could connect at least with WhatsApp calls to the works. To cope with the limitations imposed by the pandemic, the workshop structure had three phases: *Real Life Web Lab*, to facilitate discussion with community residents; *Design Group*, to coordinate discussion and processing activities within the design groups; *Experts' Round Tables*, to receive input from experts and to reflect with them on the issues of the colony. Each of these three parts had strong thematic and methodological relationships with the other two, creating a continuous flow of information and project inputs that fostered a base of knowledge and sensitivity towards contextual and global themes fundamental to develop the project proposals.

23.1.1 Real Life Web Lab

It was an online platform to share in comfort and following the indications of social isolation during the pandemic, the life experiences of the actors involved, to interact from the Paso Del Norte neighborhood with designers, local and international experts. It aimed to allow the members of the community to express themselves about their needs through a detailed description of the real life in El Paso Del Norte community. The web-lab allows a participation close as possible to the experience of in-person workshops. The need of the tool lies in the undisputed importance of listening, observing, and learning from the community members during an analysis process and community-based participatory research (Suarez-Balcazar, 2020). The *Real Life Web-Lab* collected successfully a series of videos, documents, and proposals that arise along the workshop, in the meetings between residents and their groupmates.

23.1.2 Design Groups

The working groups were interdisciplinary teams of professionals, community members, and civil society who worked together to define problems and develop ideas, solutions, and proposals. Field discussions and design consisted mainly of a design workshop for students of Tecnológico de Monterrey, supported by meetings with community members and classes of experts. The goal was to experiment with community design solutions based on surveys and conversational interviews. *Design for Vulnerables* works were organized through five main activities:

- *Mapping Paso Del Norte* (week 1): based on the knowledge of the territory and the comparison with the residents, each group had to analyze and map some characteristics of the community (colors, sharing, history, landscape, and production).
- *Here is where we live* (week 2): community members presented where they live and the public spaces where they spend their free time, what they do, what their jobs and skills are.
- *Paso Del Norte nowadays* (week 3): various analyses on the current state of the colony.
- *Paso Del Norte tomorrow* (week 4): considerations on the state of the colony in the expected future in the short, medium, and long term, based on the same analysis tools as the previous point, in order to highlight the aspects in which more is needed to intervene.
- *Let's do it* (weeks 5–6): formal development of an intervention proposal (urban, architectural, economic, etc.).

The goal was to provide participants with a clear and precise methodology to obtain material as similar and homogeneous as possible, with the aim of being able to compare the results and draw useful conclusions to answer the questions of the research project.

23.1.3 Experts' Round Tables

Discovering needs, wishes, or solutions also involves the ability to understand the challenges that vulnerable communities will have to face in the coming years: what environmental, social, and technological changes will influence the way of life of the community. Observing these phenomena through the lens of other disciplines is relevant to develop a complete understanding and thinking of new scenarios and build resilient communities. To get out of the limiting vision of pure architecture, the organization of the workshop involved the participation of various international experts on the subject of vulnerability in various fields of knowledge. This allowed to highlight some aspects, typical of different disciplines of architecture, which otherwise could not have been easily observed. Furthermore, the multidisciplinary approach, when working in the complexity of contemporary territories and in particular with vulnerable communities, is essential (Papanek, 1972). International experts came from various universities in America, Europe, and Asia: Columbia University (USA); Tongji University (China); Tecnológico de Monterrey (México); Artic University (Norway); University of Pavia (Italy); École Urbaine de Lyon (France); Instituto Mora, (México); Huddersfield University (UK); Bahia Federal University (Brazil); Westminster University (UK); ENSAP Bordeaux (France); Texas Tech University (USA); and Boston university (USA). These meetings were the occasion for the teams to reflect on methodologies, topics, and strategies to face the challenges of the coming years in vulnerable communities. The four round tables were:

- *The meaning of vulnerability.* This round table aimed to highlight the issues that make vulnerable a community. Contributions helped to understand which elements and situations are responsible to bring a community in a situation of vulnerability. The presentations focused on meaning, causes, and solutions.
- *Strategies.* A round table to highlight the strategies to solve situations of vulnerability. The contributions aimed to understand which policies, decisions, and actions could be taken according to different situations. The presentations focused on a specific context or on a specific topic.
- *Global to local.* This round table deepened global phenomena that are impacting or will impact vulnerable communities. Contributions helped to understand the risks and the opportunities for design in vulnerable communities. The presentations focused on the global phenomena and on the community scale.
- *Social entrepreneurs.* It was an occasion to present social entrepreneurship solutions in vulnerable contexts, with examples from China and Mexico.

As mentioned, the importance of the workshop lies in the fact that research on the concept of vulnerability must fit as much as possible into the real context of research, to understand what are the characteristics of the built environment and the social fabric that participate in the creation of this state of vulnerability. Therefore, it is important now to present the Paso Del Norte colony, where the workshop took place.

23.2 Paso del Norte, Chihuahua

Paso Del Norte is a neighborhood considered "outskirts" of the city of Chihuahua for many years, although it is less than 4 km from the city center. It is located on the banks of the Sacramento River, on a slope of the Sierra de Nombre de Dios (see Figs. 23.1, 23.2 and 23.3). This location is the cause that the colony conserves semi-urban and semi-rural features for more than 40 years. The founding families of Paso Del Norte received the land on which they built their homes as a donation from Mr. José Fierro. The heads of the family were Mr. Fierro's workers at that time. At the time, these families lived in a natural, clean, and pleasant environment. The Sacramento River carried water and was not as polluted with sewage and garbage as it has been for several years, two decades at least. There was only one paved street, *Calle Nube*, which later changed its name to *José Fierro*. Fruit trees were planted in the land near the houses and chickens raised for family consumption.

Currently, the Paso Del Norte neighborhood preserves much of its original environment; the city does not reach it with public services, equipment, and improvements in works at the community service, but with harmful effects due to the contamination of water, air, and soil. The area in which some fifty houses are located is surrounded by clandestine dumps of rubble and garbage. Every day a large number of trucks go up to these landfills with all kinds of waste that affect the environment in all aspects, constituting a permanent threat to the health of families. The

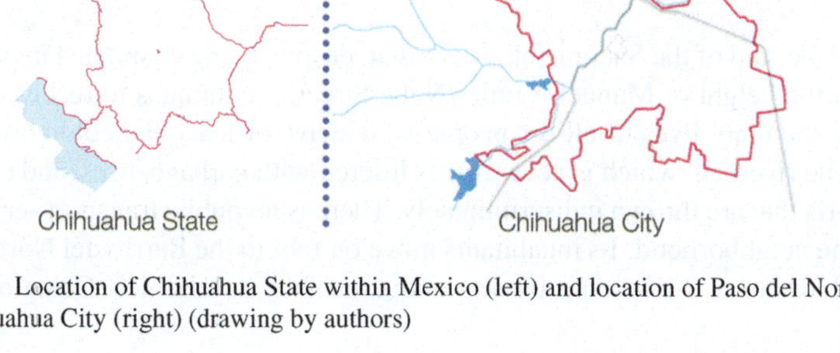

Fig. 23.1 Location of Chihuahua State within Mexico (left) and location of Paso del Norte within the Chihuahua City (right) (drawing by authors)

Fig. 23.2 Pedestrian entrance to Paso del Norte and square of Calle Farallón (picture by authors)

Fig. 23.3 Paso del Norte within its natural and urban environment (picture by authors)

same can be said of the Sacramento River that, despite being channeled in other latitudes, at the height of Manuel Aguilar/Nube Street, it continues to receive sewage, invading the daily lives of all the people who arrive or leave the colony with fetid odors. The riverbed, which is never dry, is littered with garbage, tires, and construction debris that are thrown indiscriminately. There is no public transport service that enters the neighborhood. Its inhabitants move on foot to the Barrio del Norte neighborhood where a truck route passes, and especially to Heroico Colegio Militar

Avenue. The journeys they make on foot are through areas without any protection from the inclemency of cold or heat that are extreme in the city.

In addition to pollution problems, almost no communication with the rest of the city, a difficult mobility, Paso Del Norte has few and deficient public spaces that allow families to relax, especially children and young people. Young people, especially teenagers, are not seen in the colony. Many of them have left their studies and carry out some economic activity to help the family. Young women become mothers at a very young age, often in their teens. Boys and girls play in the free spaces close to their houses, but which are often full of garbage.

One of the activities they enjoy the most is riding bicycles, but these do not last long since they are not special for mountains and the glass and stones continually destroy the tires. Another social problem in Paso Del Norte is that young mothers cannot work because there are no daycare centers that are accessible to them; they depend on the neighbors to take care of their children, but this supportive cure does not last long enough. The opposite case is that of mothers who take second and third shifts in a *maquiladora* (factory owned by foreign companies) and during that time children play in the street under the "care" of their older brothers or other children.

These problems, common to many other neighborhoods (*colonias* and *barrios*) in the city of Chihuahua, but with the own characteristics that the geographic environment gives to Paso del Norte, are the reasons why the Acciónética initiative decided to start its environmental education activities, in particular, giving support with the care of children in a safer space through games and recreational activities since 2018.

23.3 Research by Design

The teams worked on the five activities throughout the workshop in contact with the residents of the community to find the most relevant issues in the community. To standardize the work, each activity had a deliverable. All these results can be seen on the workshop website, as well as all other discussion activities between teams and experts (www.designforvulnerables.com). At the end of the workshop, the participants submitted their intervention proposals for the neighborhood, highlighting the main objectives to achieve the dreams and resolve the concerns of the residents. Figures 23.4–23.12 show some examples of intervention proposals developed by the teams that attended the workshop.

The delivery format for the final proposals consisted of these points, here presented with the instructions.

1. Title
2. Keywords
3. General Description (300 words): Resume the most relevant parts and aspects of your proposal: objectives, reasons, characteristics. Use this part to grab the interest and convince of the importance and quality of your project.

Fig. 23.4 Location of key places in the community (image by team)

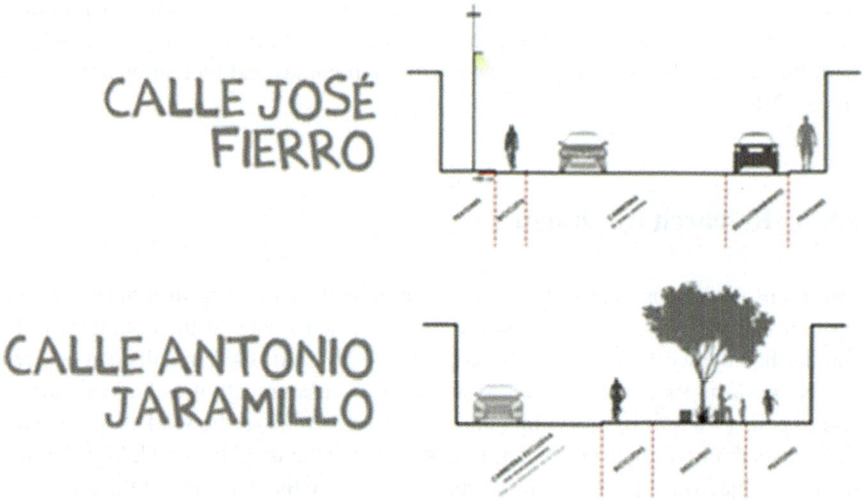

Fig. 23.5 Street section of the team proposal at Jose Fierro and Antonio Jaramilla Streets (image by team)

4. Goals (300 words): List and describe the goals that you aim to reach with the project.
5. Reasons (300 words): List and describe the reasons why your project is so relevant and why it should be supported.

Fig. 23.6 Render of proposal at Jose Fierro Street (image by team)

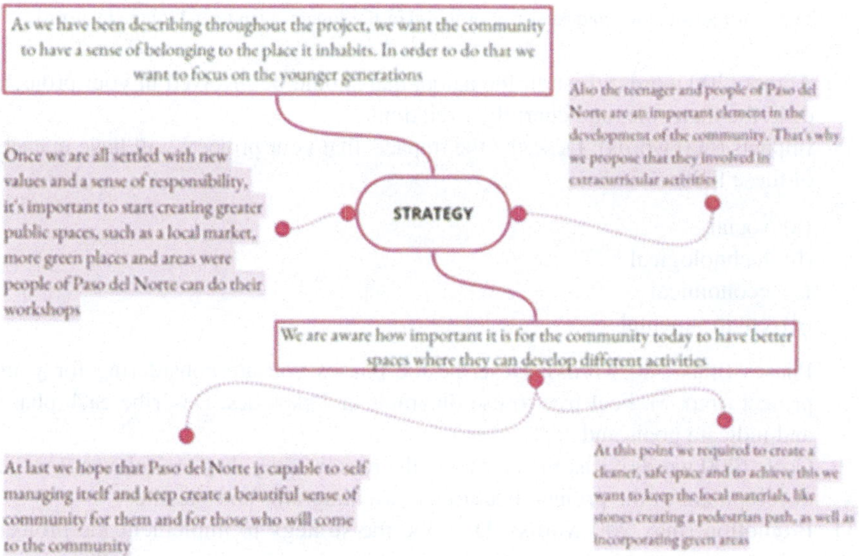

Fig. 23.7 Mind map of team strategy proposal (image by team)

6. Methodology (300 words): Describe how you imagine the project development. Refer to any methodology you are considering. You are free to use schemes and sketches.

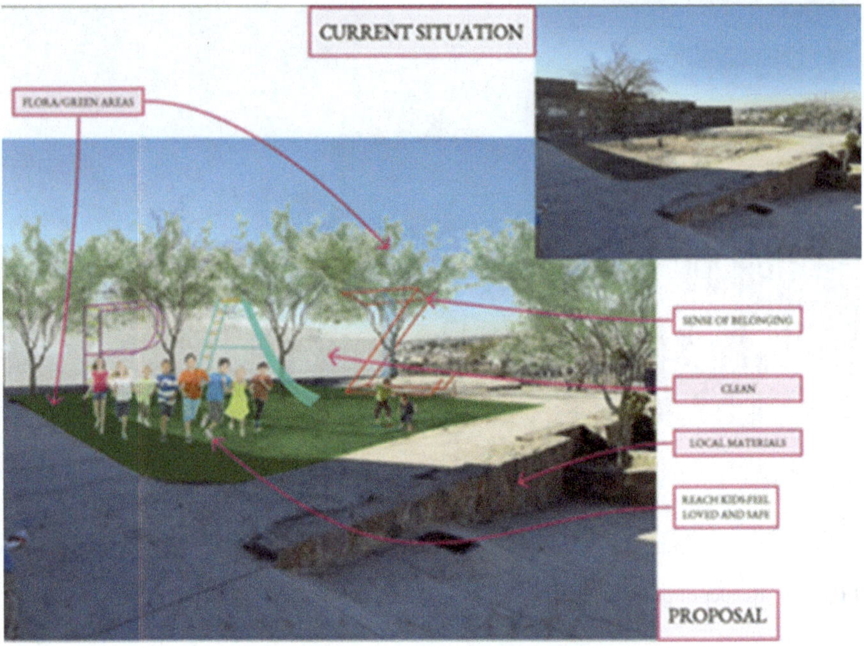

Fig. 23.8 Current situation and render of proposal of the square (picture and image by team)

7. Actors (200 words): List all the actors that should be involved in your project, describe their role, and chart their relations.
8. Impacts (500 words): Describe the impacts that your projects will have in each of these fields.
   (a) social
   (b) technological
   (c) economical
   (d) environmental
9. Phases (max 500 words): Describe the phases you are considering for your project (max 5). Feel free to use diagrams and sketches. Describe each phase and indicate goals and costs.
10. Costs (200 words): List the costs and the reasons why they are relevant for the development of the project. Feel free to use diagrams and sketches.
11. Promotion (max 200 words): Describe the strategy to implement the project (whom we should contact, which material we should prepare, etc.). Please, use words, sketches, etc.

The objective of this structure was to get a product easily presentable to local actors so that possible realizations of these projects could be taken into account, and to make comparisons between the teams. Like all the material produced during the workshop, these final projects are also available on the "Design for Vulnerables"

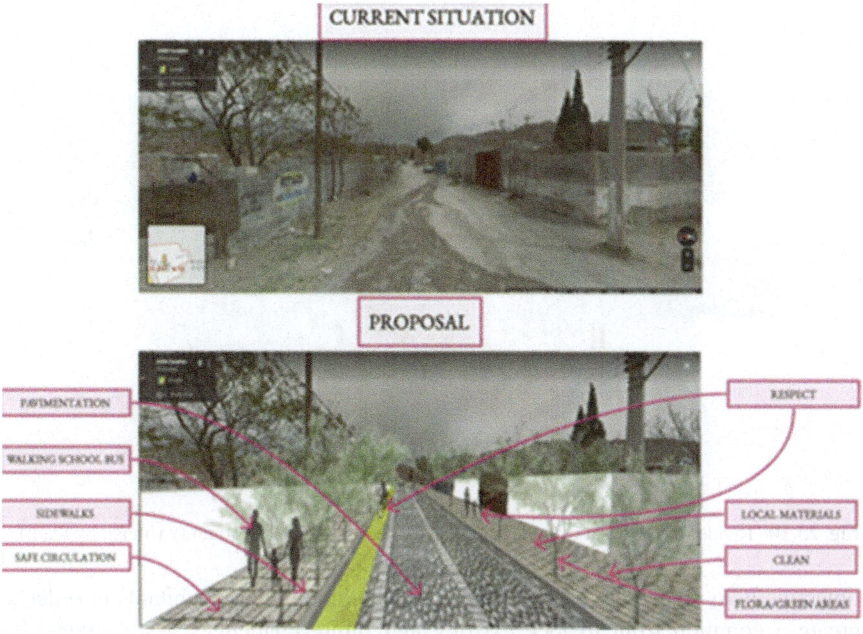

Fig. 23.9 Current situation and proposal for a typical street of Paso del Norte (picture and image by team)

website or they can be requested from the authors. Next, we present the most relevant results of each team, highlighting the themes that emerged from each discussion.

23.3.1 TEAM 1[1]: Networks and Nodes

The creation of the Paso Del Norte community was the result of a need for housing of the first settlers. Over the years, it has become a consolidated community thanks to the relationships forged with new residents or members of the same families living in the colony. This does not exclude them from the problems shared with other communities, such as pollution and insecurity, which have recently weakened the bonds between members. The latter results in a detriment to other aspects such as public spaces; the less relationship there is among the members of the community, the greater the waste of common spaces within the neighborhood, such as parks and courts. The responsibility for their care is avoided and therefore they fall into uses other than those for which they were designed or into disuse. Therefore, the main

[1] Ing. Joel Ramírez Durán, TEC Faculty, Chihuahua, Mexico; Luisa Fernanda Lira Chávez, student, Chihuahua, Mexico; Ing. Matilde Sessi, Universitá di Pavia, Italia; Nadia Sofia Medina Issa, TEC student, Chihauhaua, Mexico; Arq. Renata Enríquez Alatriste, designer, Mexico City, Mexico.

Fig. 23.10 Render of proposal of park and vehicular access to the community (image by team)

objective is to regenerate a sense of community among the inhabitants in order to create a common front to face existing and future challenges. As a strategy to achieve this, we will seek to revalue public spaces that encourage the interaction of members. This is the basis of the "Networks and Nodes" proposal, to improve the urban quality of public spaces (streets, squares and urban fronts) providing universal access to them and generating a community identity.

23.3.2 *TEAM 2[2]: Paso Del Norte, New Generations and Sense of Belonging*

Based on an observation exercise, the team focuses its proposal on a segment of the community. Although it may be one of the most affected by different problems, the solution may also lie in these people. The basic idea of this proposal is that young people have an active role in the community. Responsibility for the maintenance of public spaces and cleaning of polluted areas are some of the activities that the young people are expected to develop. This requires working with them to develop skills such as responsibility and empathy with the objective of achieving a common good, ranging from improving common spaces to the organization of community members to seek projects of greater impact on the neighborhood. This strategy aims to

[2] Alejandra Concha Contreras, TEC student, Chihuahua, Mexico; Ing. Caterina Pietra, PhD student, Università di Pavia, Italia; Arq. Lora Fahmy, designer, Ai-o Spacefactory, Bacelona, Spain; Arq. Rey Fernando Montes Trevizo, Misiones Coloniales de Chihuahua A.C., Chihuahua; Tania Carolina Cuilty Tassia, TEC student, Chihuahua, Mexico.

Fig. 23.11 Render of proposal of Recreative Route at Jose Fierro Street (image by team)

make young people feel valued by their community, this will generate a sense of belonging that in the future will motivate them to become agents of change that will lead and promote improvements in the lives of the members of the neighborhood.

23.3.3 TEAM 3[3]: Restructuring of the Neighborhood Council and Community Pavilion Paso Del Norte

This proposal presents two projects that, although they can be developed independently, both have a space-user relationship that seeks to ensure the success of the proposal. The first project consists of reorganizing the neighborhood council, which is the body in charge of making decisions in the community. The purpose of this reorganization is to ensure that more members of the community feel that they are taken into account and that more people have an active role in this group in order to generate a greater number and diversity of ideas that will be reflected in projects for the good of the community.

The second project is to generate the suitable spaces for the debate of ideas, proposals, positions and conciliations that must be achieved in any community project. These spaces shall comply not only with these purposes but also with that of improving the life of the inhabitants of the neighborhood, serving as a meeting point to develop diverse activities. As well as with external organizations that intend to collaborate with the community, and that find a plural, consolidated neighborhood council with the spaces that are required for the generation of ideas.

[3] Arq. Arturo Fuentes González, designer, Chihuahua, Mexico; Erika Paola Flores Legarreta, TEC student, Chihuahua, Mexico; Jesús Alfonso Vargas gonzález, TEC faculty, Chihuahua, Mexico.

Fig. 23.12 Render of proposal for an eco-tourism area at the community (image by team)

23.3.4 TEAM 4[4]: Paso Del Norte Ecological Corridor

This proposal takes advantage of the geographical location of the neighborhood and the natural limiting elements to generate a radical change in how the community is perceived and how to reinforce the community's identity. The natural barrier between the city and the colonia, the Sacramento River, would become an attribute of the community if it were considered in a territorial scale project, among other works, to reduce the segregation it generates in its current state. Likewise, the mountain range of Nombre de Dios can be a detonating element of commercial activities by becoming an ecological corridor that favors a tourism that seeks contact with nature. Tourism generates needs to cover by the community, so it would not only be the landscape background that it is today but a destination for people inside and outside the colony. In addition to this, and in order to provide a good service to visitors, it is necessary to create pleasant spaces that enrich the experience, one of them being a space that functions as a center of operations and tours, among other services for visitors, as well as a meeting and organization point for neighbors.

[4]Arq. Avril Díaz Arreola, designer and professor, Chihuahua, Mexico; Arq. Ernesto Ramírez García, EXATEC and professor, León, Mexico; Arq. Javier Omar Olivas López, EXATEC, Chihuahua, Mexico; María Fernanda Hernández Jáuregui, student, Chihuahua, Mexico; Arq. Paulina Grajeda Castillo, designer and professor, León, Mexico.

23.3.5 TEAM 5[5]: Paso del Norte Route

This project begins with the proposal to create a hiking area, taking advantage of the physical and geographical characteristics of the Nombre de Dios mountain range, as the first activity that will trigger adventure tourism. As a second element and taking advantage of the fact that the main access to the mountain range is José Fierro Street, to turn it into a commercial corridor that offers different products and services to tourists. All this organized and directed within the same community and supported by social networks that promote eco-tourism activities and the visit of the area in order to consume what the community offers, thus achieving an economic spill for the colony and an improvement in their quality of life.

The proposals offered in this first phase of the project have been presented on various occasions to relevant actors in the city of Chihuahua so that these ideas can be considered as a first basis for concepts and proposals for a sustainable regeneration of the Paso Del Norte. In fact, one of the research's goals is precisely to bring some design issues to the attention of key stakeholders (municipality, companies, entrepreneurs, etc.) so that they can look at the most vulnerable realities of our cities as key places in which to intervene to promote sustainable territories.

23.4 Conclusions

The interaction between the theoretical results of the round tables, with their interdisciplinary visions, and the more practical results of the design exercises could highlight some issues that seem interesting as the conclusion of this phase of the project. First of all, the "world of vulnerability" is a very vast world, full of features and interpretative possibilities that make it difficult (or even misleading) to try to give a single reading to vulnerable realities.

Without claiming to be exhaustive, the authors tried to highlight various aspects related to the theme of vulnerability, approaching various fields of knowledge. Reflections emerged on educational themes, technological themes, as well as subjects of enhancement of the architectural cultural heritage. For a complete observation of the arguments that emerged, the reader can refer to the material produced during the workshop, which is openly available on the research project web page: www.designforvulnerables.com. Alongside these transversal issues, various indications have also emerged regarding project issues which, based on the results of the workshop and the discussions with experts, should be taken into consideration to reduce the vulnerability of communities.

[5]Daniel Hernández Carreón, student, Chihuahua, Mexico; Arq. José Israel Conde Centeno, designer, Chihuahua, Mexico; Arq. Karina Manríquez Moreno, designer, Chihuahua, Mexico; Arq. Nayeli Alejandra González Anaya, EXATEC, Chihuahua, Mexico.

Enhancing the characteristics of the landscape as resources for the enrichment of the community is an important step to encourage the sense of belonging and to rediscover the relationship with a natural environment that can offer so much to those who find themselves in a state of vulnerability. In fact, it is not difficult at all to observe how these conditions of vulnerability are often generated by attitudes of indifference towards the environment, the landscape, and the context.

Working to foster a sense of belonging to a community is a second aspect that emerges as a priority for project interventions: stimulating participatory events to weave social relationships, as well as proposing shared urban spaces where to do activities with neighbors are some of the architectural strategies to achieve this goal.

At the same time, it is equally essential to propose urban-architectural interventions that ensure an optimal infrastructure network, capable of guaranteeing the basic services for a dignified life, as well as allowing an effective integration with the formal city.

During the workshop, it has been proved as equally important being able to ensure the protection of the ecosystem, intended both as a natural resource and as a social environment, capable of catalyzing economic, associative, and recreational activities.

Finally, as highlighted on several occasions, the preservation of the existing cultural and environmental heritage is an important approach for the regeneration of the existing social environment: discovering a shared heritage, both tangible and intangible, can be the promoter of a social and communitarian regeneration.

Design is therefore increasingly responsible for approaching these marginalized realities, be they urban or rural, to fulfill the task to contribute to the generation of a more equitable and sustainable society, capable of facing the challenges of future. In these most vulnerable realities, the essence of the architectural discipline is revealed, and the value of a designer is proved. In fact, the vulnerable context is very rich in facts that must be analyzed in order to understand the risks, the problems, and the expectations that contribute to creating this decision-making stalemate characterizing a vulnerable community. Here, the good designer demonstrates his ability to understand and to interpret the context, picking up the more or less hidden messages that the vulnerables transmit. The good designer intends which strategies to adopt to generate processes of involvement and empowerment of the community and which design issues should be addressed in such a difficult context. The good designer captures from the context those elements rich in potential and transforms them into catalysts of urban and social regeneration. Here, the good designer demonstrates his responsibility and humility, both necessary to let the community taking the ownership of the regenerative process and taking the responsibility for its own future.

Undoubtedly, for all the variables involved and the complexity of the existing phenomena, it is not easy to guide a successful intervention in a vulnerable community. There are so many risks and possibilities that it is very difficult to be able to grasp and manage all the variables that intervene in these realities. For this reason, it is more than ever necessary to approach work in a multidisciplinary way, letting the perspective of various disciplines contributing to form a more complete vision of the vulnerable community. In this way, the specific approach of each discipline helps also to form a range of actions which, integrated, can guarantee the success of an intervention.

Never as it is now, for this complexity, that characterizes the urban and social phenomena of a vulnerable community; it is clear how fundamental is establishing good communication and interaction with the residents, making sure that they become the protagonists of the design process. This, however, goes much further than the simple participatory design processes that are increasingly taking place. Making the residents of a vulnerable community become the protagonists of the design process requires a paradigm shift: moving from "Design for Vulnerable Communities" to "Design with Vulnerable Communities." Thus, the architect does not limit himself to proposing solutions by working for and with vulnerable communities, but he promotes the process of involving the community itself, stimulating critical analysis, discussion, and creativity within the community.

Acknowledgments We thank the external reviewers for providing critical and helpful comments on earlier versions of this chapter and on the book proposal. This work was funded by Fondo de Investigación 2020–2021 del Observatorio de Ciudades del Tecnologico de Monterrey (Mexico), through the research project "Design for vulnerables," project duration from October 2020 to December 2021.

References

Cattaneo, T., Giorgi, E., & Ni, M. (2018). Landscape, architecture and environmental regeneration: A research by design approach for inclusive tourism in a rural village in China. *Sustainability, 11*(1), 128. https://doi.org/10.3390/su11010128

European Association for Architectural Education. (n.d.). *The changing roles of the architect*. European Association for Architectural Education. Retrieved 27 September 2021, from https://www.eaae.be/eaae-academies/education-academy/themes/changing-roles-architect/.

Giorgi, E. (2020). Sharing as cultural revolution. In *The co-housing phenomenon* (Urban Book Series) (pp. 27–58. Scopus). Springer. https://doi.org/10.1007/978-3-030-37097-8_2

Giorgi, E., Bugatti, A., & Bosio, A. (2020). Social and spatial experiences in the cities of tomorrow. *Societies, 10*(1), 9. https://doi.org/10.3390/soc10010009

Kuhlicke, C., Scolobig, A., Tapsell, S., Steinführer, A., & De Marchi, B. (2011). Contextualizing social vulnerability: Findings from case studies across Europe. *Natural Hazards, 58*(2), 789–810. https://doi.org/10.1007/s11069-011-9751-6

Nast, C. (2019, September 17). *The future of design. Architectural digest*. Retrieved from https://www.architecturaldigest.com/story/future-of-design-2019.

Papanek, V. (1972). *Design for the real world: Human Ecology and social change* (2nd ed., pp. 14–15). Academy Chicago Publishers., 2005, (first published 1972).

Suarez-Balcazar, Y. (2020). Meaningful engagement in research: Community residents as co-creators of knowledge. *American Journal of Community Psychology, 65*(3–4), 261–271. https://doi.org/10.1002/ajcp.12414

Index

E. Giorgi et al. (eds.), *Design for Vulnerable Communities*, The Urban Book Series, https://doi.org/10.1007/978-3-030-96866-3

O

P

W

The manufacturer's authorised representative in the EU is Springer Nature Customer Service Centre GmbH, Europaplatz 3, 69115 Heidelberg, Germany. If you have any concerns regarding our products, please contact ProductSafety@springernature.com

Printed and bound by CPI Group (UK) Ltd, Croydon, CR0 4YY
28/09/2026
02224067-0006